# ESSENTIALS
*of*
# ORGANIC CHEMISTRY

B. HEINZ
*Palomar College*

Kendall Hunt
publishing company

This book has been printed from author prepared copy.



Kendall Hunt
publishing company

www.kendallhunt.com
*Send all inquiries to*:
4050 Westmark Drive
Dubuque, IA 52004-1840



ISBN 978-1-4652-1887-2



Printed in the United States of America
10 9 8 7 6 5 4 3 2 1

# Contents:

# LIST of EXPERIMENTS (Organic Chemistry)

## Ch. 1 Introduction to Organic Chemistry

Properties of organic compounds – flammability, solubility, conversion points; melting point and boiling point determination; selection of functional group tests; introduction to laboratory techniques as demonstrations (distillation, extraction, vacuum filtration)

## Ch. 2 Alkanes and Cycloalkanes

Molecular modeling, types of isomerism; reactions, photo-halogenation reactions of different saturated aliphatic compounds with respect to relative reaction rates

## Ch. 3 Alkenes and Alkynes

Distinction between saturated and unsaturated compounds; electrophilic addition with $Br_2$, oxidation with $KMnO_4$ to glycols

## Ch. 4 Arenes, Hetarenes and Alkaloids

Properties (appearance, solubility, odor); electrophilic substitution reaction; nitration of toluene; synthesis of a diazo dye, synthesis of phenolphthalein and fluorescein out of phenol and resorcinol resp. caffeine sublimation out of instant coffee as natural hetarene and alkaloid

## Ch. 5 Alcohols, Ethers and Thiols

Properties of alcohols (appearance, solubility, odor), pH measurements; Group identification (Jones' reagent), oxidation, reactions according to degree of substitution (Lucas test); Acidity of aliphatic and aromatic alcohols (phenols); ferric chloride test for enols ($FeCl_3$); borax test with poly-valent alcohols (glycol, glycerol, xylitol from gum, sugars); properties of thiols (appearance, solubility, odor); identification with lead acetate; oxidation to disulfides ($I_2$/NaOH)

## Ch. 6 Aldehydes and Ketones

Properties (appearance, solubility, odor); oxidation, identity tests – Benedict, iodoform test; commercial aldehydes and ketones; test on dog and cat repellant (methyl ketone and iodoform reaction)

## Ch. 7 Carboxylic Acids and Esters

Properties of carboxylic acids (solubility, appearance, odor); salt formation with alkali (neutralization), pH measurements, group identification ($AgNO_3$ reaction); synthesis of esters (esterification); hydrolysis of esters (saponification); synthesis of aspirin and oil of wintergreen, thinlayer chromatography (TLC)

## Ch. 8 Amines, Amides, and Alkaloids

Properties of amines (solubility, appearance, odor); neutralization with acids; formation of salts (hydrochlorides); pH measurements, odor quenching with lemon juice (forensics); comparative reactions of primary, secondary, tertiary and quaternary amines, Hinsberg test; synthesis of acetaminophen (analgesic drug); hydrolysis of an amide drug and TLC, color test for amides

## Ch. 9 Stereochemistry

Polarimetry, determination of specific angle of rotation of chiral compounds (D-glucose, L-alanine); Identification of unknown sugar; study of racemization, analysis of α and β pinene from turpentine; molecular modeling

## Ch. 10 Organic Halogen Compounds

Photohalogenation reactions, Friedel-Craft reaction using alkyl and acyl halides, nucleophilic substitution ($S_N$) with halogens as leaving groups, synthesis of fluorescein and eosin

## Ch. 11 Reactions of Organic Compounds

Electrophilic addition reaction ($A_E$), cycloaddition (Diels Alder 4+2π), electrophilic substitution ($S_E$) of aromatic compounds

## Ch. 12 Separation Techniques

Distillation of two miscible liquids, steam distillation of essential oils and their extraction, column chromatography of pigments, thin-layer chromatography (TLC) of printer inks from cartridges, TLC of analgesic drugs, extraction of caffeine from instant coffee via sublimation, gas chromatography (GC) of turpentine and pinenes

## Ch. 13 Analytical Methods

Synthesized products (aspirin, Diels-Alder adduct) and natural products obtained by steam distillation or by direct extraction with organic solvents and their structure elucidation via nuclear magnetic resonance (NMR) ultraviolet absorption spectroscopy (UV/UV-vis), infrared spectroscopy (IR, FTIR), gas chromatography (GC), mass spectroscopy (MS, GC-MS or HPLC-MS) if available; GC of an α and β pinene mixture from turpentine, refractometry

# INTRODUCTION

## *Message to instructors, Students and Laboratory Personnel*

The lab manual was designed to accompany the text "Essential Topics in Organic Chemistry" (chapters 1-11). The experiments follow the theoretical material covered in the chapters. The lecture portion can be covered extensively in the assigned lecture period, the labs are a maximum of about three hours long and are ideally scheduled after the lecture to ensure optimal continuity. Some labs can be covered in less time; some require a bit more time and can be extended to the next lab period to complete. However, since the lab sessions involve 3-5 individual experiments, not all experiments need to be performed. The instructor can feel free to drop certain labs, perform them as a class demonstration or assign a group of students to carry out selected experiment in order to save time. Also, individual student groups can do different experiments in a different sequence from the manual, if chemicals or equipment should be scarce or fume hoods are not abundantly available. Another strategy is to assign different experiments to different student groups and later discuss the results at the end of class, so that all students have results for their individual lab reports. Students are more or less efficient in carrying out the labs, so some might want to leave earlier. As long as the time frame is within reason it is alright to have them leave once finished, otherwise have them help others that might need a bit more time. The instructor can deal with this type of situation as seen fit.

Usually a theoretical introduction by the instructor at the beginning of the lab-session does not need to be given, since the material should have already been covered during the lecture period. The idea is to save time for the lab itself as students learn by doing. Certain practical guidelines and safety measures, however, should be pointed out.

The laboratory time is only there to collect data that can be entered into the lab manual directly. The lab report is to be written by each student on separate paper with all the results, observations, answers to the questions etc. and handed in one week after completion of the lab session pertaining to a particular chapter. Answers should be given in complete sentences. The report can be hand-written or typed. All tables and graphs must be made with the help of a ruler.

Of course, the instructor is available mainly to answer general questions concerning procedures as well as clarify the questions in the lab manual, if needed. The instructor is free to carry out the labs to his or her own discretion as an exercise of academic freedom. The main idea is to correlate the labs as much as possible with the text.

Each lab will have adequate information for the preparation of the experiments for the laboratory personnel. In the future a special manual will be devised for lab personnel to help in ordering chemicals and prepare the required reagents and solutions.

# INTRODUCTION

## *Laboratory Safety and Precautionary Measures*

Chemistry is an aggressive and possibly dangerous science. However, although things can happen they usually don't. Still, it is important to point out certain protective measures for the individual, other lab occupants and the learning institution as a whole. Many people have dogs and cats as pets, for instance, even though they can be killers if they want to be. A dog's teeth or canines can exert a force almost 300 times as strong as a human bite. As humans we have learned to live with this potential danger and learned to control it. We have respect for the animal's power. The same attitude should be applied to chemicals. We understand their potential havoc and guard ourselves by understanding their behavior and develop a possible control mechanism to keep them in check.

**Individual bodily protection:** The most obvious protective measures start with the body of the lab participant performing the experiments. Going from head to toe, the first thing to keep in mind is to tie all longer hair with a rubber band or hair barrettes to the back. Long hair can get caught in equipment and lead to "scalping" of a person. The next item involves protection of the eyes with the help of protective goggles of the type Z87. They shouldn't be cumbersome, slipping of the nose or cloudy and scratched. Although labs supply glasses, it is better to purchase one's own at the bookshop or hardware shops.

They should be worn at all times, even when only reading or doing paper work. The best and most comfortable glasses are those that students forget to take off even when leaving the lab environment, impressing their friends and loved ones with fashionable, all protective visors. Caustic chemical splashes can cause the eyes to turn into raisins, establishing the need and extra expense to purchase a glass eye substitute. When wearing contact lenses it is certainly a must to have protective goggles at all times. Good goggles can fit comfortably over prescription glasses, if they are not too big to begin with. The next item for personal protection is to wear a good cotton lab-coat. Nylon or other fibers are too flammable for safety. Chemicals can damage personal clothing creating stains or holes that may appear at a much later time. If a lab coat is not available an apron can be used. Shorts or long dresses or gowns can become a hazard by exposing the legs to chemical spills or the long material becoming a tripping hazard.

Shoes should be solid and protect the feet and tows from chemical spills or falling objects such as glass or metal pieces of equipment. Flip flops or sandals, even canvas athletic shoes should be avoided.

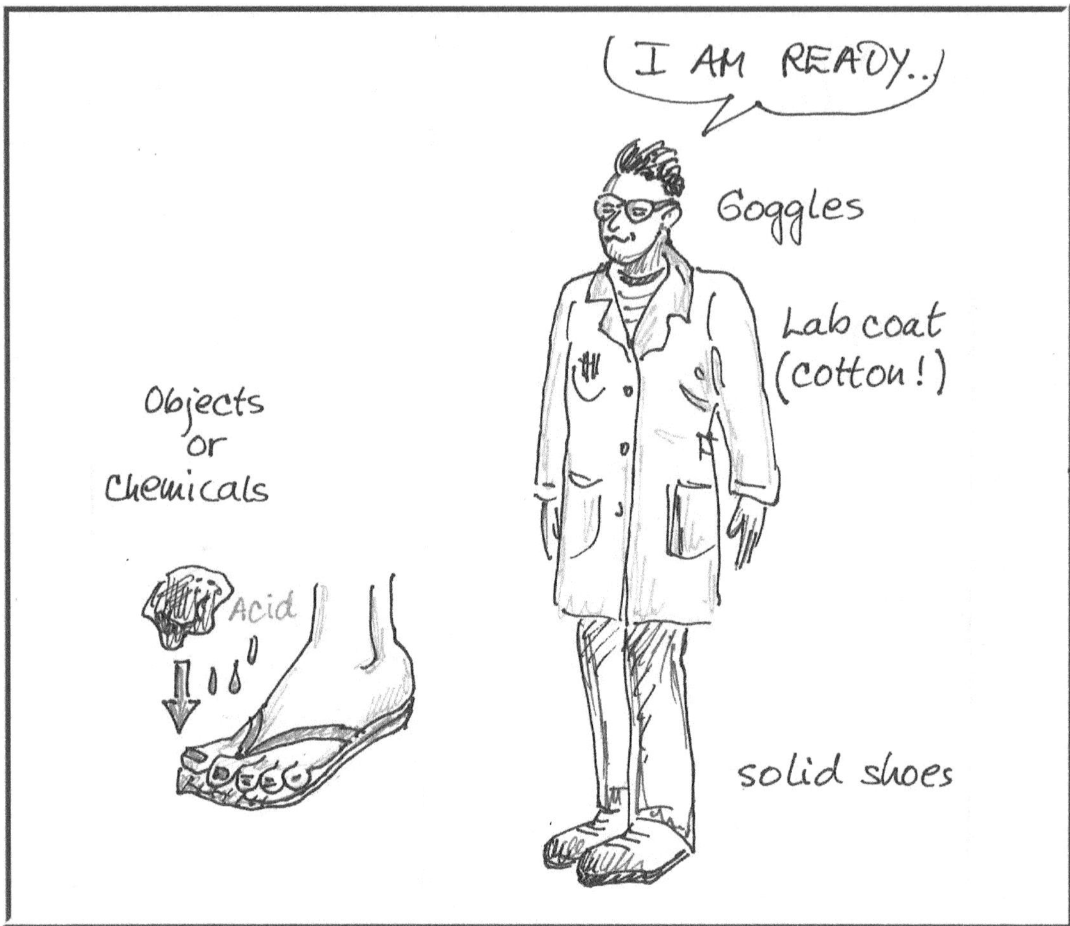


Gloves are available but do need to be worn. They can interfere with the dexterity of handling objects and the dispensing of chemicals. However, hands should be thoroughly washed after each lab session. It is obvious that one shouldn't drink or eat in the laboratory environment. Chemicals can adhere to the hands and contaminate the food or beverage.

**Possible laboratory hazards:** The most dangerous event in a lab is the start of fires. Fires, however, always start small. One should always be alert and on the look-out that open flames are always controlled and turned off when not needed. A fire once started will require oxygen to sustain itself. Therefore any material that prevents oxygen from feeding the fire is of value when placing it over the fire, provided it doesn't burn itself such as a ceramic pad.

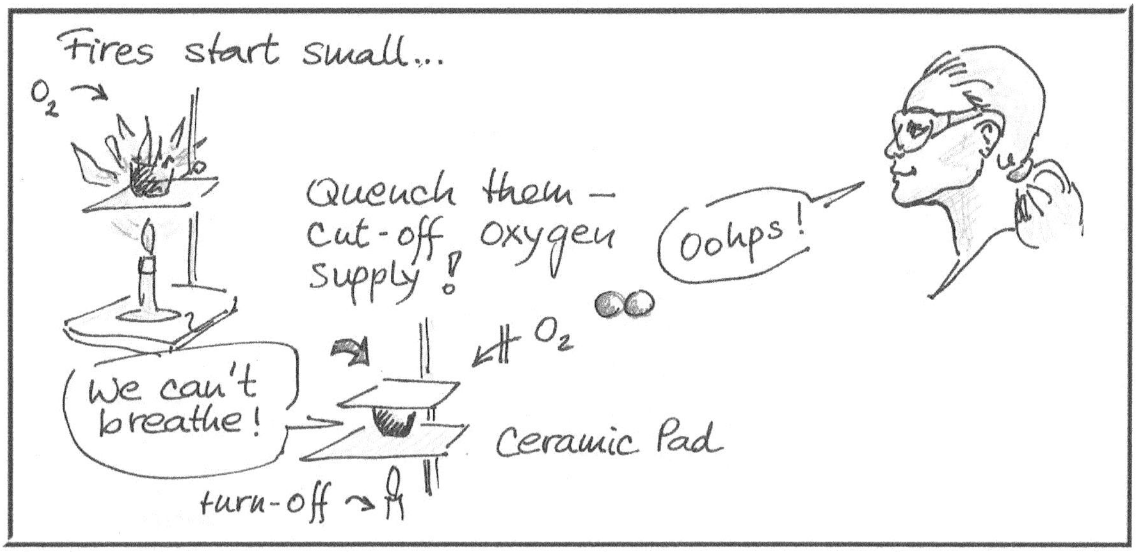


If the fire gets out of control by not being quenched adequately enough, the use of a fire extinguisher will help. These red canisters are usually mounted at various sites of the laboratory. Some are in special cabinets outside of the laboratory in the hallway. The cabinet door must be opened and the fire extinguisher removed for action. One should familiarize oneself with the use of a fire extinguisher, directions posted on the extinguisher itself, even before a fire ever starts.

Modern labs are equipped with smoke and fire detectors which will go off by ringing an alarm. If this happens the students must evacuate the premises leaving their belongings behind. A technical crew will investigate the situation and clear the area for re-entry.

Every student has to be careful not to lean with the back on the lab counter while a Bunsen burner is still burning. Distractions can be dangerous. If the student catches on fire the move under a shower close to the exit door can quench the fire by pulling a handy lever.

When carrying out experiments, one must use caution when dealing with test tubes. The opening should always be facing away from a person, either oneself or a lab-partner. If chemicals splatter into someone's eyes, the eye douche situated at waste level close to the exit doors is of a handy relief, squirting water into the eyes when leaning over the faucets and pushing down a lever.

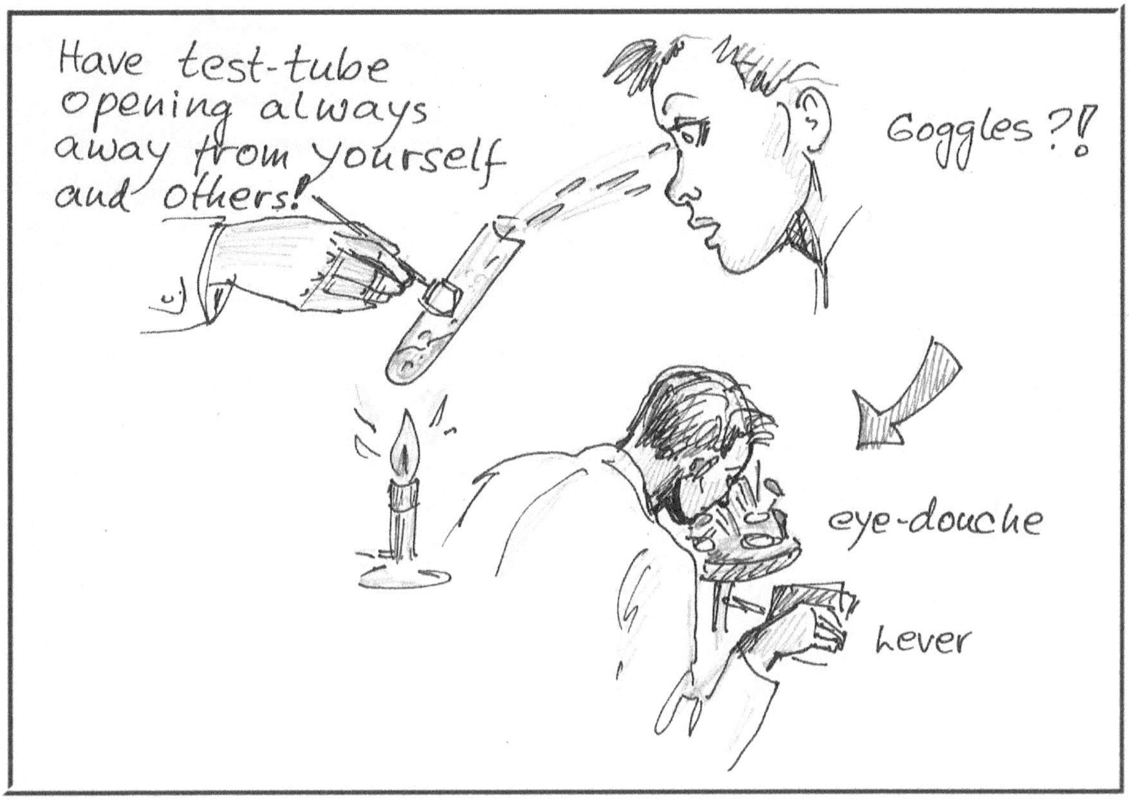


When the laboratory exercise is finished for the day, one should clean the table tops, balances and any other used equipment and make sure that all the gas spigots are turned off by pushing the handles into a perpendicular position.

In the event of glass breakage, the shattered pieces should be carefully mopped up, transferred to a dust bin and emptied into a special broken glass container. All waste chemicals are emptied into provided waste containers, one for liquid wastes and one for solid wastes.

When noxious, flammable or other irritating chemicals are used for an experiment the fume hoods offer protection from vapors and fumes. They also offer protection, if the sashes are pulled down, from the effects of possibly explosive or combustible material.

The instructor will inform the students of any hazards that might be involved with certain experiments.

A pregnant student should receive written permission from the personal physician to engage in chemical experiments. The same should be observed by any student with certain health concerns or disabilities.

However, usually the exposure is too short-lived for many chemicals to cause any harm. With expertise and the necessary precautions, the student will have a memorable experience working in a chemistry lab and will leave healthy and unharmed like thousands of other students before.

# Periodic Table of The Elements



| 1A (1) | 2A (2) | 3B (3) | 4B (4) | 5B (5) | 6B (6) | 7B (7) | (8) | 8B (9) | (10) | 1B (11) | 2B (12) | 3A (13) | 4A (14) | 5A (15) | 6A (16) | 7A (17) | 8A (18) |
|---|---|---|---|---|---|---|---|---|---|---|---|---|---|---|---|---|---|
| hydrogen<br>1<br>H<br>1.0079 | | | | | | | | | | | | | | | | | helium<br>2<br>He<br>4.0026 |
| lithium<br>3<br>Li<br>6.941 | beryllium<br>4<br>Be<br>9.0122 | | | | | | | | | | | boron<br>5<br>B<br>10.811 | carbon<br>6<br>C<br>12.011 | nitrogen<br>7<br>N<br>14.007 | oxygen<br>8<br>O<br>15.999 | fluorine<br>9<br>F<br>18.998 | neon<br>10<br>Ne<br>20.180 |
| sodium<br>11<br>Na<br>22.990 | magnesium<br>12<br>Mg<br>24.305 | | | | | | | | | | | aluminium<br>13<br>Al<br>26.982 | silicon<br>14<br>Si<br>28.086 | phosphorus<br>15<br>P<br>30.974 | sulfur<br>16<br>S<br>32.065 | chlorine<br>17<br>Cl<br>35.453 | argon<br>18<br>Ar<br>39.948 |
| potassium<br>19<br>K<br>39.098 | calcium<br>20<br>Ca<br>40.078 | scandium<br>21<br>Sc<br>44.956 | titanium<br>22<br>Ti<br>47.867 | vanadium<br>23<br>V<br>50.942 | chromium<br>24<br>Cr<br>51.996 | manganese<br>25<br>Mn<br>54.938 | iron<br>26<br>Fe<br>55.845 | cobalt<br>27<br>Co<br>58.933 | nickel<br>28<br>Ni<br>58.693 | copper<br>29<br>Cu<br>63.546 | zinc<br>30<br>Zn<br>65.39 | gallium<br>31<br>Ga<br>69.723 | germanium<br>32<br>Ge<br>72.61 | arsenic<br>33<br>As<br>74.922 | selenium<br>34<br>Se<br>78.96 | bromine<br>35<br>Br<br>79.904 | krypton<br>36<br>Kr<br>83.80 |
| rubidium<br>37<br>Rb<br>85.468 | strontium<br>38<br>Sr<br>87.62 | yttrium<br>39<br>Y<br>88.906 | zirconium<br>40<br>Zr<br>91.224 | niobium<br>41<br>Nb<br>92.906 | molybdenum<br>42<br>Mo<br>95.94 | technetium<br>43<br>Tc<br>[98] | ruthenium<br>44<br>Ru<br>101.07 | rhodium<br>45<br>Rh<br>102.91 | palladium<br>46<br>Pd<br>106.42 | silver<br>47<br>Ag<br>107.87 | cadmium<br>48<br>Cd<br>112.41 | indium<br>49<br>In<br>114.82 | tin<br>50<br>Sn<br>118.71 | antimony<br>51<br>Sb<br>121.76 | tellurium<br>52<br>Te<br>127.60 | iodine<br>53<br>I<br>126.90 | xenon<br>54<br>Xe<br>131.29 |
| caesium<br>55<br>Cs<br>132.91 | barium<br>56<br>Ba<br>137.33 | 57-71<br>La-Lu | hafnium<br>72<br>Hf<br>178.49 | tantalum<br>73<br>Ta<br>180.95 | tungsten<br>74<br>W<br>183.84 | rhenium<br>75<br>Re<br>186.21 | osmium<br>76<br>Os<br>190.23 | iridium<br>77<br>Ir<br>192.22 | platinum<br>78<br>Pt<br>195.08 | gold<br>79<br>Au<br>196.97 | mercury<br>80<br>Hg<br>200.59 | thallium<br>81<br>Tl<br>204.38 | lead<br>82<br>Pb<br>207.2 | bismuth<br>83<br>Bi<br>208.98 | polonium<br>84<br>Po<br>[209] | astatine<br>85<br>At<br>[210] | radon<br>86<br>Rn<br>[222] |
| francium<br>87<br>Fr<br>[223] | radium<br>88<br>Ra<br>[226] | 89-103<br>Ac-Lr | rutherfordium<br>104<br>Rf<br>[261] | dubnium<br>105<br>Db<br>[262] | seaborgium<br>106<br>Sg<br>[266] | bohrium<br>107<br>Bh<br>[264] | hassium<br>108<br>Hs<br>[269] | meitnerium<br>109<br>Mt<br>[268] | ununnilium<br>110<br>Uun<br>[271] | unununium<br>111<br>Uuu<br>[272] | ununbium<br>112<br>Uub<br>[277] | | ununquadium<br>114<br>Uuq<br>[289] | | | | |

| | | | | | | | | | | | | | |
|---|---|---|---|---|---|---|---|---|---|---|---|---|---|
| lanthanum<br>57<br>La<br>138.91 | cerium<br>58<br>Ce<br>140.12 | praseodymium<br>59<br>Pr<br>140.91 | neodinium<br>60<br>Nd<br>144.24 | promethium<br>61<br>Pm<br>[145] | samarium<br>62<br>Sm<br>150.36 | europium<br>63<br>Eu<br>151.96 | gadolinium<br>64<br>Gd<br>157.25 | terbium<br>65<br>Tb<br>158.93 | dysprosium<br>66<br>Dy<br>162.50 | holmium<br>67<br>Ho<br>164.93 | erbium<br>68<br>Er<br>167.26 | thulium<br>69<br>Tm<br>168.93 | ytterbium<br>70<br>Yb<br>173.04 |
| actinium<br>89<br>Ac<br>[227] | thorium<br>90<br>Th<br>232.04 | protactinium<br>91<br>Pa<br>231.04 | uranium<br>92<br>U<br>238.03 | neptunium<br>93<br>Np<br>[237] | plutonium<br>94<br>Pu<br>[244] | americium<br>95<br>Am<br>[243] | curium<br>96<br>Cm<br>[247] | berkelium<br>97<br>Bk<br>[247] | californium<br>98<br>Cf<br>[251] | einsteinium<br>99<br>Es<br>[252] | fermium<br>100<br>Fm<br>[257] | mendelevium<br>101<br>Md<br>[258] | nobelium<br>102<br>No<br>[259] |

**Chapter 1  Introduction to Organic Chemistry**

1.1 Woehler's experiment and the birth of organic chemistry

1.1.1 The difference between inorganic and organic chemistry

1.1.2 Naturally occurring and synthesized organic compounds

1.2 Hydrocarbons with and without hetero-atoms

1.3 Types of organic compounds

1.3.1 Acyclic

1.3.1.1 Branched and un-branched

1.3.1.2 Saturated and unsaturated

1.3.2 Cyclic

1.3.2.1 Carbocyclic (aromatic and non-aromatic)

1.3.2.2 Heterocyclic (aromatic and non-aromatic)

1.3 Functional Groups and hetero-atoms

1.4 Introduction to isomerism

1.5 Bond arrangements of the most important atoms (C, H, N, O, S,  and Halogens)

1.5.1 Lewis dot structures of organic molecules

1.5.2 Resonance and formal charges

1.5.3 Arrow formalism

1.6  Formula types (molecular, expanded structural, condensed, abbreviated or line-angle)

**Chapter 1 Introduction to Organic Chemistry**

The first documented, "organic" molecule artificially produced out of an "inorganic" substance was discovered by the German chemist Friedrich Wöhler 1800-1882). He heated Ammonium Cyanate, an inorganic salt [$NH_4^+$][$NCO^-$], and surprisingly found the organic compound Urea in his reaction flask. Urea is a major compound found in the urine of living systems, such as humans and most animals.

**Fig. 1.1**

The rearrangement of the atoms under the influence of heat is shown in the models below:
**Fig. 1.2**

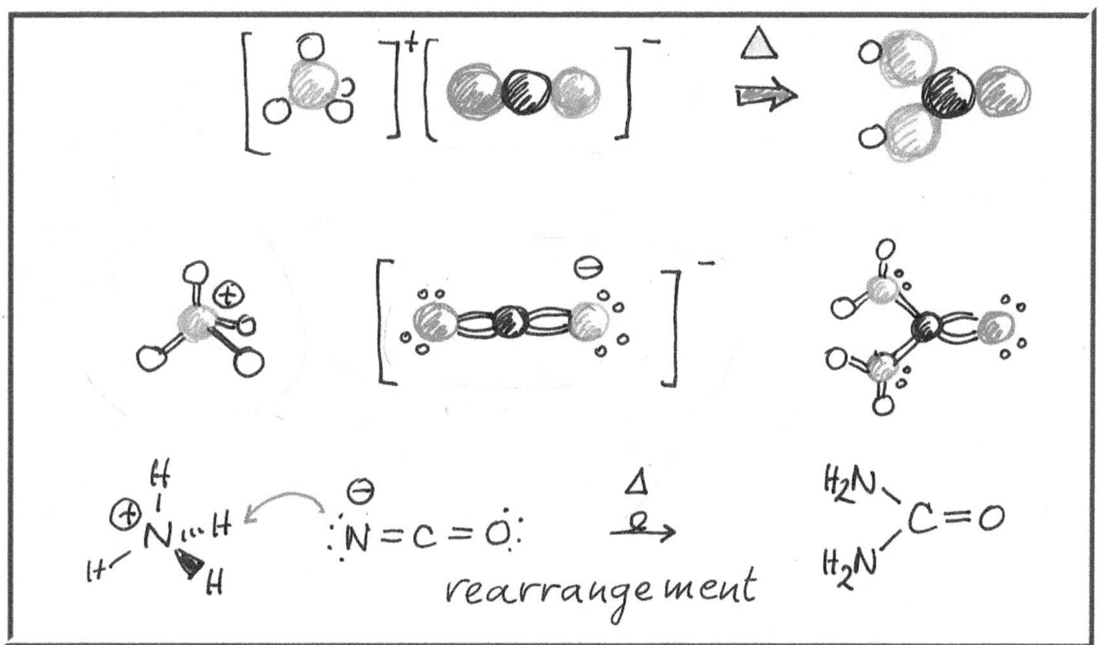

Therefore, Urea was considered an "organic" compound – life-based, attributed to "organs" and organisms. The name stuck and a new branch of chemistry, "Organic Chemistry," was born. Although many so-called organic chemicals are not obviously connected to living systems, found in nature, or are being newly discovered, they are still considered to be part of the organic spectrum, as long as they are mainly carbon-based molecules. Petroleum and coal, for instance, are a mix of many organic chemicals thought to be derived from long dead living systems. However, inorganic deposits of the Earth's crust are also derived from the decay of living materials, such as silicas ($SiO_2$-compounds derived from many plants and marine organisms) and lime stone ($CaCO_3$ deposits of marine life). Many geologic deposits have come about due the activity of various life forms. A branch called bio-inorganic chemistry involves the study of these relationships.

Today organic chemistry is the vast and endless field of mostly covalent compounds having a core-structure of a carbon framework that is more or less saturated with hydrogen atoms (hydrocarbons) , occasionally interlaced with hetero-atoms, leading to new derivatives (alcohols, amino acids, nucleotides, etc.).

Only carbon and hydrogen can group together in endless arrays such as the hydrocarbons. These frameworks can be substituted by the so-called **hetero-atoms** (O, S, N, P, halogens, etc.). The hetero-atoms can become part of the main framework or certain attachments, either by themselves or in special group arrangements, called **functional groups**. These variations lead to certain structural relationships that allows the classifications into different families, such as the **alcohols** (-OH groups), **Aldehydes** ( -CHO groups ), **Ketones** ( -CO- groups), **Nitriles** ( -CN groups), etc. , as shown in the following table.

**Table. 1.1**

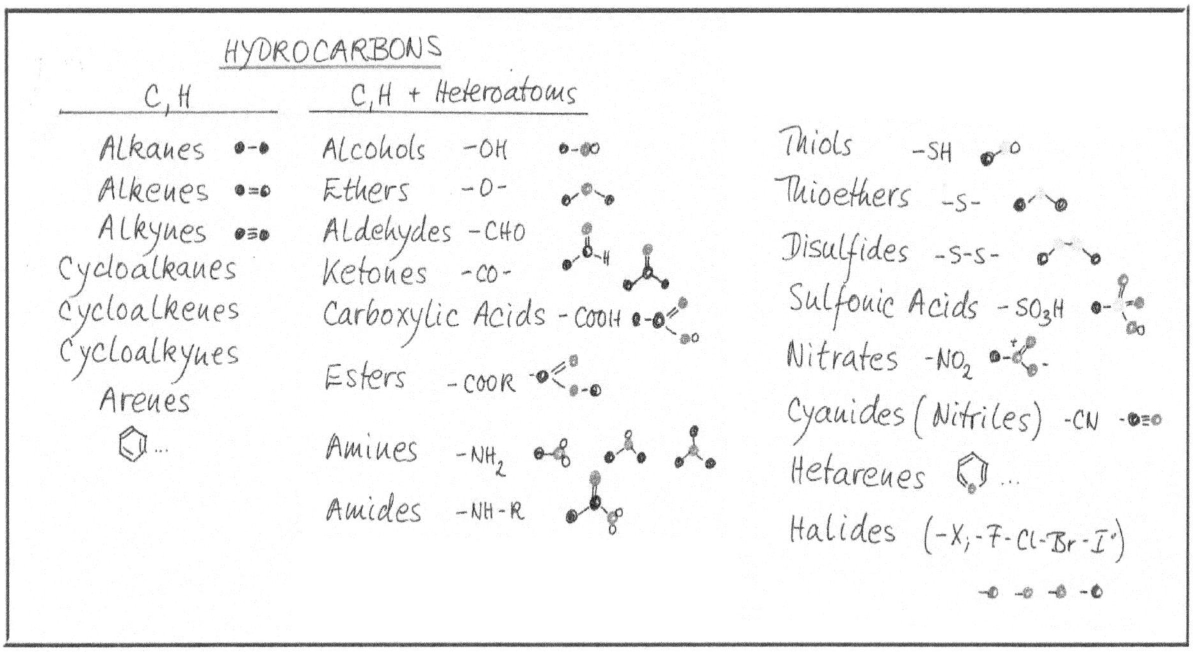


The main skeleton for these molecular derivatives is always based on the C/H framework.

**Fig. 1.3**

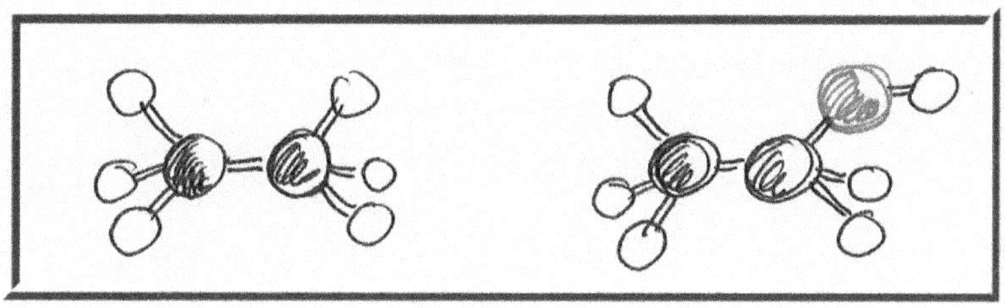

**Example problem 1.1**
**Identify the functional group in the following molecules**

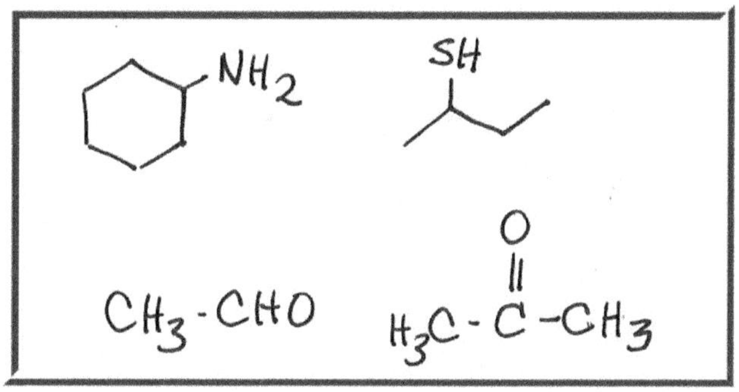


Carbon is the central and key element in all organic structures. It has special and diverse bonding characteristics and can bond with itself in unlimited arrays and varieties. It can also make multiple bonds with itself and other elements (mainly O, N, S, and P). By knowing the specific bonding characteristics ("bondedness" or valence) of the main elements involved, correct structural formulas can easily be derived.

**Fig. 1.4**

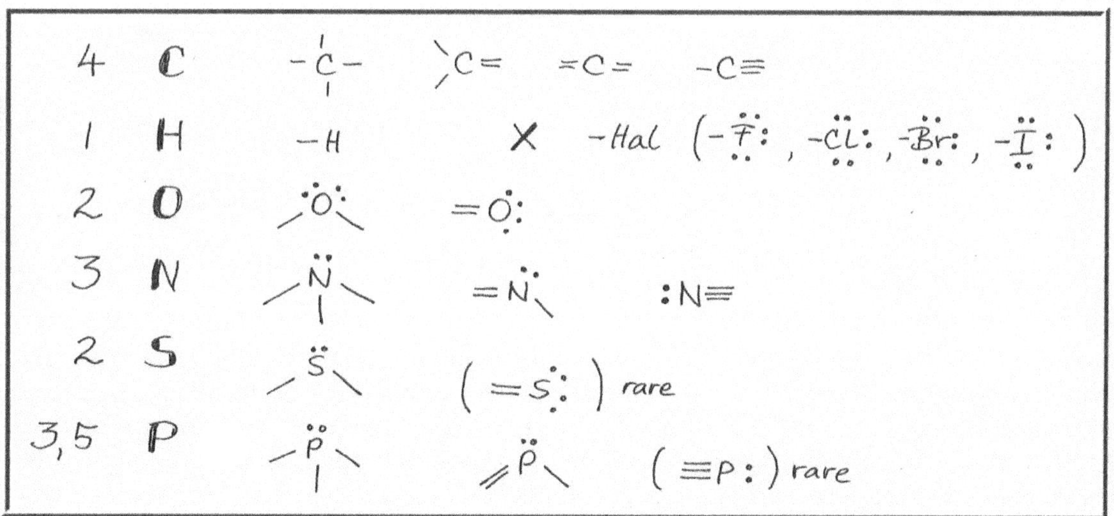

**Example problem 1.2**
**Which of the following molecules have wrong structures due to incorrect bonding? Make the necessary changes to receive correct structural formulas, if possible.**

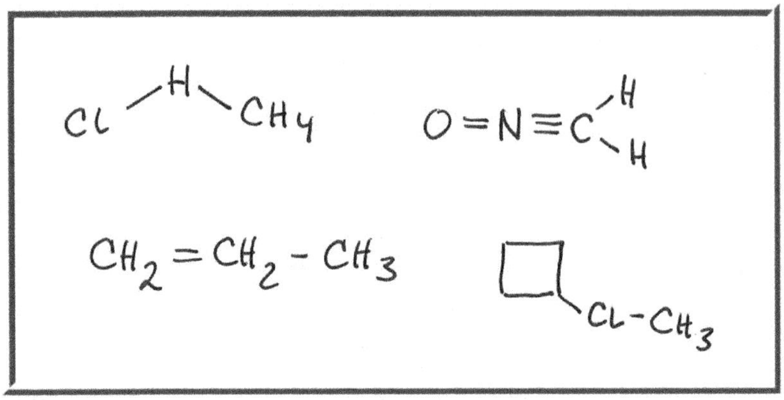

Exceptions to these bonding rules are only found in charged species, where the **formal charges** need to be included in the structure.

**Fig. 1.5**

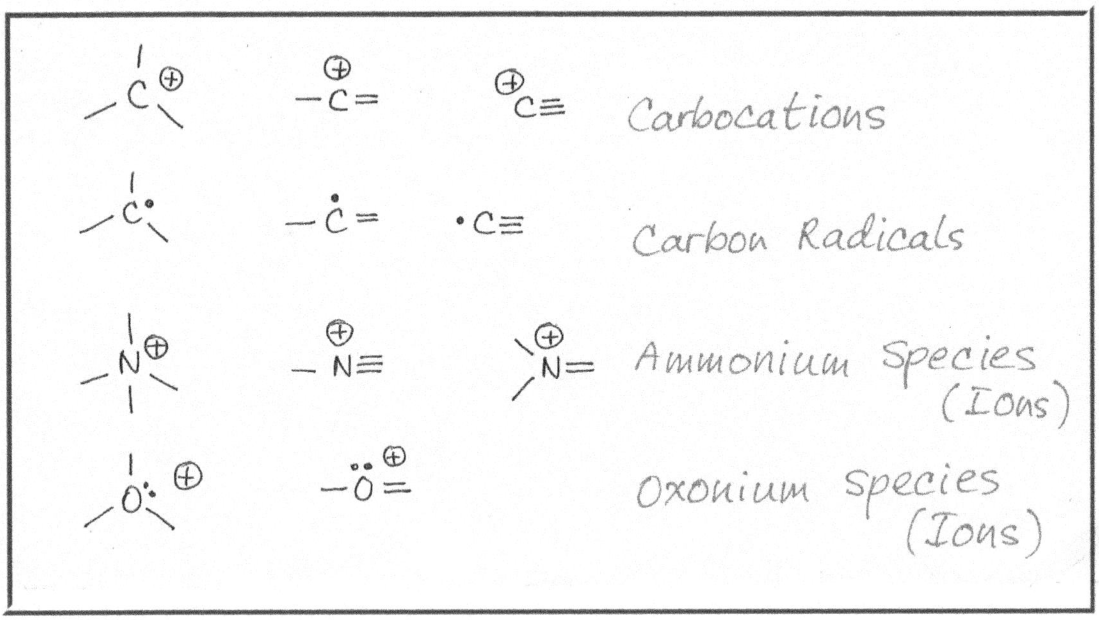


In organic chemistry, lone and bonded electron pairs are often interchangeable, leading to **resonance structures**, expressed by the different Lewis structures for the respective resonance contributors Even though the phenomenon of resonance implies oscillations between two or more extremes, the actual status quo is more of a "spread-about" condition, best exemplified by the **resonance hybrid.** For the study of reaction mechanism, so common in organic chemistry, the individual resonance contributors are much more useful.

**Fig. 1.6**

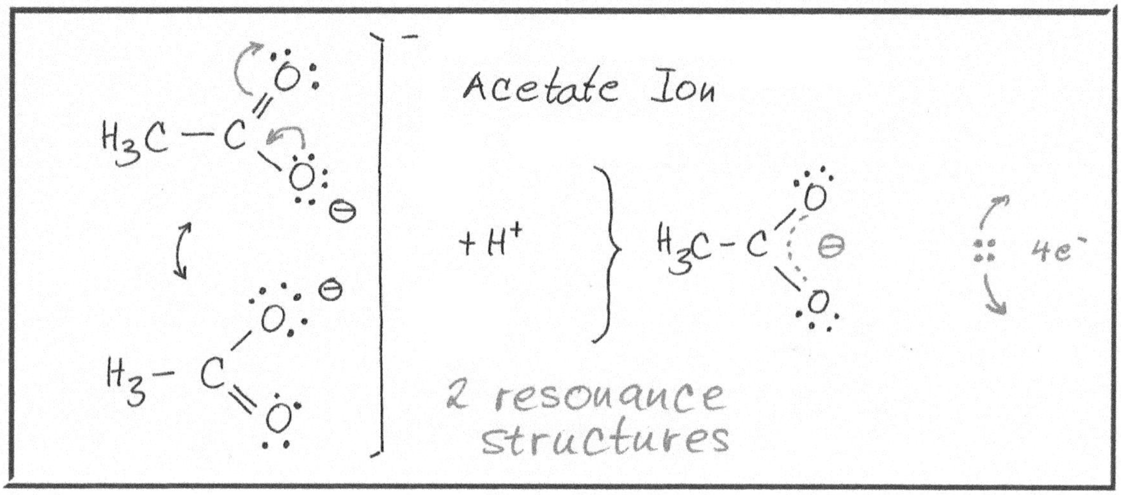


**Fig 1.7**

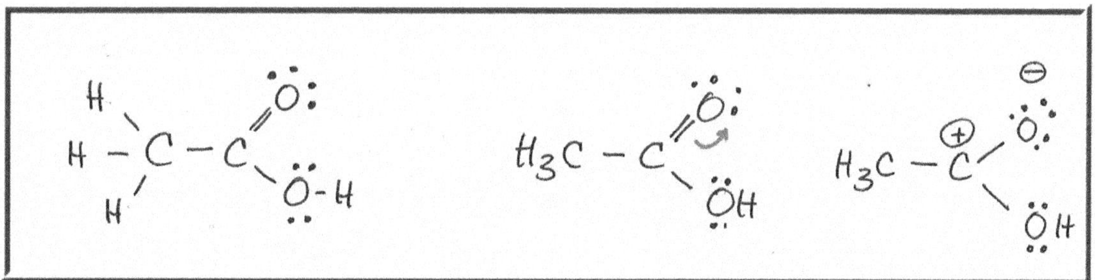

Formal charges are introduced, when “should-have” (number of original valence electrons) and “have” (number of valence electrons plus or minus a shared one) are no longer the same.

**Fig. 1.8**

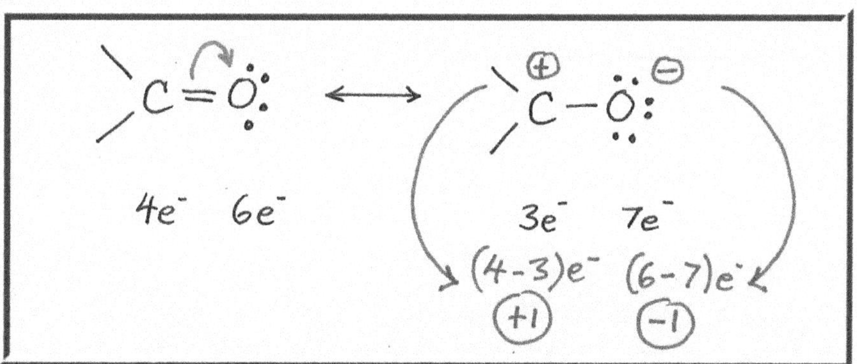


The more electronegative atom within the bond usually gets the negative formal charge. “Formal” in this sense means “arbitrary” and serves as a construct to help understand the bonding situation reflected in electron density shifts and final distribution.

Resonance hybrids – the actual state of electron distribution, can be thought of as the result of superimposed individual resonance structures. They are illustrated by dotted lines tracing the path in which the electrons fluctuate between lone pair and bonded pair existences within the molecule or molecular ion.

**Fig 1.9**

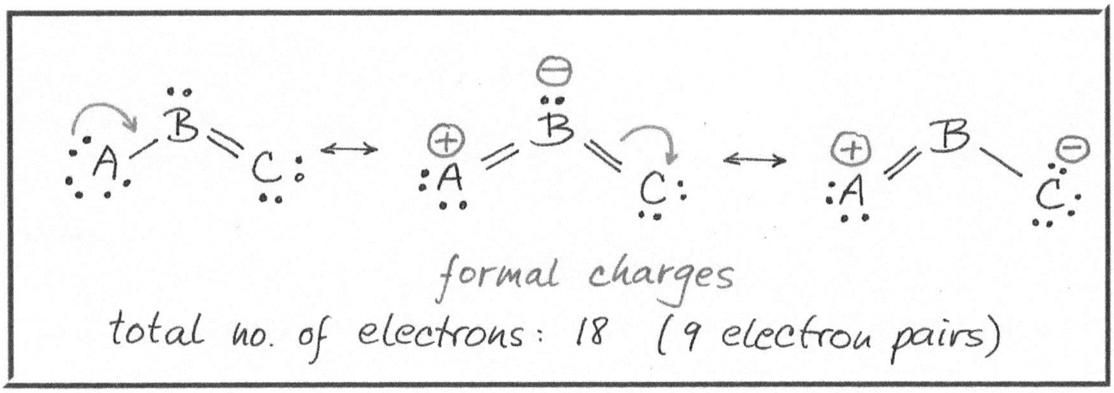


Formal charges are no longer needed and the hybrids are usually used in cases where more than 2 atoms are involved.

**Fig 1.10**

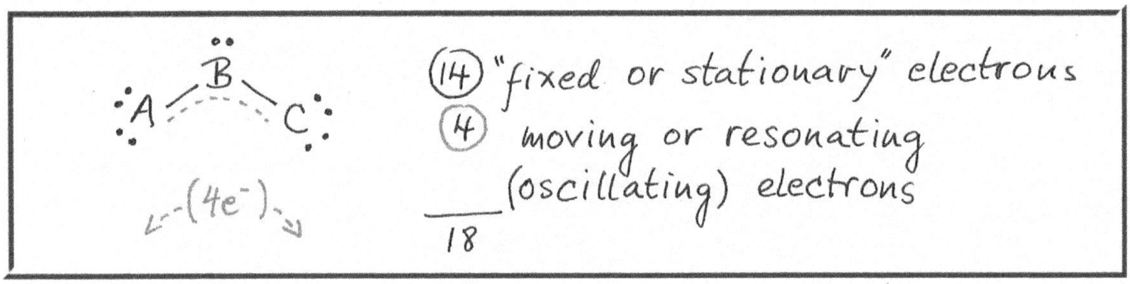


**Example Problem 1.3**
**Fill in the missing formal charges for the following structures:**

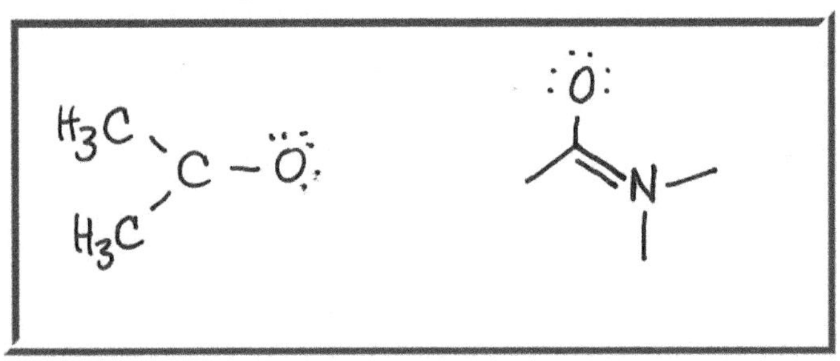

**Curved arrows** are often used to show the movement of **electron pairs**, usually from the status of a lone pair to a bonded pair (or vice versa) or individual **fish-hook arrows**, that symbolize the move of single electrons. The fish-hook arrow is an arrow with one half of its tip removed, resembling a fish hook, hence the name. This is only used when a bond is **homolytically** cleaved or separated, producing two equal parts, namely two **radicals,** each having an unpaired electron. When one bond of multiple bond situation is cleaved in such a way, a **biradical** is produced – the two lone electrons still situated on the original molecule.

**Fig. 1.11**

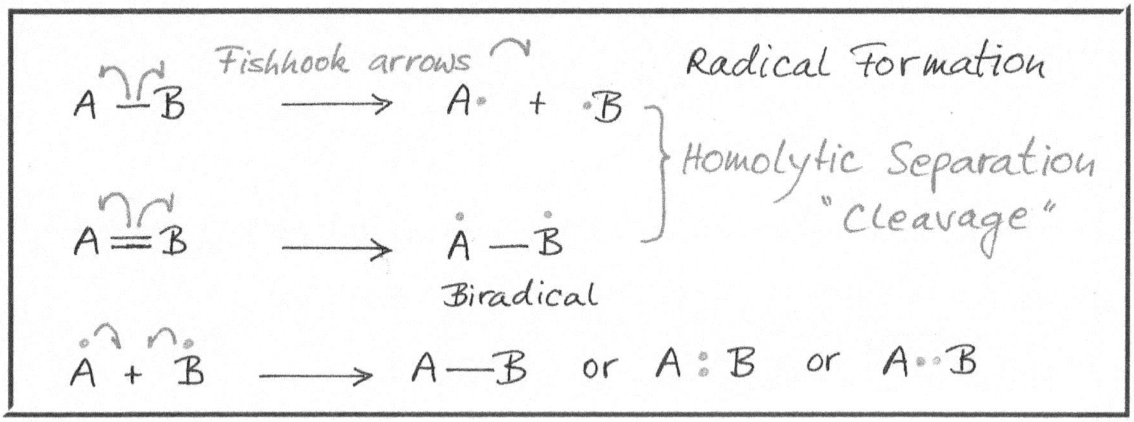


Double headed arrows are curved and have a complete arrow tip. They always designate the movement of a whole electron pair within a molecule. Often, formal charges are needed for the new species being formed to balance an electron deficiency or a surplus within the structure.

**Fig. 1.12**

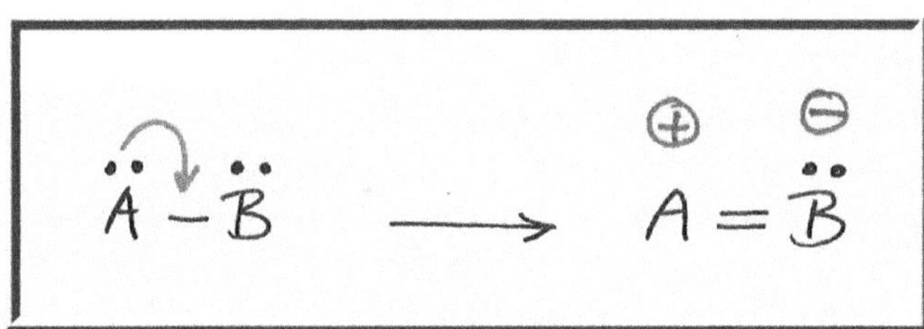


When single bonds are being moved, separate ionic species are formed. This process is known as **heterolytic** cleavage or separation.

**Fig. 1.13**

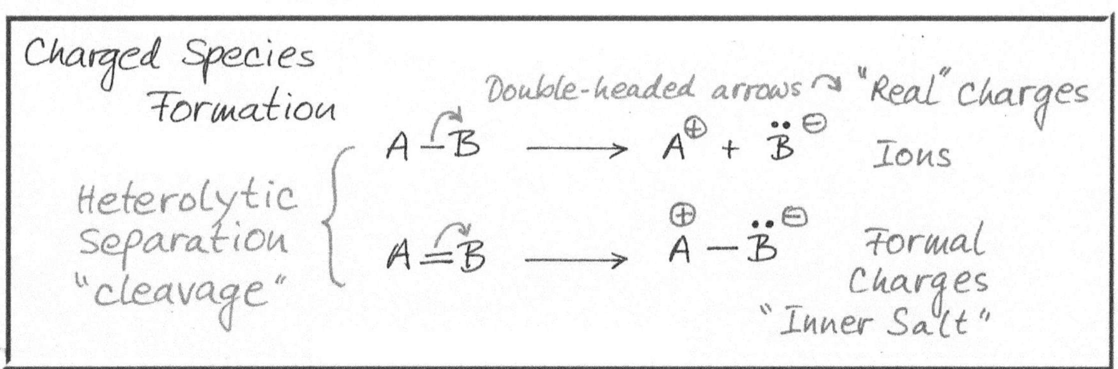

When an electron pair of a multiple bond is moved, or a lone pair is moved to make a bonded pair, species with two opposite formal charges are being formed. This resembles a type of "inner salt" or **betaine.**

**Example Problem 1.4**
**Draw the new structures formed by following the indicated arrow movements for the following species:**

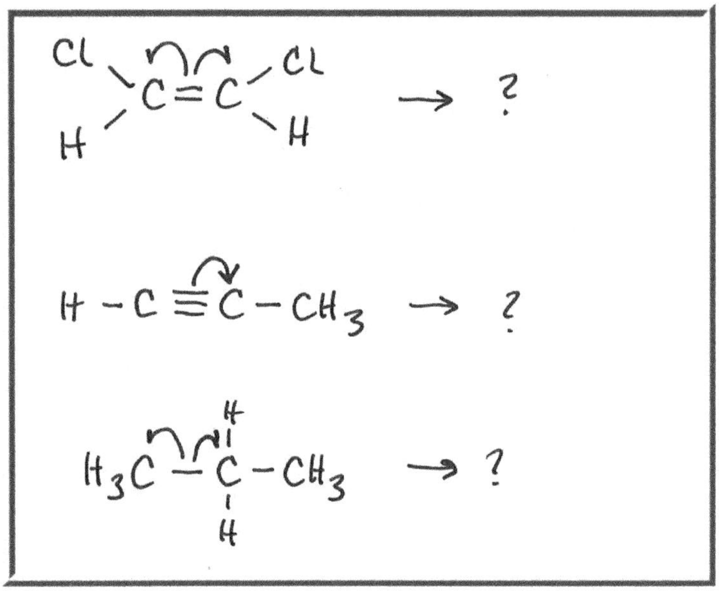

Organic compounds are identified by a variety of formulas. The **molecular formula** (MF), represents a line-up of the involved atoms in the order of C, H, N, O, and Halogens. If Phosphorous and/or Sulfur atoms are present, then they are usually written before the O. The numerical subscripts indicate the number of atoms that are involved per molecule, such as $CH_4N_2O$. The molar mass can easily be calculated from the molecular formula by adding up the atomic weights (masses) of each element, multiplied by the corresponding subscript.

**Fig. 1.14**

$CH_4N_2O$

| | | |
|---|---|---|
| C | = | 12 |
| 4 H | = | 4 |
| 2 N | = | 28 |
| 1 O | = | 16 |
| | | 60. |

$$M_{CH_4N_2O} = 60 \frac{g}{mol}$$

However, the MF does not give any information regarding the structure (constitution) or the actual alignment pattern of the atoms. This is given by the **structural formula** (SF).

**Fig. 1.15**

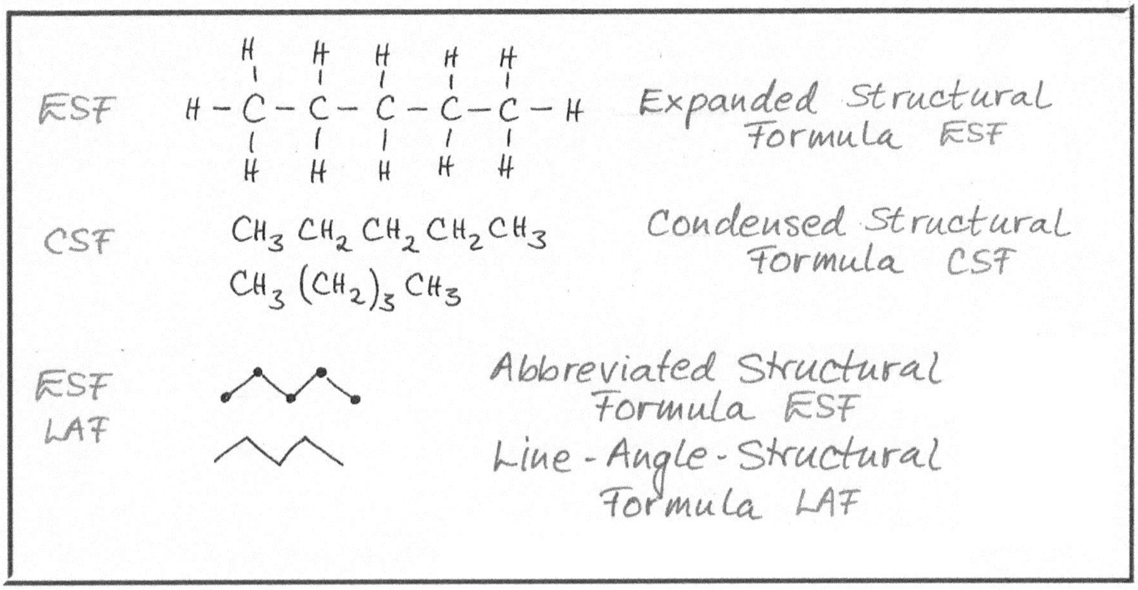


The one shown above is the **expanded structural formula** (ESF), in which every bond is accounted for. A **condensed structural formula** (CSF) compacts the formula to a "one-liner", originally conceived to save money and paper for printing space, leaving out some of the bonds without obscuring the actual structure.

**Fig. 1.16**

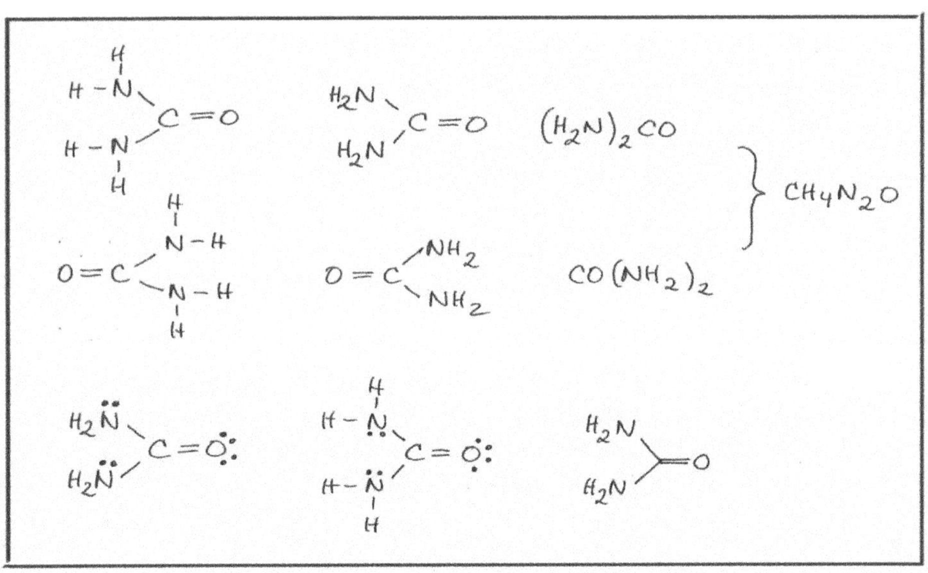

The **Lewis dot structure** would leave the entire "skeleton" intact, since it depends on the exact location of all electrons, shared or unshared.
Functional groups are often semi-condensed, just enough to become simplified but still in recognizable form. The hydroxyl group for alcohols is written –OH and not –O-H.

**Example Problem 1.5**
**Draw the complete, expanded formula for the following condensed formulas:**

$CH_3CH(CH_3)CH(CH_3)CH_3$

$CH_3CH(Cl)(CH_2)_2CH_3$

To make things even simpler, chemists have adopted the **abbreviated structural formulas** (ASF), sometimes also referred to as the **line-angle formulas** (LAF), to streamline the molecular patterns into the most abstract form. This makes it much easier to recognize structural forms and helps prevent the notion of "not seeing the forest due to all the trees." Urea thus becomes as shown in the figure above.

In these formulas, carbons become points or angles and the attached hydrogen atoms are not written but assumed.

**Example Problem 1.6**
**Convert the following formulas into another type of formula**

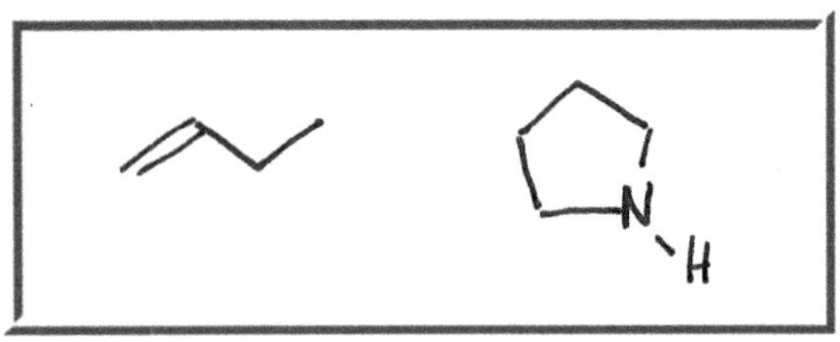


These formulas will be the preferred form for the remainder of this text. Hydrogen atoms attached to hetero-atoms that make up the functional groups are always expressed explicitly, as is the case for the $-NH_2$ group, the so-called Amino group, or the -OH group, known as the hydroxyl group, found in alcohols. The MF for ethyl alcohol (the normal drinking alcohol), known also as Ethanol, is $C_2H_6O$. However, another chemical, namely diethyl ether, has the same MF as ethyl alcohol.

Any molecules having the same molecular formula (MF), but a different spatial arrangement of the atoms, are called **isomers**. If the structural formula (SF) differs, then they are a specific type of isomer, namely a **constitutional** or **structural isomer**. This is the case for ethanol and dimethyl ether – the atoms are linked together in a different pattern.

**Fig. 1.17**

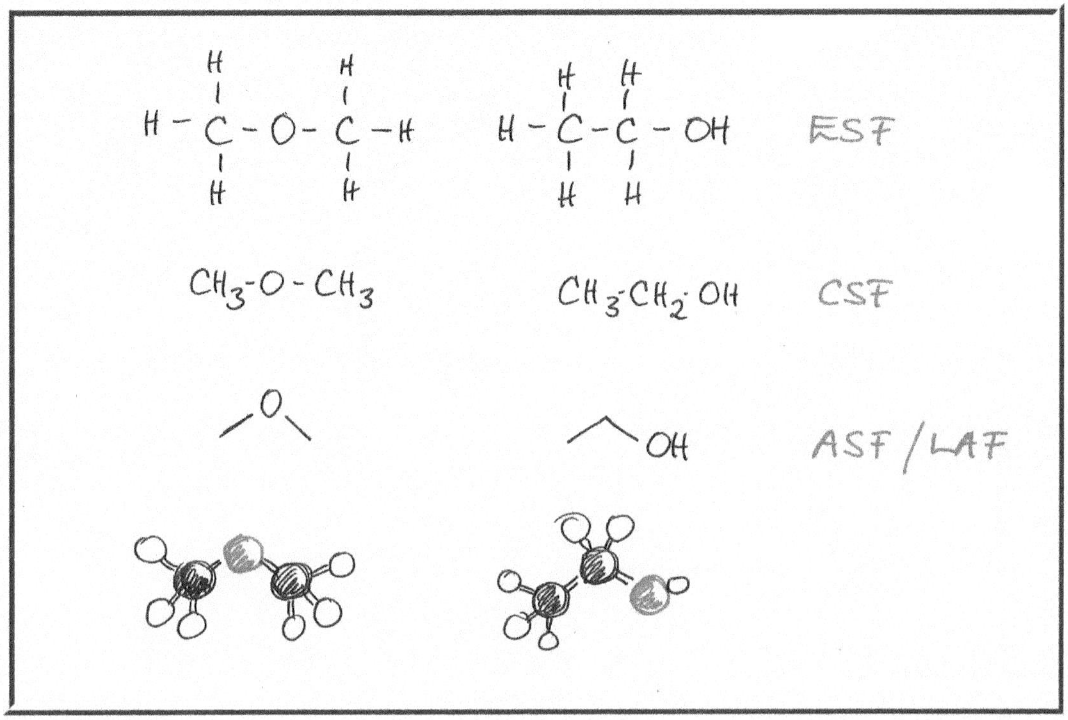


All molecules that have the same MF but different names are definitely structural (constitutional) isomers. There are two more types of isomerism that will be discussed in later chapters:

- **configurational isomerism** (stereoisomerism, geometric isomerism – isomerism based on same general structure but different spatial orientations or geometry)

**Fig. 1.18**

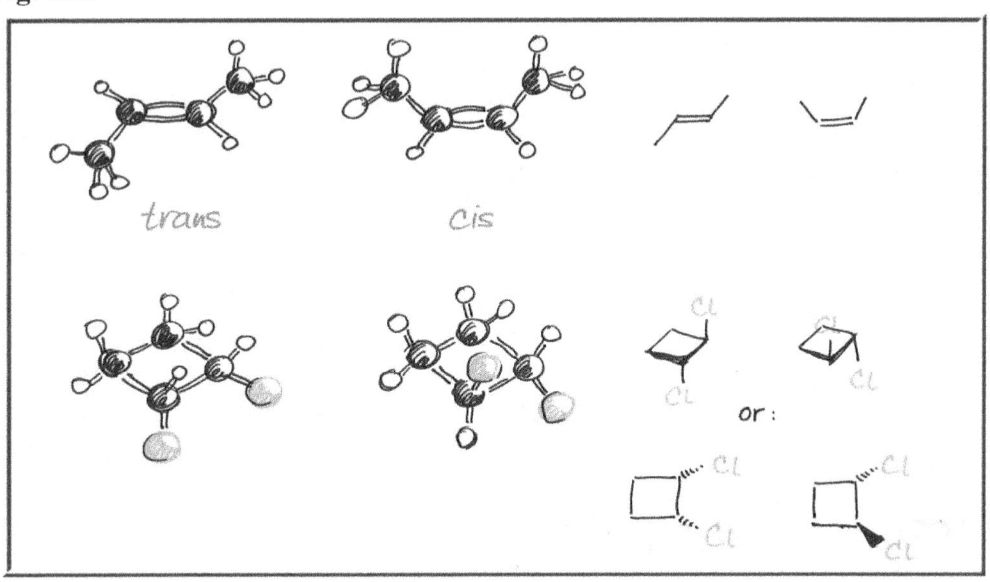

⅄ **conformational isomerism** (isomerism based on free rotation of the bonds)
**Fig. 1.19**

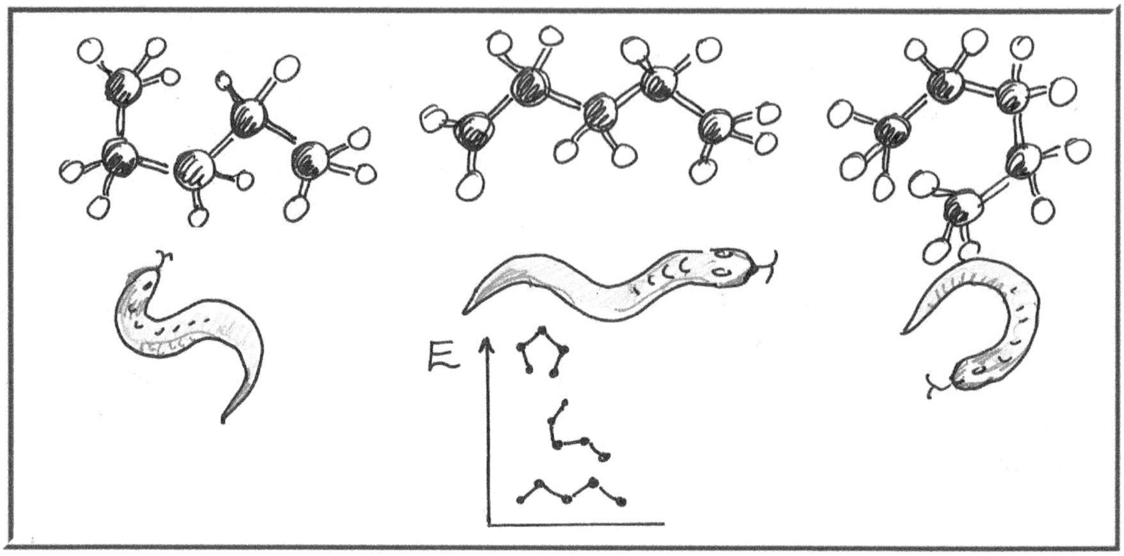


**Example Problem 1.7**
**Classify the following sets of two molecules as either being constitutional, configurational, or conformational isomers:**

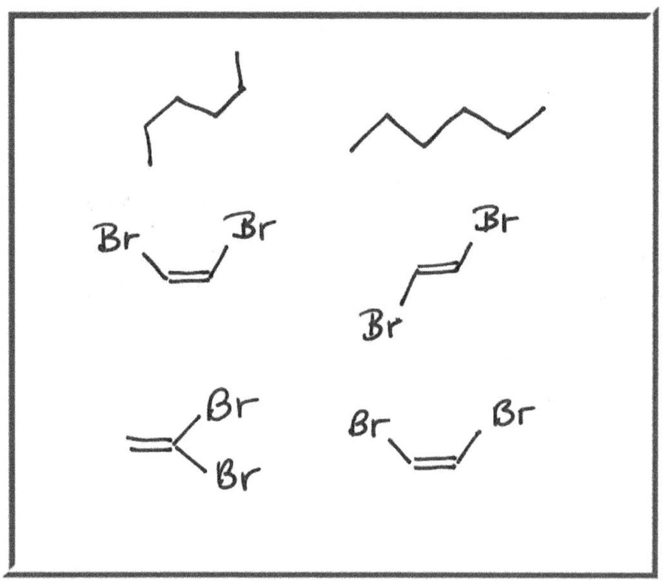


**Example Problem 1.8**
**To each of the following structures draw an additional constitutional, configurational and/or conformational isomer, if possible**

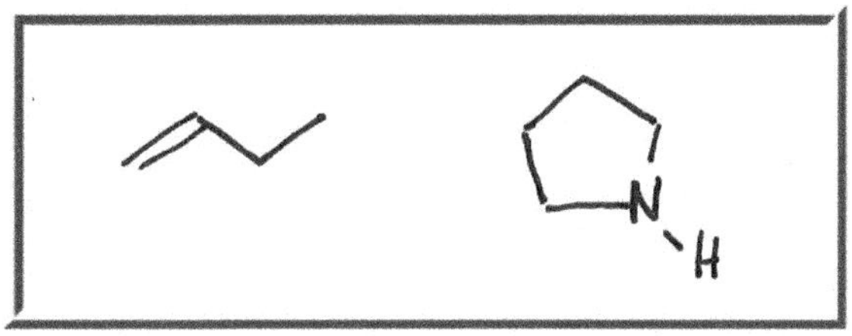

**Example Problem 1.9**
**Draw possible structures (2 or more) for the following molecular formulas**

$C_5H_{10}$ $C_4H_8Br_2$

All organic chemicals are divided into two major groups: the **acyclic and cyclic compounds.** The **acyclic** compounds comprise all open-chained structures with or without hetero-atoms, branched or un-branched. The open-chained pure hydrocarbons as well as the non-aromatic cyclic molecules are often referred to as **aliphatic** compounds.

The **cyclic** compounds subdivide themselves into two groups: the **carbocyclic** compounds, with rings made up of only C-atoms, and the **heterocyclic** compounds, where the rings can incorporate one or more, same or different, hetero-atoms. Carbocyclic compounds also go under the name **alicyclic** compounds.

Both cyclical types can be broken down even further into the so-called **aromatic or non-aromatic** carbocyclic or heterocyclic molecules, depending on whether the ring is made up of only carbon atoms or interspersed with hetero-atoms. The aromatic representatives are referred to as **arenes** (carbocyclic) and **hetarenes** (heterocyclic).

**Example Problem 1.10**
**Classify the following structures as acyclic or cyclic**

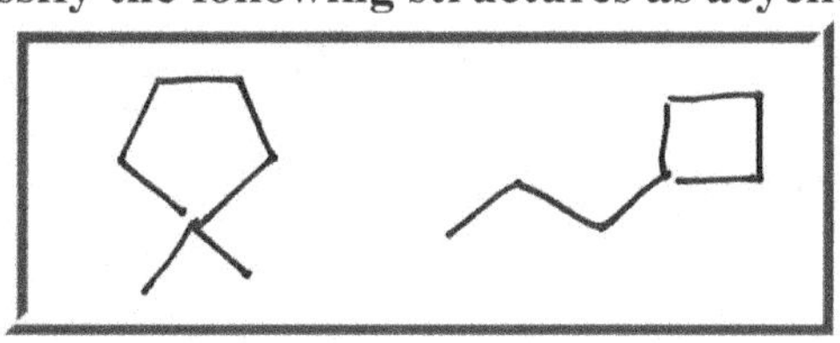

**Example problem 1.11**
**Group the following compounds as either being carbocyclic or heterocyclic**

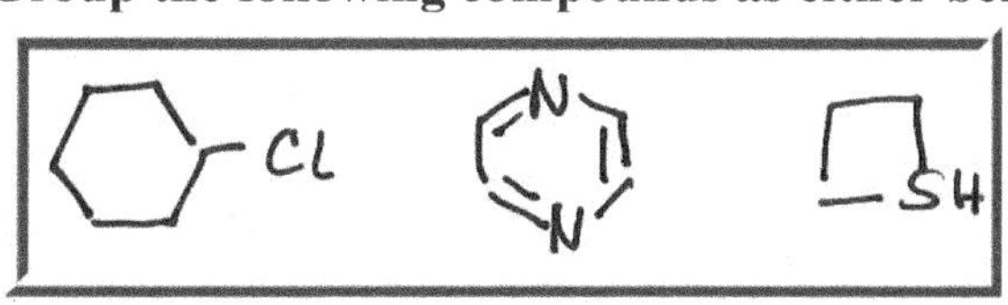


**Example Problem 1.12**
**Which of the following cyclic structure could be classified as being alicyclic?**

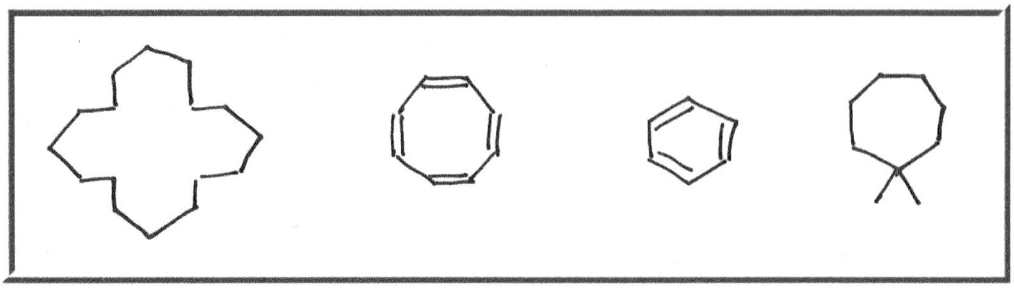

**Example problem 1.13**
**Which of the following molecules are aliphatic?**

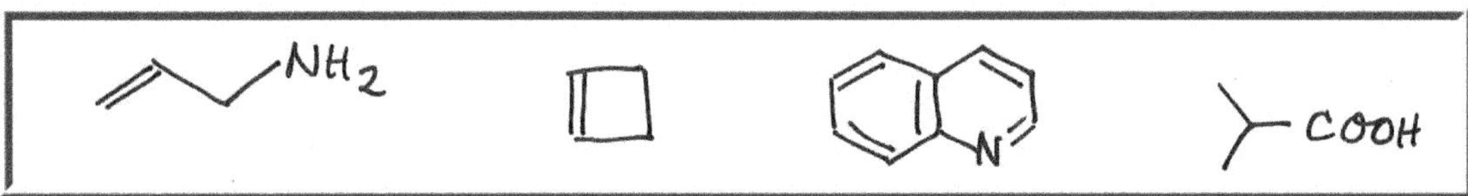


The **aromatic ring system** originally involved "aromatic," well-smelling molecules, but today represents a ring system with a special type of bonding that gives the molecule a significant energy gain and thus extra stability. The energy gain is the result of a special circular resonance of **conjugated double bonds** (2 or more multiple bonds each separated by a single bonds) or p-electrons, all in phase to overlap each other for an extended circular pi-bond, often referred to as a pi-sextet, producing a type of "ring current." Not all such bond arrangements qualify as aromatic, since they have to obey an empirical rule of **4n + 2 pi** electrons (n being any whole number: 0, 1, 2, etc.), known as the **Hűckel Rule** (Erich Hűckel, German (1896-1980)
The prototype for the aromatic system is the benzene molecule.

**Fig. 1.20**

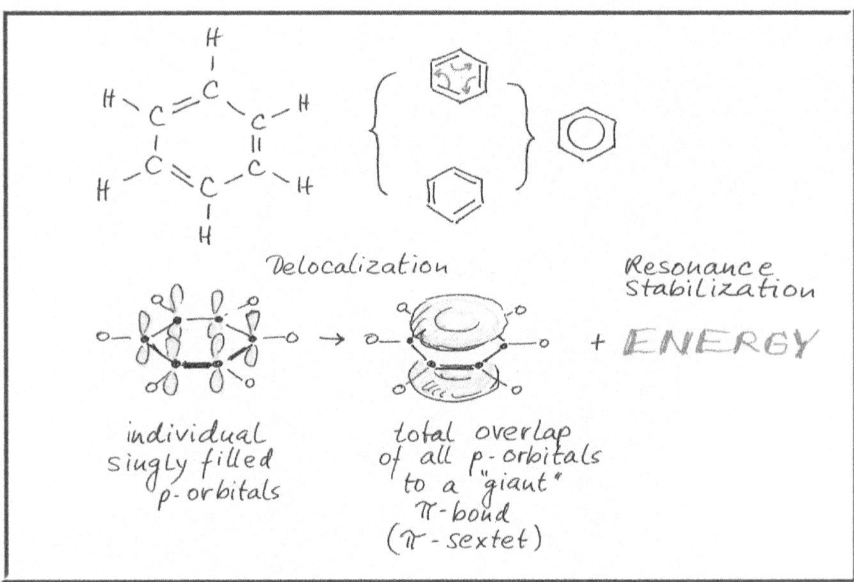

The double bonds (localized or de-localized) come about by an overlap of p-orbitals that have to be in-plane, like poles of a picket fence, to make **pi-bonds (π).**

**Example Problem 1.14**
**Which of the following cyclic structures with conjugate double bonds (2 pi electrons per double bonds) fits the Hűckel-Rule and is aromatic?**

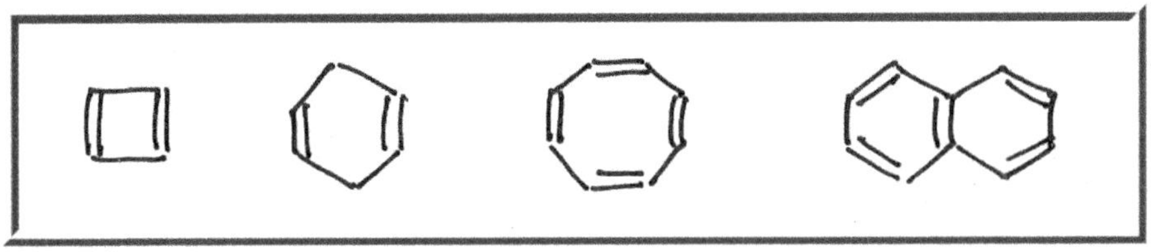

The following table gives a compact overview of some major classes of organic compounds that will be addressed in the context of this course.

**Table 1.2**

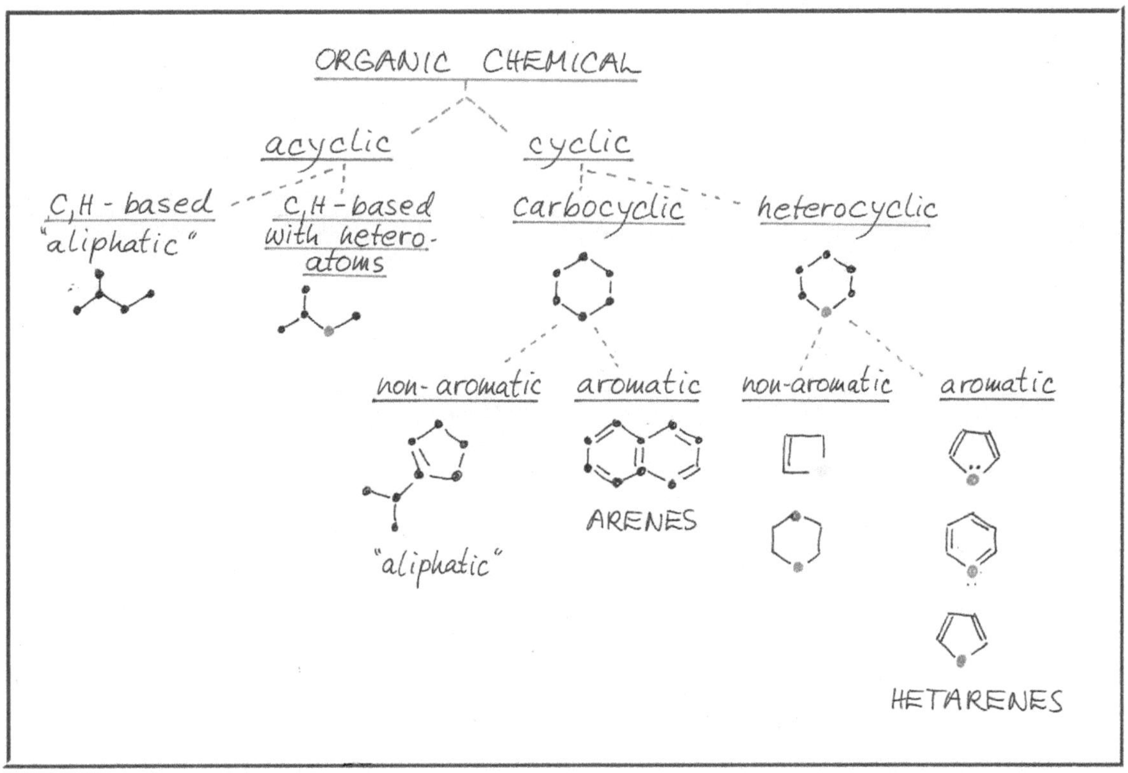


The **Lewis dot structures** of organic molecules involve all the bonded and lone pairs (if present).

The molecule can be simplified by a semi-condensed structure.

By moving lone pairs into bonded positions or vice versa, resonance structures can be produced. Electron pairs are usually moved to the more electronegative element.

The more resonance structures are possible – where the normal structure is included – the more stability can be achieved by the molecule or molecular ion.

Acetic Acid is a proton donor and the resulting acetate ion is resonance-stabilized as follows:

**Fig. 1.21**

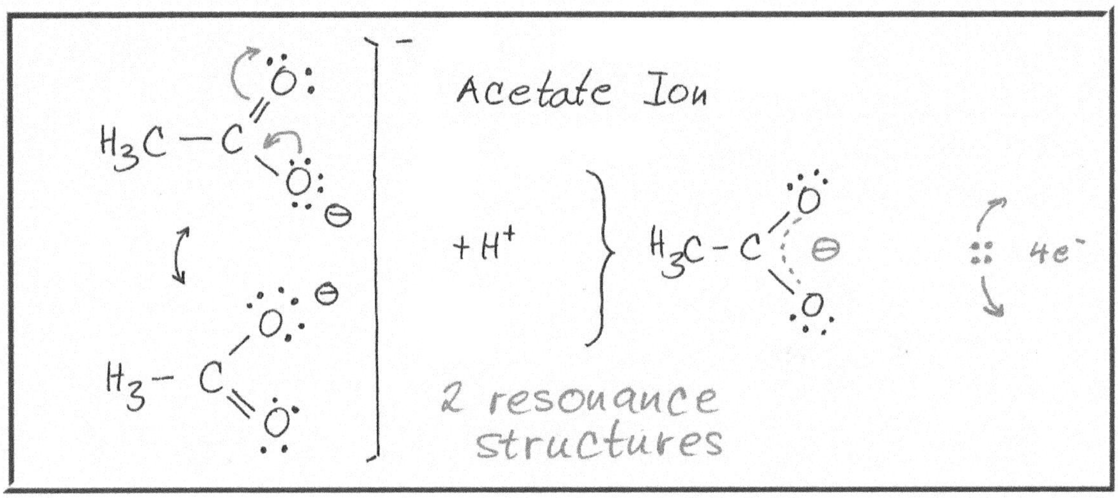


4 Electrons oscillate between the two oxygen atoms as is indicated by the dotted line. The negative charge then appears between the 2 oxygen atoms as well, since an ion is involved. Hence, the charge is also "de-localized." Sometimes a subdivided charge is distributed among the involved atoms as fractions

**Example Problem 1.15**
**Draw a Lewis dot structure for the following molecules or ions:**

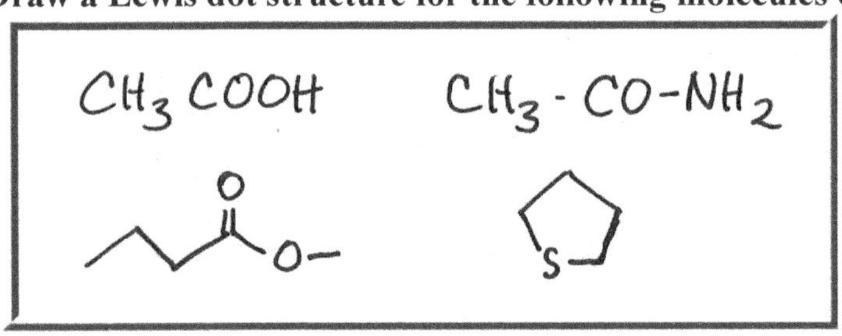


**Example Problem 1.16**
**To each of the shown structures below draw a possible resonance structure.**

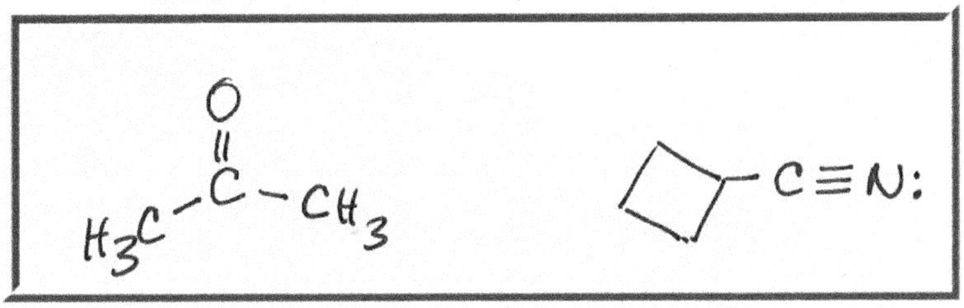

**Example Problem 1.17**
**Draw a resonance hybrid structure for the given resonance structures for each example given below:**

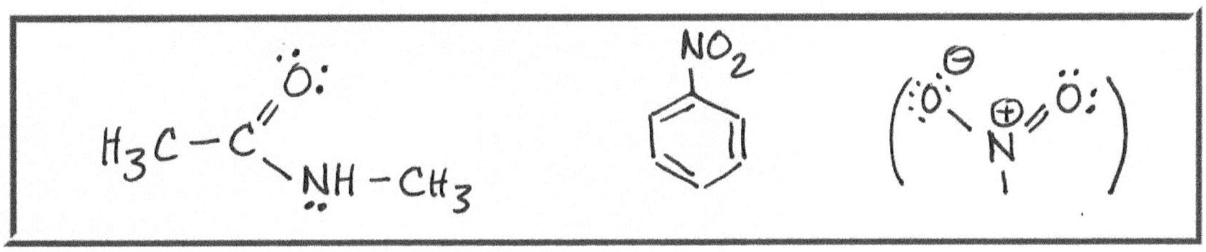

If this chapter is well studied and the "rules of the game" internalized as much as possible, all the following chapters in organic as well as biochemistry can be mastered with much more ease.

According to older classification systems all organic compounds can be divided into **aliphatic** and **aromatic** compounds. Aliphatic substances can be cyclic or acyclic, with or without heteroatoms N, O, S etc.), but must be, if they happen to consist of ring structures, strictly non-aromatic. Aromatic are all cyclic compounds with or without heteroatoms that have a specific energy gain referred to as resonance energy provided by interacting circular pi-systems. This "ring current" should follow the Hückel rule (4n+2 pi electrons required, n= 0.1.2…) as is already discussed above, but which will be covered in more detail in chapter 4. Aliphatic cyclic compounds can be saturated, having only single bonds, but aromatic rings must be unsaturated, meaning all multiple bond must have their pi systems in plane. The following tables show an overview of the classes and types of organic compounds with specific examples:

**Table 1.3**

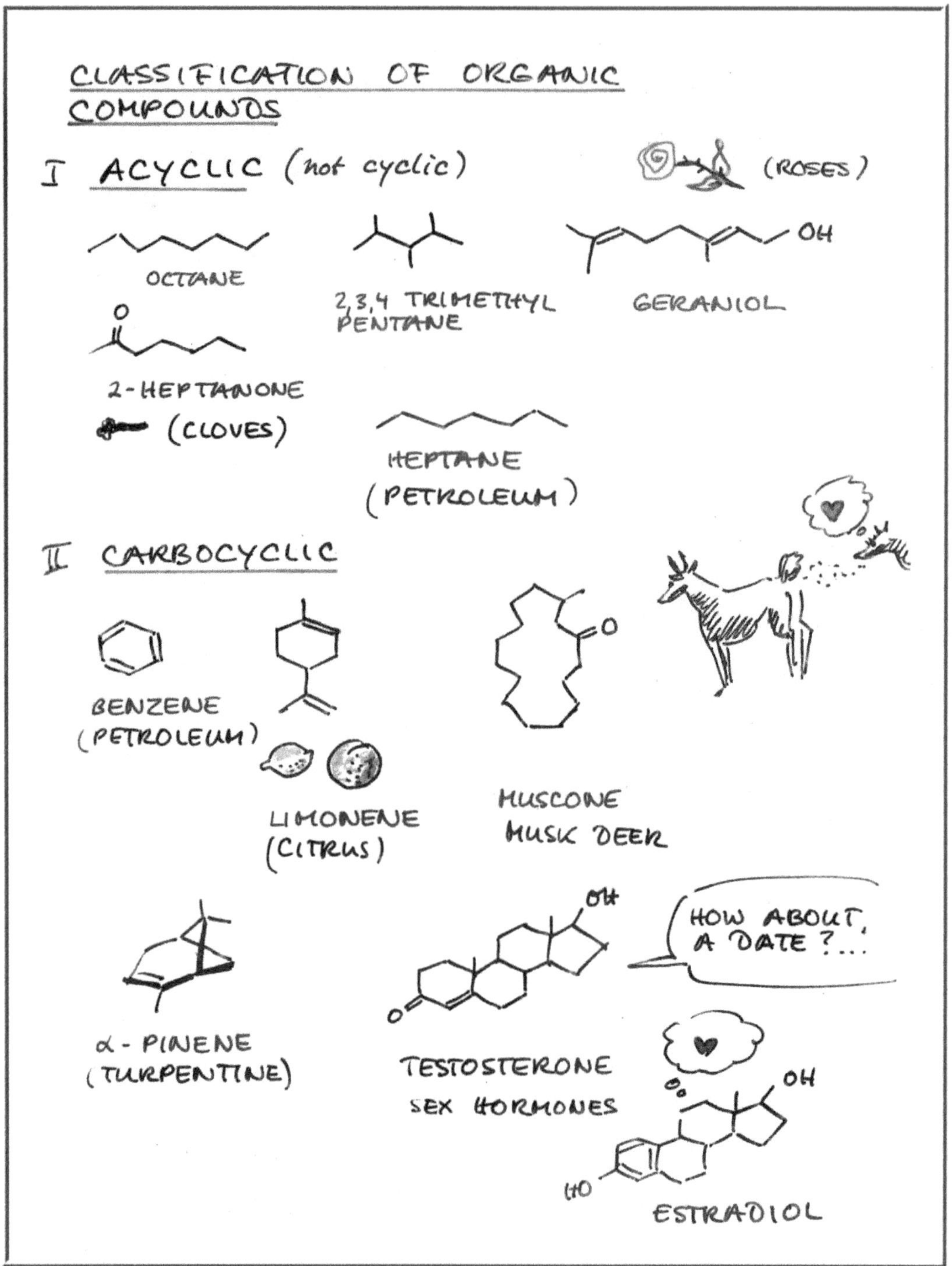
CLASSIFICATION OF ORGANIC COMPOUNDS
I ACYCLIC (not cyclic)
(ROSES)
OCTANE
2,3,4 TRIMETHYL PENTANE
GERANIOL
OH
O
2-HEPTANONE
(CLOVES)
HEPTANE
(PETROLEUM)
II CARBOCYCLIC
BENZENE
(PETROLEUM)
LIMONENE
(CITRUS)
O
MUSCONE
MUSK DEER
α-PINENE
(TURPENTINE)
OH
O
TESTOSTERONE
SEX HORMONES
HOW ABOUT A DATE?...
OH
HO
ESTRADIOL

**Table 1.4**

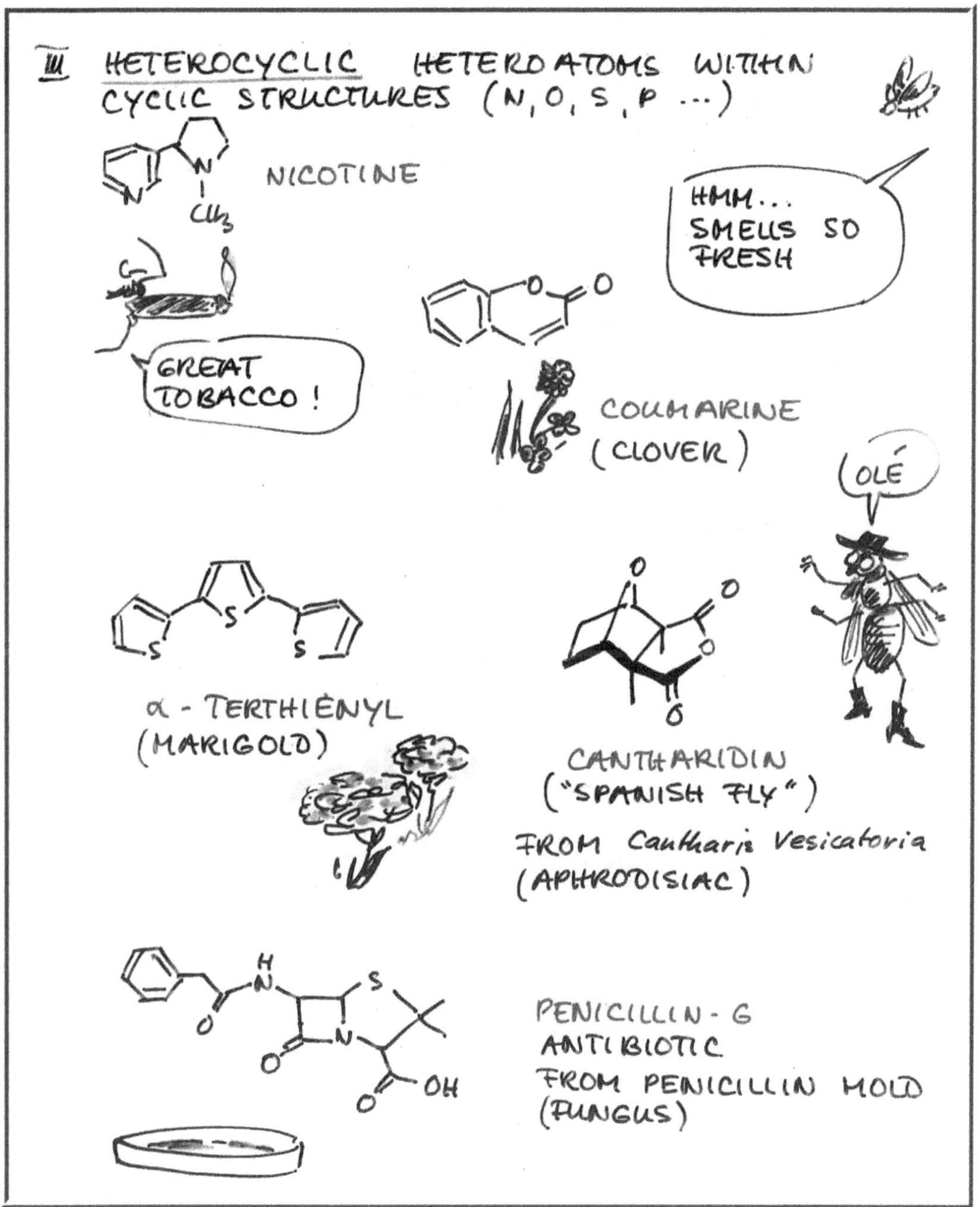


In table below gives a sample of many structural possibilities for natural products, which can be aliphatic or aromatic. The formula symbolism for organic molecules will be laid out in the following chapters.

**Table 1.5**

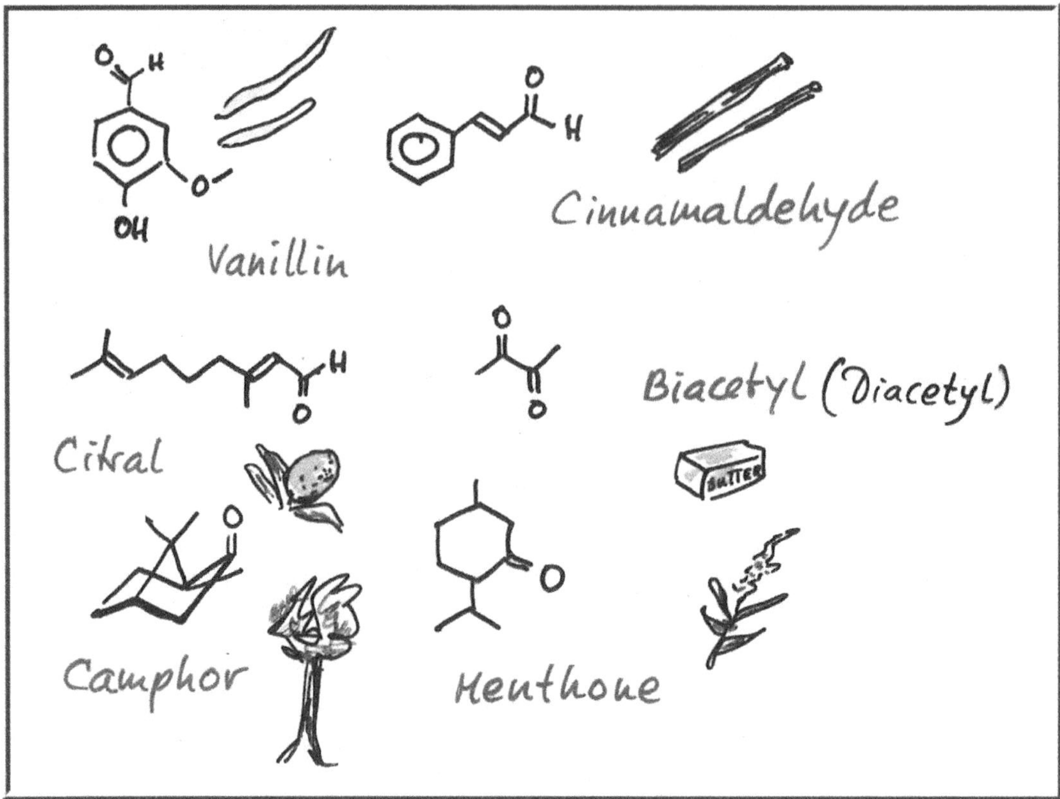
Vanillin
Cinnamaldehyde
Citral
Biacetyl (Diacetyl)
BUTTER
Camphor
Menthone

# EXPERIMENTS – Ch. 1

## *Introduction to Organic Chemistry*

**Key Terms:** Hydrocarbon compounds, saturated and unsaturated compounds, functional groups, non-polar and polar covalent compounds, hetero atoms, solubility and miscibility, conversion points (Mp, Bp), types and classes of organic compounds, aliphatic, aromatic, cyclic, acyclic, carbocyclic, heterocyclic.

### I Wöhler's Experiment – Thermal Conversion of an Inorganic Compound into an Organic Compound.

**Chemicals:**

*Ammonium cyanate ($NH_4NCO$)*
*Ammonium thiocyanate ($NH_4SCN$)*
*Biuret reagent ($CuSO_4$, KOH)*

**Equipment and Materials:**

*Test tube (large, pyrex), clamp or holder*
*Bunsen burner (hot flame)*
*Test tubes (medium) with rack*
*Electrode set-up with light bulb and small beakers*
*Infrared spectrometer (IR or FTIR)*

**Procedure:**

1. Place 2-3 spatula tips of $NH_4NCO$ in a large pyrex test tube. Attach a test tube clamp or holder and swerve to and fro through a strong Bunser burner flame. Heat a few minutes, then set aside to cool.
2. Scrape out the reaction material and place on a watch glass.
3. Dissolve a bit of unreacted $NH_4NCO$ in water (small beaker) and test the solution for conductivity by placing the solution under the set of electrodes. Do the same for the reacted material and compare the conductivities.
4. To the two solutions in the small beakers from above, add a few drops of the biuret reagent and note any color change. Copper (II) ions form a violet colored complex with amide or peptide groups ($-CONH_2$ and -CONH-). Biuret is a dimer of urea that gives the reaction its name.
5. If available, test a small solid sample of the reaction product in the IR-spectrometer to search for the characteristic absorption bands of the carbonyl (-CO- ) group and the amide ($-CONH_2$) groups, typical of any formed urea as organic compound, $(H_2N)_2CO$ (this can be demonstrated or carried out by the instructor)
6. Repeat the reaction with $NH_4SCN$ which should lead to thiourea as organic compound, $(H_2N)_2CS$.

## Observations and Results:

| Observations/Tests: | $NH_4NCO$ vs. $(H_2N)_2CO$ | $NH_4SCN$ vs. $(H_2N)_2SO$ |
|---|---|---|
| • **Conductivity**<br>• **Biuret reaction**<br>• **IR spectrum** | | |

## Questions and Problems:

1. Explain any changes in conductivity.
2. Discuss the reactions with the biuret reagent.
3. Interpret the IR spectrum as far as any of the typical absorption bands were observed for urea. Look up "infrared spectrum for urea" in the internet and compare with the one obtained experimentally.
4. How would an IR spectrum for thiourea differ from that of urea?

# II Classification of Organic Compounds by Type and Physical Properties

## Chemicals:

*Selection of sample bottles of various organic compounds (solid and liquid):*
*Hexane, dichloromethane, cyclohexane, cyclohexene, benzoic acid, nitraniline, ethylacetate, azabenzene, naphthalene and others*

## Equipment:

*None*

## Procedure:

1. Inspect each sample and classify according to appearance, color, odor, physical state.
2. Classify each sample according to the organic molecular type (hydrocarbon, alcohol, aldehyde, ketone, amine, ether, ester, thiol etc.) Look up the formula and do some research on the web or in the literature. Sometimes more classifications apply.
3. Record all observations in the table below:

## Observations and Results:

| Name of Chemical: | Color: | Odor: | Physical state, under normal conditions: | Type of compound: (*hydrocarbon, alcohol etc.)* |
|---|---|---|---|---|
| | | | | |

| | | | | |
|---|---|---|---|---|
| | | | | |
| | | | | |
| | | | | |
| | | | | |
| | | | | |

## III Conversion Point Determination (Mp., Bp.) for Organic Compounds

### Chemicals:

*Acetylsalicylic acid*
*Hexane*
*Ethyl acetate*

### Equipment:

*Melting point device*
*Capillary tubes (closed end), glass rod (long)*
*Hot water bath, used as steam bath*
*Test tube, thermometer*
*Metal stand and clamps, hot plate*

### Procedure:

1. Fill a capillary tube about 2-3 mm high in a capillary tube, by dapping the open end of the tube in the finely powdered sample. To ensure filling the tube to the bottom and tight packing of the crystals, send the capillary tube with open end facing upwards down a vertically held long glass tube. The bouncing of the capillary tube will lead the crystals to settle tightly at the bottom.
2. Place the tube in the proper slot of the melting point device and heat the metal block via a setting between 2 and 3. Insert the mercury thermometer into its provided slot and observe the heating process until the crystals collapse into a liquid. Read the temperature off the thermometer.
3. If the melting process is drawn out (wider Mp. range), record the beginning and the end of the melting process and determine the average.
4. Set up the metal stand and attach a medium sized test tube and a thermometer at a higher elevation using a slit, one-holed rubber stopper to surround the thermometer as support.
5. On a hot plate below test tube prepare a steam bath of boiling water in a beaker.
6. Place 2-3 mL of the liquid chemical (hexane and later repeated with ethyl acetate) into the test tube. Position the tip of the thermometer over the surface of the liquid (3-4 mm). Lower both equally into the steam bath (2-3 cm from the boiling water surface).
7. When the liquid sample begins to boil, its vapors will hit the tip of the thermometer and the temperature reading will mark its boiling point (Bp.).

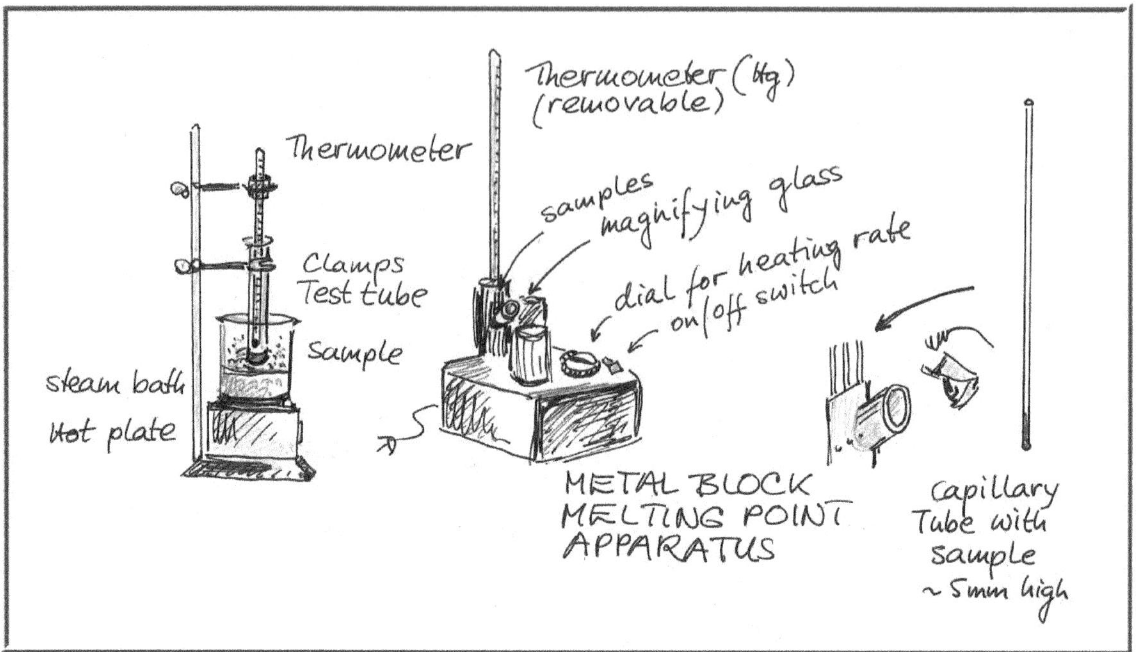


## Observations and Results:

| Chemical: | Mp. (°C) (theoretical): | Mp. (°C) (practical): | Bp. (°C) (theoretical): | Bp. . (°C) (practical): |
|---|---|---|---|---|
| **Acetylsalicylic acid** | 135 | | | |
| **Ethyl acetate** | | | 77.1 | |
| **Hexane** | | | 68 - 69 | |

## Questions and Problems:

1. What was the percentage error for the observed melting point (Mp.) compared to the theoretical value?
2. What was the percentage error for the measured boiling point (Bp.) of ethyl acetate compared to the theoretical value? Explain.
3. How is the melting point determination affected, if the crystals within the capillary tube are too crude or not grinded down enough? Explain.
4. How is the melting point measurement affected, if the capillary tube is packed too high?
5. If an organic liquid boils at 120°C, would a steam bath be adequate? Explain.

## IV Solubility and Miscibility of Organic compounds

### Chemicals:
***Solid solutes***: *Urea,*
*cinnamic acid, naphthalene , phenol*
***Liquid solutes***: *toluene, acetone, ethyl acetate,*
*ethanol*
***Solvents***: *toluene, acetone, ethyl acetate, ethanol, water*

### Equipment:
*Test tubes with rack, spatula*
*Droppers with bulb*
*Hot water bath*
*Hot plate*

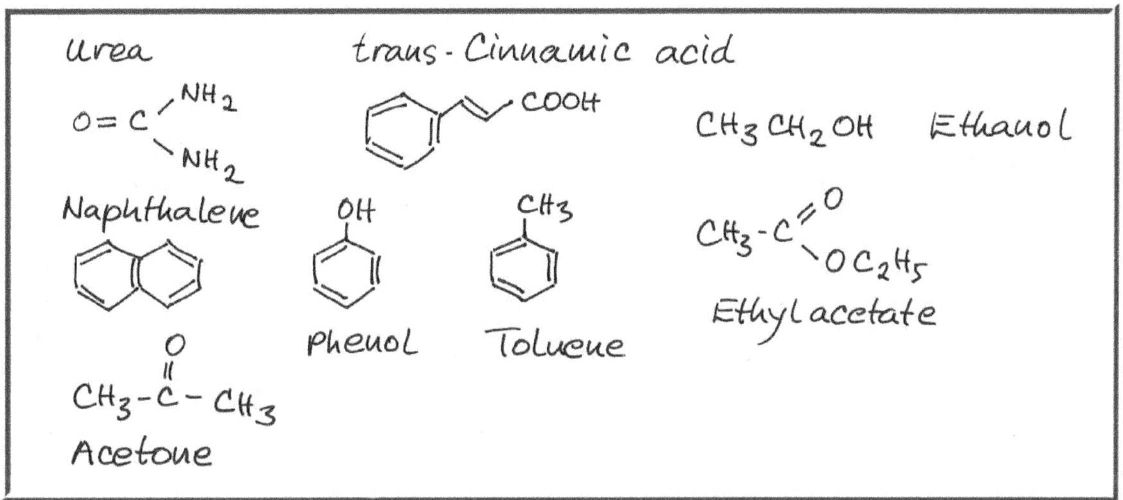


### Procedure:
1. Prepare the hot water bath in the fume hood.
2. Set up four test tubes in a rack.
3. For testing **solubility** of the solid chemicals in different liquid solvents, place a spatula tip of each solid solute separately in one of the test tubes.
4. Add a few drops of the first liquid solvent into each sample, swirl contents, add more solvent and/or heat in hot water bath, if necessary. Only aqueous solutions (containing water as a solvent) should be heated in a hot water bath! Organic solvents might catch fire or evaporate too quickly!
5. Clean out the test tubes and repeat the procedure for the other four solvents and record observations in the table below.
6. Clean out the four test tubes and fill about 1 cm high separately with each of the liquid solute samples to test **miscibility** of different liquids.
7. Add an equivalent the first solvent to each sample and swirl. Observe if one phase forms or the two liquids form two phases (phase separation).
8. Clean out the test tubes and repeat the procedure for the other four solvents in the list and record miscibility results in the table below.

## Observations and Results:

Evaluate solubility according to very soluble (++), slightly soluble, heat needed (+), and insoluble (-):

| Sample (solid solute): | Toluene: | Acetone: | Ethyl acetate: | Ethanol: | Water: |
|---|---|---|---|---|---|
| Urea | | | | | |
| Cinnamic acid | | | | | |
| Naphthalene | | | | | |
| Phenol | | | | | |

Evaluate miscibility with (+) for miscible and (-) for immiscible:

| Sample (liquid solute): | Toluene: | Acetone: | Ethyl acetate: | Ethanol: | Water: |
|---|---|---|---|---|---|
| Toluene | | | | | |
| Acetone | | | | | |
| Ethyl acetate | | | | | |
| Ethanol | | | | | |

## Questions and Problems:

1. Evaluate the different solubility and miscibility behavior by comparing relative polarities and the ability to form hydrogen bonds (as acceptors and/or donors).
2. Which of the solvents are **protic** (able to make hydrogen bonds as donors via hydroxyl groups) and which are considered **aprotic** (not able to act as hydrogen bond donors) organic solvents?
3. How does water solubility of organic compounds in general compare with that of inorganic compounds? Explain.
4. Why are some organic solutes such as acetic acid, ethanol, and sodium acetate very well soluble in water?
5. Many drugs are sparely soluble in water such as amines. To make them soluble as a pill taken orally, they are converted to hydrochlorides, indicated as drug name ------- hydrochloride (HCl). This doesn't mean that the drug contains hydrochloric acid. What type of chemical has actually been produced?

**Chapter 2 Alkanes and Cycloalkanes**

2.1 Homologous Series of the alkanes

2.1.1 General formula

2.1.2 Structures, geometry, and formulas

2.1.3 Constitutional (structural) isomerism by branching

2.2 Nomenclature

2.2.1 IUPAC nomenclature rules

13.2.2 Common names

2.2.3 The alkyl groups

2.2.4 Degree of substitution

2.3 Conformational (Rotational) isomerism by rotation of single bonds

2.4 Physical properties

2.5 Chemical properties and reactions

2.5.1 Combustion

2.5.2 Photohalogenation

2.5.3 Polymerisation

2.6 Cycloalkanes

2.6.1 Ring strain and relative stability

2.6.2 Ring-puckering and conformations (boat, chair for cyclohexane

2.6.3 Configurational isomerism (cis, trans)

2.6.4 Nomenclature

2.7 Natural resources

2.7.1 Natural gas (earth gas)

2.7.2 Petroleum Products

2.8 Uses and applications

2.8.1 Fuels

2.8.2 Anti-cracking additives

2.8.3 Lubricants and machine oils

2.8.4 Medicinal salves and balms (Baby oils, Vaseline etc.)

**Ch. 2 Alkanes and Cycloalkanes**

Alkanes are saturated, acyclic Hydrocarbons. The smallest alkane is **methane** ($CH_4$), the 2-carbon alkane is **ethane** ($C_2H_6$), the 3-carbon alkane is **propane** ($C_3H_8$), and on and on, making a **homologous series**

**$CH_4$**
**$CH_3CH_3$**
**$CH_3CH_2CH_3$**
**$CH_3CH_2CH_2CH_3$**
**$CH_3CH_2CH_2\,CH_2CH_3$**
**$CH_3CH_2CH_2\,CH_2\,CH_2CH_3$**
**Etc.**
**$CH_3\,(CH_2)_nCH_3$**

**Fig. 2.1**

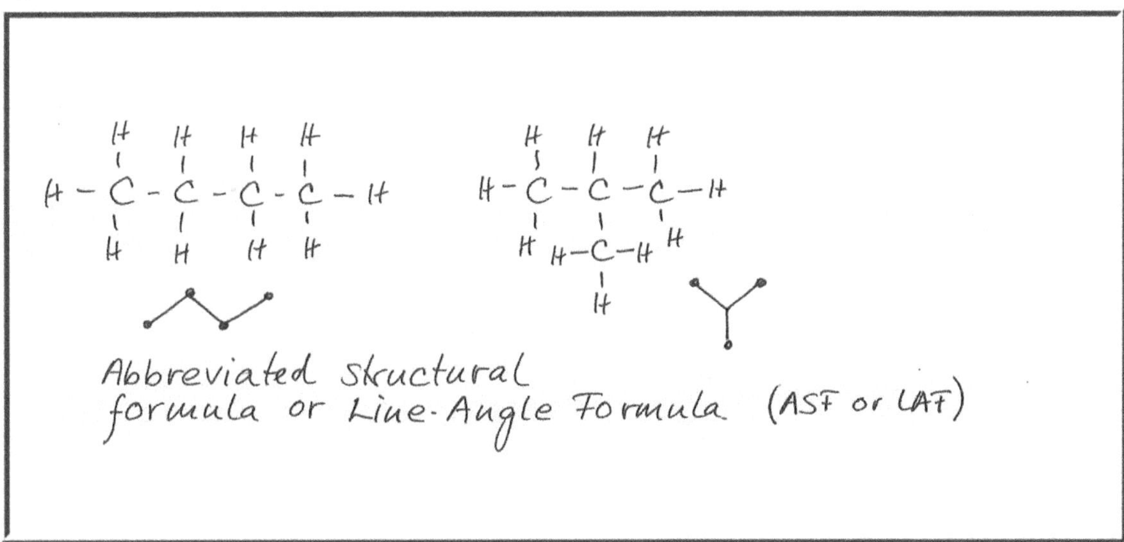


Each higher alkane has an extra **methylene group (-$CH_2$-)** compared to the previous one
The names of the alkanes all end with **-ane**. The first four representatives have their own original names ending with the 4-carbon-alkane **butane**. Every alkane with 5 or more C-atoms has its name based on the Latin numbering system (… penta, hexa, hepta, etc.). So the higher homologs are named pentane, hexane, heptane, etc.
The formulas for all alkanes follow a certain algorithm and therefore the alkanes, as a family, are considered a homologous series. With the help of the **general formula $C_nH_{2n+2}$**, the molecular formula of any alkane can easily be derived.

**Example Problem 2.1**
**What is the molecular formula for alkane with 32 Carbon atoms?**

$C_{32}\ H_{2(32)+2} = \underline{C_{32}H_{66}}$

**Example Problem 2.2**
**An alkane has 44 hydrogen atoms. What is the corresponding number of Carbon atom?**
**$C_xH_{44} = C_{(44-2)/2}\ H_{44}\ = \underline{C_{21}H_{44}}$ (21 Carbon atoms)**

This formula works for straight-chain as well as branched alkanes, which are constitutional isomers with the same molecular formula.
For molecules butane and beyond, the carbons can be linked together in various branched arrangements.

**Fig. 2.2**

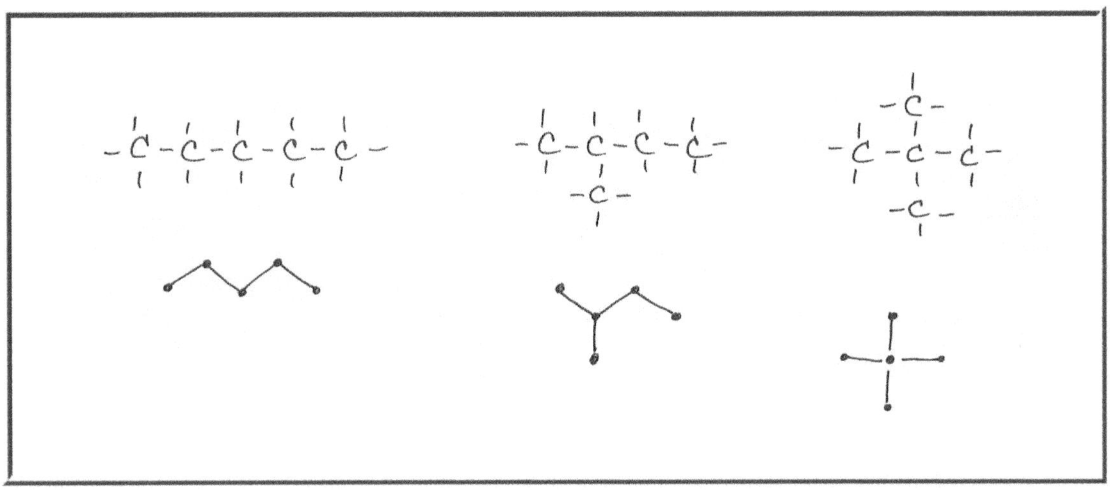

The different possible arrangements are referred to as **constitutional or structural isomers**. To convert one isomer into another, bonds must be broken and the atoms realigned and connections reformed. The **energy barrier** for isomeric inter-conversions is the highest in comparison with the other two major types of isomerism – configurational and conformational isomerism, already introduced in chapter 1.

.

**Fig. 2.3**

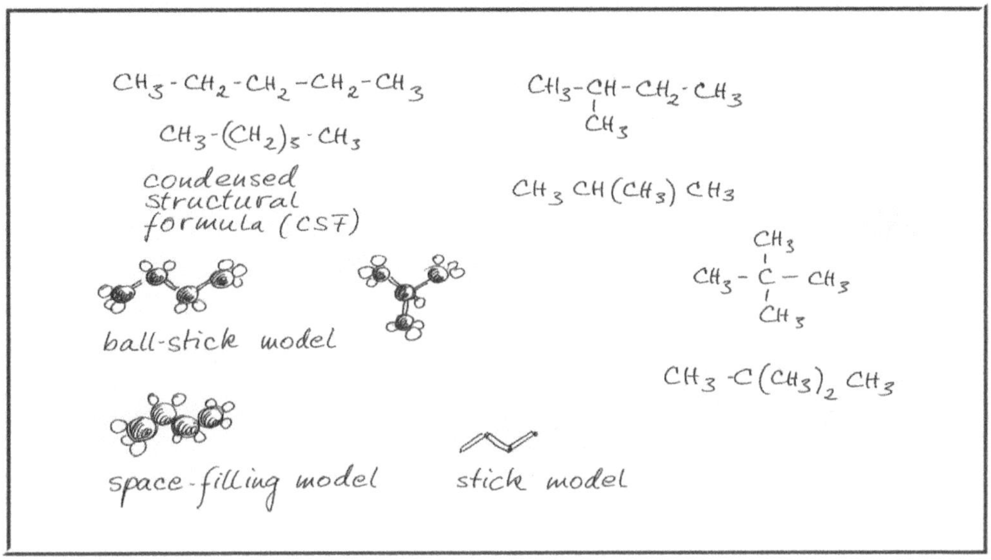


The pentane isomers still have their own names. Hexane and all the higher alkanes need systematic names for differentiation. The standard system for nomenclature of organic compounds is governed by the **IUPAC** organization – the **I**nternational **U**nion for **P**ure and **A**pplied **C**hemistry. The number of isomers increases significantly with increasing number of Carbon atoms, as is shown in the following table:

**Table 2.1**

No. of possible constitutional isomers

| No. C | Isomers | |
|---|---|---|
| 4 | 2 | \|\| |
| 5 | 3 | \|\|\| |
| 6 | 5 | \|\|\|\|/ |
| 7 | 9 | \|\|\|\|/ \|\|\|\| |
| 8 | 18 | \|\|\|\|/ \|\|\|\|/ \|\|\|\|/ \|\|\| |
| 9 | 35 | \|\|\|\|/ \|\|\|\|/ \|\|\|\|/ \|\|\|\|/ \|\|\|\|/ \|\|\|\|/ \|\|\|\|/ |
| 10 | 75 | |
| 11 | ' | |
| 12 | ' | |
| 13 | ' | |
| 14 | ' | |
| 15 | ' | |
| 16 | ' | |
| 17 | ' | |
| 18 | ' | |
| 19 | | |
| 20 | $3.66319 \times 10^5$ | |

Nonacosane, a straight-chain alkane with 29 carbon atoms ("nona" = nine, "cosa" = twenty), is used as a pheromone by many insects including the mosquito (*Anopheles stephensi).* It can be arranged in 1,590,507,121 different structural (constitutional) isomers. How this heavy molecule can act as a volatile sexual attractant over great distances is still a mystery. Besides activating chemical sensors within the insect's body such as antennae, it could also act via vibrational signals.

Another long-chain alkane with 24 carbon atoms known as tetracosane, found in minerals such as "evenkite", has the possibility of 14,490,245 isomeric structures. It is detected by its yellow-orange fluorescence.

Plants also harbor long-chained alkanes such as hentricontane or unitriacontane with 31 carbon atoms, allowing the possibility of 10,660,307,791 constitutional isomers. This molecule occurs in peas but also in the resin gum Arabic and in bees wax

The following illustration shows the type of branched structural isomers of the lower alkanes. The straight chain representative, indicated by *n* infront of the name such as *n*-hexane, derived from *"normal"*, also counts as one of the isomers.

**Fig. 13.4**

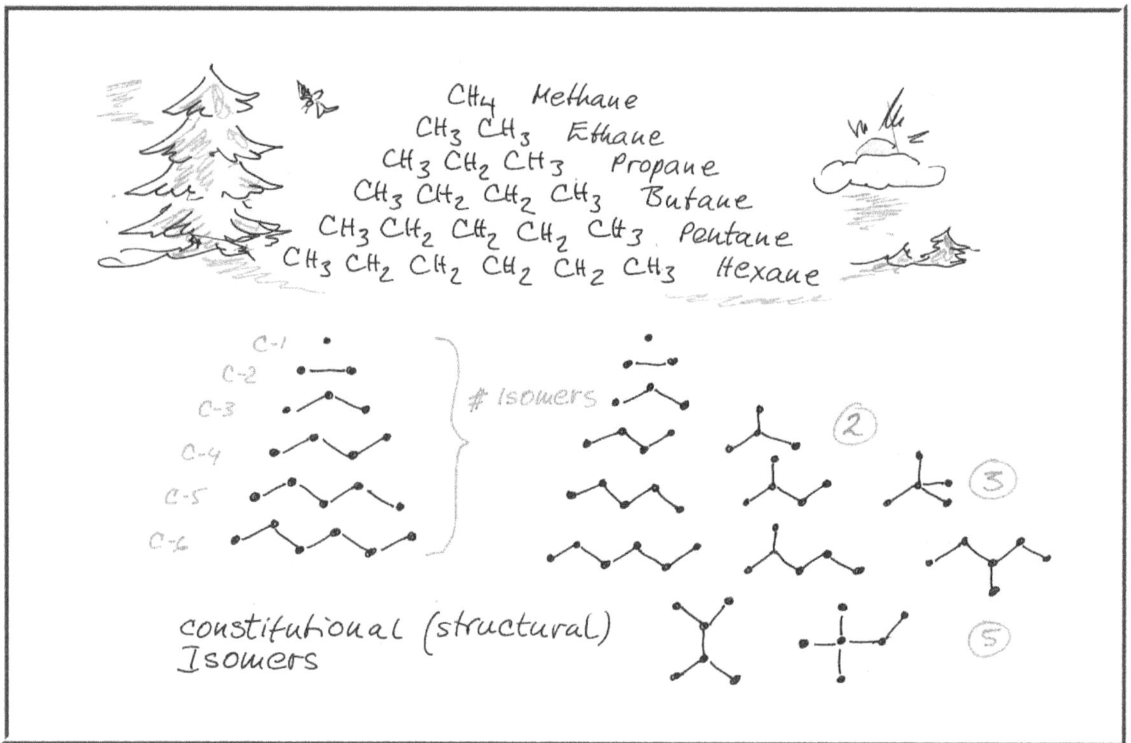


The individual names of the lower alkanes are referred to as "common names," that are often derived from Greek names, referring to their sources or origin. The standard rules for nomenclature follow specific rules that are very practical for branched forms of the parent alkanes. For this, the side-chains or branches are considered to be **alkyl** groups and, depending on the number of C-atoms involved, are named **methyl** (- $CH_3$ ), **ethyl** ( -$CH_2CH_3$ ), **propyl** ( -$CH_2CH_2CH_3$ ), **isopropyl** ( - CH $(CH_3)_2$ ), etc. The original ending of "ane" is replaced by

suffix **"yl"** indicating that an R-group (radical group or organic Rest) is implied, meaning the group cannot exist by itself and must be attached to something. A radical is strictly speaking a species with an unpaired electron ($^{\cdot}CH_3$), so incomplete by itself.
The primary rule for naming is finding the longest chain and naming it as the **parent**. The next step is to find and identify the substituents or side-chain groups ("the children"). By numbering the longest chain, the position ("home address") of the side-chains can be identified. The idea is to keep numbers low or finding the lowest number combination if more groups are involved. The parent is written last in the name and the substituents are enumerated and named, in alphabetical order, in front of the parent.

**Fig. 2.5**

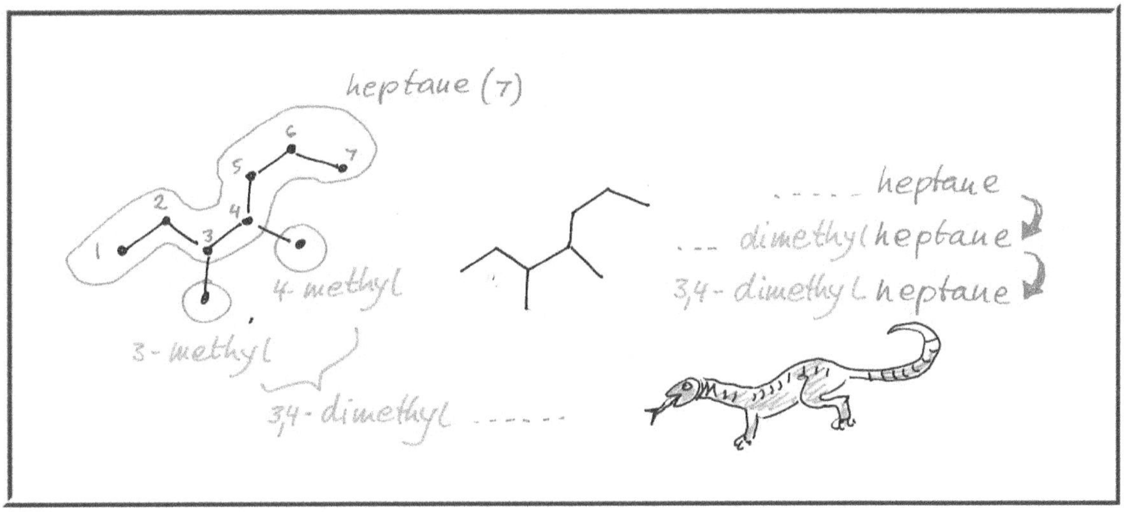


Punctuation rules apply. Numbers are separated by commas whereas numbers and names are separated by hyphens or dashes. Letters are written together, not separated, with the exception of **italicized prefixes**, such as *cis-*, *trans-*, *sec-*, *tert-*, etc.

**Fig. 2.6**

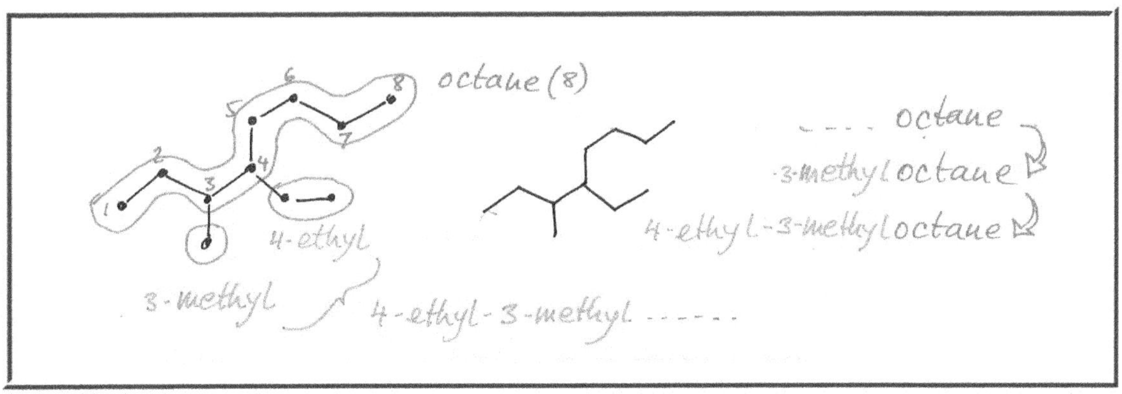

There are certain rules for writing branched alkanes in a condensed molecular formula. Side-branches are placed in parenthesis usually to the right of the carbon they are attached to.

1 2 3 4
**$CH_3CH(CH_3)CH_2CH_3$**

The linking element $–CH_2$- is called the **methylene group**. A formula can be simplified by placing repeating groups also in parenthesis

**$CH_3C(CH_3)_2\ (CH_2)_4CH_3$**

One should be careful not to confuse linker groups such as $–CH_2$- with terminal (side-chain) groups such as $–CH_3$ (methyl), $–C_2H_5$ (ethyl) or other.
When more structural isomers exist, the un-branched, straight chain representative is supplied with a small **n** hyphenated to the parent name – **n-alkane**, in order to distinguish it from the other branched isomers. For example, hexane that can exist in five possible structural types would be named n-hexane in its linear form. The prefix "n-" is derived from "normal" as to the normal, un-branched variety.

**Example Problem 2.3**
**Name the following molecules:**

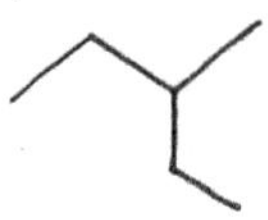

$CH_3\ CH_2\ CH_2\ CH_3$

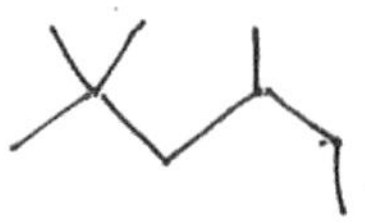

$CH_3-CH(CH_3)-CH(CH_3)\ CH_3$

**Example Problem 2.4**
**Give the structure for the following names:**
***2,3-Dimethylhexane***
***2,2-Dichlorobutane***
***3,3,4-trimethylheptane***
***3-Ethyl-4,4-dimethyloctane***

**Cycloalkanes** are closed ring structures that require the removal of two hydrogen atoms for the possibility of ring closure, which is done by connecting the resulting end groups (minus a hydrogen atom each).

**Fig. 2.7**

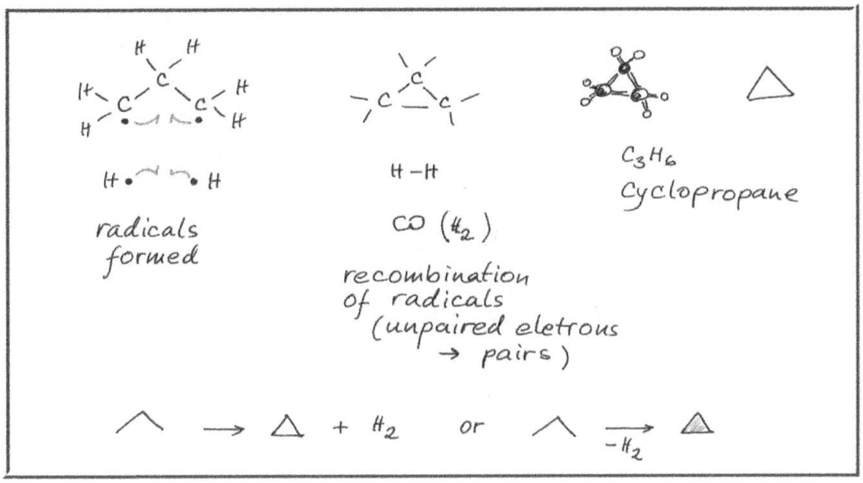


The general formula for the **homologous series** of cycloalkanes is $\mathbf{C_nH_{2n}}$. Some representatives are:

**Fig 2.8**

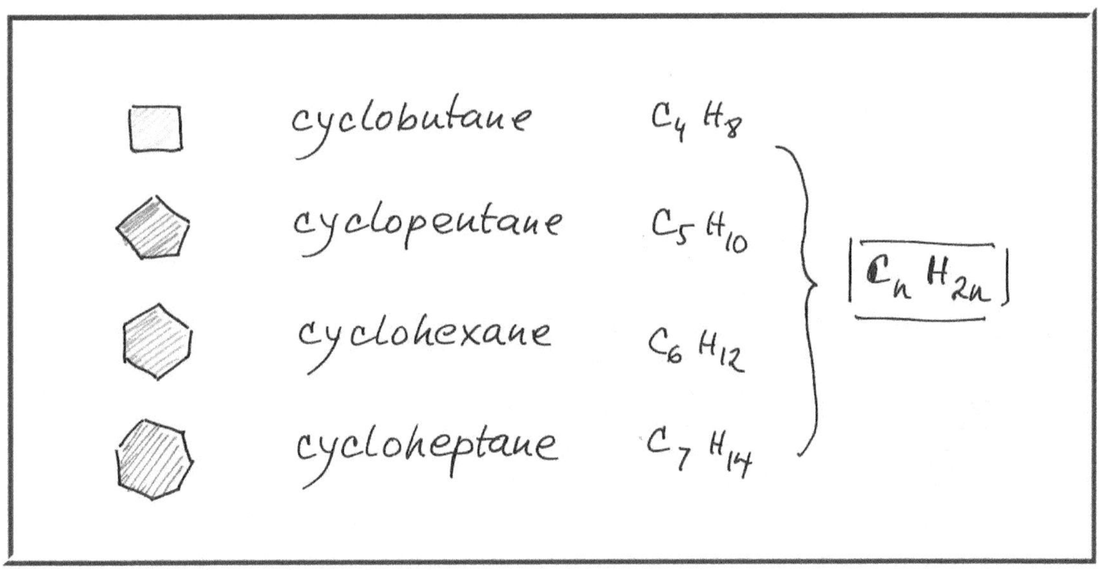


The best formulas for their representation are the abbreviated or line-angle formulas. Condensed formulas are no longer much in use.

The nomenclature rules are similar to the acyclic alkanes. The parent here is always the ring itself. The positions or carbon atoms are numbered if more than one side-chains or other substituents (groups) are present, keeping in mind that numbers are kept low and that the alphabetical listing of the groups prevails.

**Fig. 2.9**

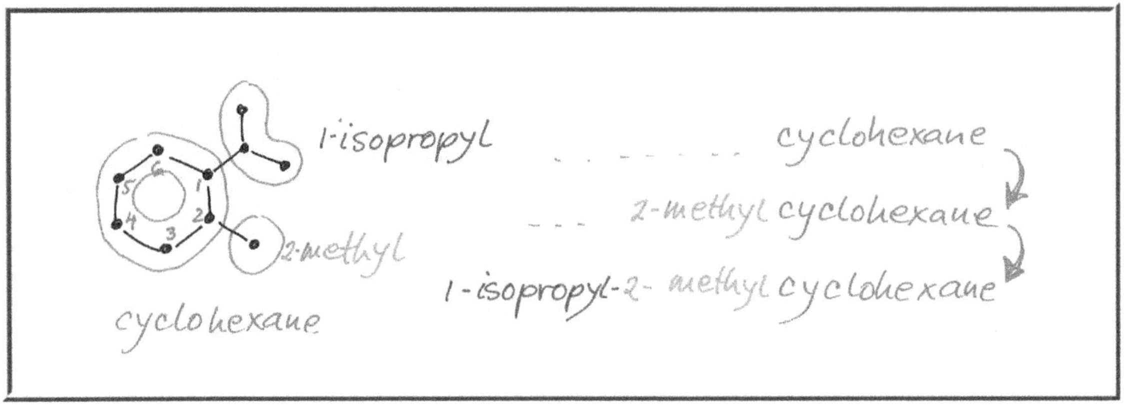


For multiples of the same group that are attached to the same molecule, then the Latin prefixes **di, tri, tetra**, etc., must be used. Each substituent, even if sharing the same carbon, must have its own number or **locant.**

**Fig. 2.10**

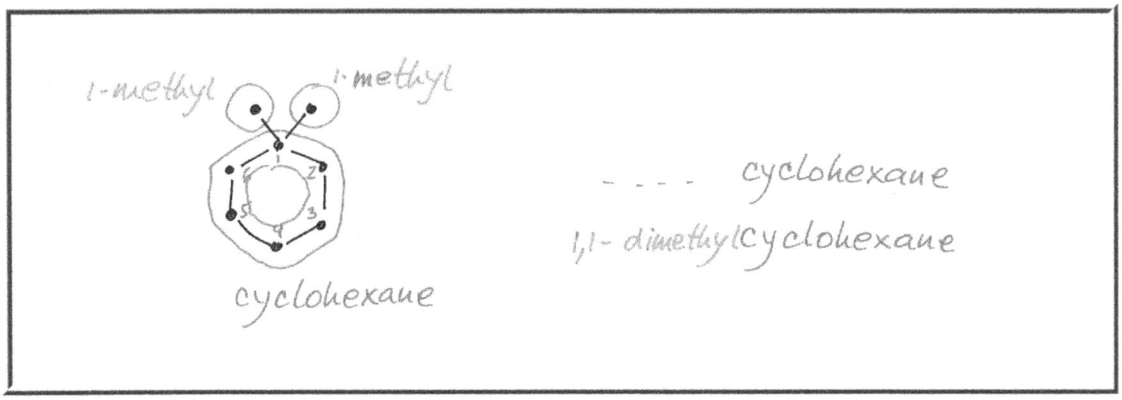

**Example Problem 2.5**
**Name the following molecules:**

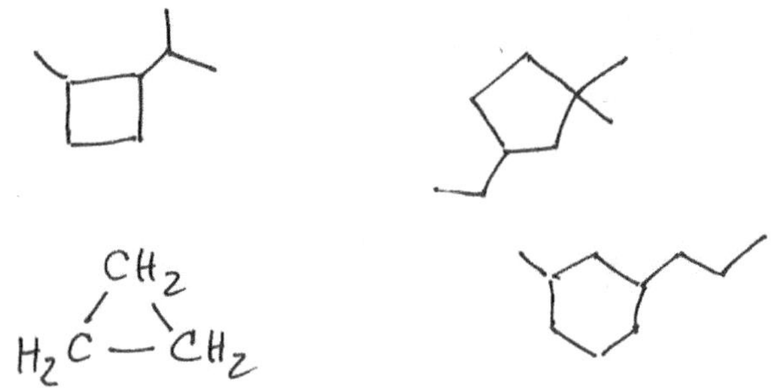


**Example Problem 2.6**
**Give the corresponding structures (ASF) for the following names:**
***1,2-dimethylcyclopentane***
***Cis-1-chloro-2-bromocyclohexane***
***Isopropylcyclobutane***
***1,1,4-Triethylcyclohexane***

In saturated hydrocarbon molecules, Carbon is surrounded by 4 bonded electron pairs. According to the VSEPR theory (valence-shell-electron-pair-repulsion) the four density areas will seek maximum distance from each other, thereby leading to a tetrahedral, geometric arrangement. The optimal, maximum distance for the for the electron pairs, in order to minimize the repulsion of like charges, is best achieved by each pair pointing into the four directions of a tetrahedron with an ideal bond angle of 109.5° between each of the two bonds.

**Fig. 2.11**

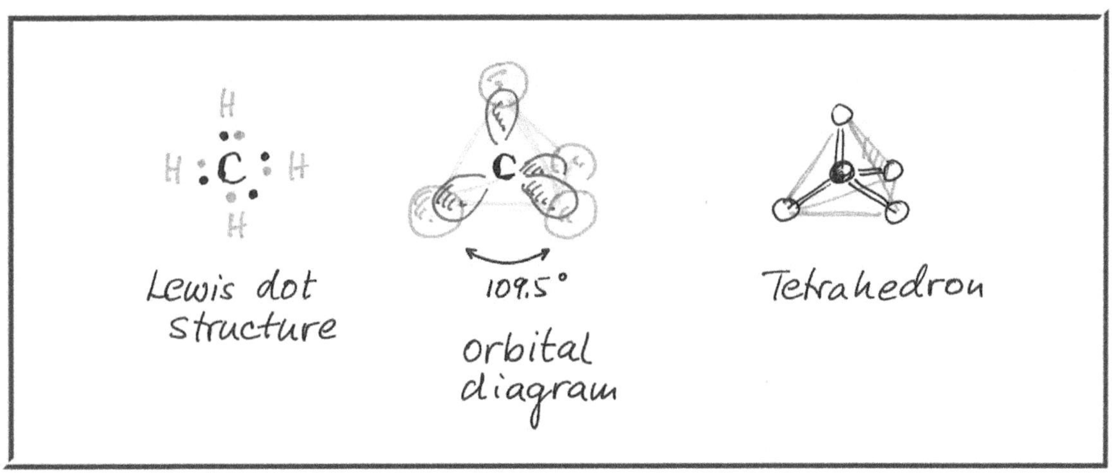


All structures try to preserve this angle. The line angle or abbreviated formulas best represent the actual bonding arrangements.

**Fig.2.12**

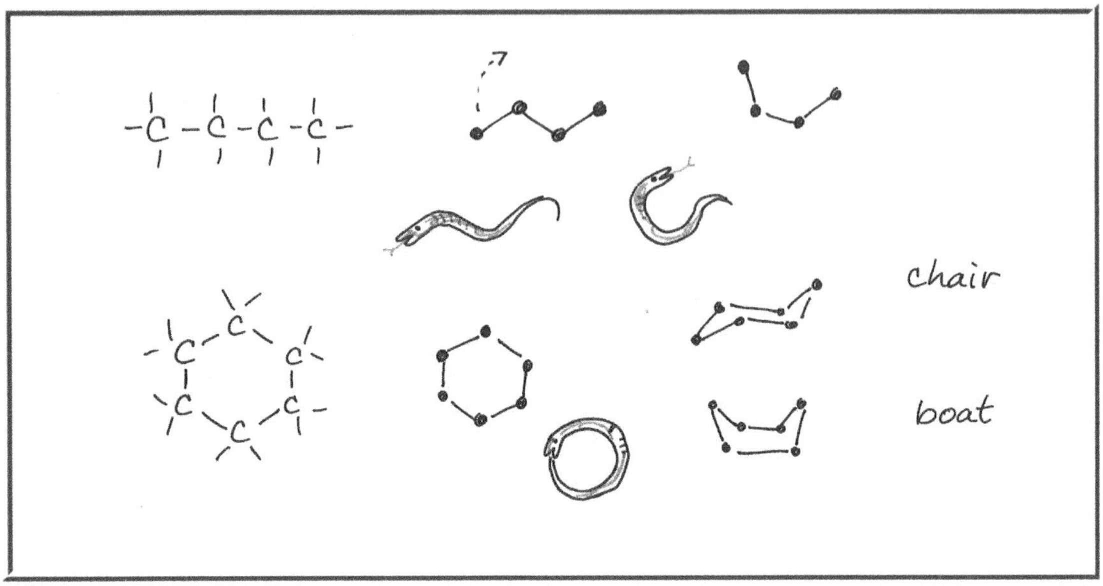


Carbon chains can be linear, branched or in cyclic form. All these are considered **aliphatic compounds. Carbocyclic** rings as long as they are not aromatic are also **called alicyclic.**
In cyclohexane, all tetrahedral angles are preserved, making the molecule not flat but puckered. The lower rings still have some **puckering** but it is obvious that the smaller the ring gets the more the tetrahedral angle is compressed, producing considerable strain on the ring, which results in a decrease of molecular stability. This type of ring tension is called **Pitzer tension** (Kenneth S. Pitzer, USA). A perfect example is cyclopropane, in which the tetrahedral angle is reduced to almost $60^{o}$, leading to considerable **angle strain**. Large rings experience another type of instability since they become too wobbly. This tension is known as the **Baeyer tension** (Adolf v. Bayer, 1835-1917, German). Cyclohexane is the midpoint for all cycloalkanes having zero angle strain.
Large rings also experience another type of strain called trans-annular or **Prelog** tension (Vladimir Prelog, 1906-1998, Croatian), where the large rings behave like a loose rubber band, some areas getting to close to each other.

**Fig. 2.13**

The snake-like moves of an acyclic alkane and the puckering types of the cycloalkanes are the result of free movement or rotate-ability of the single bond. This leads to a new type of isomerism, the **conformational or rotational isomerism**, based on the free rotation around the carbon atom as a pivot point or joint, similar to rotating an arm around the shoulder. The energy differences are only slight when interconverting on form into another. Keeping arms stretched overhead is more strenuous than having them dangling beside the hips can be considered two types of conformational "body" isomers with different energy content and stability.

**Fig. 2.14**

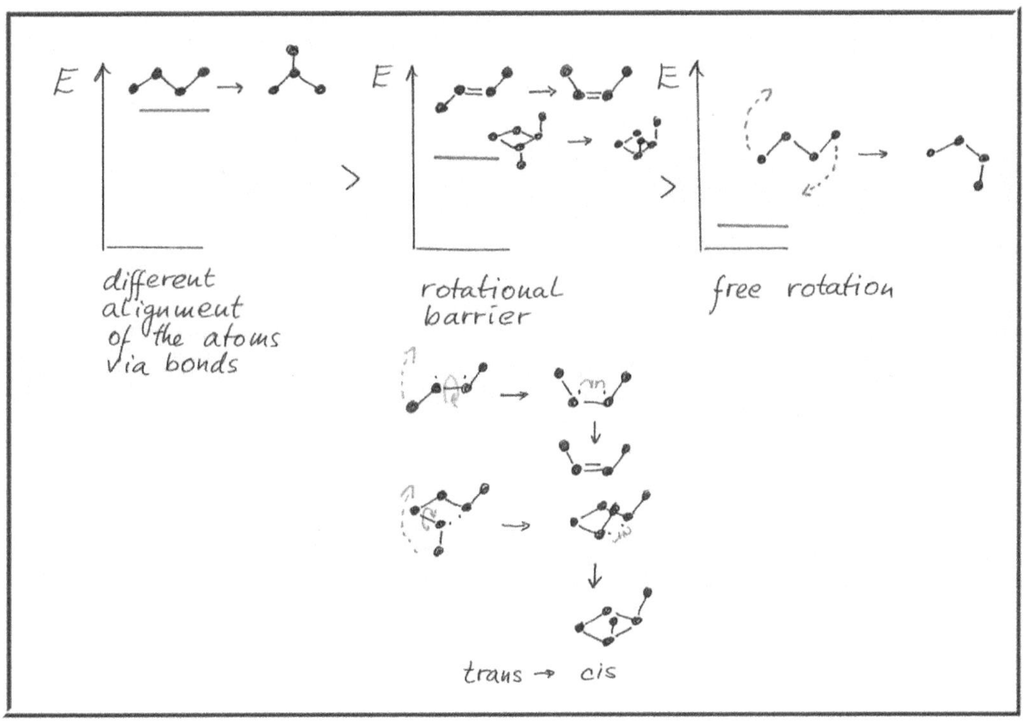

Acyclic compounds have no rotational barriers. However, cyclic hydrocarbons can only be twisted so far, the plane of the ring itself representing a rotational barrier. When substituents are attached to different C-atoms, they can be either on the same side of the ring – in the **cis** position, or on opposite sides of the ring - in the **trans** positions. Here another type of isomerism is encountered – the **configurational or geometric isomerism**, a type of **stereoisomerism** consisting of two opposites ("stereo"). This type requires the presence of a rotational barrier such as a ring or double bond. If the barrier is temporarily disrupted or opened up, one isomer can convert into the other, which is "frozen" into the new position upon renewed ring closure or reestablishment of the barrier.

**Fig. 2.15**

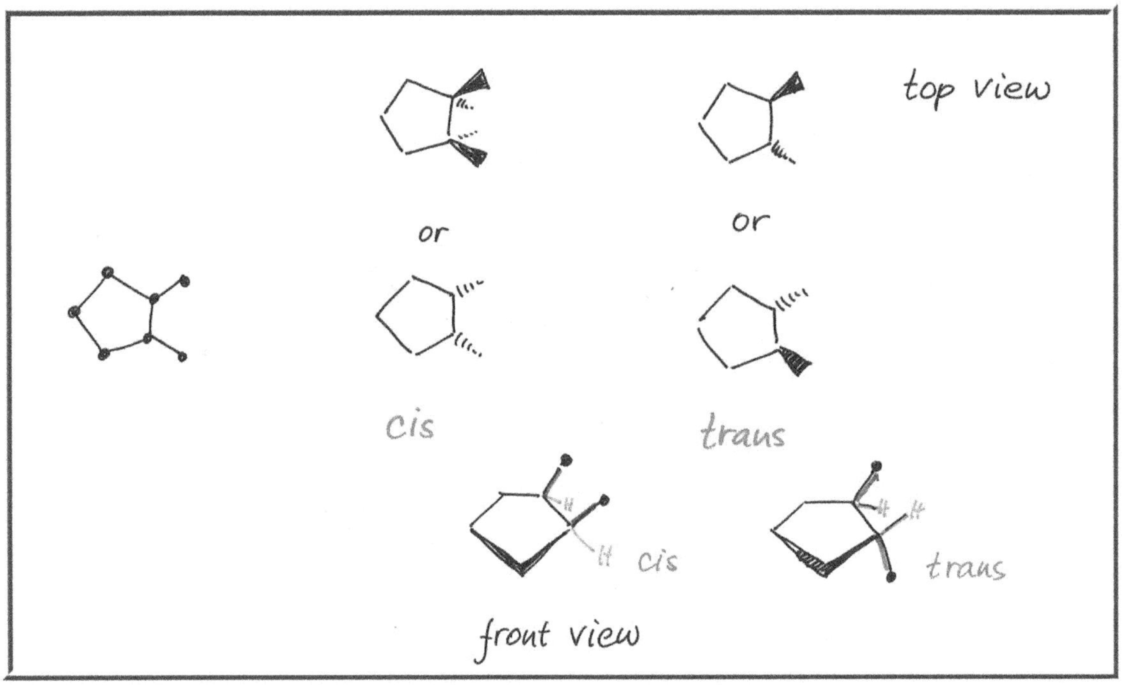


**Example Problem 2.7**
**Identify the type of isomerism involved for the following selected pairs of molecules:**

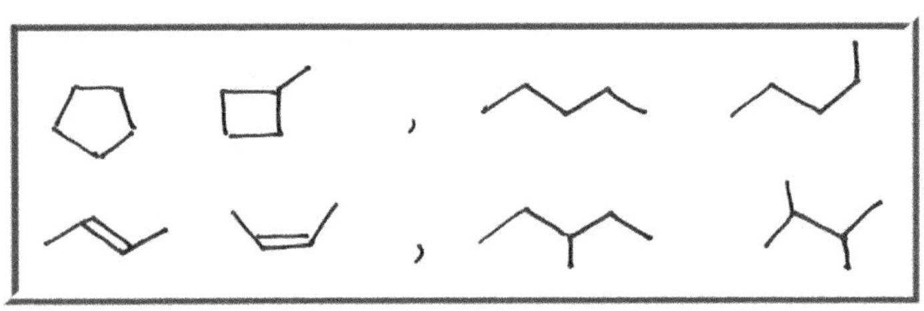

When the molecule is looked upon face down or when it is drawn on a planar surface, a new type of bond formalism has to be introduced, namely bonds expressed by the **dash and wedge symbolism**, that allow a two-dimensional (2D)formula to be converted into a three-dimensional structure (3D)– **dash bonds going into the plain** and **wedge bonds coming out of the plane**. The tetrahedron can be expressed in 3D this way as well.

**Fig. 2.16**

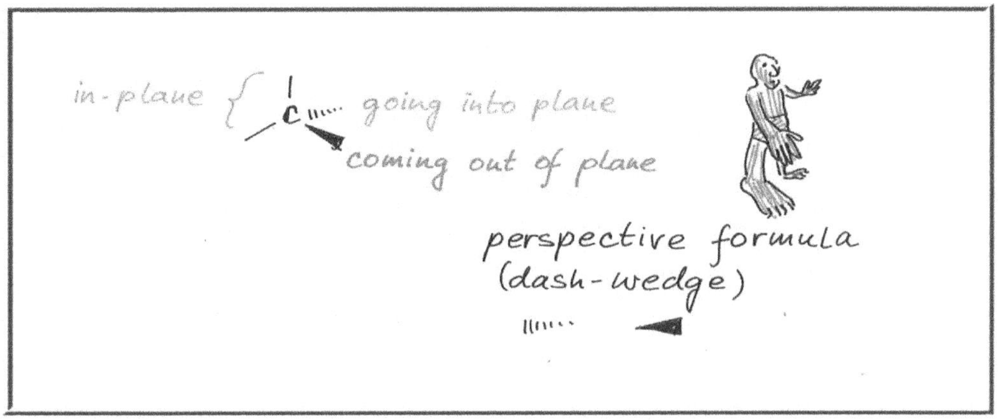


**Configurational Isomerism has an energy barrier** between the highest, observed by constitutional isomers, and the lowest, found in the conformational isomers.

**Fig. 2.17**

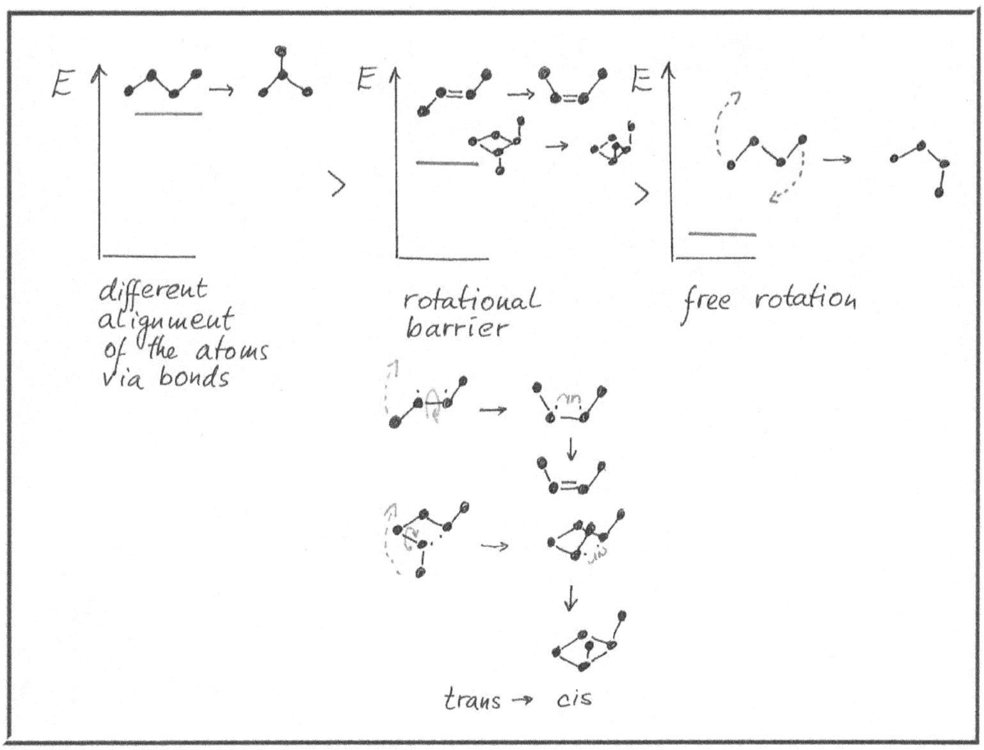

Of special importance are the two extreme conformational isomers of cyclohexane - the **boat** and **chair** conformation, a result of different puckering.

**Fig. 2.18**

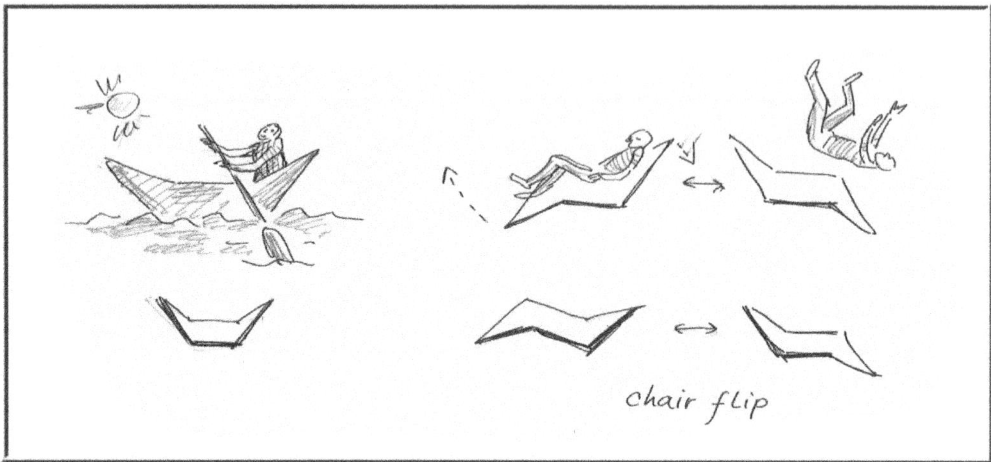


When groups come too close together, **steric hindrance** or **steric repulsion** leads to a strained condition, reflected by a higher, less favorable energy state.

**Fig. 2.19**

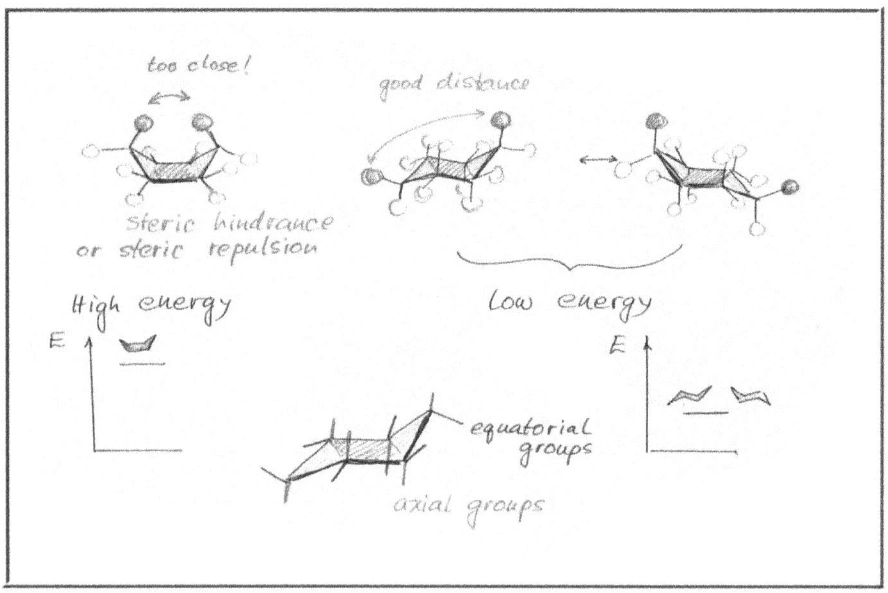

All alkanes and cycloalkanes are colorless, without odor, and with more or less high boiling points. The lower representative ($C_1$ - $C_4$) are gases, the middle ones ( $C_5$ - $C_{10}$ ) are liquids under normal conditions, the ones higher molecular weights are wax-like solids with a consistency of Vaseline. They are all non-polar, with the liquid ones making excellent solvents for non-polar materials. Chemically they are, for the most part, not very reactive, except when reacted with oxygen, which leads to a highly exothermic reaction known as **combustion**. Controlled oxidation leads to heat and light, which makes the hydrocarbons excellent energy molecules. Propane gas is mainly used for heating water and gas furnaces in households and industry. Butane gas tanks are sold mainly for camping and outdoor cooking. Methane is found in the earth as "earth gas" and used for gas stoves and laboratory Bunsen burners.

**Complete combustion** of any hydrocarbon in the presence of sufficient amounts of oxygen always leads to the same products - $CO_2$ and $H_2O$. When Oxygen is limited, **incomplete combustion** leads to a whole array of other compounds, ranging from carbon monoxide to alcohols.

**Fig. 2.20**

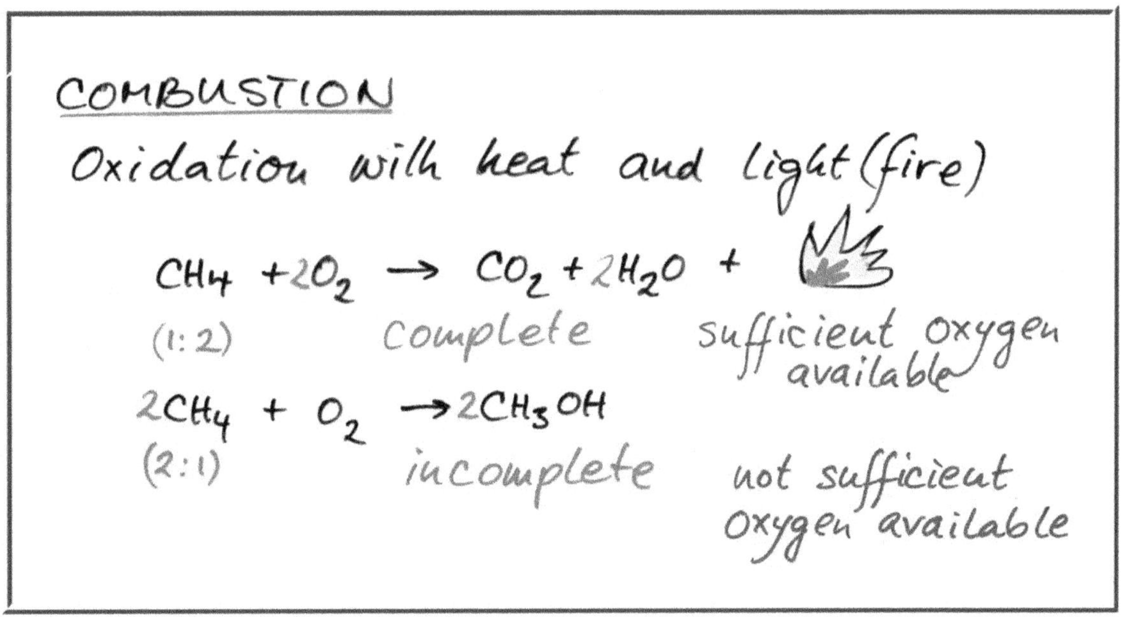


**Example problem 2.8**
**Complete and balance the following combustion reactions:**

**a)** $C_2H_6 + O_2 \rightarrow$ **.... + ....** *(complete combustion)*
**b)** $2C_2H_6 + 1O_2 \rightarrow$ **....** *(incomplete combustion)*

The blue flame observed coming out of the valves on a gas stove is given by the excited state of $C_2$ molecules that have a more probable existence under high energy conditions.
Another reaction for alkanes is the light catalyzed **photo-halogenation**, initiated by the radicals formed during homolytic cleavage of diatomic halogen molecules, which propagate a radical chain reaction by knocking off electrons from the alkane molecules, making them turn into

radicals and thus more reactive themselves.

**Fig. 2.21**

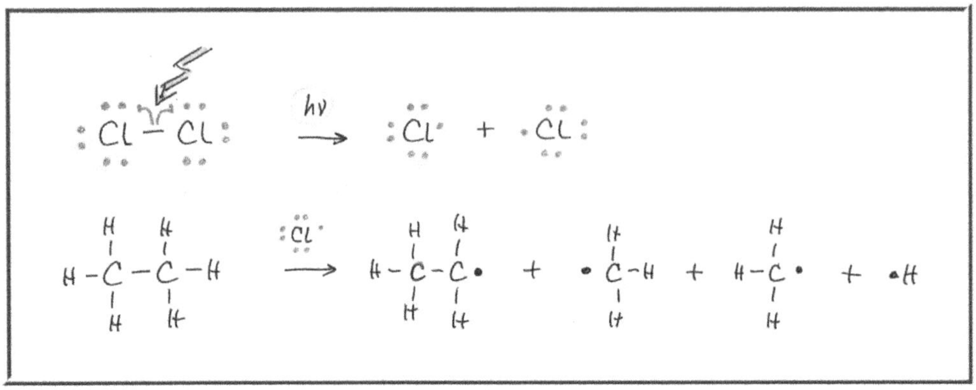

The free halogen radicals cause "havoc" by inducing radical formation in the hydrocarbon molecules.

**Fig. 2.22**

The various radicals can link up to various combinations once the influence of light is removed and the radical "chain reaction" thereby quenched. Possible new recombination products are chloromethane, chloroethene, propane, octane and hydrogenchloride.

**Example Problem 2.9**
**Pentane is treated with $Br_2$ in the presence of light. Give 5 examples of possible products that can be forms by this photo-halogenation reaction.**

The ease of halogenation depends on the **degree of substitution** of the particular C radical involved. The degree of substitution accounts for the level of dehydrogenation of a certain carbon atom, meaning how many attached hydrogen atoms where replaced (substituted) by something else, usually in the form of other carbon groups or functional groups that can contain hetero-atoms (both types often symbolized as R groups). For carbons surrounded by four single bonds, the sequence goes from unique, primary (1° ), secondary (2°), tertiary (3°) up to quaternary (4°). **Carbon radicals** ($C^{\cdot}$) or **carbocations**, carbon groups with a positive charge ($C^+$), both
being three-bonded species, only go up to the tertiary degree.

**Fig. 2.23**

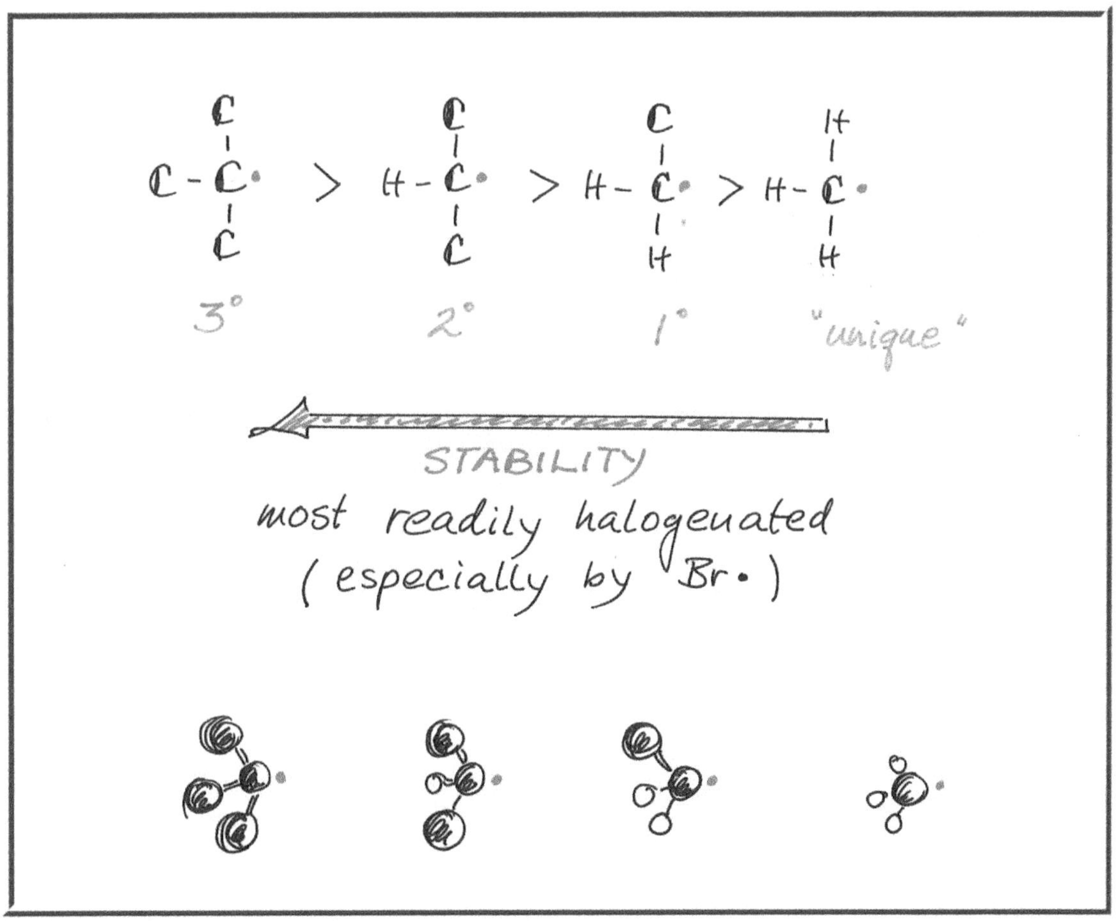

**Example Problem 2.10**
**Determine the degree of substitution of the following species:**

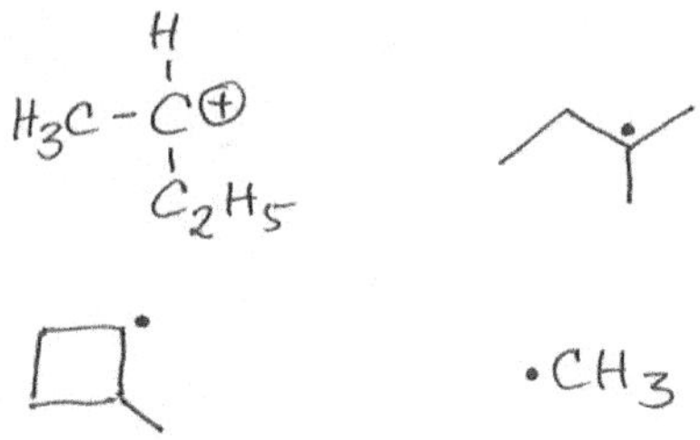

The same mechanism, but without the photo-induced halogen radicals initiating the chain-reaction, can be induced with the help of radical-producing catalysts such as superoxide molecules as **initiators**. These will lead to the formation of hydrocarbon radicals by way of the **initiation step** to carry on the run-away reaction of joining more and more radicals together to larger more complex molecules. If **polymerization** is the desired outcome, then this process is called the **propagation phase**. When the reaction needs to be stopped a radical quenching molecule is added that "kills" all the free radicals. This is called the **termination step**. The individual subunits joined together to make long chained macromolecules or **polymers** are the **monomers**

**Fig. 2.24**

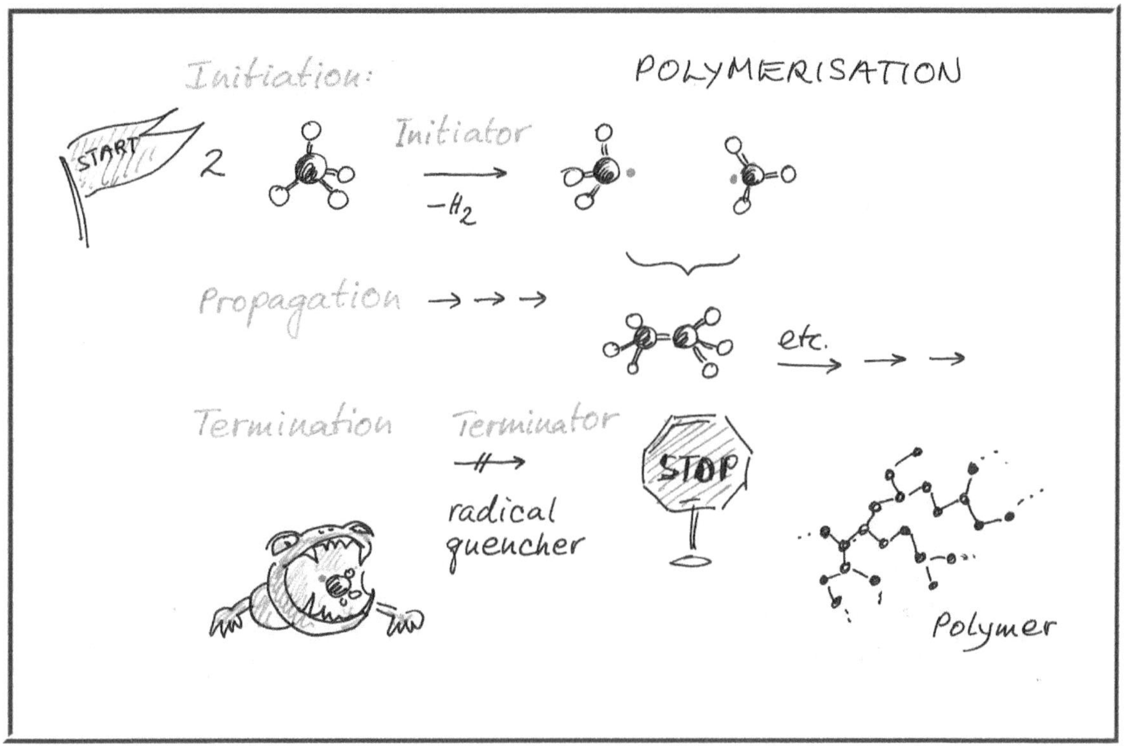


Besides earth gas containing mainly methane, most hydrocarbons are found in the petroleum deposits of the earth. Fractional distillation breaks down the crude, black oil into light boiling (gasoline), to higher boiling fractions (Kerosene, Diesel) such as motor oils and machine oils as lubricants, viscous to wax like liquids to solids (mineral oils, Vaseline, etc.) all the ways up to the black tar as remaining residue. Oil is a most valuable commodity, not only as an energy source but also as a starting material for the synthesis of all other organic molecules in the chemical-pharmaceutical industries.

Gasoline contains many structural isomers. It seems that so-called **isooctane** , although there are many different branched varieties of octane, has the best anti-knock properties that prevent a bumping type of ignition ("knocking") of the internal combustion engine. It ensures optimal oxidation and smooth driving. Isooctane has been attributed the octane rating of 100. A higher the octane rating makes for better, more efficient fuel for the internal combustion engine, providing higher activation energy for combustion. In comparison n-heptane has the isooctane rating of zero.

**Fig. 2.25**

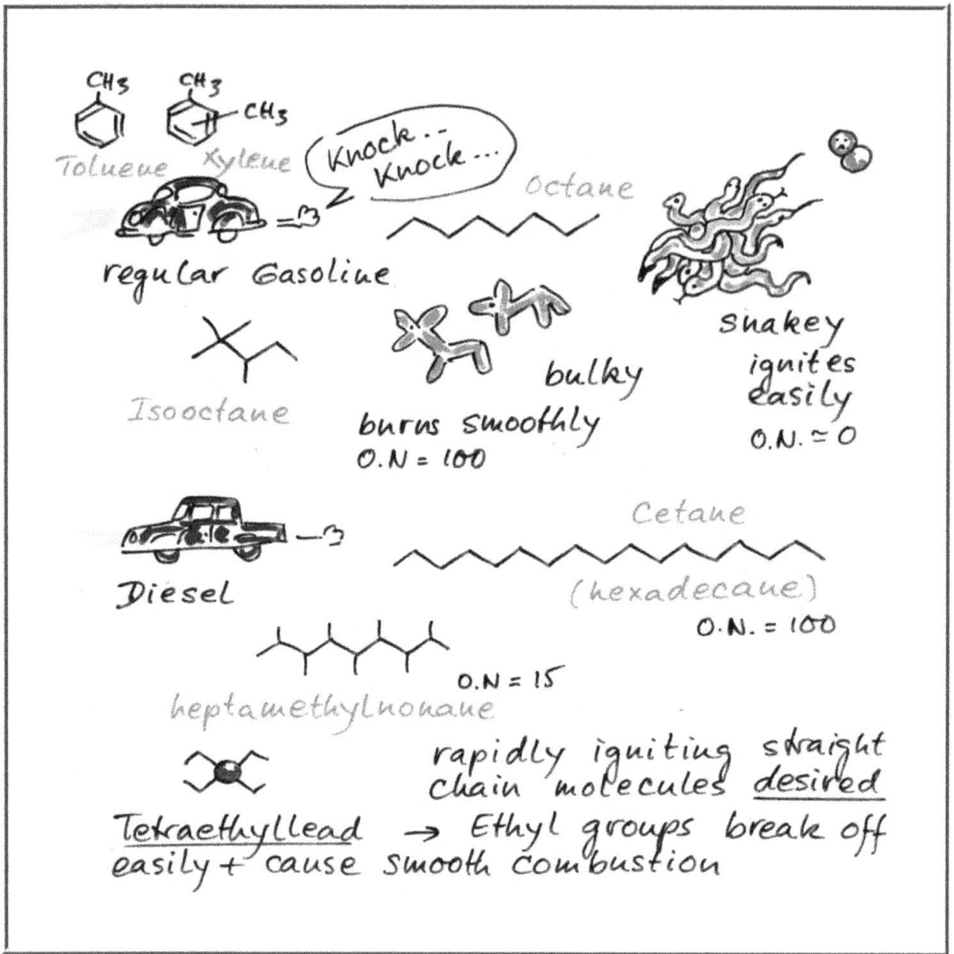


Some molecules, including other branched hydrocarbons, have an even higher isooctane rating. In the table below some chemicals are compared as to their isooctane ratings:

**Table 2.2**

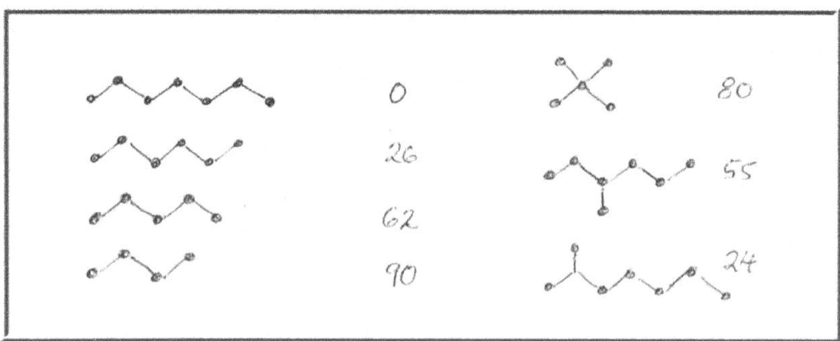


Methane is also known as **marsh gas,** since it appeared oozing out of marshes, as a result of decaying organic matter (plants, animals) Anaerobic bacteria convert $CO_2$ by anaerobic (oxygen-free) respiration with the help of a reducing enzyme into $CH_4$ and $H_2O$. They are found in the

ground, in land-fills and trash dumps but also in the gut of ruminating animals (cattle, goats etc.) as well as cockroaches. Even human beings release some methane as a by-product of flatulation due to natural and unnatural digestive conditions.
Methane breath testing can help identify certain conditions related to the problems of the digestive tract. About 50% of healthy adults excrete methane in their breath. This leads to the assumption that there might be some truth in the fire-spewing dragons of myth. Some dinosaurs of prehistoric times might have been able to exhale flammable gases as a deterrent for their enemies, but not harmful to themselves.

**Fig. 2.26**

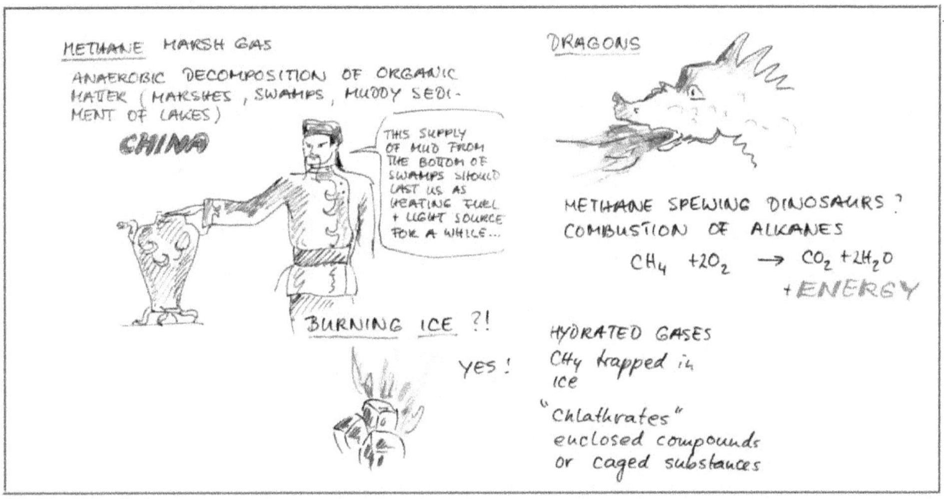


Methane gas monitors are always situated close to landfills and municipal trash dumps as to measure the degree material breakdown and microbial activity.
The ancient Chinese used special porcelain vases for petroleum-like lamps using marsh gas as fuel.
An interesting phenomenon is "burning ice", a **clathrate** or an encapsulation compound**.** Methane gas is entrapped in the interstices (hollow parts) of the ice crystals, which can be set on fire.

# EXPERIMENTS –Ch. 2

## *Alkanes and Cycloalkanes*

**Key Terms:** Linear and branched alkanes, homologous series ($C_nH_{2n+2}$), constitutional (structural and conformational (rotational) isomers, nomenclature, cycloalkanes ($C_nH_{2n}$), fused and linked ring systems, boat, chair conformations, Newman, sawhorse and perspective formulas, dash/wedge bond formalism for 3D representations, syn/anti conformations, configurational (geometric) isomerism, rotational barrier, *cis/trans* isomers, axial (a) and equatorial (e) group conformations on cyclohexane chairs, conversion points (Mp. /Bp. ), molecular stacking and intermolecular forces, reactions (combustion, photo-halogenation) haloalkanes and halocycloalkanes, ring-puckering, formulas: expanded structural (ESF), condensed structural (CSF), abbreviated structural (ESF) or line-angle structural (LAF), molecular (MF); homolytic and heterolytic cleavage of bonds and radical/ionic formation, formal charge concept, radical addition, poly-addition reactions ($A_R$ –mechanism): initiation, propagation and termination stages.)

### I Molecular Modeling of Alkanes

**Chemicals:**

*None*

**Equipment and Materials:**

*Molecular modeling kit (Molymod)*

### Procedure:

1. Build a pentane molecule starting with a five carbon chain (black spheres) and saturating all remaining bonds with hydrogen pieces (white spheres). Number the carbon atoms 1-5 in the un-branched, n-pentane chain.
2. Rotate the bonds in different patterns and draw these **conformational isomers** using abbreviated structural formulas (ASF) in the table below.
3. Break off the C-5 atom group (-$CH_3$, methyl group) and substitute it with a hydrogen atom on C-2. The hydrogen atom attach to the empty slot on C-4. This produces the branched constitutional isomer of isopentane.
4. Break off the methyl group (-$CH_3$) on C-4 and substitute for the other hydrogen atom on C-2. This produces the model for neopentane.
5. Revert to the straight, un-branched form of pentane, remove one hydrogen atom each from C-1 and C-5 and join C-1 with C-5 to build the cyclopentane ring. Rotate the bonds a bit to observe ring-puckering.
6. Substitute a hydrogen atom on C-1 with a chlorine group (green sphere) and another hydrogen atom on the adjacent C-2 group with another chlorine atom. Observe how the ring is a rotational barrier and that two possible orientations in reference to the ring plane occur which are **configurational isomers**, in this case known as ***cis*** and ***trans*** types.

7. Widen the ring by inserting a **methylene group (-$CH_2$-)** and construct a cyclohexane ring. Rotate the bonds to fashion a chair and a boat conformation. Leave the chlorine groups in place.
8. When looking at the chair sideways, flip the C-1 from its original position up or down. Do the same with C-4. This is known as a **chair flip**. Notice that the orientation of the groups in the chair conformation can either be equatorial (e, in plane) or axial (a, out of plane or perpendicular to the ring plane). When a chair flips,**e** becomes **a**, and **a** becomes **e**.

## Observations and Results:

Record the information and structures in the following table:

| **Constitutional isomers for Pentane:** | **IUPAC Name:** | *MF* | *Abbreviated structural formula:* | *Dash-wedge formula:* | *Conformational Isomers for n-pentane* |
|---|---|---|---|---|---|
| **n-pentane** | | | | | |
| **Isopentane** | | | | | |
| **Neopentane** | | | | | |

Draw all structure concerning cycloalkanes in the following table;

| **Cycloalkane** | **MF** | **Abbreviated structural formula (ASF):** | **Dash-wedge formula:** | **Perspective chair formula:** | **Chair flip:** $^{1}C_{4} \leftrightarrow {}^{4}C_{1}$ |
|---|---|---|---|---|---|
| Cylopentane | | | | | |

| | | | | | |
|---|---|---|---|---|---|
| 1,2-Dichlorocyclopentane | | | | | |
| *Cis-1,2-*Dichlorocyclopentane | | | | | |
| *Trans-1,2-*Dichlorocyclopentane | | | | | |
| Cyclohexane | | | | | |
| 1,2-Dichlorocyclohexane | | | | | |
| *Cis-1,2-*Dichlorocyclohexane | | | | | |
| *Trans-1,2-*Dichlorocyclohexane | | | | | |

(darkened areas remain blank)

## Questions and Problems:

1. Look up the boiling points for the three pentane structural isomers and relate the different values to the strength of intermolecular forces between the individual molecular types (steric hindrance, bulkiness, stacking possibilities).
2. Correlate the energy barriers between the individual constitutional isomers with those existing between the different conformational isomers when converting one form into another (inter-conversion energy)
3. Are all the compounds studied aliphatic (non-aromatic acyclic or cyclic) hydrocarbons?
4. Does a chair flip interconvert ***cis*** and ***trans*** isomers? Explain.

# II Physical Properties of Alkanes and Cycloalkanes

## Chemicals:

*Hexane*
*Cyclohexane*

## Equipment and Materials:

*Test tubes and rack*
*Metal stand, clamp*
*Large beaker for hot water bath, used as steam bath, hot plate*
*Thermometer with holed stopper*
*Top-loading balance*
*Graduated cylinder*

## Procedure:

1. On a hot plate below prepare a steam bath of boiling water in a beaker.
2. Place 2-3 mL of hexane into a test tube. Position the tip of the thermometer over the surface of the liquid (3-4 mm). Lower both equally into the steam bath (2-3 cm from the boiling water surface).
3. When the hexane begins to boil, its vapors will hit the tip of the thermometer and the temperature reading will mark its boiling point (Bp.).
4. Repeat the procedure for cyclohexane in another fresh, clean test tube.
5. Determine the density for both substances by determining mass and volume. Tar out the mass of a graduated cylinder on a top-loading scale and pour in a random amount of the liquid. Determine the mass and read off the volume. Density is mass/volume in g/mL.
6. Mix the chemicals with each other and determine whether they are mutually miscible.

## Observations and Results:

Record the results in the following table:

| **Alkane/ Cycloalkane:** | **MF:** | **Molar mass (g/mol):** | **Bp.(°C)** (practical): | **Bp. (°C)** (theoretical): | **Odor:** | **Miscibility** (with each other): | **Density (g/mL)** (pract.): | **Density (g/mL)** (theor.): |
|---|---|---|---|---|---|---|---|---|
| **Hexane** | | | | 68-69 | | | | 0.66 |
| **Cyclohexane** | | | | 80.74 | | | | 0.78 |

## Questions and Problems:

1. Why would cyclohexane have a higher boiling point and higher density than hexane although it has almost the same molar mass?
2. What is the typical odor of the alkanes and cycloalkanes?
3. If combined with water, would they make up the lower phase or the upper phase?

## III Chemical Properties and Reactions of Alkanes and Cycloalkanes (Combustion and Photo-halogenation)

### Chemicals:

*Hexane*
*Cyclohexane*
*Bromine ($Br_2$ in $CH_2Cl_2$)*
*Dichloromethane ($CH_2Cl_2$)*

### Equipment and Materials:

*Test tubes with rack, droppers with bulb*
*Light bulb (high wattage, 150 W), stop watch*
*Hot water bath (beaker, 250 mL), hot plate*
*Graduated cylinder (10 mL)*

*Evaporating dish (porcelain)*
*Wooden splint, matches*

## Procedure:

1. Prepare a warm water bath, which is warm to the touch.
2. Prepare two test tubes and fill each with 4 mL dichloromethane (methylene chloride).
3. To the first add 1mL hexane and to the other tube 1 mL cyclohexane.
4. To each add 1mL of the $Br_2/CH_2Cl_2$ solution and mix.
5. Start timing immediately the discoloration of the solutions at room temperature (RT) and daylight.
6. Empty the tubes, refill with the two samples and heat the two liquids in a warm water bath.
7. Transfer the warmed tubes to a dark room and switch on the incandescent light bulb.
8. Add the bromine reagent, swirl the contents and record the time for discoloration at about 10 cm away from the light source.
9. In the fume hood, place about 5 drops of hexane in an evaporating dish and ignite it with a glowing wooden splint. Repeat the procedure with cyclohexane.

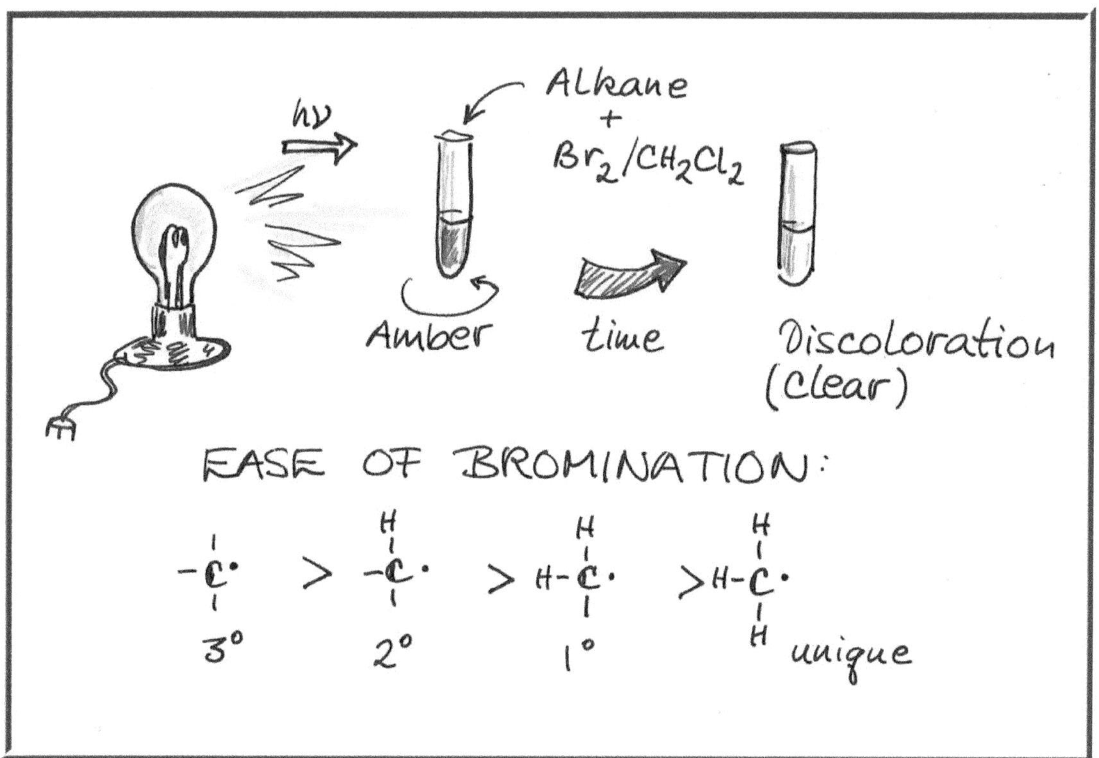

## Observations and Results:

Tabulate the observed results and add the other information required according to the following table:

| Alakane/cycloalkane: | Bromination time (sec or min) at RT and daylight: | Bromination time (sec or min) at elevated temp. and incandescent light (150W): | Combustion observations (color of flame, smoke, residue): |
|---|---|---|---|
| Hexane | | | |
| Cyclohexane | | | |

## Questions and Problems:

1. The halogenation reaction proceeds via photo induced radical formation of the halogen molecule, which in turn produces radicals of the hydrocarbon molecule. Draw some possible radicals that can form for each starting molecule, hexane and cyclohexane.
2. What are some of the termination products that result from pairing off of the radicals in different combinations, once the energy source is removed?
3. If the light source would be UV light, how would the reaction speeds become affected?
4. Write the two balanced equations for the combustion of hexane and cyclohexane with the oxygen of the air to carbon dioxide and water (complete combustion)
5. Why is initial ignition with a glowing or burning splint required if the reaction is exothermic anyway?

## IV Distillation of Hexane (Class Experiment)

### Chemicals:

*Hexane*

### Equipment and Materials:

*Distillation apparatus, heating mantle*

*Round bottom flask, boiling stones*

*Graduated cylinder*

### Procedure:

1. In the fume hood and with the help of the instructor set up the distillation apparatus according to the image below.
2. In a small round-bottom flask (100mL) place about 60mL of hexane and heat the contents with the help of the heating mantle until the boiling point of hexane is reached. Prepare a heating curve by plotting temperature ($^{o}$C) against recovery volume collected in the

receiver (graduated cylinder). The recovered liquid can be used again by other students.

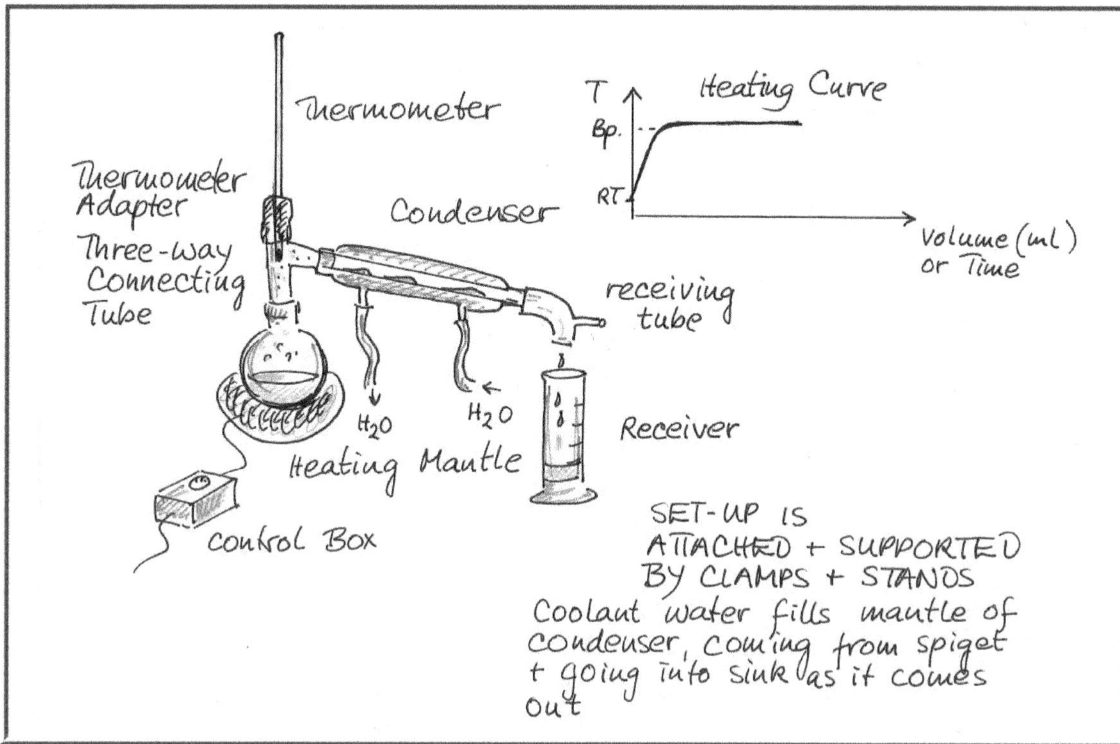


## Observations and Results:

| Temp. (°C) | Volume (mL) |
|---|---|
| 25 | 0 |
| | |
| | |
| | |
| | |

## Questions and Problems:

1. Describe the general pattern of the heating curve.

2. Why does the curve flatten out (known as a plateau) when the boiling point of the liquid is reached?
3. Why shouldn't the reaction vessel run completely dry during the distillation process.
4. How can the distillation of a mixture of two or more miscible liquids be used to separate them according to their different boiling points. How many plateaus would be observed if hexane and cyclohexane are separated by distillation?
5. What is understood by fractional distillation?

## V Hexane as an Extraction Solvent

### Chemicals:

*Hexane*

*Iodine solution ($I_2$ in aqueous KI)*

### Equipment and Materials:

*Large test tube with fitting rubber stopper*

### Procedure:

1. Place 2-3mL of the aqueous iodine solution in a large test tube.
2. Add an equivalent amount of hexane using a dropper or Pasteur pipette.
3. Vigorously shake the tube by placing the stopper tightly over its opening.
4. Remove the stopper and let the two phases separate out.

### Observations and Results:

Record any color changes:

| **Color of $I_2$/KI before extraction:** | **Color of hexane extract:** |
|---|---|
| | |

### Questions and Problems:

1. What happened to the color of the aqueous layer after the iodine was extracted?
2. What happened to the appearance of the organic layer after the extraction process?
3. Which of the two solvents was the denser phase?
4. Why is extraction only possible for non-miscible liquids?
5. Why is extraction of a certain solute from one solvent into another even possible?

**Chapter 3 Alkenes and Alkynes**

3.1 Homologous series of alkenes and alkynes

3.1.1 General formulas

3.1.2 Structures, geometries, and formulas

3.2 Nomenclature rules and bond order priorities

3.2.1 IUPAC names and common names

3.2.2 The vinyl and allyl groups

3.3 Configurational isomerism of the double bond (cis, trans)

3.4 Physical properties

3.5 Chemical properties and reactions

3.5.1 Oxidation and combustion

3.5.2 Addition reactions

3.5.3 Poly-addition to various polymers

3.5.4 Acidity of terminal alkynes (R-C≡CH)

3.6 Natural resources and functions

3.6.1 Ethylene (ethene) as ripening hormone in plants

3.6.2 Vision process

3.6.3 Biological signaling processes

3.7 Uses and applications

3.7.1 Acetylene (ethyne) blow torch for welding

3.7.2 Monomers for various commercial polymers (plastics etc.)

## Ch. 3 Alkenes and Alkynes

Unsaturated hydrocarbons are classified as either belonging to the alkene or alkyne family, depending on whether a double bond or a triple bond is involved. They are both homologous series and their molecular formulas follow a certain pattern, just like the family of alkanes, only that the two and four Hydrogen atoms are removed, respectively. The general formula for alkenes with only one double bond is $\mathbf{C_nH_{2n}}$ and for the alkynes with only one triple bond it is $\mathbf{C_nH_{2n-2}}$. The names use the same stem as the alkanes ("eth," "prop," etc.), but a different ending. The suffix **–ene** is used for the alk**enes**, and **-yne**, for the alk**ynes**.
The first molecule for each series is of course the 2C compound, since a multiple bond involves 2 Carbon atoms.

**Fig. 3.1**

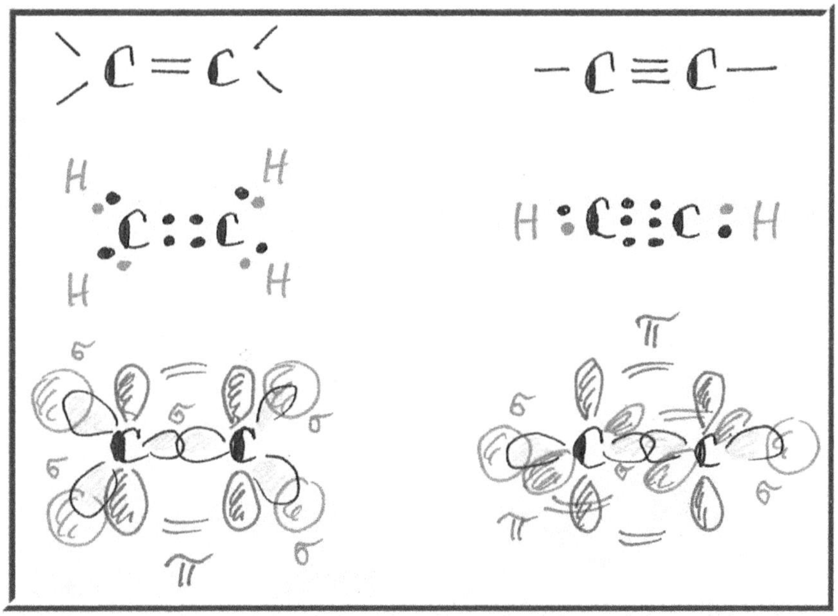

The multiple bond changes the shape of the atomic arrangement in its electronic environment. According to the VSEPR concept, the maximum distance between electron density areas (the multiple bond counting as one, even though it involves more electrons, is in the shape of a flat triangle called a "**trigonal planar"** geometry. The angle between the three atoms surrounding the central carbon atom is 120°.

The triple bond and the single bond surrounding the central C-atoms account for 2 electron density areas and the maximum distance is 180°, making the molecular environment around the triple bond **linear** in shape.

So, in review, single bonds lead to tetrahedral, double bonds to trigonal planar, and triple bonds to linear atomic arrangements.

**Fig. 3.2**

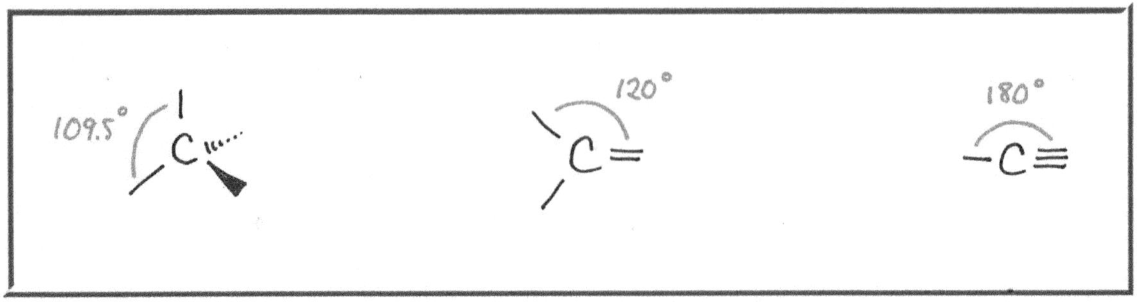


Nomenclature is a bit more complicated, since now the position of the multiple bond within the parent has to be accounted for. Here the C-atoms of the chain are numbered as usual and the multiple bond designated by a number, indicating from which part of the chain it originates, keeping locant numbers low. Thus, even though the multiple bond spans between two carbons, only one number is needed.

**Fig. 3.3**

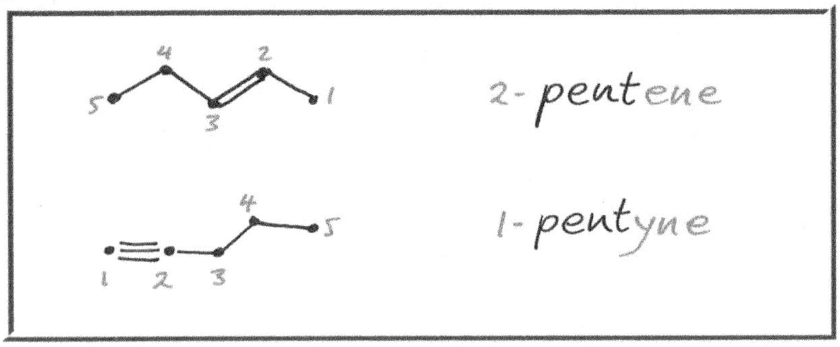


When substituents are involved, such as alkyl groups that make up for branches, the longest chain incorporating the multiple bond must be used for the parent molecule.

**Fig. 3.4**

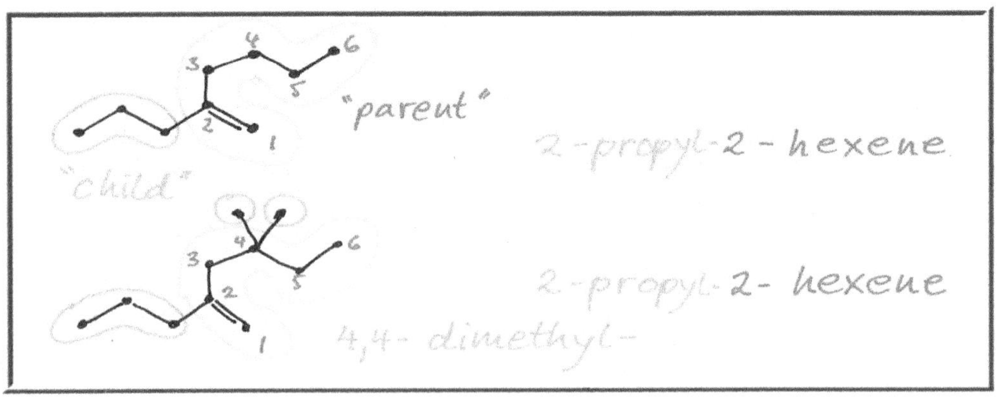

The longest chain with the double in this case is 1-hexene. The "child" or group attached is propyl. The complete name is therefore 2-propyl-1-heptene. If two or more groups are attached, they must be (just like for the alkanes) listed in alphabetical order.
Cyclic alkenes are possible even for small rings. However the triple bond, being linear in shape, cannot fit in the smaller rings, producing excessive ring strain. Carbocycles need to be larger than the 6-membered ring, in order to incorporate triple bonds as well.
If alone in the ring, the double bond automatically counts as no. 1 and does not have to be numerically mentioned in the name. Any attached substituents have to be numbered in reference to where the double bond originates or with the flow of the double bond.

**Fig. 3.5**

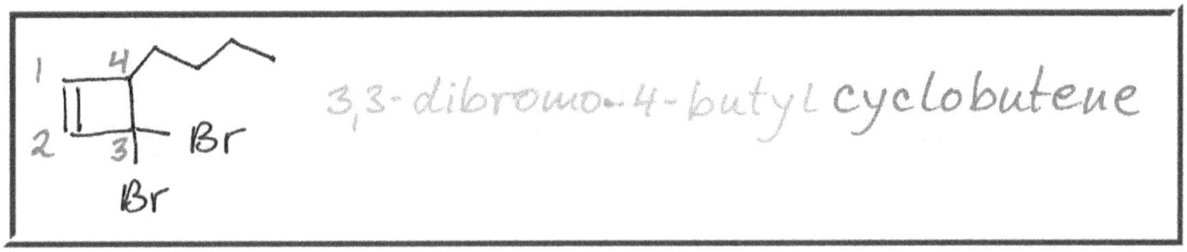


**Example Problem 3.1**
**Name the following molecules:**

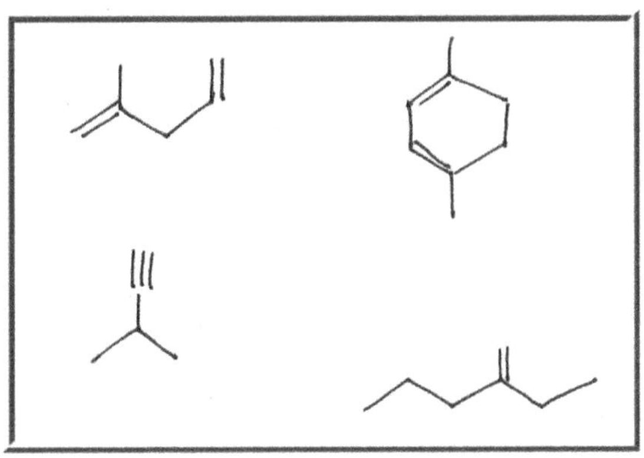

**Example Problem 3.2**
**Give the structural formulas (ASF/LAF) for the following names;**
***3-Methyl-2-pentene***
***2-Sec-butylcyclohexene***
***1,2,3-Trichlorocyclopentene***
***3-prpyl-1-butyne***

It must be remembered that the Latin numerical prefixes do not contribute to the name (di, tri, tetra, etc.) when listing the groups in alphabetical order. When more multiple bonds are involved, the double bond always has priority over the triple bond, mainly because the name is easier to pronounce.

**Fig. 3.6**

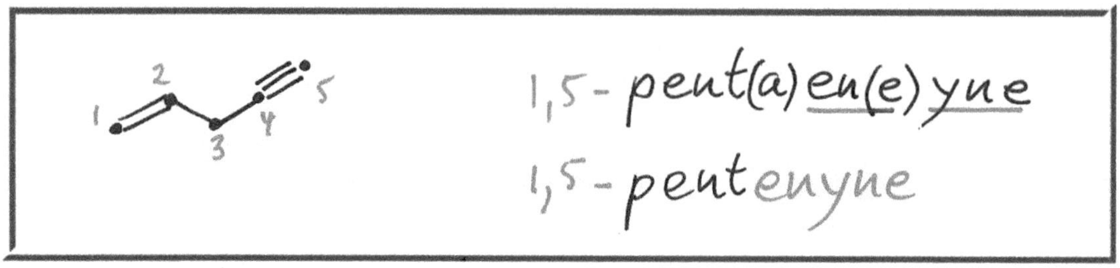


The number can be separated to give the following name as well: 1-pentene-5-yne.
To prevent two similar sounding vowels, one is usually dropped (ae → e, ye → e). When more of the same multiple bond are involved, an ***ene*** becomes a ***diene, triene, ... polyene***, or an *yne* becomes ***diyne, triyne, ... polyyne***. Here the vowels are expressed separately and not aligned. The multiple bonds are treated like functional groups, requiring their own locant numbers and proper prefixes if occurring in more than one.

**Fig. 3.7**

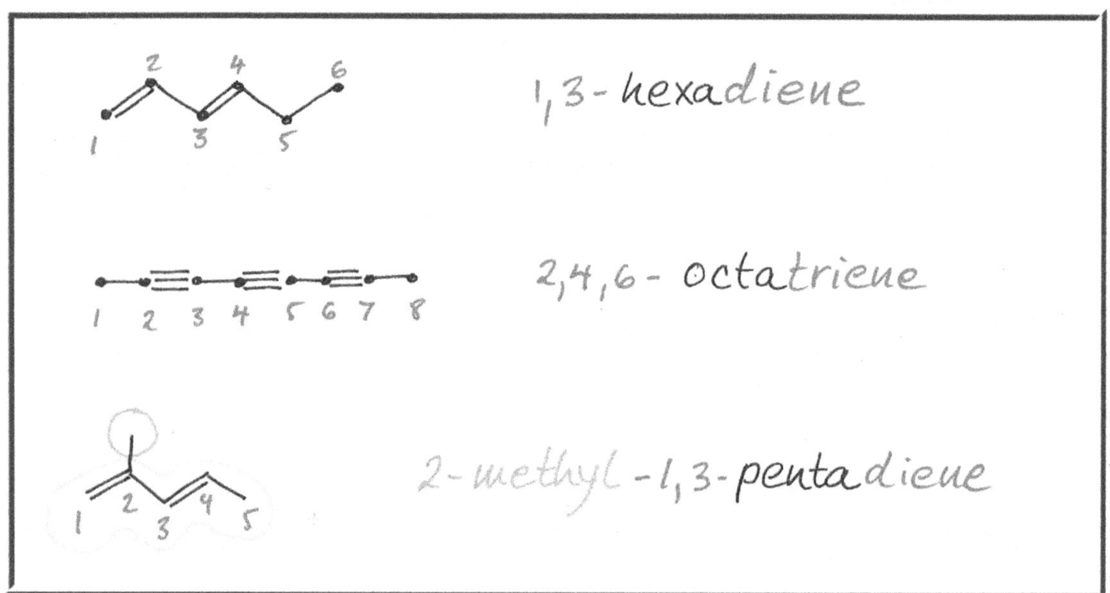


The homologous series are generalized by the family names.

**Alkadienes / Alkadiynes**
**Alkatrienes / Alkatriynes**
.
.
.
**Alkapolyenes/ Alkapolyynes**

If substituents are present, the naming proceeds in the usual way: finding the parent (including all multiple bonds), numbering the chain with lowest possible numbers, and naming and numbering the groups according to alphabetical order.
For rings containing more than one double bond, one double bond will start as carbon no.1, the others named within sequence and flow of the primary double bond.

**Fig. 3.8**

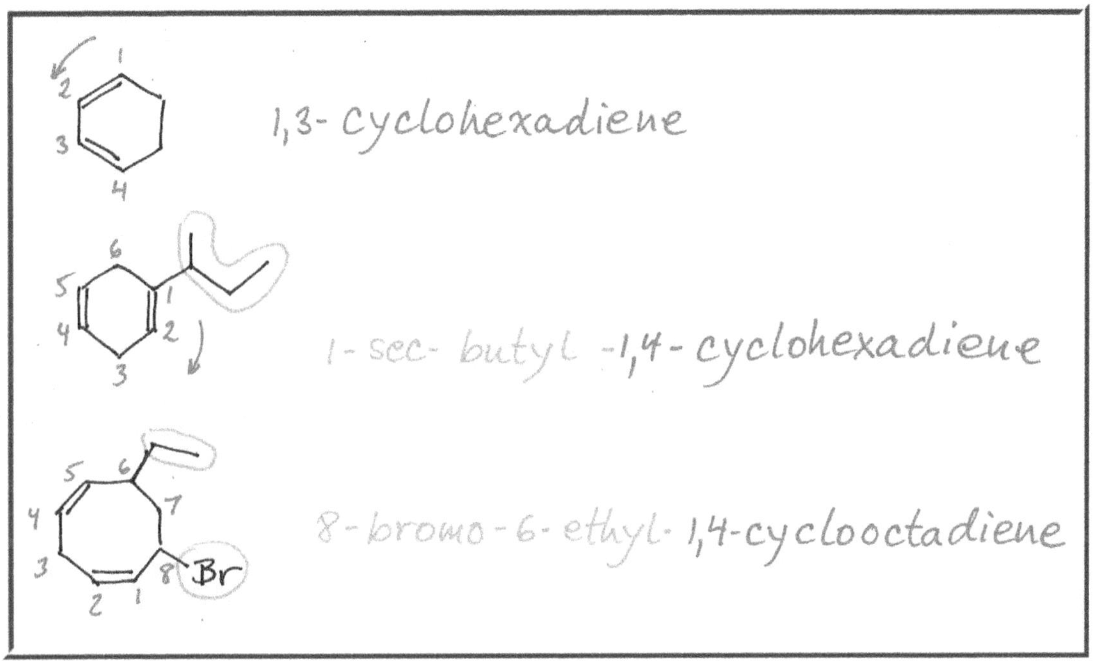


**Example Problem 3.3**
**Name the following structures:**

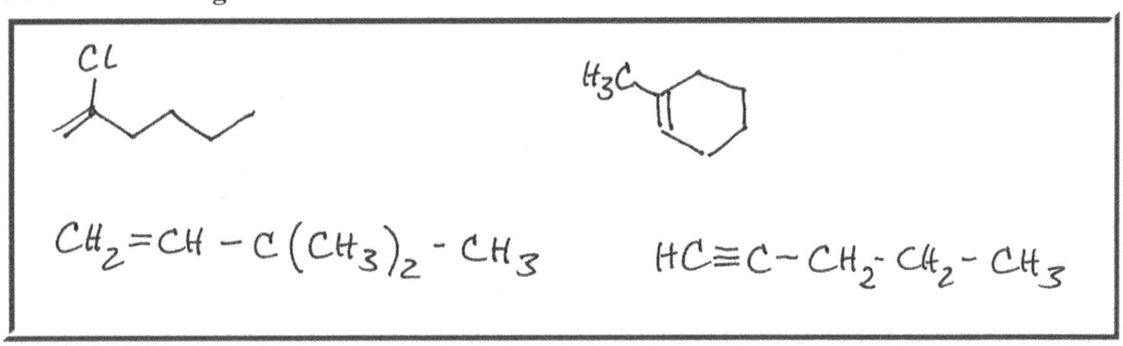


**Example Problem 3.4**
**Write the formulas (ASF/LAF) for the following names:**
***1,2-Butadiene***
***3-Ethyl -5-chloro-1,3-cyclohexadiene***
***3,4-Dimethyl-1-hexene-5-yne***
***2,3,5-Octatriene***

More multiple bonds within a molecule can be arranged in different patterns. Three types of specific **multiple bond arrangements** are distinguished as **isolated, conjugated, and cumulated.**

**Isolated** bond arrangements mean that there are more than one single bond between two individual multiple bonds.

**Fig. 3.9**

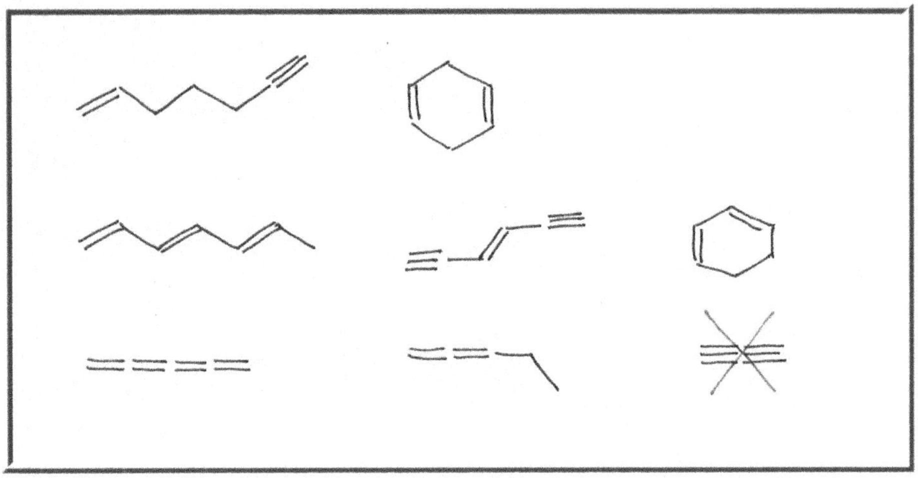

**Conjugated** bond arrangements mean that each multiple bond is separated by one single bond.

**Cumulated** is only possible for double bonds, since carbon can only have four bonds. Here the double bonds are directly adjacent to one another.

The structure for 2 cumulated double bonds is also linear, since the central carbon atom is surrounded by two electron density areas (VSEPR theory).

**Fig. 3.10**

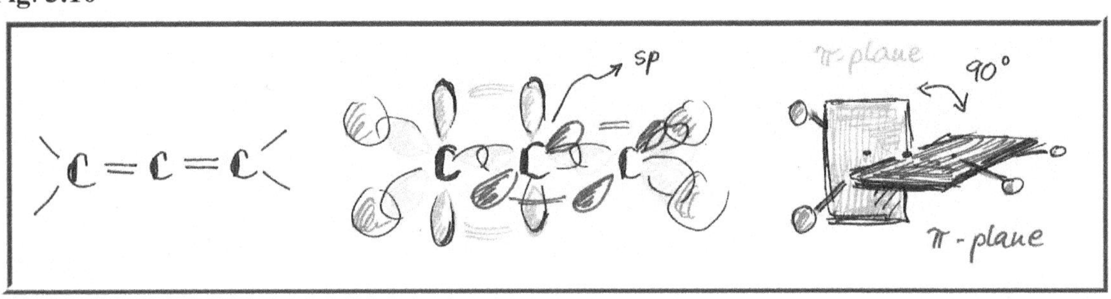

**Example Problem 3.5**
**Identify the type of multiple bond arrangements for the following molecule, comparing two at a time:**

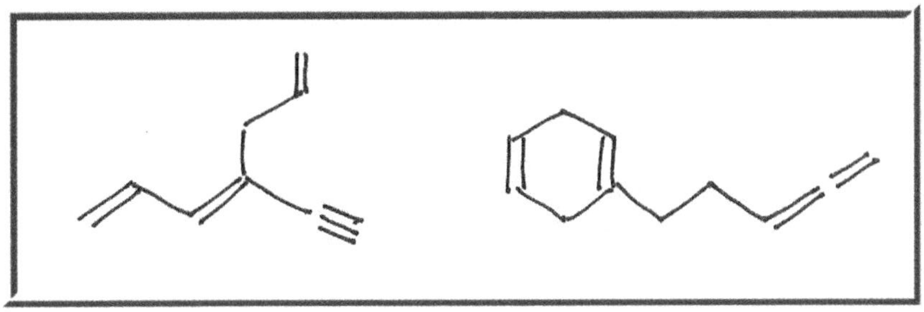

**Example Problem 3.6**
**Draw a molecule consisting of two isolated, 3 conjugated, and 3 cumulated multiple bond arrangements. Two of the multiple bonds should be triple bonds.**

The **physical properties** for alkenes and alkynes are similar to the alkanes. They are colorless, odorless, and highly flammable. The low-molecular representatives are gases.
**Chemical properties** involve a higher reactivity than the saturated counterparts. The multiple bond being an electron-rich area is highly susceptible to **addition reactions**. Small molecules, for instance, can easily be added, mostly with the help of a catalyst, creating an opportunity to synthesize a variety of substances or **derivatives**. Four important addition reactions are as follows:

1) **Hydrogenation (addition of Hydrogen $H_2$ )**
2) **Halogenation (addition of a halogen molecule, $F_2$, $Cl_2$, $Br_2$, $I_2$)**
3) **Hydration (addition of water $H_2O$)**
4) **Hydrohalogenation (additions of hydrogenhalides HF, HCl, HBr, HI)**

Of course there are many more such as sulfonation, nitration, hydroboration etc., but discussion of these would exceed the scope of this publication.

The alkene or alkyne is considered the **substrate**, the molecule being acted upon, and the adding molecule is considered the **reagent**, that performs the active attack. The multiple bond is considered an electron rich area and the reagent must therefore be an **electrophile** – a species that "loves" electrons. For this reason, the reaction category is considered an **$A_E$ reaction** – an addition reaction proceeding via the electrophilic mechanism.
The reagent can be separated into an A and a B part (A – B) and then placed above the double bond to quickly determine the product that can be formed by rearranging the bonds (shown by the curved arrows) as follows:

Fig. 13.11

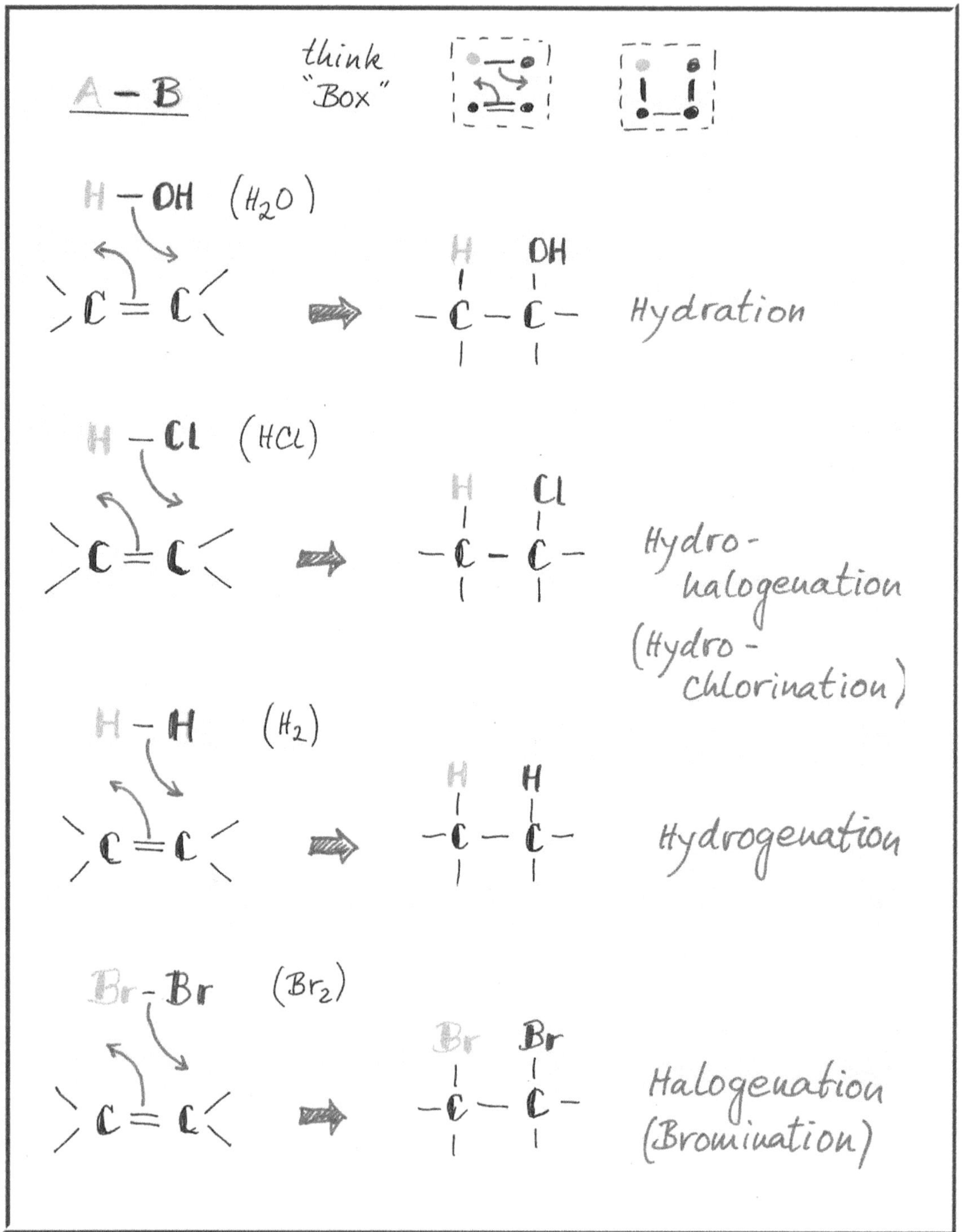

A – B
think "Box"
H – OH ($H_2O$)
H OH
Hydration
H – Cl (HCl)
H Cl
Hydro-halogenation (Hydro-chlorination)
H – H ($H_2$)
H H
Hydrogenation
Br – Br ($Br_2$)
Br Br
Halogenation (Bromination)

When an asymmetric alkene is the substrate for an asymmetric reagent, the addition is usually **stereoselective**, meaning that of the two possible **adducts** (addition products), one is preferred over the other. According to **Markovnikov's Rule**, the Hydrogen will go to the Hydrogen. This means the hydrogen will favor that part of the double bond that already has the most hydrogen atoms attached.

**Fig. 3.12**

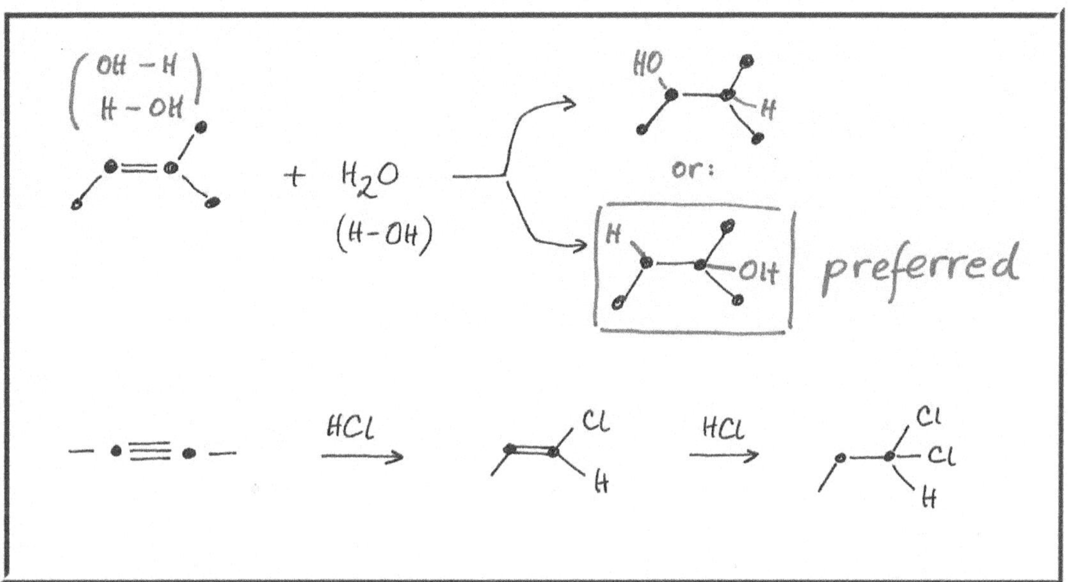


The actual driving force for such stereoselectivity is the relative strength of the **intermediate carbocation** ($3^o > 2^o > 1^o$) occurring during the **two-step mechanism**, which is of minor significance to the scope of this course.

Alkynes can also undergo addition reactions and under equal molar reaction conditions lead first to the corresponding alkenes. If more reagent is supplied, it can be saturated all the way up to the alkanes.

**Example Problem 3.7**
**Complete the following reactions:**

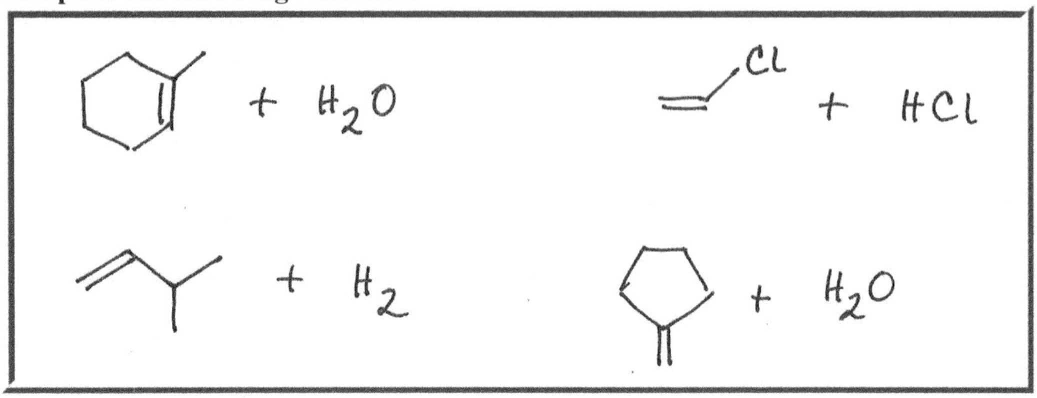


**Example Problem 3.8**
**By what addition reaction could the following products be made?**

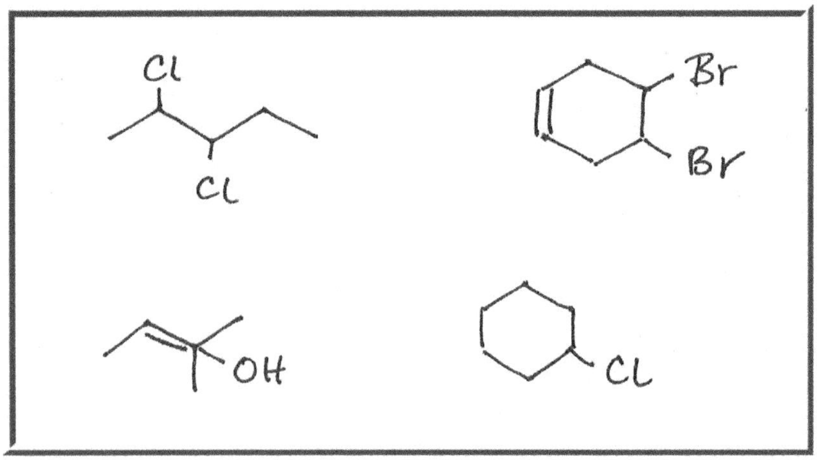


Another important reaction for alkenes is the **oxidation** with $KMnO_4$, which leads to **glycols**, divalent alcohols with two adjacent hydroxyl groups.

The alkene is oxidized and the $MnO_4^-$ ion reduced from a magenta colored solution to $MnO_2$ – a brown, precipitating solid. This is an excellent test for the identification of an unsaturated hydrocarbon. Alkenes can also be oxidized to epoxides, a three-membered cyclic ether.

**Fig. 3.13**

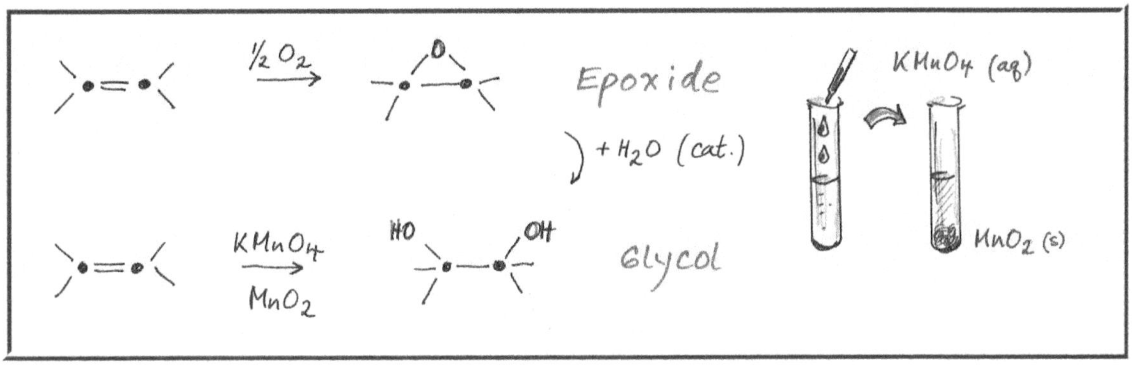


Another popular laboratory test is the discoloration reaction of a yellow to orange, aqueous bromine solution. The presence of an alkene or cycloalkene is indicated by the turning of a yellow $Br_2$ -reagent to colorless solution, when added.

**Example Problem 3.9**
**Which of the following molecules would give positive reaction with the Bromine reagent ($Br_2$) and Potassium permanganate ($KMnO_4$) reagent?**

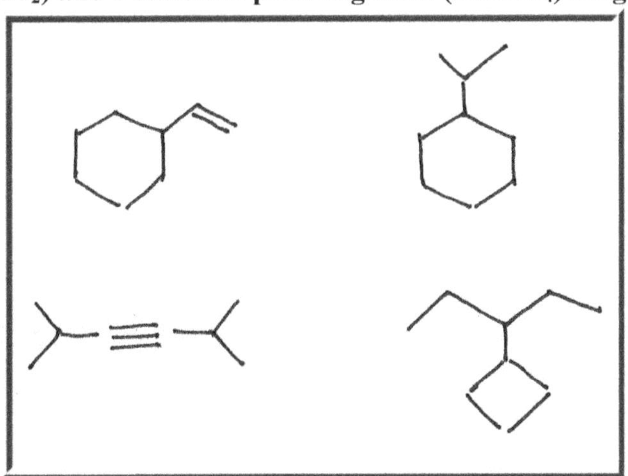

**Example Problem 3.10**
**Complete the following reactions:**

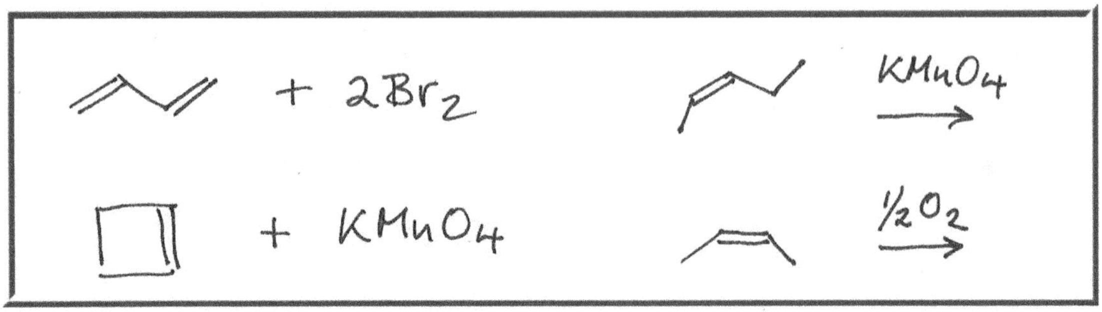


The multiple bond also lends itself well to repeated addition reactions – **polyaddition** – producing a variety of polymers (macromolecules).

**Fig. 3.14**

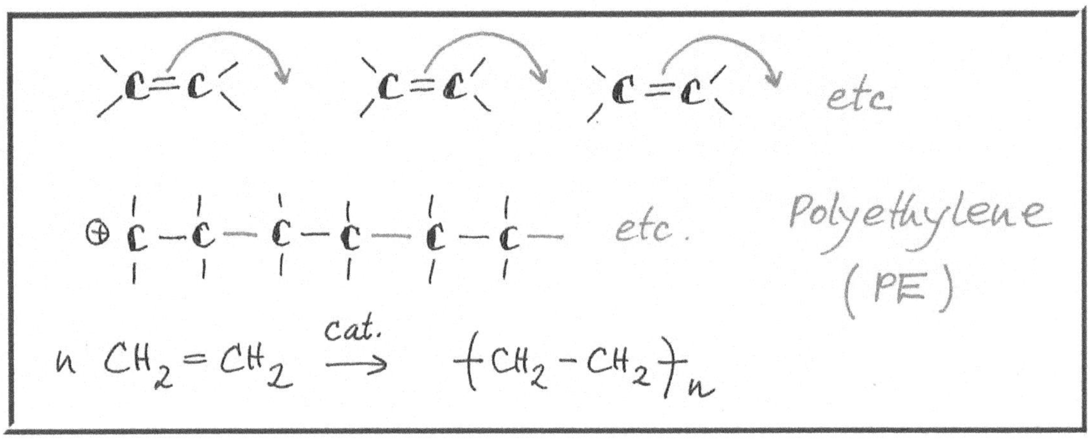


The polymer names are still based on the common names of the monomeric units such as ethylene, propylene, vinylchloride, vinylbenzene (styrene), etc. The vinyl group is the unsaturated 2-carbon group (ethenyl). In the same manner Polyvinylchlorine (PVC), polyvinylalcohol (PVA), Teflon (PTFE), Polystyrene (PS), and many others can be produced. The three steps are also as important here as with the radical mechanism of alkanes, namely

- **Initiation** by an unstable molecule that starts the process (often peroxides like Benzoylperoxide).
- **Propagation** – the continued process of chain elongation.
- **Termination** – a controlled stop to the continued process by a quenching molecule or "terminator"

**Fig. 3.15**

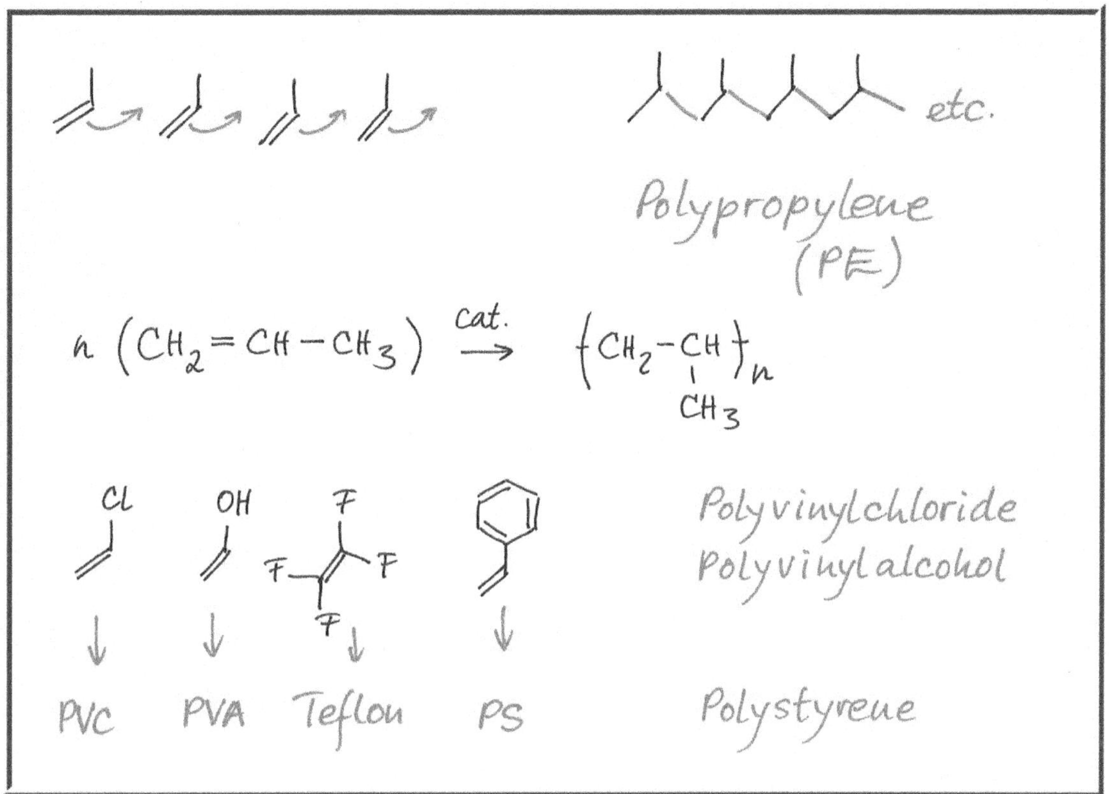


**Example problem 3.11**
**What polyaddition products can be formed using the following monomers:**

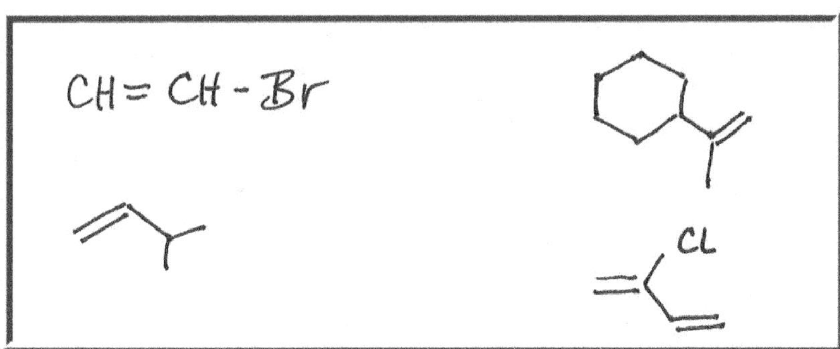


An interesting phenomenon is the effect of plastic grocery bags (usually polyethylene – PE) on the ripening of fruits and vegetables. Decomposition of polymers can lead back to the original monomers. It happens to be the case that ethylene (ethene) is a **ripening hormone** for plants, as gas if signal "ripening," which of course can lead to earlier spoilage for the consumer. This condition let to the implementation of a consumer ripening prevention product, consisting of various types of containers containing $KMnO_4$, which, as noted above, oxidizes the ethylene (ethene) gas molecules to glycols.

**Fig 3.16**

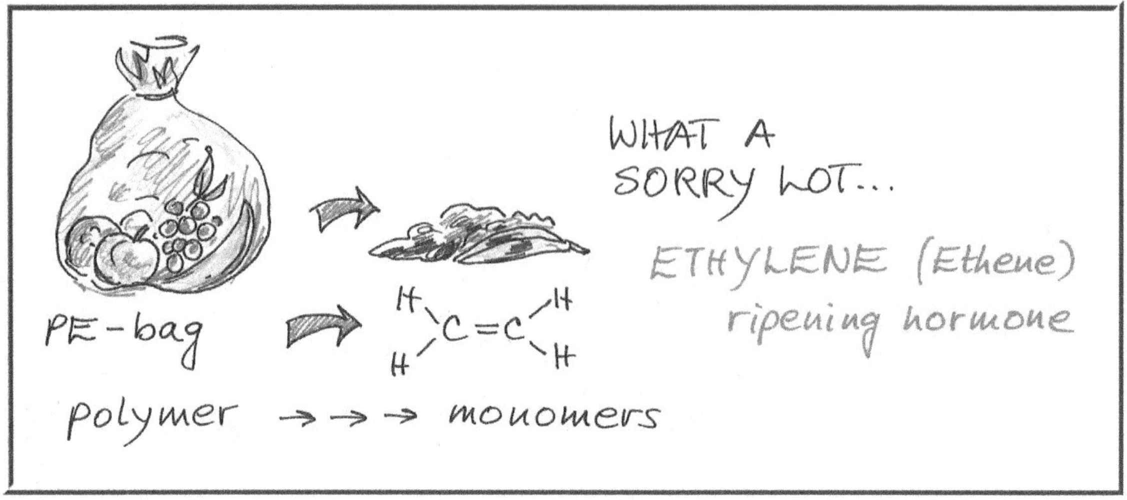


Plants use ethylene and other alkenes for many physiological processes, including signaling processes. Certain proteins have receptors for these molecules. Cis- and trans- configurations of the alkene can act as on or off switches. This is especially apparent for the **vision process**, which is based on the isomerization of 11-cis retinal to all-trans retinal in the presence of light. The former is covalently bonded to the protein opsin and released as the photo-induced conversion into the trans-isomer occurs, simultaneously releasing an electrical signal to the brain (visual photo-transduction).

**Fig. 3.17**

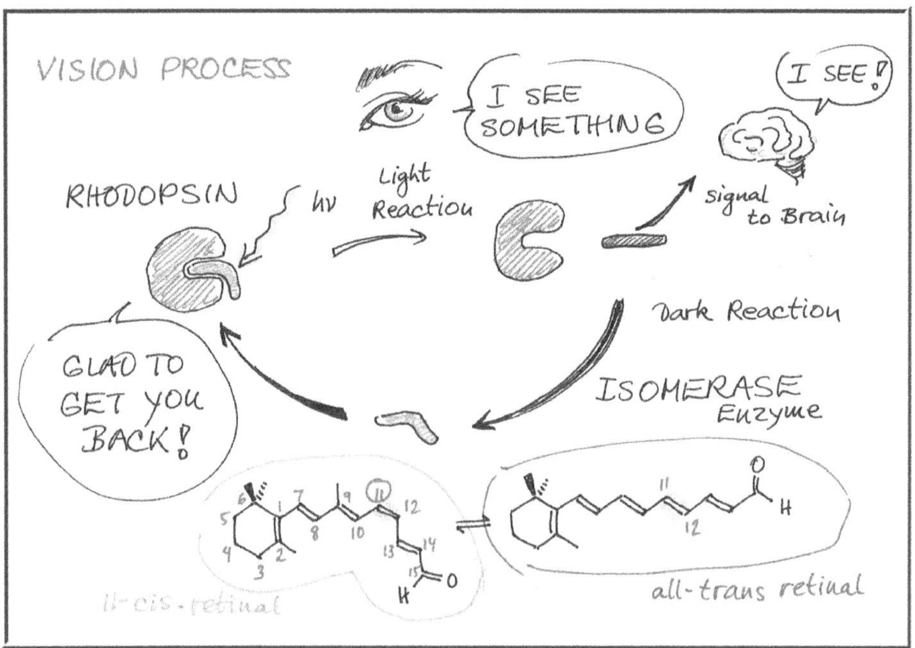

Alkenes and alkynes are widely spread in nature. They usually derive from the monomeric building block, the **isoprene** unit. Of particular importance is the possible of the configurational isomers *cis* and *trans* that can act for various signaling processes as is shown above. Many insect pheromones rely on only one configurational isomer for their biological activity. A sequence of conjugated multiple bonds are often a source of color for many plants and animals. The long stretch of delocalized pi-systems allows the molecule to interact with visible light, making them also more colorful, as is the case when the big chemical class of the carotenoids is involved.

**Fig. 3.18**

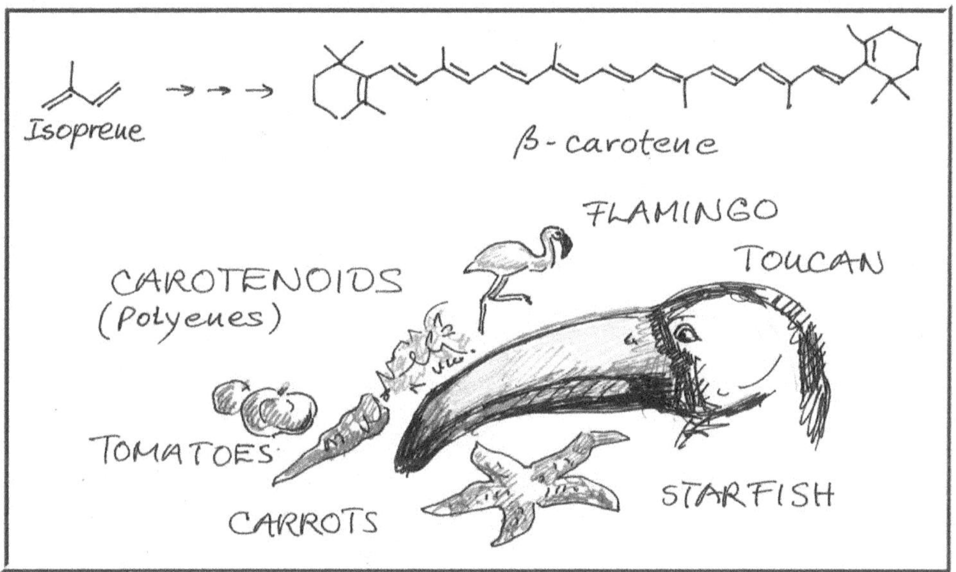


Isoprene is the monomer for natural rubber, a polyprene harvested from the rubber tree. An important artificial rubber is neoprene, polymerized out of 2-chloro-1,3-butadiene or chloroprene, an analog of natural isoprene made by the chemical company DuPont.

One of the most important alkynes is acetylene (common name for ethyne). Due to the strong negative **induction effect** (electron withdrawing tendency) of the electron- rich triple bond, acetylene and other H-containing alkynes have an acidic behavior. It can form salts when reacting with bases called **acetylides**.

Fig. 3.19

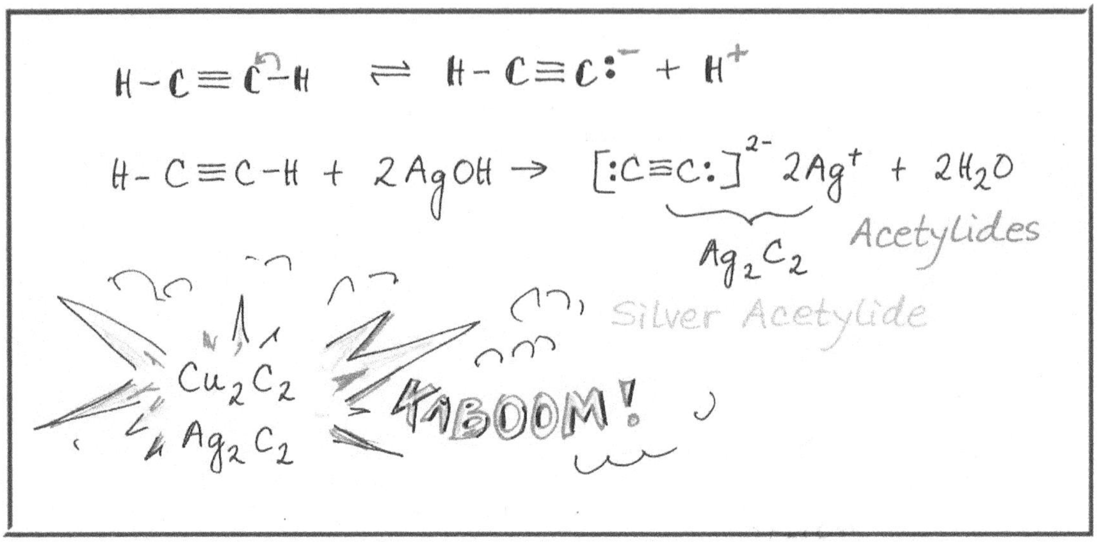


Di-substituted acetylides, such as $Ag_2C_2$ and $Cu_2C_2$, are powerful explosives. These are related to the carbides (compounds with carbon as anion) and can be considered dicarbides, although the bonding is strictly covalent.

Acetylene is also used as gas component for the **oxyacetylene blow torch**, producing sufficient energy for welding, cutting, and soldering metals. As a gas mixture it produces the highest possible temperatures (4000° C – 5000° C).

Fig. 3.20

All alkenes and alkynes are also valuable building blocks and starting materials for a variety of more complex organic molecules.

Alkenes are usually prepared by dehydration reactions of alcohols with the help of a condensation agent such as phosphoric acid. Dehydrogenation of aliphatic alkenes can lead to the corresponding alkyne.

**Fig. 3.21**

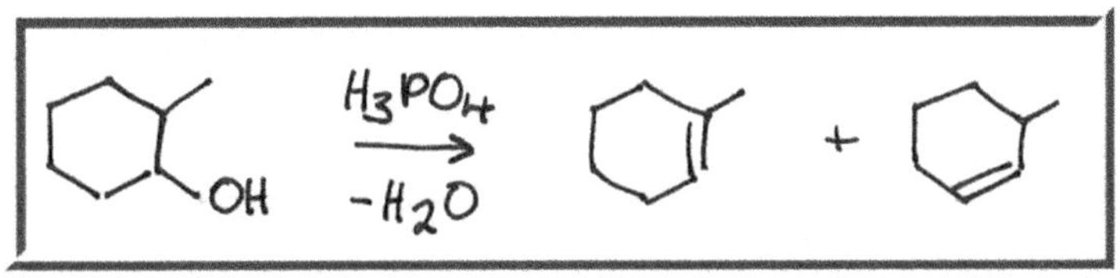

# EXPERIMENTS – Ch. 3

## *Alkenes and Alkynes*

**Key Terms:** Multiple bonds, pi (π) orbitals and bonds, homologous series, alkenes or ( $C_nH_{2n}$), alkynes or ($C_nH_{2n-2}$), nomenclature rules, relative positions of more than one multiple bond (cumulated for only double bonds, conjugated, isolated), saturated and unsaturated molecules, addition reactions ($A_E$), electrophiles, Markovnikov Rule, *cis-trans* isomerism, extended nomenclature for more multiple bonds within one molecule (alkadienes, alkadiynes, alkapolyenes, alkapolyynes, alkaenynes), molecular symmetry (point, plane and rotational symmetry), theoretical and practical yields.

### I Molecular Modeling of Unsaturated Molecules

**Chemicals:**

*none*

**Equipment and Materials:**

*Organic chemistry modeling kits (Molymod)*
*C (black), O (red), H (white), Cl (green), Br (orange),*
*Single and double bond pieces*

### Procedure:

1. Build the molecule for 1-butene first and then move the double bond into the position for making 2-butene.
2. Draw the different structures in the grid shown below using the different formula types (MF, ESF, CSF and ASF/LAF).
3. Identify your model as either being in the *cis* or *trans* configuration and name as such, using the prefixes *cis* or *trans*. Is the molecule symmetrical or asymmetrical? Is there a point, plane of symmetry (mutually exclusive) and does the molecule have a rotational symmetry? Convert the molecule into the other isomer and analyze accordingly.
4. Build a molecule of $H_2O$ and add to the double bond and draw the possible addition product(s) for the hydration reaction in the grid shown below. Use the ESF and the ASF/LAF types of formulas.
5. Build the molecule 1-butyne and 2-butyne and discuss in the same way as for the butenes, studied above.
6. Add a molecule of $H_2O$ and draw and discuss the possibilities.
7. Build a molecule having two multiple bonds within the four-carbon structure, name and draw all the possible constitutional and configurational isomers.

## Observations and Results:

| Name of Molecule: | Formulas: | | | | Hydration (+ $H_2O$ or H-OH) reaction product(s) |
|---|---|---|---|---|---|
| | MF | ESF | CSF | ASF/LAF | |
| | | | | | |

## Questions and Problems:

1. Which is the electrophile in the addition (hydration) reactions?
2. For what reactant molecules does the Markovnikov Rule apply?
3. What type of compound is produced by the addition of water to the unsaturated molecule(s)?

## II Analytical Tests for the Distinction of Saturated and Unsaturated Compounds (Excluding Aromatic Compounds)

## Chemicals:

*Bromine solution in $CH_2Cl_2$ ($Br_2$, amber to brown colored)*
*Cyclohexane*
*Cyclohexene*
*Unknown A*
*Unknown B*
*Unknown C*
*Potassium permanganate ($KMnO_4$), 1 % aq. Solution*

## Equipment and Materials:

*Test tubes with rubber stoppers*
*Droppers with bulb*

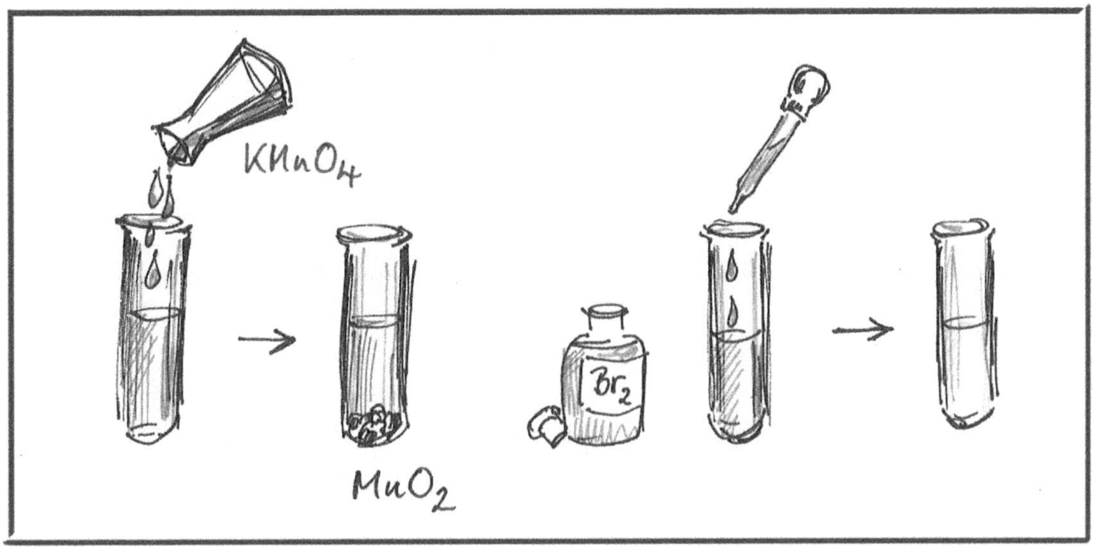


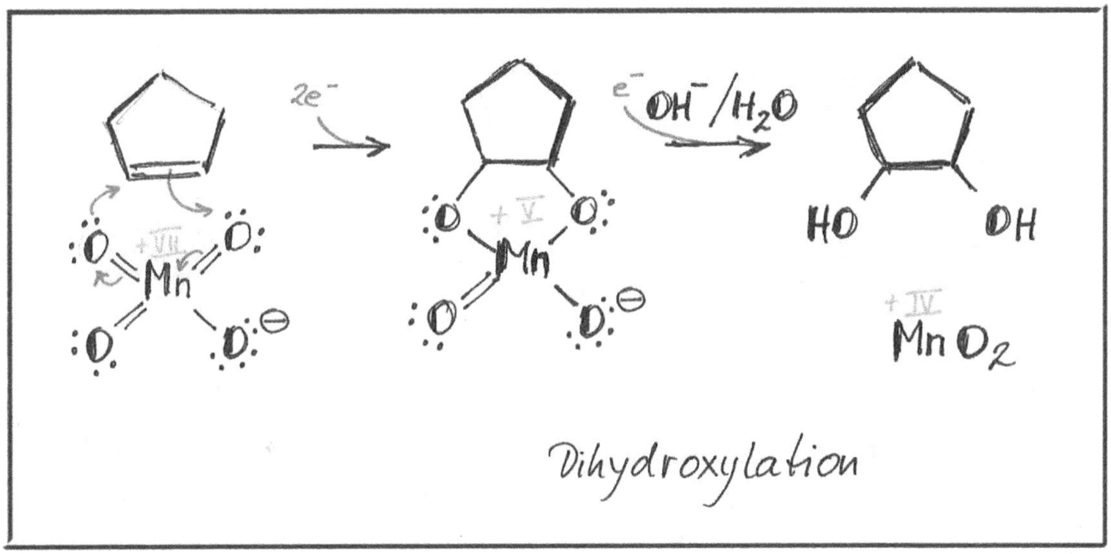


**Procedure:**

1. Prepare five test tubes in a rack.
2. In the first test tube, place the saturated compound cyclohexane, in the second the unsaturated compounded cyclohexene and in the third, fourth and fifth, respectively the unknowns A, B, and C. Each should be filled to a level of 1-2 cm.
3. To each tube add a few drops of bromine solution. Stopper the tube and shake vigorously. Observe and record color changes.
4. Clean out the tubes and fill again with new sample solutions of the five compounds and test each with about 15 drops of $KMnO_4$ solution. Watch for a change of the clear magenta solution (Mn VII) into a brown precipitate (Mn IV). As an oxidizing agent

$KMnO_4$ attacks double bonds, which the other strong oxidizing agent, $CrO_3$, does not tend to oxidize.

## Observations and Results:

| Compound: | Reaction with $Br_2$: | Reaction With $KMnO_4$: | Saturated: | Unsaturated: |
|---|---|---|---|---|
| • **Cyclohexane**<br>• **Cyclohexene**<br>• **Unknown A**<br>• **Unknown B**<br>• **Unknown C** | | | | |

## Questions and Problems:

1. What happens when an unsaturated compound reacts with bromine solution?
2. What type of reaction is involved?
3. Why does water need to be added to make the diatomic bromine molecule a good electrophile?
4. What is the product formed when cyclohexene reacts with bromine? Give the name and draw its structure.
5. In the case of the $KMnO_4$ reduction, the magenta colored solution produced by Mn (VII) is reduced to Mn(IV) oxidation state, appearing in the form of the brown $MnO_2$ precipitate. How many electrons were transferred here?

## III Hydration ($A_E$) of Fumaric Acid to Malic Acid (Model Enzymatic Reaction for Fumarase in the Krebs-Cycle and oxidation to Oxaloacetic acid).

### Chemicals:

*Fumaric acid*
*Demineralized water*
*$H_2SO_4$ or $H_3PO_4$ (concentrated, as catalyst)*
*Jones Reagent ($CrO_3$ in dil. $H_2SO_4$/acetone, yellow/orange)*

### Equipment and Materials:

*Erlenmeyer flask (small), glass rod, graduated cylinder*
*Hot plate and hot water bath with stir bar*
*Test tubes*
*Top-loading digital scale/balance*

### Procedure:

1. Prepare a hot water bath over a hot plate with stirring function.

2. Weigh out 2g of fumaric acid in a small Erlenmeyer flask, dissolve in 2-3 mL water and add a few drops of the concentrated acid catalyst (goggles!). Add a stir bar to the solution.
3. Attach the flask to a stand with a clamp and lower into the hot water bath.
4. Heat for 15 -20 min under constant stirring.
5. Remove the flask, take out the stir bar with clean forceps and pour a portion into a clean test tube.
6. To test the formation of an alcohol, add a few drops of Jones reagent. A color change will indicate oxidation to an aldehyde or ketone, whereby Cr (VI, yellow/orange) is reduced to a lower oxidation state, Cr (III, blue).

## Observations and Results:

| **Reaction product:** | **Structure:** | **Jones reagent + or –** |
|---|---|---|
| **Malic acid** | | |

## Questions and Problems

1. What is the equation for the formation of malic acid via hydration of fumaric acid?
2. Calculate the theoretical yield for malic acid if 2.00g of fumaric acid are used.
3. If a student obtained 0.80 g, what would be the practical or experimental yield in %?
4. What is the purpose for using an acid catalyst? What happens to the water being by itself a weak electrophile?
5. How does maleic acid differ structurally from fumaric acid? What type of isomerism is involved?
6. Draw the structure for oxaloacetic acid that forms when malic acid is oxidized with Jones reagent.

## Chapter 4 Aromatic Compounds (Arenes and Hetarenes)

4.1 The aromatic ring system

4.1.1 Introduction to sigma and pi bonds

4.1.2 Pi electrons and their delocalization

4.1.3 Conjugated multiple bonds

4.1.4 Resonance energy and the Hückel rule

4.2 Natural distribution and discovery of aromatic compounds

4.3 Nomenclature

4.3.1 Mono-substituted benzene derivatives as parent molecules

4.3.2 *Ortho-, meta-, para-* prefixes for di-substituted benzene derivatives

4.4 Fused aromatic ring systems

4.4.1 Carbocyclic compounds

4.4.2 Heterocyclic compounds

4.5 Physical and chemical properties

4.5 Reactions

4.5.1 Substitution on the ring

4.5.2 Side chain reactivity

4.6 Uses in medicine and industry

4.7 Relative toxicity

**Chapter 4  Aromatic Compounds (Arenes and Hetarenes)**

Aromatic compounds derive their name from being originally “aromatic,” as they were first discovered and recognized as a distinct organic class. The prototype is the **Benzene** molecule, with its regular, hexagonal ring structure first recognized by August **Kekulé** (German, 1829 – 1896) interpreting his dream of a snake biting into its tail - the “ouroboros”. Although the molecular formula ($C_6H_6$) or percentage composition was known, the highly unsaturated molecule gave way to much speculation as to its molecular structure. Kekulé spent so much time investigating this mystery that the problem manifested itself into his dreams as he explained that it motivated him to come up with the two postulated structures of benzene, which are now called the **Kekulé structures**

**Fig. 4.1**

In today's view, benzene is a completely renonance stabilized molecule with about 34 kcal/mol or 143 KJ/mol extra energy for stabilization. This means that the two Kekule structures are only separate resonance structures and that the actual structure is somewhere in between, called the resonance hybrid, and symbolized by a ring within the hexagon.

**Fig. 4.2**

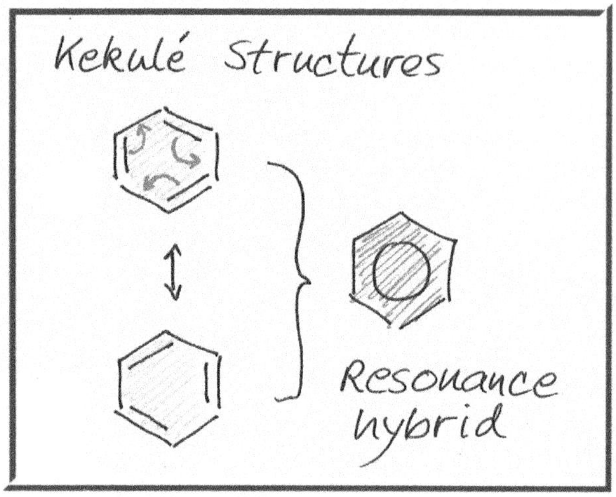


The hybrid involves the delocalization of the three conjugated double bonds to an extended pi-system ($\pi$). This is often referred to as a pi-sextet. If it would be a true cyclohexatriene, with the three double bonds in fixed position, it would look like the following representation – the double bonds being shorter than the single bonds.

**Fig. 4.3**

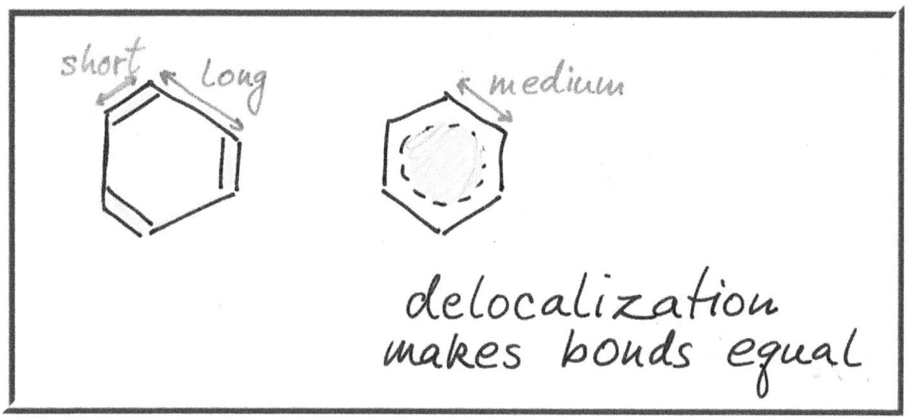


However, studies (x-ray diffraction, etc.) show that benzene is a completely flat molecule with all bonds of equal length (an average of the single and double bond lengths). This can only be possible if the three double bonds are delocalized and spread over the whole system, making the skeletal structure an even-sided planar hexagon.

The energy gain can be explained by calculating the hydrogenation energy values for each individual double bond of one of the Kekulé structures and comparing the calculated value for the three double bonds to the actual measured value of the hydrogenation energy required for the whole benzene system.

**Fig. 4.4**

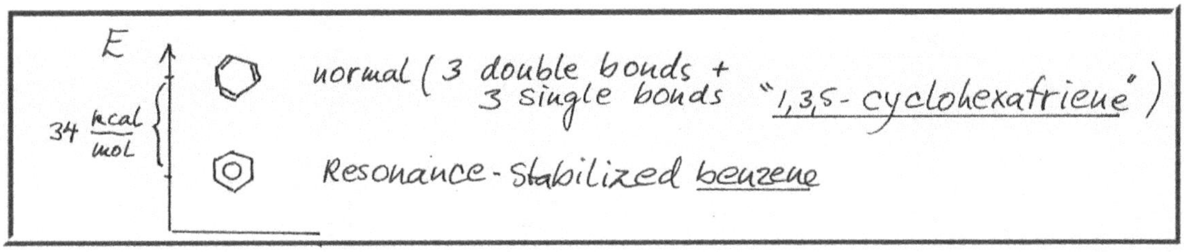


**Fig. 4.5**

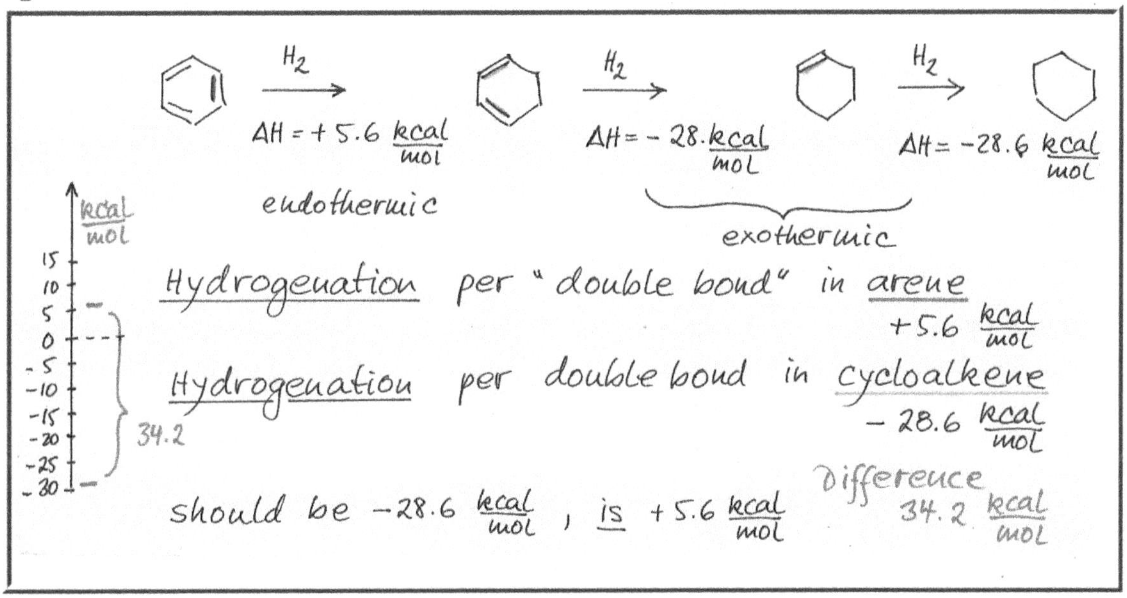


Although exceeding the scope of this course, it is still useful for some to understand the basics of the Molecular Orbital Theory (MO-theory). The atomic orbitals – understood as probability areas for electrons to move about in, with each orbital able to house a maximum of two electrons with opposite spin and often symbolized as circles or boxes with vertical arrows pointing up or down to indicate spin direction.

**Fig. 4.6**

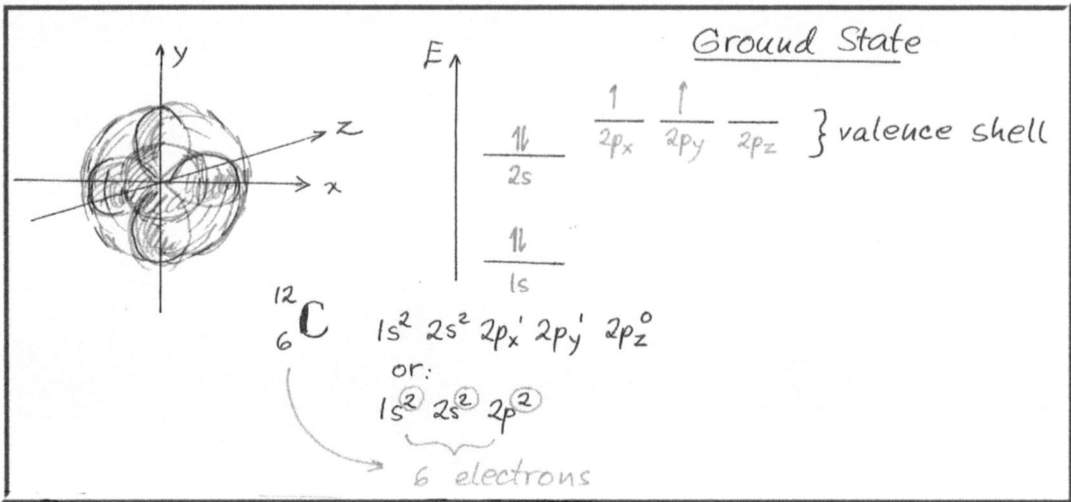


The ground state of the Carbon, being the central element for the whole field of organic chemistry as well as biochemistry, should be described here in more detail. It has a full 1st shell (n = 1) with a doubly filled *s*-orbital ($1s^2$). In the 2nd shell, it has a higher *s*-orbital, doubly filled ($2s^2$), and the 3 *p*-orbitals, 2 of which are filled with one electron each and the third remaining empty or unoccupied, according to **Hund's rule**. Hund's Rule (Friedrich Hund, German, 1896-1997) states that degenerate (equal energy) orbitals are first filled separately before they are doubled up. The configuration of the 6 electrons is therefore $1s^2\ 2s^2\ 2p^2$ or $1s^2\ 2s^2\ 2p_x^1\ 2p_y^1\ 2p_z^0$.

**Fig. 4.7**

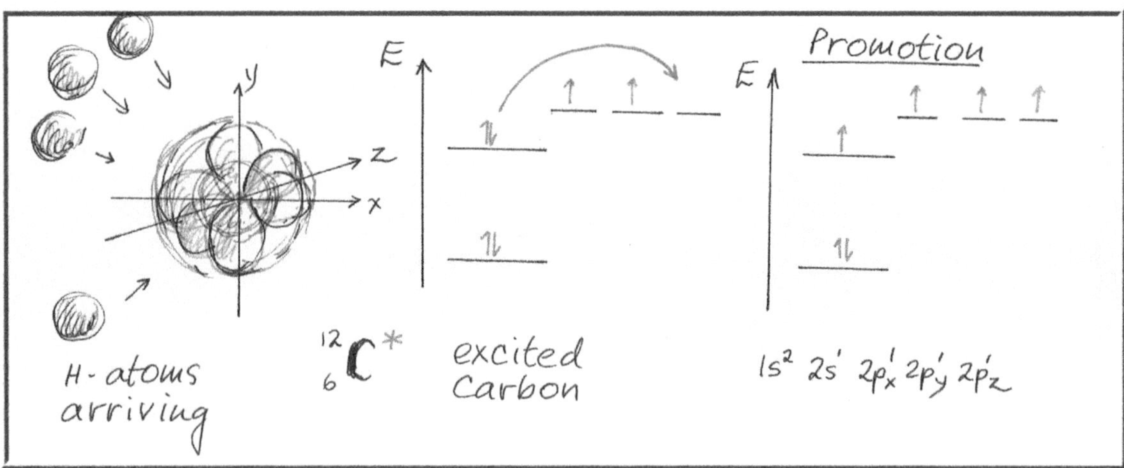


When bonding partners such as H-atoms become available, C becomes excited ($C^*$), resulting in one electron jumping from the doubly filled 1s orbital into the empty *p* orbital. This electron transition leading to all singularly filled orbitals is called the **promotion**. According to the **MO-orbital theory**, where bonds come about by atomic orbital (AO) overlap (merging of their probability spaces or volumes), the atom converts the singly filled non-degenerate (different energy levels) atomic orbitals into orbitals that all look alike and have the same energy content (degenerate), referred to as **hybrid orbitals**. The process is called **hybridization**, which literally creates a mix of the different types of orbitals into a new species, like the mixing of two or more dog breeds into a mongrel (hybrid). One *s*- orbital, when mixed with one *p* becomes an *sp*, when mixed with two *p*'s becomes $sp^2$, and when mixed with three *p*'s becomes an $sp^3$ orbital.

**Fig. 4.8**

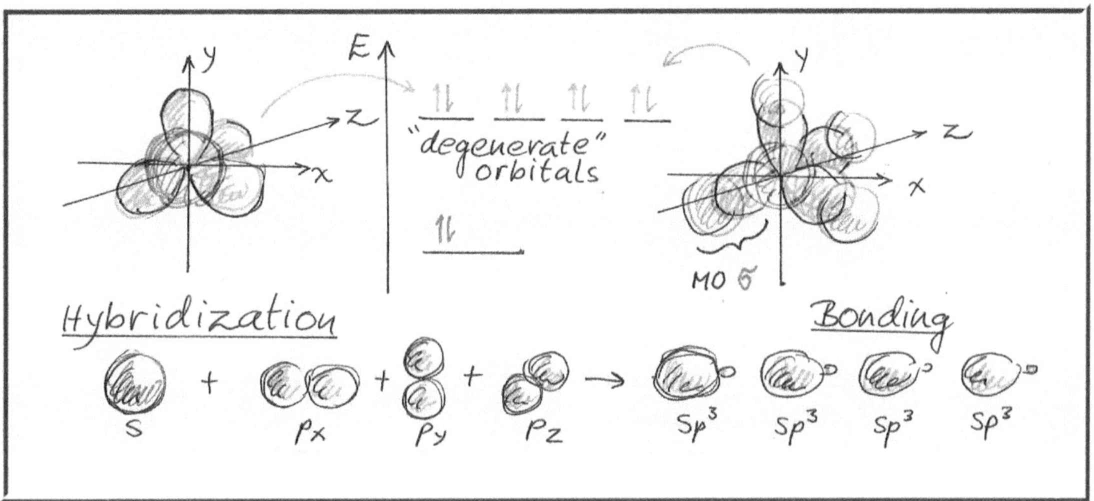


The newly formed hybridized atomic orbitals (HAO's) are much more voluminous and extend further out into space, offering more overlap area for the bonding with Hydrogen s-orbitals. 4 $sp^3$ orientate themselves into the corners of the tetrahedron and once overlapping occurs, new orbitals are formed - the **molecular orbitals** or MO's. MO theory is very helpful in understanding the 3-dimensional space alignment for molecules.

**Fig. 4.9**

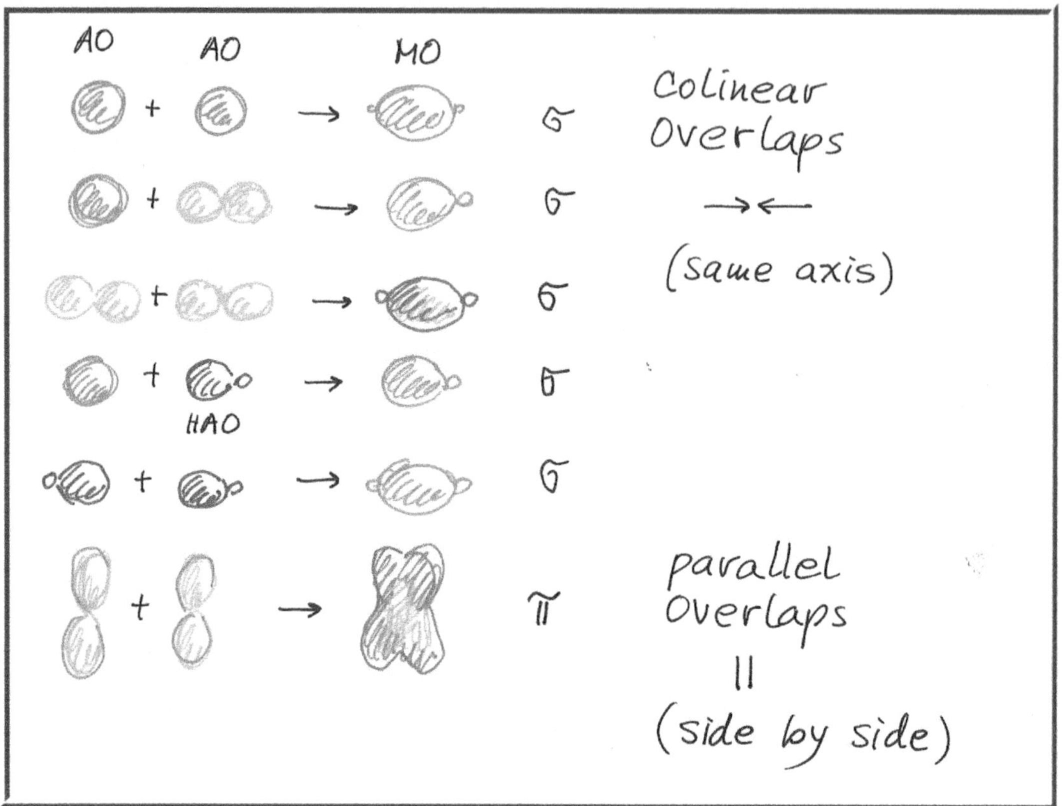


The new overlap-orbitals or MO's, formed by two atoms as bonding partners, can come about in two ways. If they result from a co-axial or co-linear overlap (head to head), they are called **sigma (σ) bonds**. Any head to head overlap, regardless of whether *s*, *p*, or hybrid orbitals are involved (like or unlike) leads to a sigma bond.

When there are not enough bonding partners available, so-called **pi-bonds (π)** are made. Un-hybridized $p_y$ or $p_z$ orbitals left out of the mixing process will align side by side and do a parallel ( || ) overlap. This bond is not as strong as the sigma bond, since the overlap space is a result of more strain and therefore less voluminous.

The **orbital-diagram,** a structural type model showing all the involved orbitals for a molecule, with ethene and ethyne as examples would be as follows:

**Fig. 4.10**

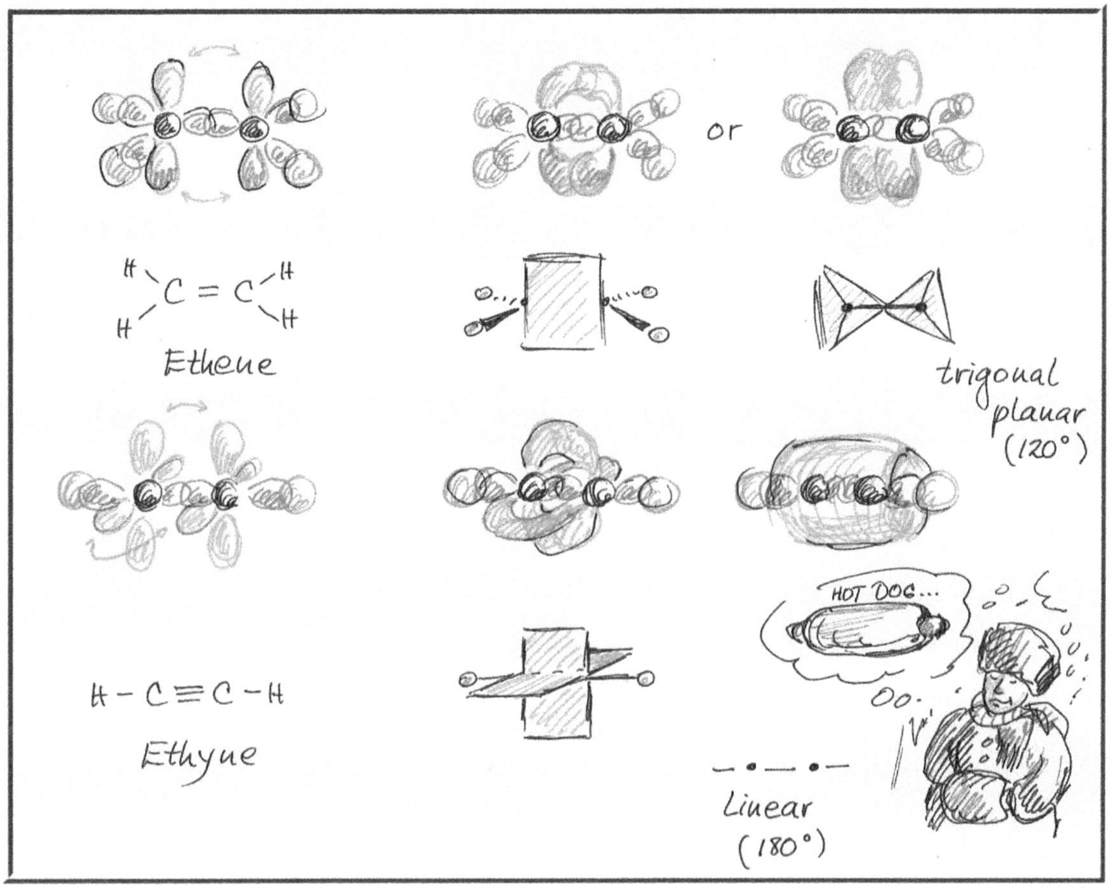


A double bond is actually a combination of a centrally located sigma bond and a top and bottom connected pi-bond ($\sigma + \pi$). A triple bond is the ($\sigma + \pi$) combination plus another $\pi$-bond connecting the front and back of the triple bond, perpendicular to the other $\pi$-bond plane, thus being a ( $\sigma + 2\pi$) combination.

When looking at benzene ring, it is now understood that the orbital diagram of the molecule should look like the following:

**Fig. 4.11**

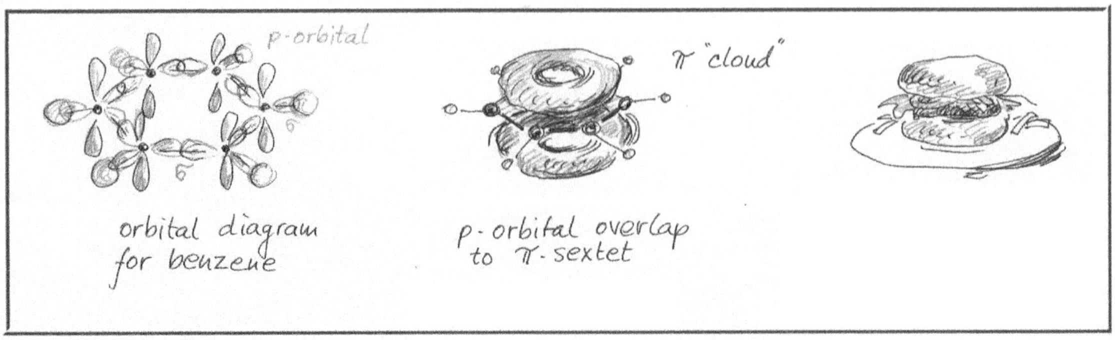


**Example Problem 4.1**
**How many sigma (σ) and pi (π) bonds are involved in the following molecules?**

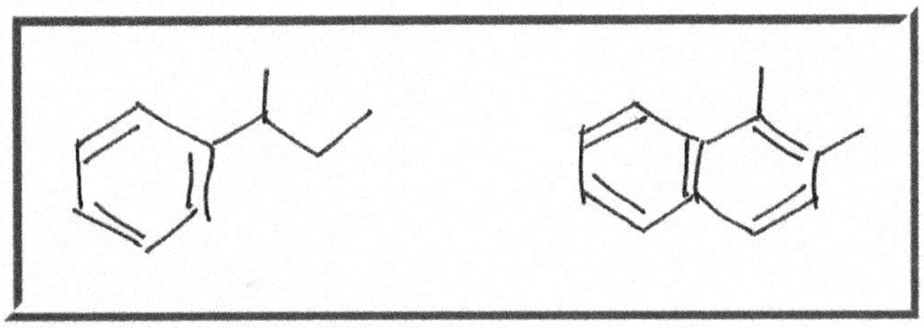

**Example Problem 4.2**
**Sketch the orbital diagram for the following molecule:**

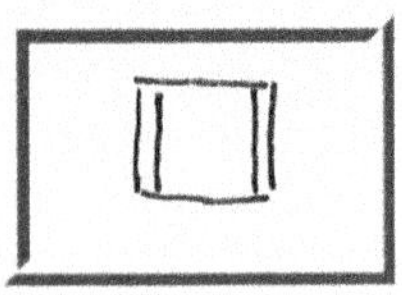

Not only can the π-bonds pair up by twos (Kekulé structure), but also make a giant π-bond or π-sextet (6 *p* electrons being shared as one delocalized system), symbolized by the resonance hybrid formula. The "pi-cloud" is layered on both sides of the ring, pushing the C-atoms of the benzene into a completely flattened out framework.

The special π-sextet gives the benzene ring its added stability (34 kcal/mol), a condition to be preserved at all costs. That is why benzene and its homologs will not add molecules, even though it is an unsaturated system. The reaction preferred is the **substitution** ($S_E$) reaction, where the rate-determining step is similar to the $A_E$ reaction, also involving an electrophile being attracted to the electron rich system. However, the ring will kick out a proton ($H^+$) and reestablish its π-sextet or **aromaticity** once the new group is attached. So a group is substituted for a hydrogen.

**Fig. 4.12**

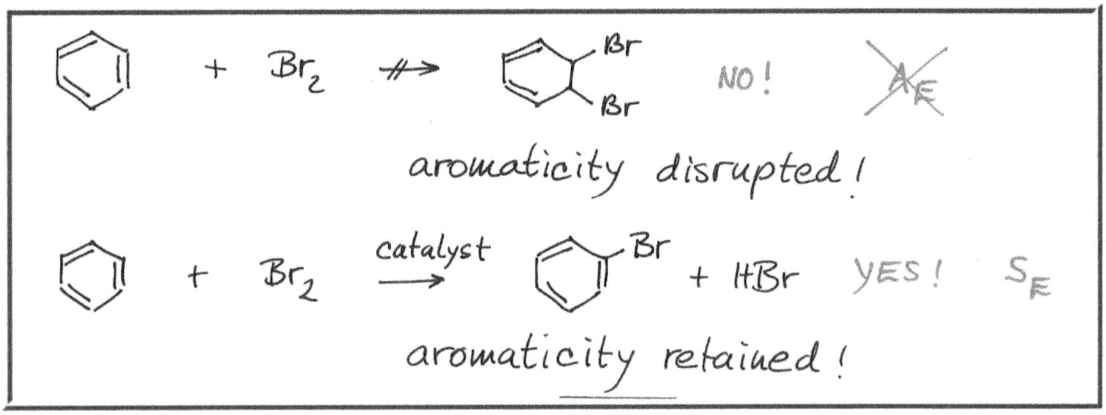


**Example Problem 4.3**
**Complete the following substitution reactions:**

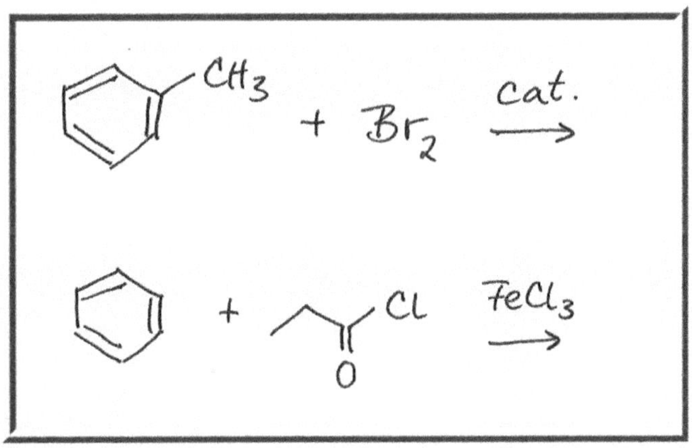


**Example Problem 4.4**
**What sequence of reactions, considering the property of o/p and m directing properties, would lead to the following molecule?**

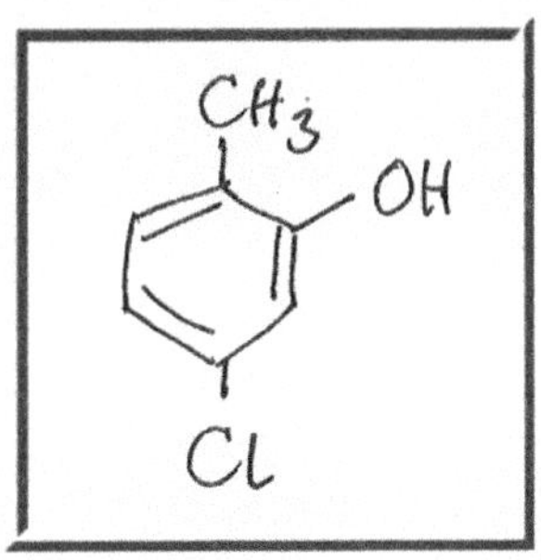


Not all cyclic compounds that have complete, conjugated double bonds are aromatic. To predict whether a compound is aromatic or not, the **Hückel Rule**, based on energy calculation for various molecules carried out by Hückel, is of help. Here the ring must contain **(4n + 2)** $\boldsymbol{\pi}$ electrons (n being any number). For benzene, the number of pi electrons is 6, so

**Fig. 4.13**

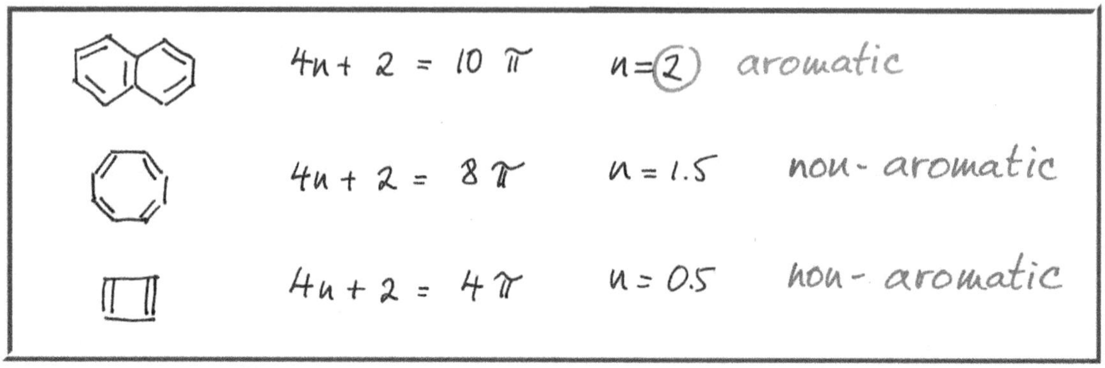


If n is a whole number the system is aromatic.

**Fig. 4.14**

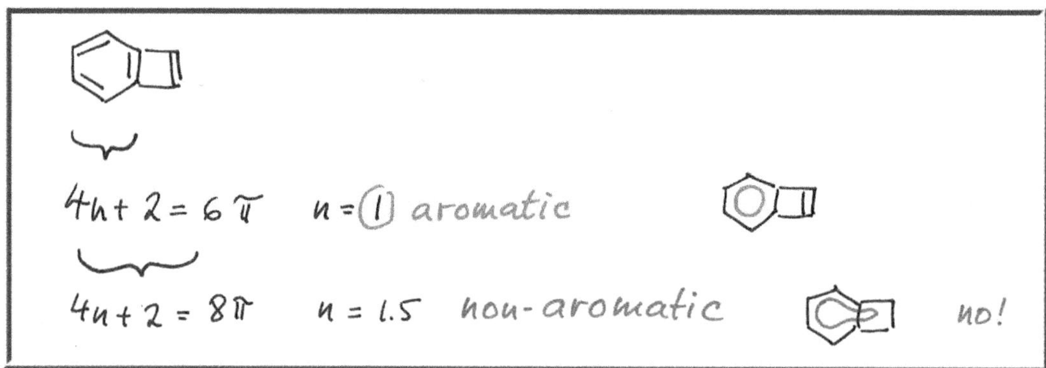


As long as part of the molecule is aromatic, the whole molecule is considered aromatic.

**Example Problem 4.5**
**Which of the following cyclic molecules are aromatic?**

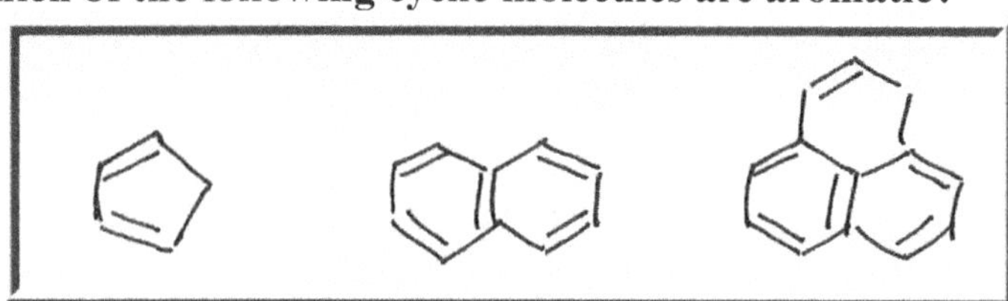

**Example Problem 4.6**
**Which of the following molecules comply with the Hückel Rule?**

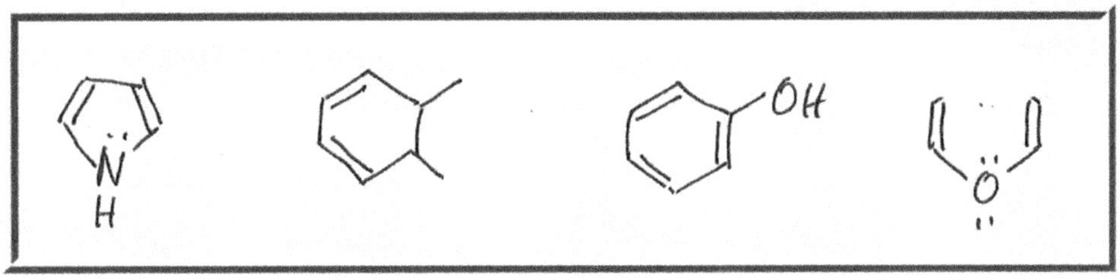


Only the ring, that has total aromaticity, can be represented by a circle in the formula.

**Fig. 4.15**

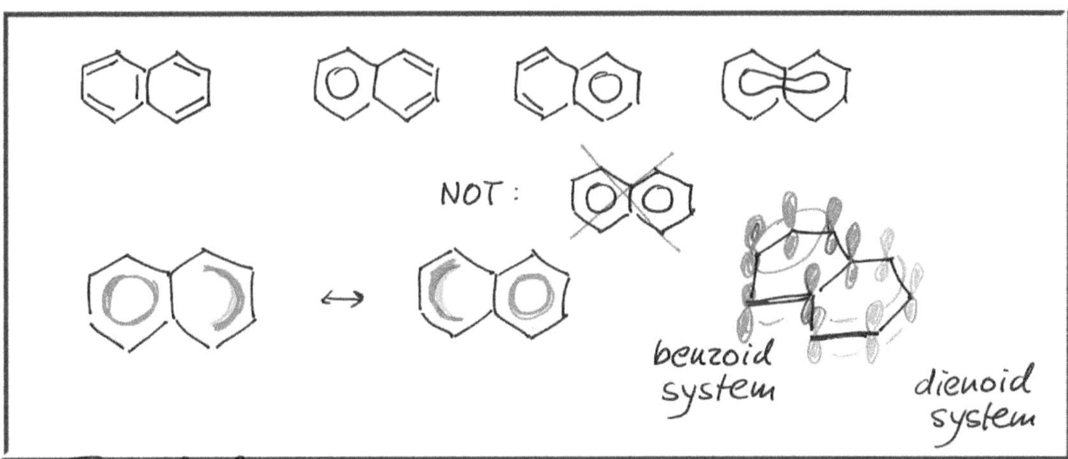


Benzene and fused arenes ("polynuclear" arenes), aromatic systems that have two or more carbons in common, such as naphthalene and anthracene, have their own homologous series.

**Fig. 4.16**

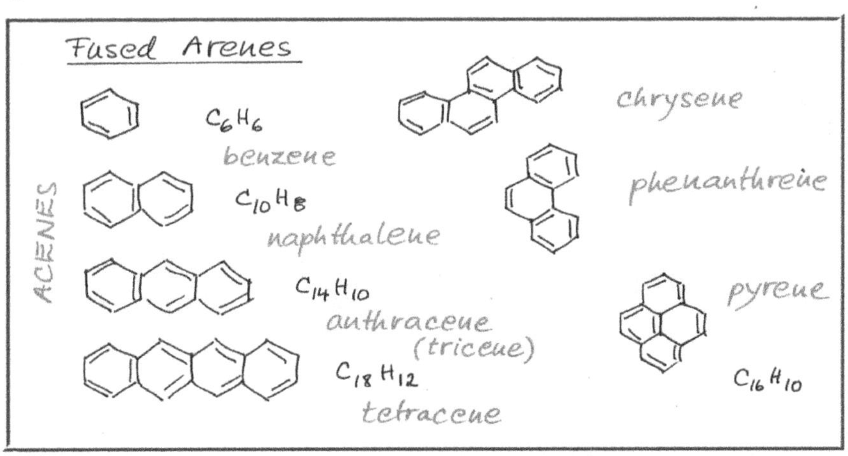

**Nomenclature** of benzene derivatives is similar to naming cyclohexane based products, only "benzene" now being the parent. The ring is numbered in such a way that the substituents have the lowest number combination. Mono-substituted benzenes do not have to be numbered, since all sites on the ring are equivalent.

**Fig. 4.17**

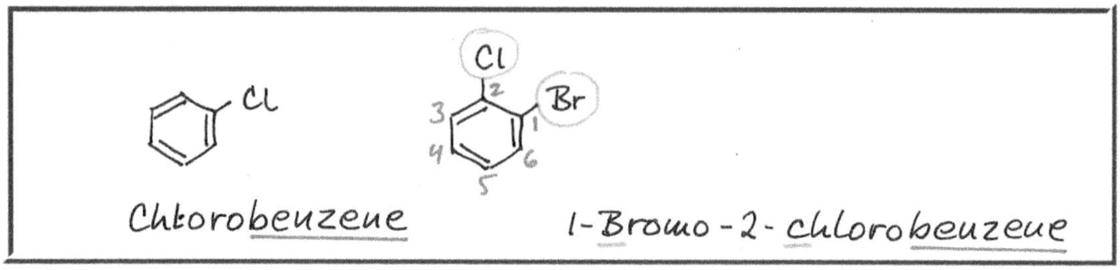


For di-substituted benzene, instead of using the numbering system, the prefixes ***ortho-*** (for 1,2-), ***meta*** (for 1,3-), and ***para*** (for 1,4-) can be used. These prefixes, when in print, are always italicized. The abbreviations ***o-***, ***m-***, and ***p-*** can also be used. Some mono-substituted benzene derivatives have the luxury of having their own name. Some important ones are

**Fig. 4.18**

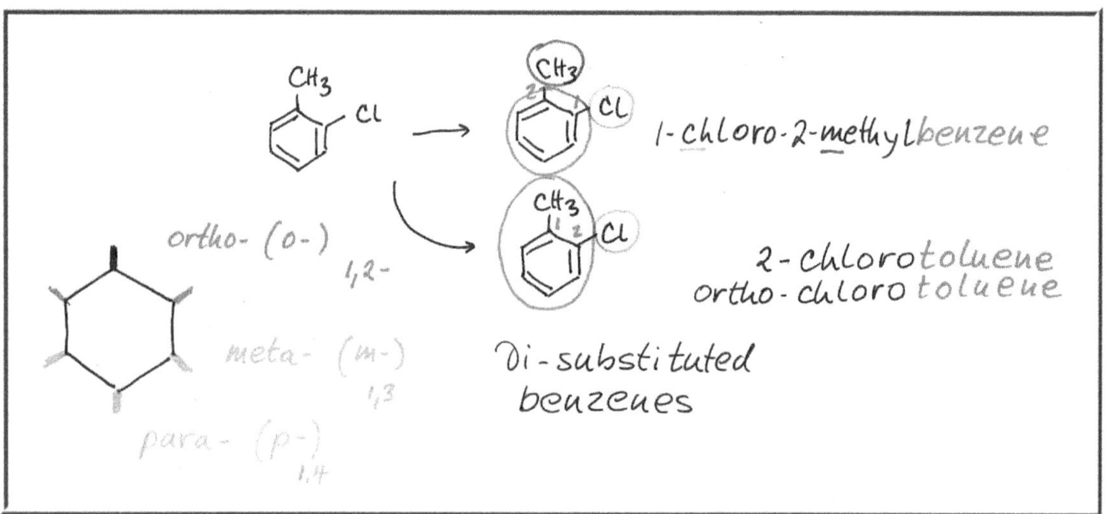


When using the named derivatives with other substituents, the one that gives the name is automatically number 1.

**Fig. 4.19**

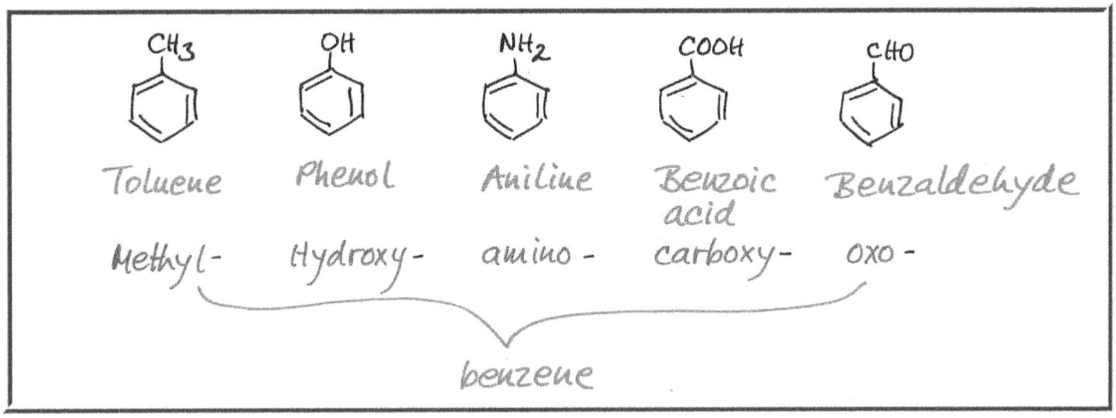


When two groups that would have their own name as benzene derivatives occur on the same molecule, only one can be used as a parent at a time. So, often many names are possible for a certain benzene derivative.

**Fig. 4.20**

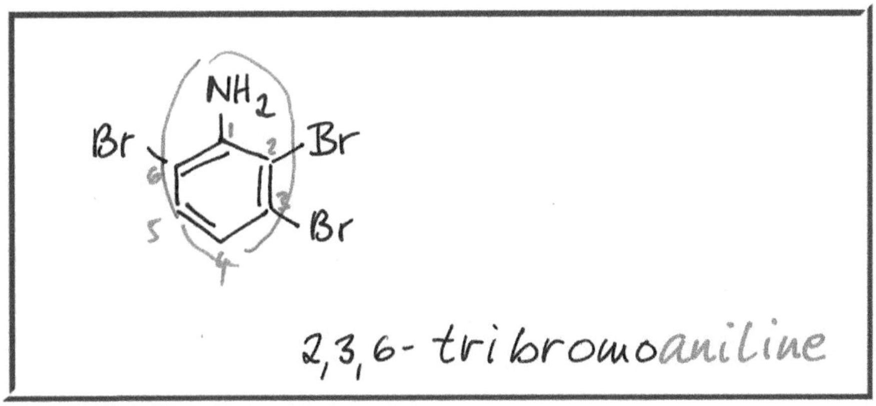


**Example Problem 4.7**
**Find two or more names for the following molecules:**

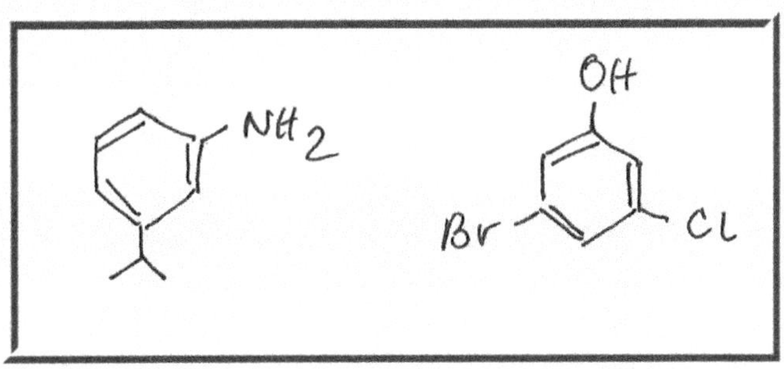

**Example Problem 4.8**
**Write the structures represented by the following names:**
***Ortho-ethylaniline***
***1,3,4 –Tribromobenzene***
***Meta-Nitrophenol***
***2,4,6-Trinitrotoluene***

Pure benzene is found in petroleum but its derivatives are widely spread in the living and non-living worlds. They are key components in many essential oils of plants – oil of cloves, bitter almonds, oil of wintergreen, and allopathic molecules such as salicylic acid, which dissolves the root system of competing plants.

The first substituent on the benzene nucleus has a big influence on the site of the second substitution. If the first substituent is an alkyl group ($-CH_3$, $-C_2H_5$, etc.) or a group that contains lone pairs in association with only single bonds (-OH, $-NH_2$, -halogen), it will direct the second substituent to the *ortho* or *para* position. These groups activate the benzene ring by contributing their electrons to the aromatic ring system and can be considered as "pushers."

**Fig. 4.21**

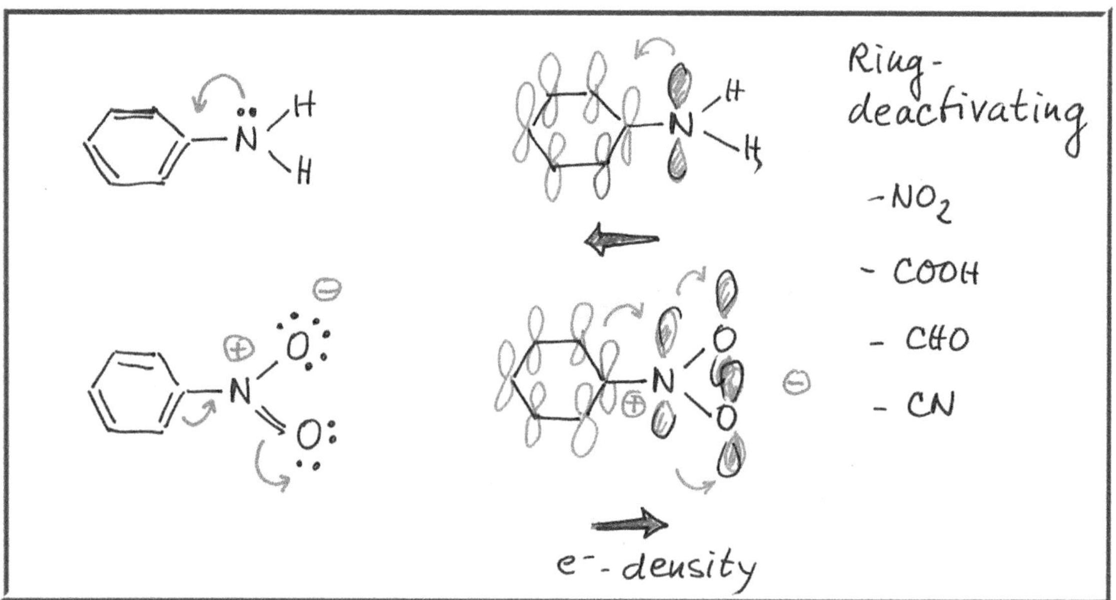


The lone pairs get "sucked" into the π-sextet, participating in the "dance," so they are electron-enriching and thus also ring-activating groups having a +M, +I, contributing, positive mesomeric (M) or inductive (I) effect. Exceptions are the halogens, which are o-, p-directors but ring deactivating. When the first substituent has its own resonance system, consisting of multiple bonds and lone pairs (π-systems), then it is considered to compete with the π-system of the benzene ring, having its own "dance" or "party." These groups are electron-withdrawing and thus ring-deactivating (-M or -I effect). They can be considered to be "pullers" and direct the second substituent to the meta position.

**Fig. 4.22**

If this would not be the case, then anybody could synthesize TNT, a powerful explosive. The nitro group is one of the most deactivating group for the benzene ring

**Fig. 4.23**

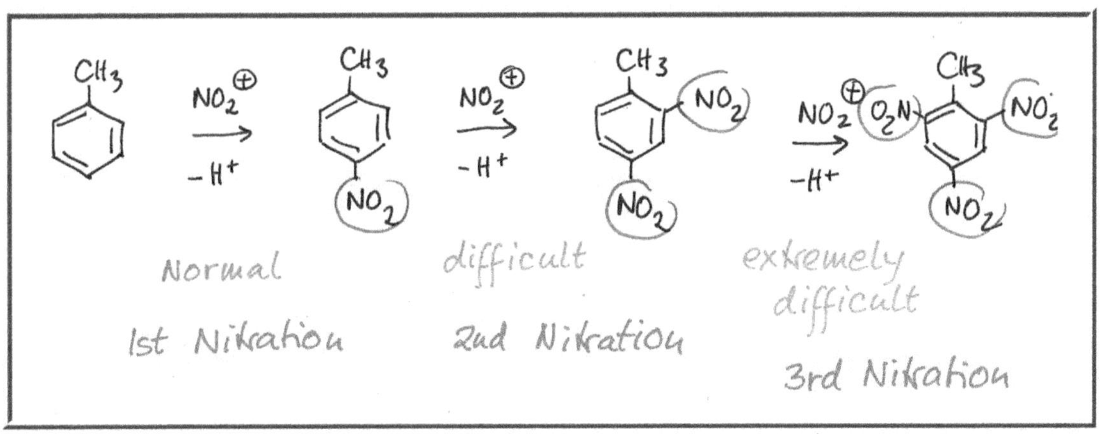


Aromatic systems without functional groups are mostly colorless, of petroleum like odor, and more or less flammable. Benzene has a high octane number and is, like some of its homologs, found in crude oil and gasoline.

It was originally discovered in the natural product gum benzoin, mainly containing benzoic acid, which can easily decarboxylate into benzene.

**Fig. 4.24**

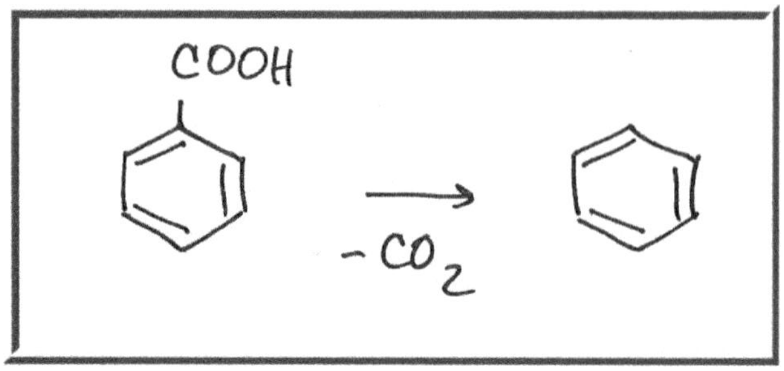


Benzene has excellent solvent properties, especially for grease removal and extraction purposes. However, since it became known that it is a powerful carcinogen (bone-marrow pathologies leading to leukemia and lymphoma, etc.) its use in industry and laboratories has been discouraged, now being replaced by toluene and higher methylated derivatives (xylenes). Benzene toxicity is mainly attributed to the formation of epoxides by oxidative metabolism. Epoxides are usually DNA damaging and cannot be easily eliminated out of the body. Toluene would oxidize in its side-chain, leading ultimately to benzoic acid, a water soluble carboxylic acid.

Benzene has great value as starting material to produce a whole array of other chemicals (drugs, pesticides, detergents, explosives, dyes, etc.) and thus a most sought after resource. It is converted to styrene (vinylbenzene or ethenylbenzene), which is the monomeric unit for the production of polysterene (PS), a polymer with unlimited uses in the plastic industry. Benzene can be alkylated (R = alkyl groups) or acylated (R = CO-Alkyl groups), via the **Friedel-Crafts Reaction**, with alkyl or acyl halides and a Lewis acid as a catalyst ($AlCl_3$ or $FeCl_3$).

**Fig. 4.25**

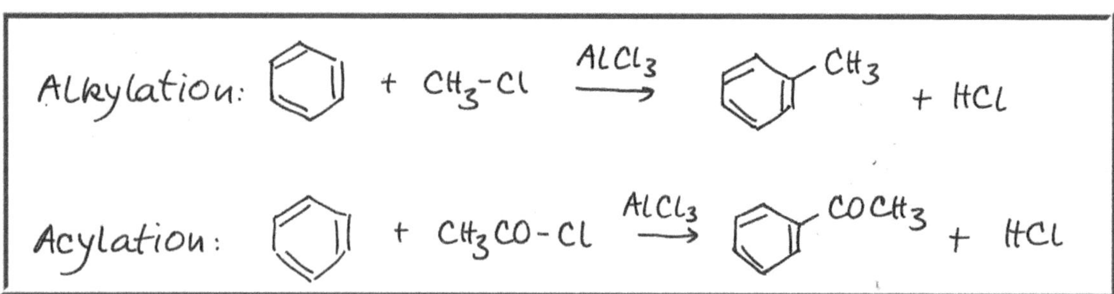


**Naphthalene**, the next higher homolog, has the typical smell of moth balls and is used as an ingredient for the prevention of moth infestation in protein based fibers such as wool. Due to its flammability it is often substituted or mixed with p-Dichlorobenzene.

**Fig. 4.26**

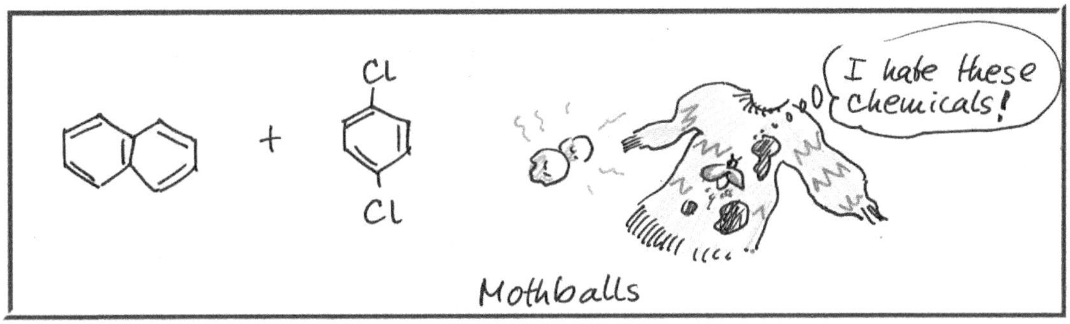


Mothballs added to gasoline could increase a car's horsepower. **Anthracene** , a rectilinear homolog of Benzene and Naphthalene, is isolated from coal-tar. It is not a carcinogen but can be environmentally damaging due to its persistence and durability. Its main use is as a starting material to make various dyes.
If the benzene molecule serves as a substituent to other, parent molecules, it is called **phenyl**, not to be confused with phenol (hydroxybenzene).

**Fig. 4.27**

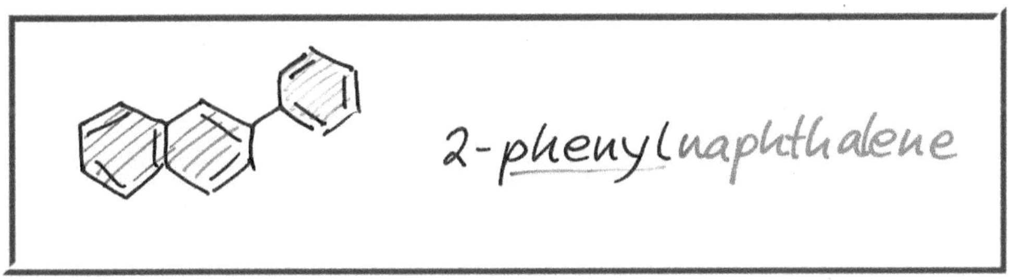


A list of many fused aromatic ring systems is found in the following table

**Benzopyrene** (Benzpyrene)is of special interest since it is a well-studied carcinogen, thought to mimic steroid hormones but also causing damage by intercalation of the DNA double helix, sqeezing between the rungs of the base pairs (see Ch.27). Along with other similar fused ring systems it is found wherever burning, roasting or combustion of organics occur such as automobile exhaust, cigarette smoke, coffee roasts, barbeque drippings reacting on the hot coals to toxic vapors. An interesting phenomenon is that the spectrum of polycyclic aromatic hydrocarbons from car exhaust is very similar to that of most organic compounds found in the interstellar media.

**Fig. 4.28**

**Bisphenol A (**BPA), an important component for making polycarbonates and epoxresins, has received some notoriety as of late, in that has become a potent endocrine disruptor, mimicking female hormones causing, similar to the phthalates (Phthalic acid derivatives, which are used as plasticisers such as PETE in water bottles), the "feminization of society".

**Fig. 4.29**

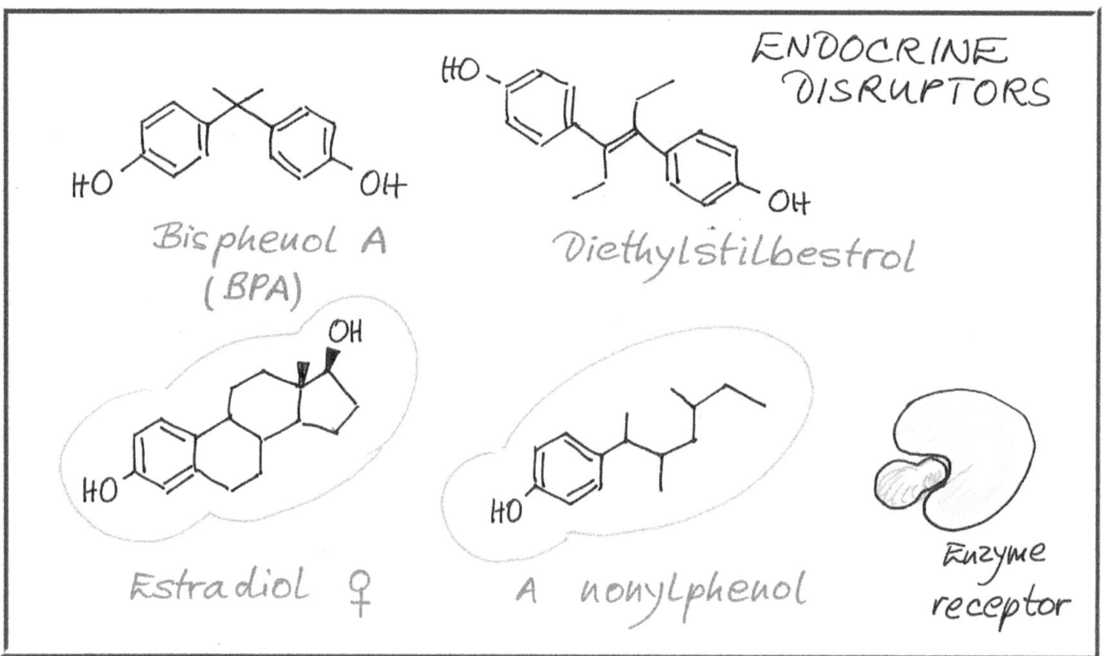

When two benzene rings are linked together the product is called **Biphenyl**, a crystalline solid of pleasant smell. It is added to many products as fungicide. An important family of halogenated derivatives is the group of **polychlorinated biphenyls** (PCB's), with to 2-10 chlorine atoms as substituents, which, although banned, have become persistent pollutants of the environment. They had many uses in technology but due to their action as endocrine disrupters and neurotoxins were discontinued.

**Fig. 4.30**

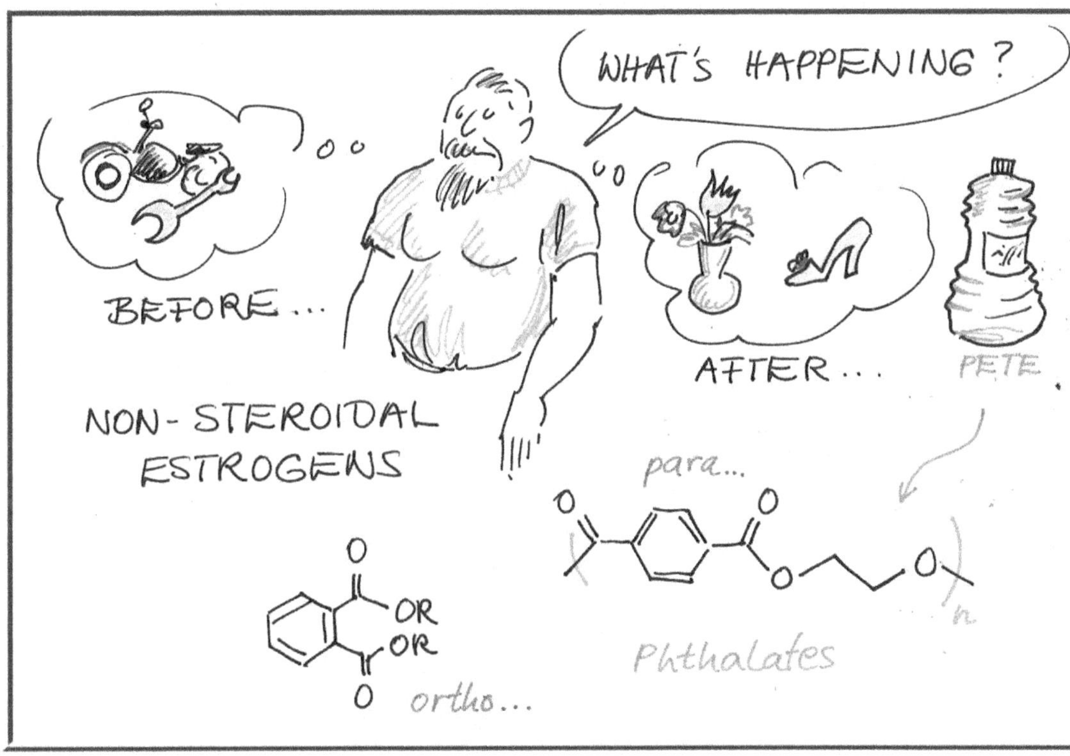


Aromatic ring systems containing one or more, same or different hetero-atoms with the ring framework are called heterocyclic compounds in general, but if they are also aromatic they are known as **hetarenes.** The group is endless, but certain families stand out, especially those that contain nitrogen and/or oxygen atoms. Sulfur containing rings are also quite abundant, so that only a short mention has to suffice. In many ways they are similar to the arenes, but have much more diverse characteristic due to the inclusion of the hetero-atoms in the parent ring structures. In the following table some important arenes are listed:

**Table 4.1**

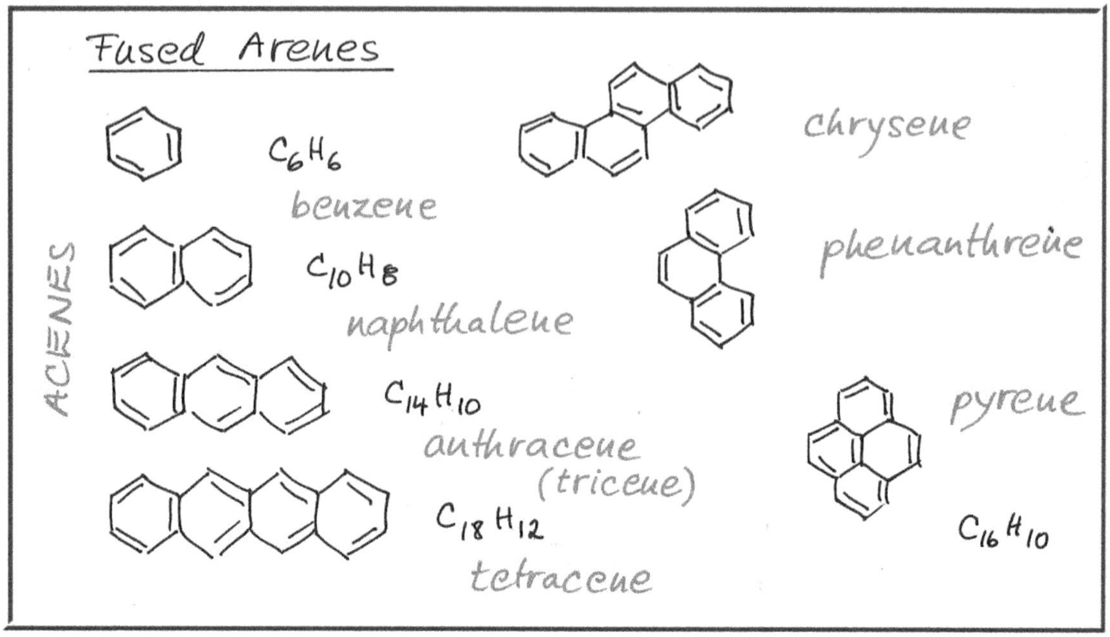
Fused Arenes
ACENES
$C_6H_6$
benzene
$C_{10}H_8$
naphthalene
$C_{14}H_{10}$
anthracene
(tricene)
$C_{18}H_{12}$
tetracene
chrysene
phenanthrene
pyrene
$C_{16}H_{10}$

# EXPERIMENTS – Ch. 4

## *Aromatic Compounds (Arenes and Hetarenes)*

**Key Terms:** Conjugated ring systems, benzene as aromatic prototype, delocalized pi systems, resonance energy, Kekulé structures, resonance hybrid, Hückel rule for aromaticity [(4n + 2) $\pi$ electron systems)], electrophilic substitution ($S_E$), two-step mechanism, carbocations, activation energy, heteroatoms , second substitution and ortho/meta and para conductors, induction effect and electron "pushers" (+I effect) and "pullers" (- I effect).

### I Molecular Modeling of Benzene and Heterocyclic N-Analogs (Azarenes):

**Chemicals:**

*none*

**Equipment and Materials:**

*Organic chemistry modeling kits (Molymod)*
*6C (black), 6H (white), 2N (blue), 2 Br (orange)*
*3 Single and 6 double bond pieces*

**Procedure:**

1. Build a benzene molecule, connecting six carbon atoms by alternating single and double bonds (3 conjugated double bonds) and attach a hydrogen atom to each carbon atom.
2. Pick one carbon atom as reference and change all single bonds to double bonds and vice versa. Draw the resulting Kekulé structure including the previous one in the grid below.
3. Build a diatomic bromine molecule and pretend it is polarized in the presence of water. Separate one atom of the bond to act as electrophile and attach to an opened up double bond piece. Make a mental note of which carbon now as a positive formal charge and remove a hydrogen atom from the four-bonded carbon to which the bromine is attached and re-establish a double bond. The hydrogen atom is then connected to the other bromine atom to which the single bond piece is attached. This represents the mechanism of an electrophilic substitution reaction.
4. Replace one carbon atom of the unsubstituted benzene ring with a nitrogen atom and draw the structure of this aromatic heterocycle (hetarene).
5. Replace a second carbon atom with a nitrogen atom at different positions in reference to the first one. Draw all the possible constitutional (structural) isomers.
6. Return to the model of benzene and save for the next modeling assignment (exp. II).

## Observations and Results:

| Molecule: (Arene / azarene) | Molecular Formula (MF) | Kekulé Structures: | Resonance hybrid: |
|---|---|---|---|
| • **Benzene**<br><br>• **Pyridine (azabenzene)**<br><br>• **Pyridazene (1,2-Diazabenzene)**<br><br>• **Pyrimidine (1,3-Diazebenzene)**<br><br>• **Pyrazine (1,4-Diazabenzene)** | | | |

### Questions and Problems:

1. What are the products of each bromination reaction ($Br_2/H_2O$), listed in the table above, proceeding via the $S_E$ mechanism?
2. Why are there more bromination possibilities for the azarenes? Think of the number of equivalent positions in order to answer this question.

## II Molecular Modeling of Benzene Derivatives (Toluene, Phenol, Aniline and Benzoic Acid) and Conception of Directing Effects for Second Substitution

### Chemicals:

*none*

### Equipment and Materials:

*Molecular model kit*
*Benzene model (from exp. I)*
*1 N, 2 H for the amino group (-$NH_2$)*
*1 O, 1 H for the hydroxyl group (-OH)*
*1 C, 3 H for the methyl group (-$CH_3$)*
*1 C, 2 O, 1 H for the carboxylic group (-COOH)*
*2 Br for $Br_2$*
*single bond and double bond pieces*

## Procedure:

1. To an already constructed model of benzene attach a methyl group by replacing a hydrogen atom. In the same manner build a hydroxyl group, an amino group and a carboxylic group and attach it to the benzene ring ("nucleus").
2. On the models of the derivatives perform an electrophilic substitution reaction with a $Br_2$ model and study the directed positioning of the second substituent by the first.

## Observations and Results:

| **First Substitution of Benzene:** | **Name of derivative:** | ***Ortho/para* or *meta* director for second substitution** |
|---|---|---|
| • **Methylation (+ $-CH_3$)** | • | |
| • **Hydroxylation (+ -OH)** | • | |
| • **Amination (+ $-NH_2$)** | • | |
| • **Carboxylation (+ -COOH)** | • | |

## Questions and Problems:

1. How many possible products can be formed by (catalytic) bromination of the four benzene derivatives above?
2. Show the equations for each bromination reaction discussed above.
3. Why don't benzene derivatives or other arenes add molecules even though they are unsaturated?

## III Nitration of Toluene

## Chemicals:

*Toluene*
*Nitric acid (conc.)*
*Sulfuric acid (conc.)*

## Equipment and Materials:

*Erlenmeyer flask (medium)*
*Droppers with bulb*
*Ice bath*
*Graduated cylinder (10 mL)*

## Procedure:

1. In the fume hood place 5.0 mL of toluene in a medium sized Erlenmeyer flask

2. Add 4-5 drops each of the concentrated $HNO_3$ and concentrated $H_2SO_4$ to the toluene. This produces the nitronium (nitrosyl) cation ($NO_2^+$) in situ (on site) as electrophile.
3. After a few minutes of reaction, place the Erlenmeyer flask in an ice bath.
4. Observe changes of the reaction mixture and record the results in the table shown below.

## Observations and Results:

Tabulate the observed results and add the other information required according to the following table:

| **Substrate molecule:** | **Structure:** | **Density (g/mL):** | **Mass (g):** | **Moles (mol):** | **Electro-phile (reagent):** | **Structure of nitronium cation ($NO_2^+$):** | **Names and Structures of the two possible products:** |
|---|---|---|---|---|---|---|---|
| Toluene | | | | | Nitronium cation ($NO_2^+$) | | |

## Questions and Problems

1. Write the equations for the two step mechanism for the nitration reaction.
2. What is the theoretical yield for the two possible products combined? Show the calculation.
3. Which isomer would have the higher stability and thus the higher probability of being formed?
4. What visible indications showed that a reaction occurred?

## Chapter 5 Alcohols, Ethers, and Thiols

5.1 The hydroxyl group in alcohols

5.1.1 Polarity of the Oxygen containing group

5.1.2 Hydrogen bonding properties

5.2 Categories and classes of alcohols

5.2.1 Valence and number of hydroxyl groups

5.2.2 Degree of substitution – primary, secondary, and tertiary alcohols

5.3 Natural distribution, occurrence, and formation

5.3.1 Simple alcohols

5.3.2 Complex alcohols – often with other functional groups

5.4 Nomenclature

5.5 Physical and chemical properties

5.6 Reactions

5.6.1 Oxidation to aldehydes and ketones

5.6.2 Dehydration to ethers and esters

5.7 Toxicity

5.7.1 Methanol and blindness

5.7.1 Ethanol and intoxication scales

5.7.2 Glycol as enzyme inhibitor

5.8 Uses and commercial significance

5.9 Ethers

5.9.1 Physical and chemical properties

5.9.2 Nomenclature

5.9.3 Significance as solvents and anesthetics

5.10 The sulfhydryl group in thiols (mercaptans, thio-alcohols)

5.10.1 Physical and chemical properties

5.10.2 Nomenclature

5.10.3 Reactions (Redox)

5.10.4 Distribution and occurrence in nature

5.10.5 Uses and commercial significance

## Chapter 5 Alcohols, Ethers, Thiols

Alcohols are all molecules distinguished by one or more **hydroxyl groups** (-OH). They can be mono-, di-, tri-, or polyvalent, depending on the number of -OH groups within a molecule or **valency.**

Monovalent alcohols begin with methanol, followed by ethanol, propanol, butanol, etc. They can be primary (1°), secondary (2°), or tertiatry (3°), depending on the **degree of substitution** of the carbon that the -OH group is attached to, as shown below:

**Fig. 5.1**

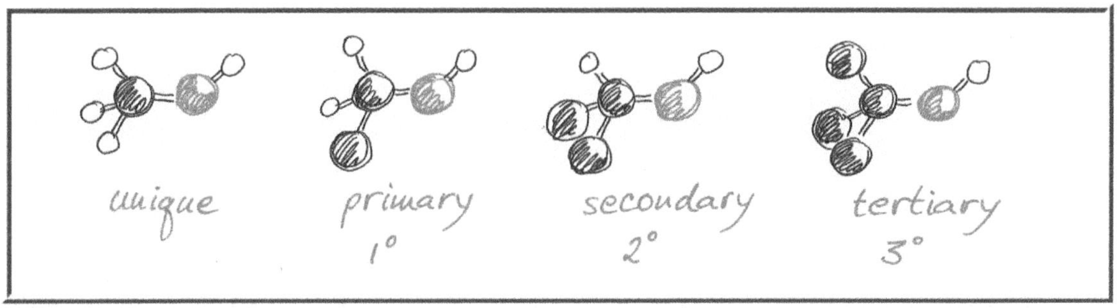


**Example Problem 5.1**
**Classify the following alcohols by their degree of substitution (1°, 2°, 3°)**

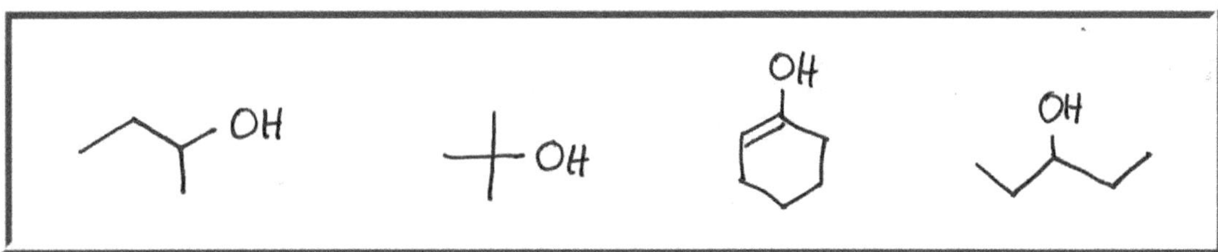


**Example Problem 5.2**
**Indicate whether the following alcohols are mono-, di-, or trivalent.**

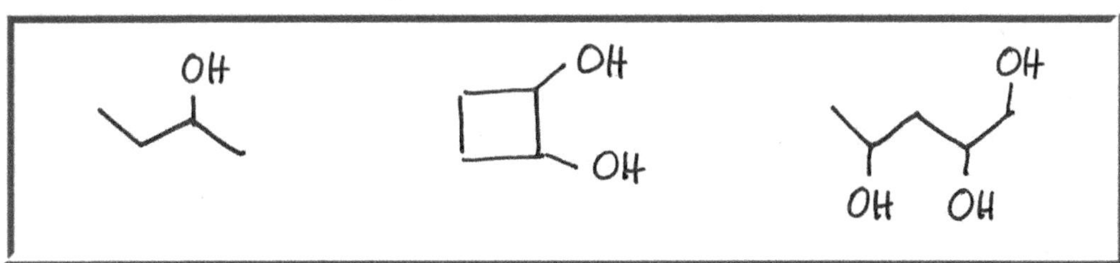

There are three ways of naming an alcohol:

1. Group name "**alkyl**" + "alcohol" for primary alcohols

**Fig. 5.2**

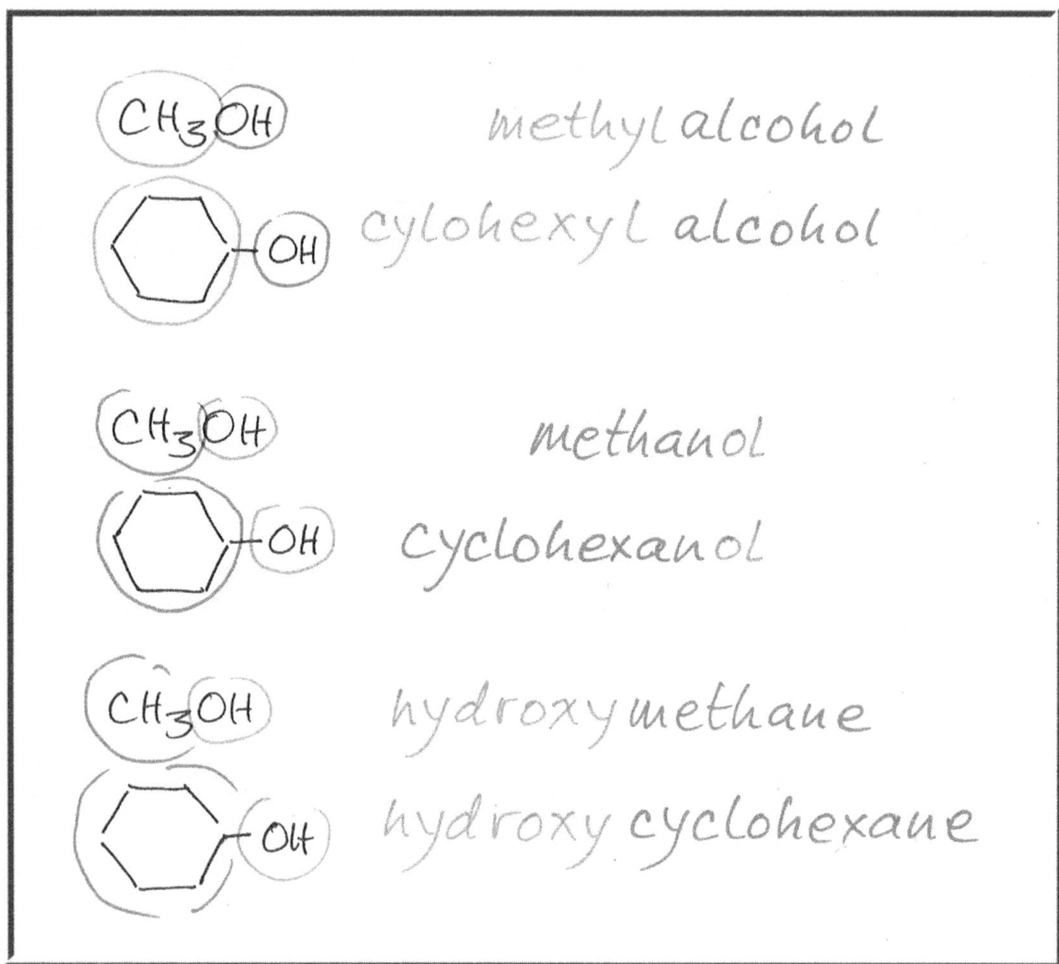


2. Hydrocarbon stem + ending or suffix **"-ol",** the location on the hydrocarbon chain or ring indicated by a number up front

3. Functional group name "**hydroxy**" (for alcohols) + the hydrocarbon as parent molecule + the location number on the chain or ring (if needed)

**Example Problem 5.3**
**Name the following alcohols in two or three different ways:**

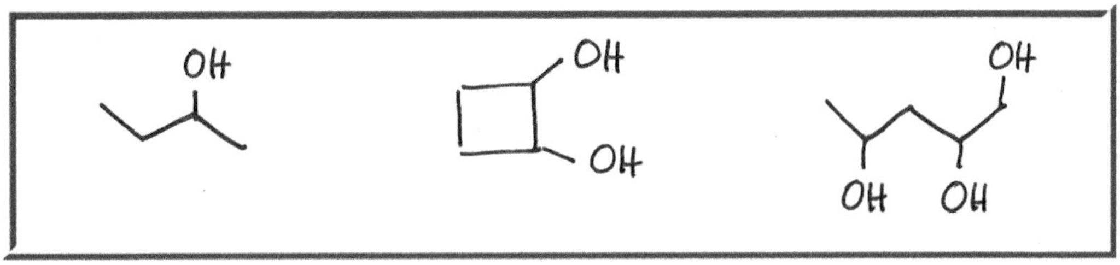

**Example Problem 5.4**
**Give the structures for the following names:**
***3-Pentanol***
***Hydroxybenzene***
***Ethyl alcohol***
***2-Hydroxynaphthaline***

More complex alcohols are often referred to by their common names but are mostly recognized by the ending **"-ol"** such as:

Eugenol
Cholesterol
Estradiol
Glycerol
Clycol
etc.

When more hydroxyl groups are involved, the number is indicated by **-diol, -triol, etc.,** as a suffix. Here the complete name of the hydrocarbon is used, followed by the appropriate suffix.

**Fig. 5.3**

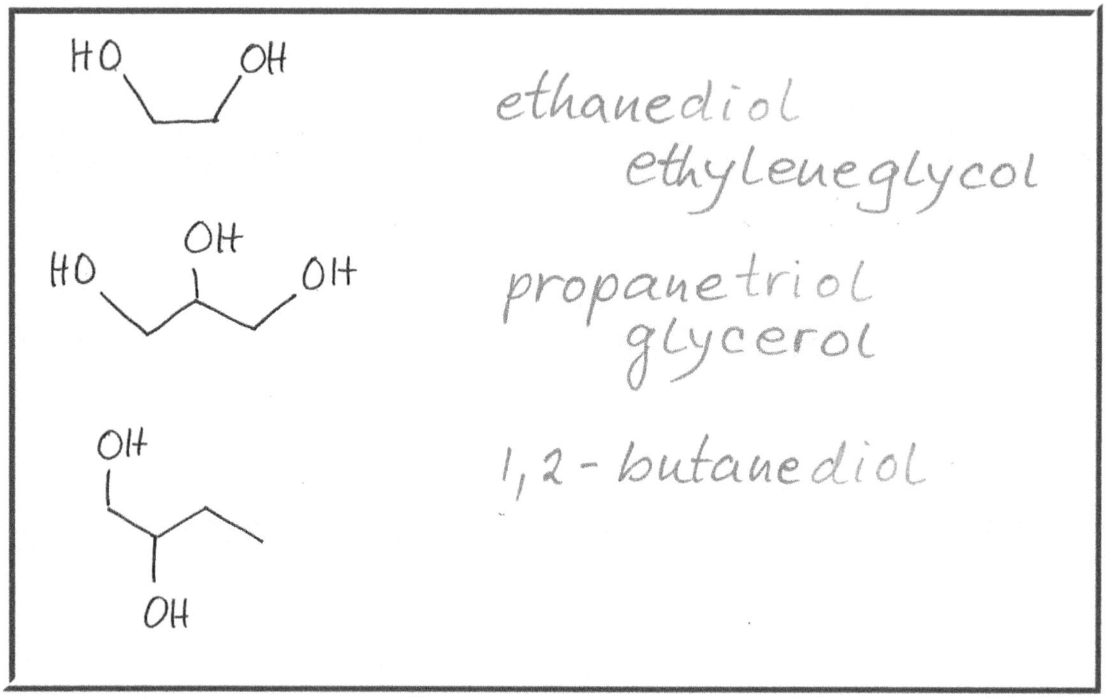


**Example Problem 5.5**
**Name the following polyvalent alcohols**

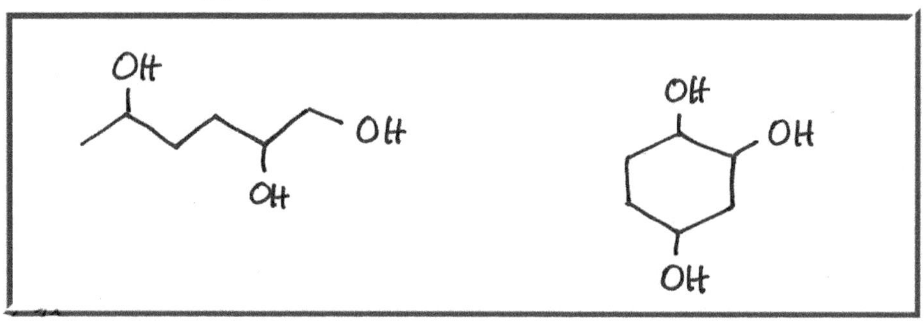


**Example problem 5.6**
**Draw the molecular structures for the following names:**
***2,4-pentanediol***
***1,2,5-Trihydroxycyclohexane***
***1,2,3-Cyclopentanetriol***
***1,2,3-Trihydroxypropane***

The aromatic alcohols are known as **phenols**, when derived from benzene, and **naphthols**, when derived from naphthalene (phenol homolog). Alpha- and beta- Naphthols are 1- and 2- substituted derivatives, resp..

Fig. 5.4

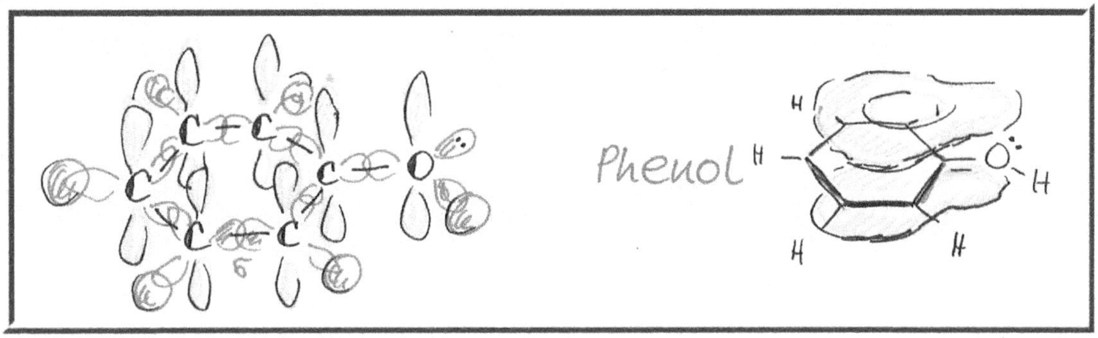


**Anthranols** are the anthracene alcohol homologues, having more isomeric possibilities.

**Fig. 5.5**

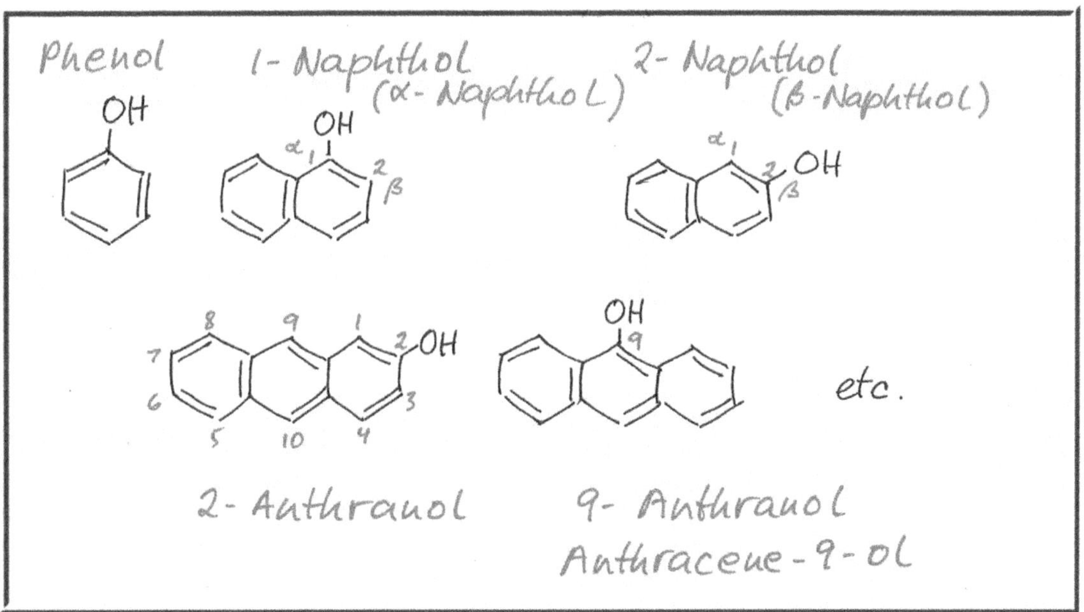


In a sense, alcohols are derivatives of water, having one H-atom substituted by an organic, usually hydrocarbon rest.
So part of the water molecule, namely the -OH group, is still there with all the polar properties intact. The highly polarized -OH bond allows the hydroxyl group to make H-bonds with other alcoholic groups, water molecules, and in general all groups involving a highly electronegative element, mostly oxygen and nitrogen to a lesser degree. It can act as a hydrogen bond donor as well as acceptor.

**Fig. 5.6**

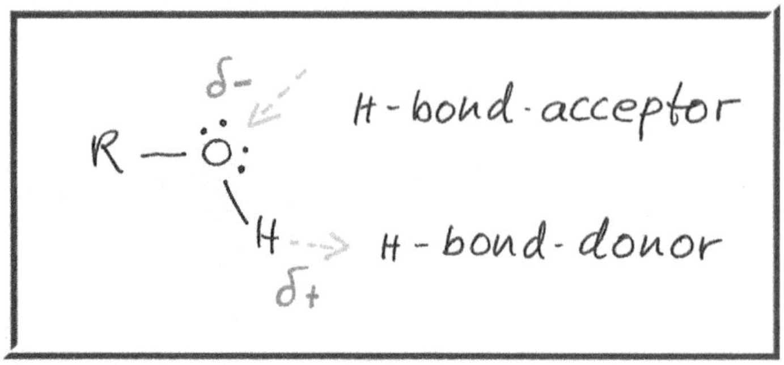


This explains the excellent solubility and miscibility properties of the lower alcohols with other polar substances. However, the higher the hydrocarbon content of the alcohol, the more hydrophobic the molecule becomes, culminating into the wax-like **fatty alcohols**. The characteristics of the hydrocarbon part of the molecule outweigh the polar properties of the hydroxyl part of the molecule.

**Fig. 5.7**

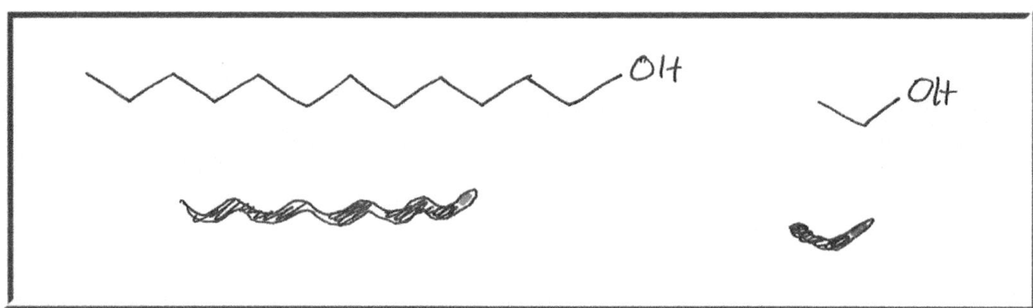


**Fig. 5.8**

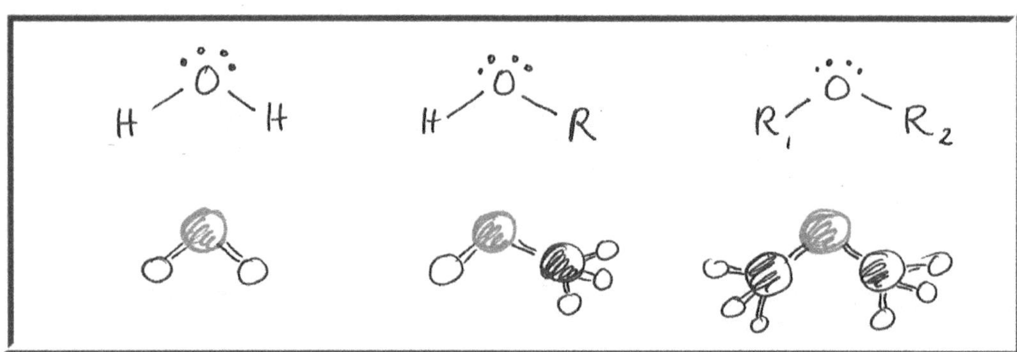

The following table shows a list of different monovalent, primary alcohols and their physical properties:

**Table 5.1**

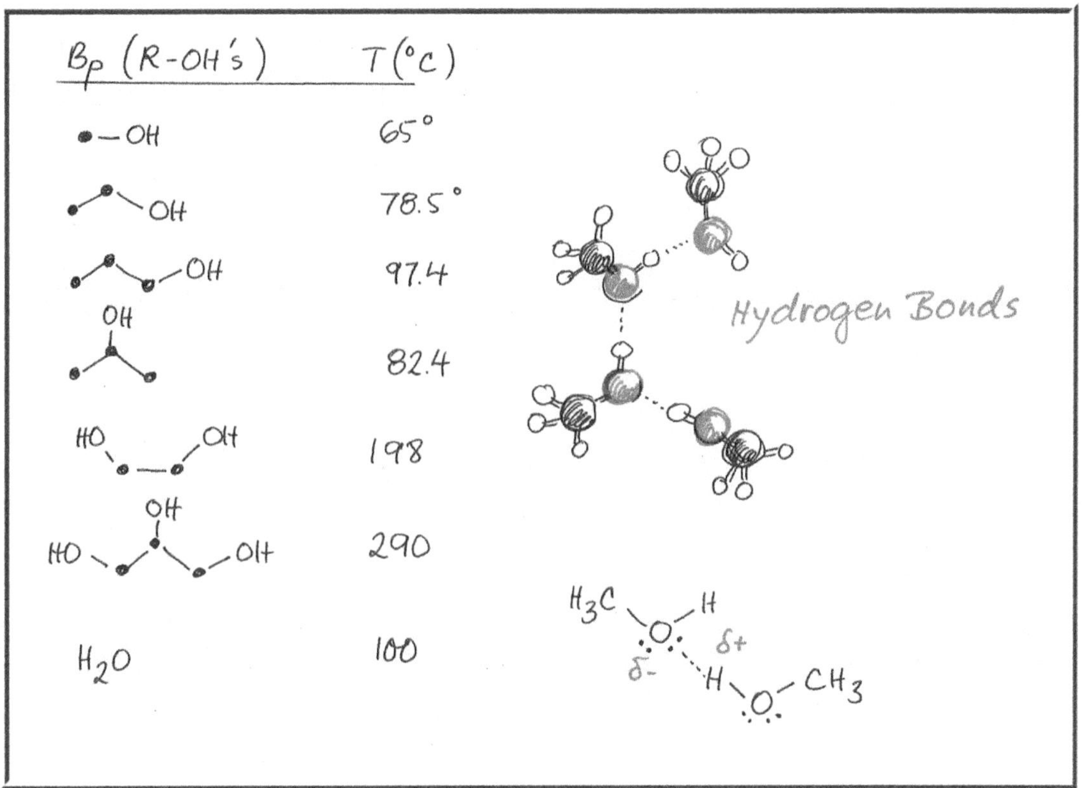


The simple alcohols are mostly liquid, with higher boiling points due to the pronounced H-bonding, holding the molecules together more lightly. Their vapor pressure is therefore relatively low.

Alcohol abuse is a big problem in our society. To monitor alcohol consumptions, especially for traffic related conditions, the Breathalyzer test was developed. The various colors of chromium compounds, related to various oxidation states of the element, can be used to differentiate between blood alcohol levels.

**Fig. 5.9**

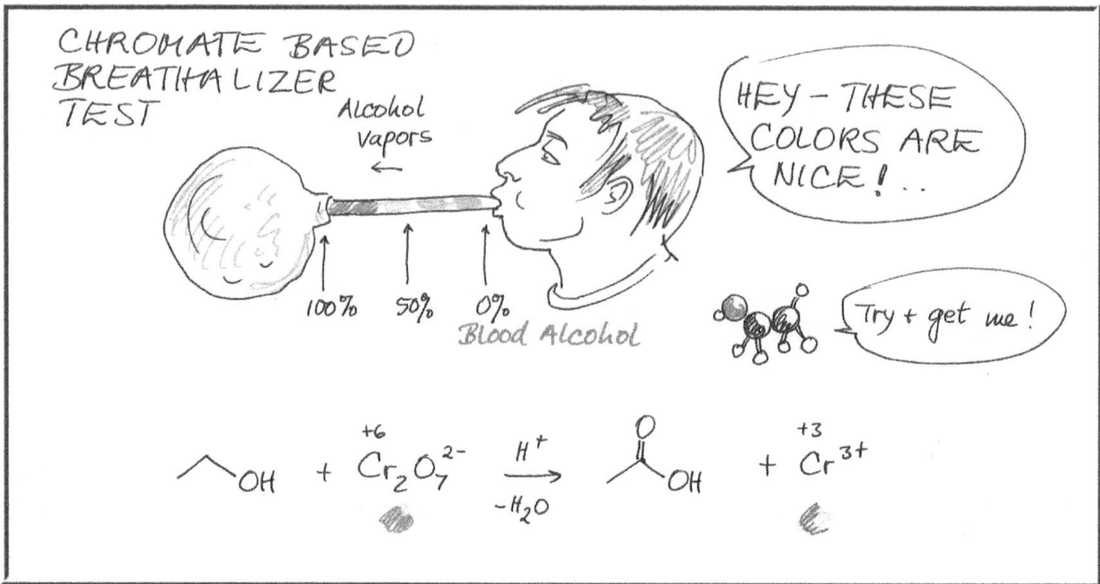


Alcohol's toxicity is mainly related to its metabolites Acetaldehyde and Acetic acid, produced by liver enzymes such as alcohol dehydrogenase. Acetaldehyde is the culprit of the typical alcohol hangovers. Methanol is not as sedative but leads to more dangerous metabolites: Formaldehyde and Formic acid (Methanal and Methanoic acid, respectively). Methanol consumption can therefore lead to blindness and death.

**Table 5.2**

ALCHOL INTOXICATION LEVELS

| Levels %(v) | Effect on Body |
|---|---|
| 0.05 - 0.15 | Lack of coordination |
| 0.15 - 0.20 | Intoxication |
| 0.30 - 0.40 | Unconsciousness |
| 0.50 | Death |

Alcohols are great disinfection agents. **Isopropanol** (2-propanol, propan-2-ol, isopropyl alcohol) is of value as "rubbing alcohol" with an excellent cooling effect on the skin. Apart from its refreshing odor, it is also a universal solvent with many applications in households and industry.

**Fig. 5.10**

Glycol or dihydroxethane  toxicity is more of a problem than ethanol. Sweet tasting, it is often licked up by dogs and leads to an agonizing death. It binds to **alcohol dehydrogenase** in the liver, but since the enzyme has a higher affinity for ethanol than glycol, regular drinking-alcohol is the best antidote for glycol poisoning, as the glycol is driven out of the binding site by the in-coming alcohol molecules.

**Fig. 5.11**

Glycol often referred to by its common name ethylene glycol is used as antifreeze, since it prevents water molecules from making ice crystals at cold temperatures, competing for hydrogen bonds. Glycerol, another important higher-valent alcohol could do the same. It is mainly used as a humectant for crèmes and lotions, mainly formed by saponification of fats and oils (triacylglycerols). It also makes the ester trinitroglycerol or glycerine, which together with an absorbant material finds use as the important explosive **dynamite.**

**Fig. 5.12**

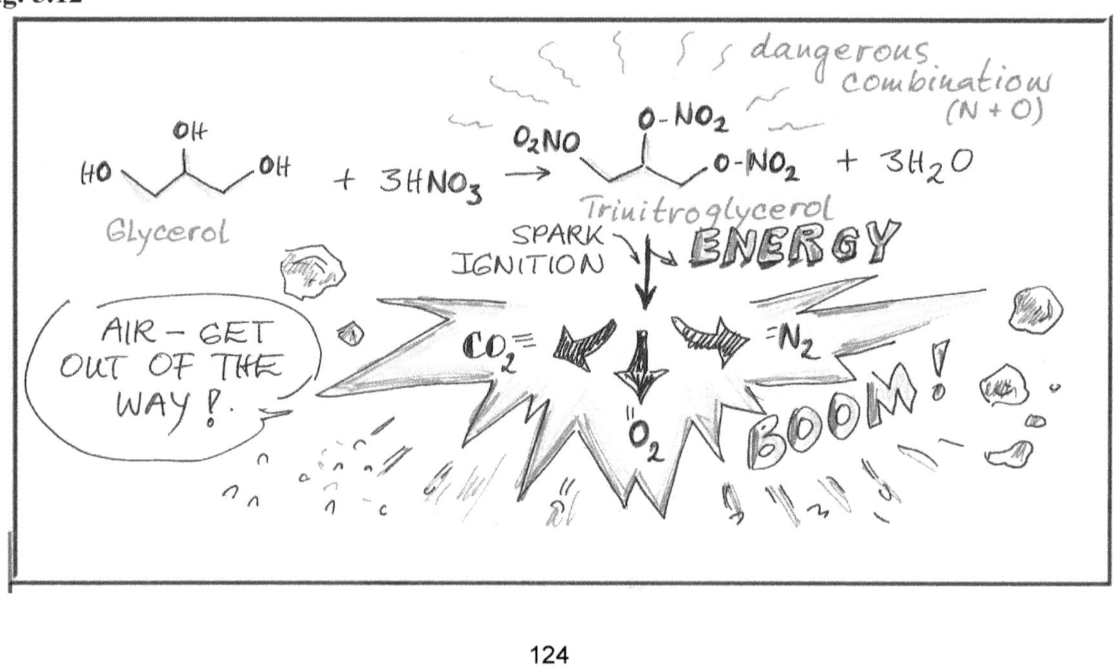

Poly-alcohols like xylitol or sorbitol are mostly of sweet taste, which makes them good sweeteners for diabetics. They are also used for sugarless gums, which can become a problem with too much use, causing diarrhea and possibly glaucoma.

The more -OH groups, the higher the boiling point (Bp) due to more hydrogen bonding possibilities. Straight-chained n-propanol boils at higher temperature than isopropanol due to closer stacking possibilities, which leads to stronger intermolecular forces to overcome.

**Fig. 5.13**

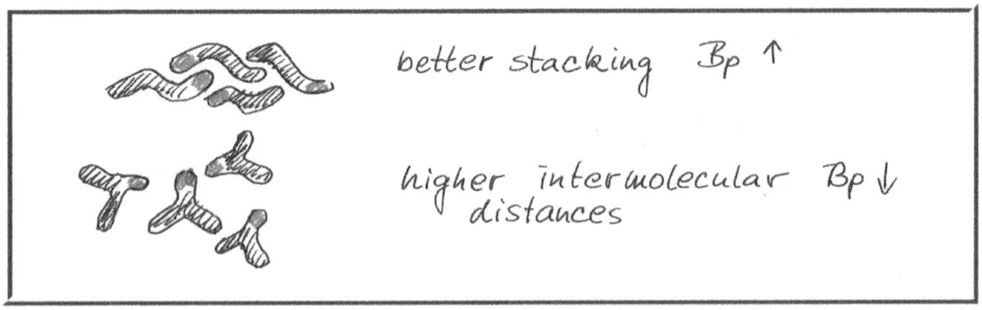


**Example Problem 5.7**
**Order the following alcohols in the order of increasing water-solubility:**

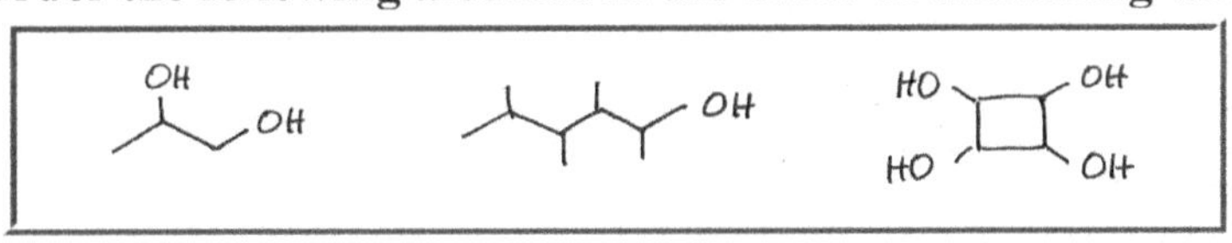


Ethanol is produced in nature and in the beverage industry by fermentation of sugars (carbohydrates), various plant juices, with the help of yeast enzymes.

**Fig. 5.14**

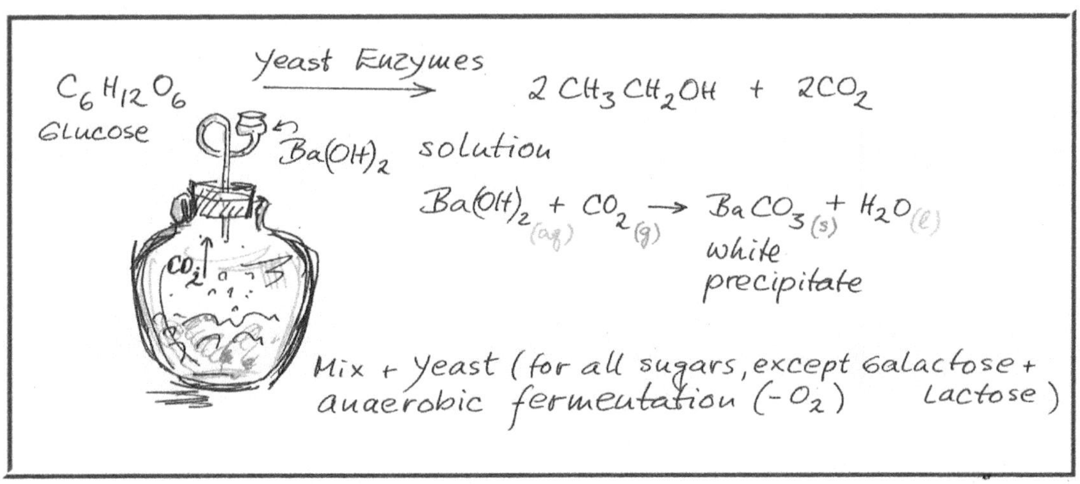

The relative alcohol content of various alcoholic beverages is listed in the following table:
**Table 5.3**

| SOURCE | | BEVERAGE | (%EtOH) | (% proof) |
|---|---|---|---|---|
| Grapes | Grape Juice | Wine | ≤ 12% | 20 - 24 |
| Barley Wheat | | Beer | 5% | 10 |
| Distilled Wine | | Brandy | 40-45 | 80 - 90 |
| Barley, Rye, Corn | | Whiskey | 45-55 | 90 - 110 |
| Molasses, Sugar Cane | | Rum | ~45° | 90 |
| Potatoes | | Vodka | 40-50 | 80 - 100 |

Physical properties of alcohols such as the dielectric constants (ε), that indicate relative polarity, based on higher valence and higher degrees of substitution are shown below:

**Table 5.4**

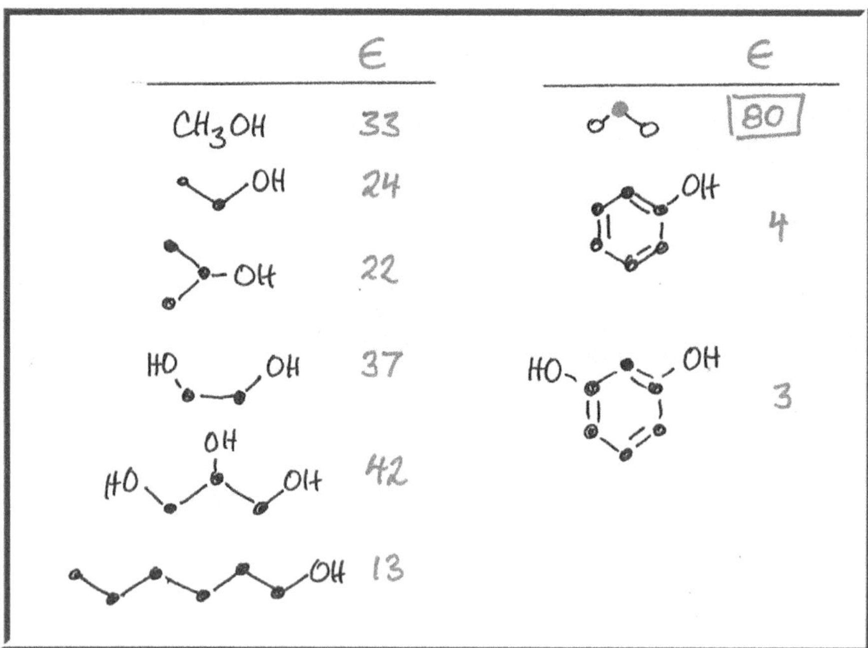

The hydroxyl functional group is one of the most important and wide-spread groups in naturally occurring products. Associated with other functional groups in a given molecule, they are involved in sugars, amino acids, lipids, steroids, etc. **Tetrodotoxin**, a cage-like, highly toxic molecule found in the Pufferfish and some other animals, has six OH- groups in its structure.

**Fig. 5.15**

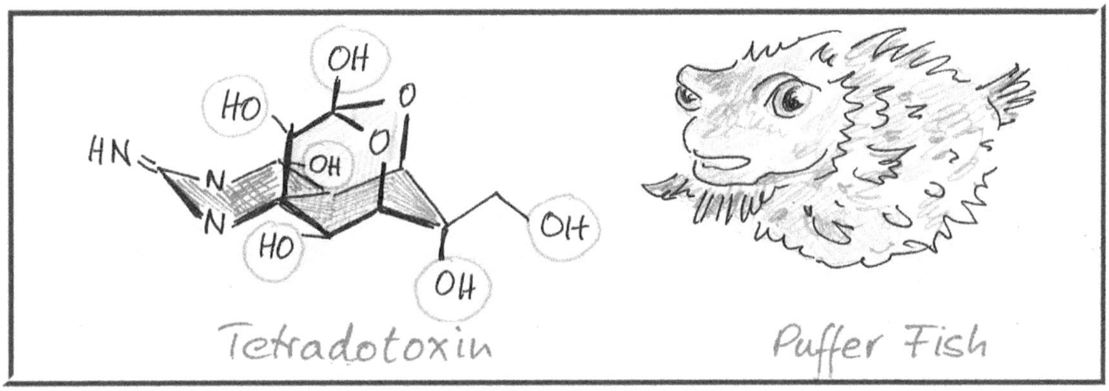


Alcohols can undergo a variety of reactions. Primary alcohols can be oxidized into the corresponding aldehydes, (R-CHO) which can be oxidized further to the carboxylic acids (R-COOH or $RCO_2H$)

**Fig. 5.16**

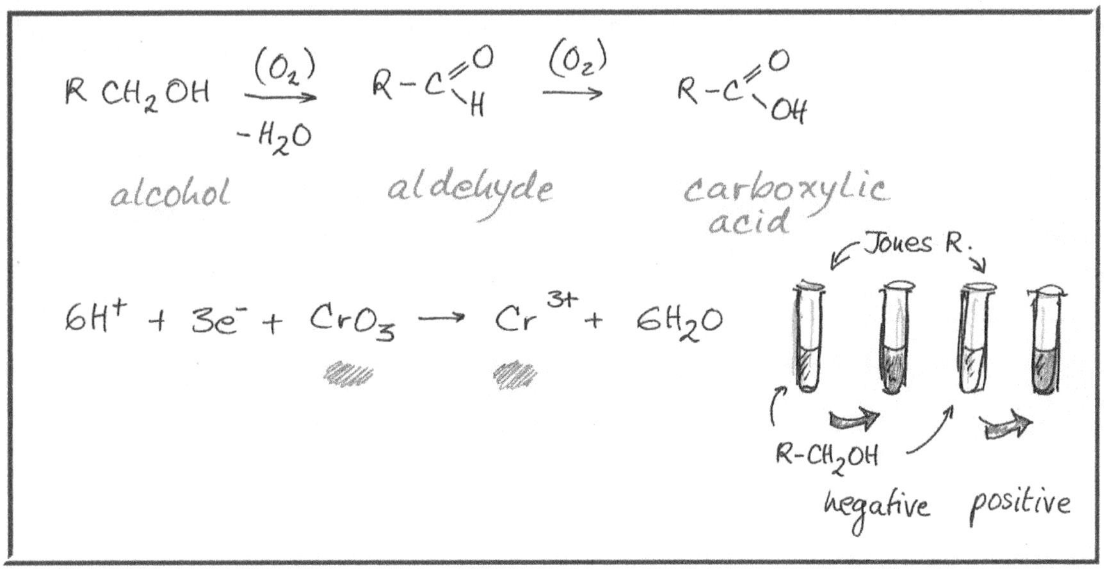


A good test for the oxidizability of 1° alcohols is the treatment with Jones Reagent – an orange $CrO_3$ solution that turns green in the presence of 1° alcohols. When adding Sodium Bicarbonate ($NaHCO_3$, also known as Sodium hydrogencarbonate) to the green solution and bubbles occur, the evolution of carbon dioxide ($CO_2$) gas indicates the formation of a carboxylic acid.

**Fig. 5.17**

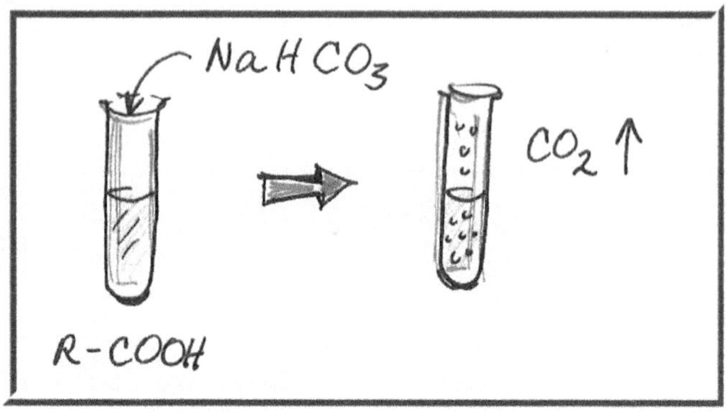


Secondary alcohols can only be oxidized to the ketones ($R_1$- CO-$R_2$), which are chemicals with the carbonyl group (-CO-) internalized, not attached to a terminal carbon and a hydrogen as is the case for aldehydes (-CHO)

**Fig. 5.18**

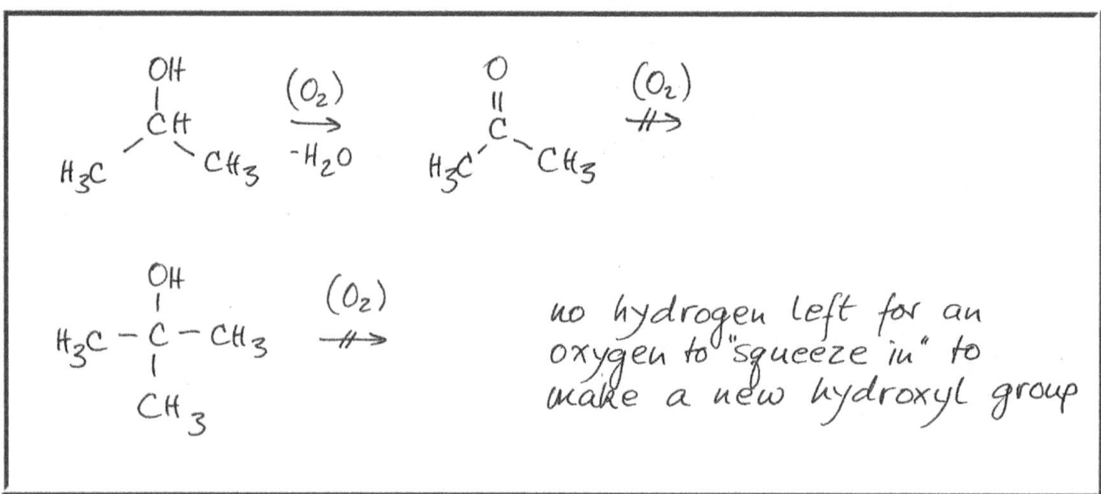


Tertiary alcohols cannot be oxidized, since they have no more available hydrogen attached to the carbon atom, between which an oxygen can squeeze in to make another –OH group, geminal (attached to the same carbon temporarily) to the hydroxyl group already in place. Two hydroxyl groups attached to the same carbon ("geminal", like twins) immediately combine to split off as water, leaving behind a carbonyl group.

**Example problem 5.8**
**Complete, if possible the following oxidation reactions:**

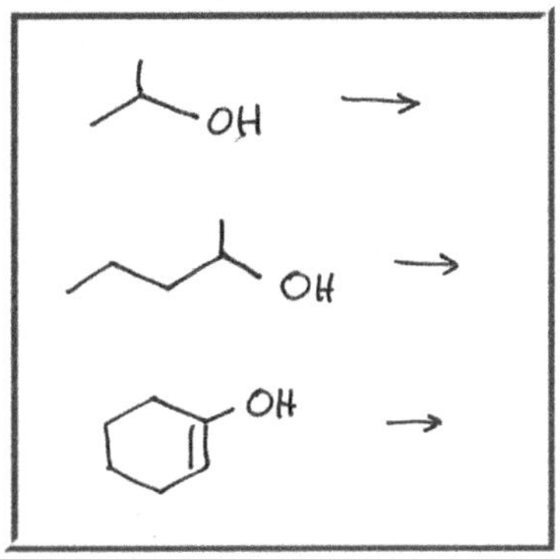


Alcohols can also polymerize, especially the unsaturated alcohol "vinyl alcohol"(ethenol) that converts to **polyvinyl alcohol (PVA)**, although it is prepared by a different route.

Alcohols can also polymerize, especially the unsaturated alcohol "vinyl alcohol"(ethenol) that converts to **polyvinyl alcohol (PVA)**, although it is prepared by a different route. It is a water-soluble polymer finding many uses such as making "slime" with Borax (sodium borate).

**Fig. 5.19**

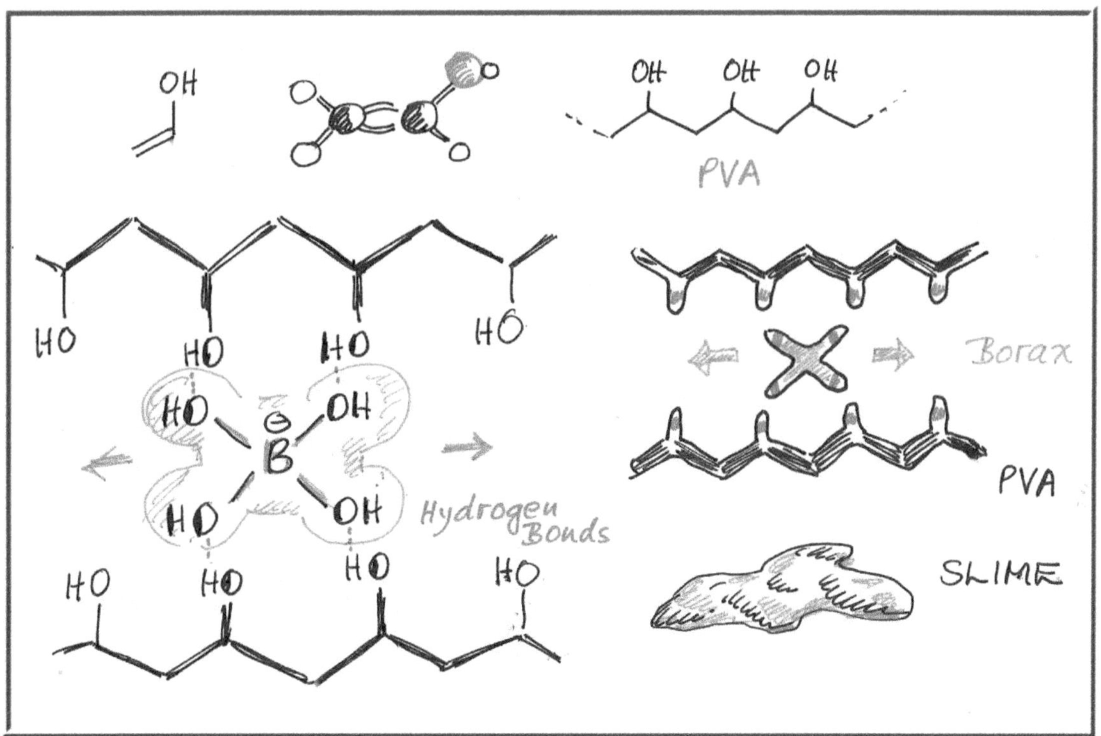

**Phenols** and their many derivatives the so-called **Phenolics** have become an important class of their own. They have a higher acidity than the corresponding aliphatic alcohols, due to the strong electron-withdrawing effect of the aromatic ring. Easier dissociation of the – O-H proton leads to $pK_a$ values in the range of $10^{-10}$ to $10^{-8}$.

Both, the aliphatic and aromatic alcohols, can form salts with strong bases. These are called **alkoxides** or **aryloxides**, resp.

**Fig. 5.20**

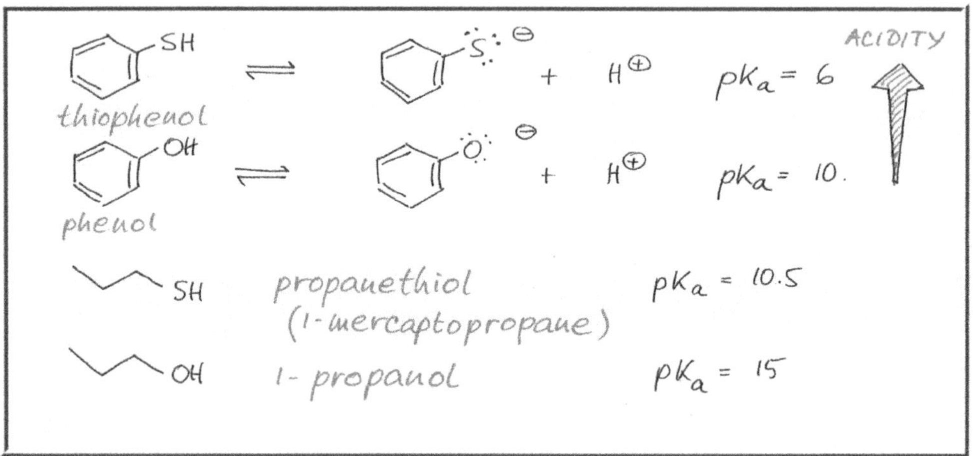


Many derivatives of aromatic alcohols (**Arenols**) are found in plants and have medicinal properties.
Phenols (Benzenols) are considered **enols**, compounds that have the hydroxyl group attached to a double bond. They can be identified by a specific color reaction when treated with a Ferric Chloride solution by coordinating as ligands to the Ferric ion ($Fe^{3+}$) on two sides to give a **chelate complex,** in which the two enols form scissor-like loops around the central metal ion. The natural amber colored reagent turns dark purple when an enol is present.

**Fig. 5.21**

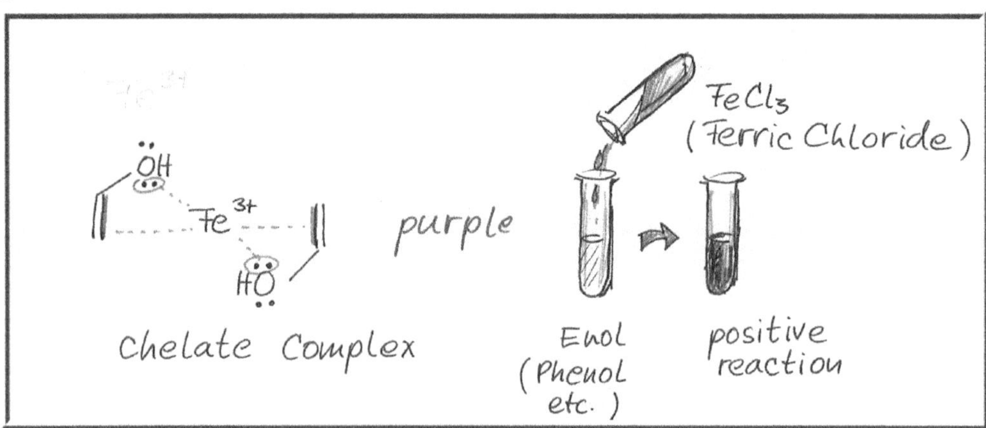

**Example problem 5.9**
**Name the following molecules:**

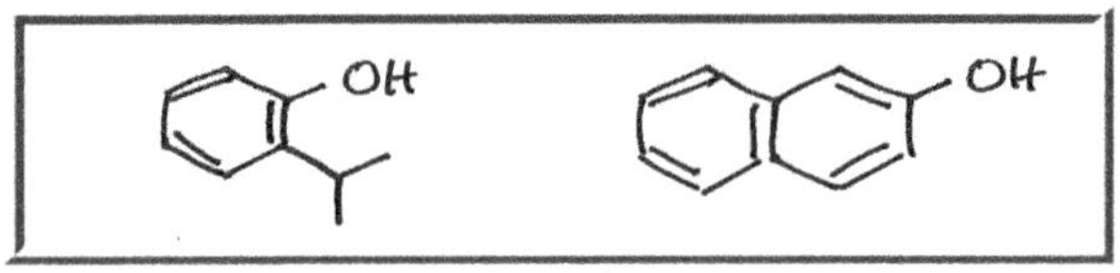


There are also benzene derivatives with more than one hydroxyl group such as the **benzenediols**, of which there are 3 isomers, depending on whether substitution is 1,2; 1,3; or 1,4. Each representative has its own common name such as catechol, resorcinol, and Hydroquinone.

**Fig. 5.22**

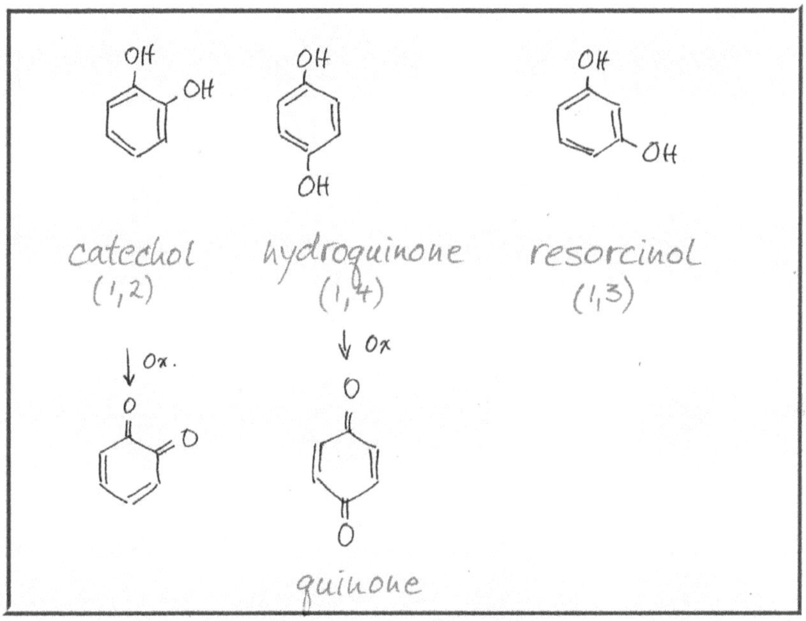


Phenols are commonly used as disinfecting agents. In industry they are used to make a variety of resins, especially with formaldehyde. **Bakelites** were the first man-made resins, used in the old radios, telephones, toys and even in jewelry, the latter having become a highly sought after collector's item.

**Fig. 5.23**

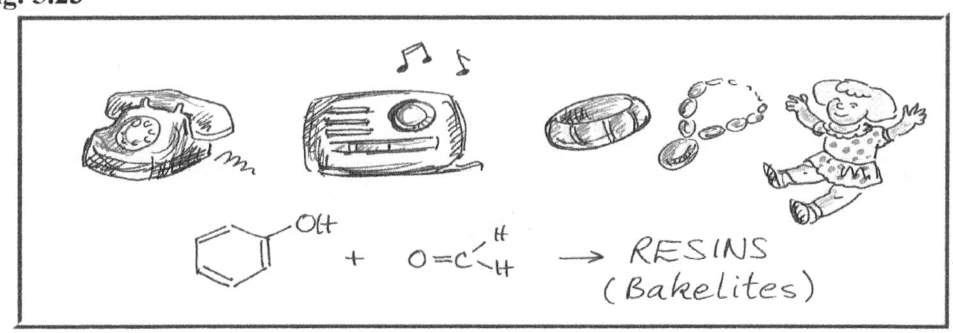

All alcohols can react with carboxylic acids to give esters. This proceeds via intermolecular dehydration (condensation) between the hydroxyl and carboxylic group. The process is specifically known as an **esterification** reaction

The functional group for esters is **$R_1$-O-CO-$R_2$** (or $R_1$-CO-O$R_2$). Esters are found in many fruits and are of distinct and pleasant aromas. They are discussed in more detail in Ch. 7 (Carboxylic acids and derivatives).

**Fig. 5.24**

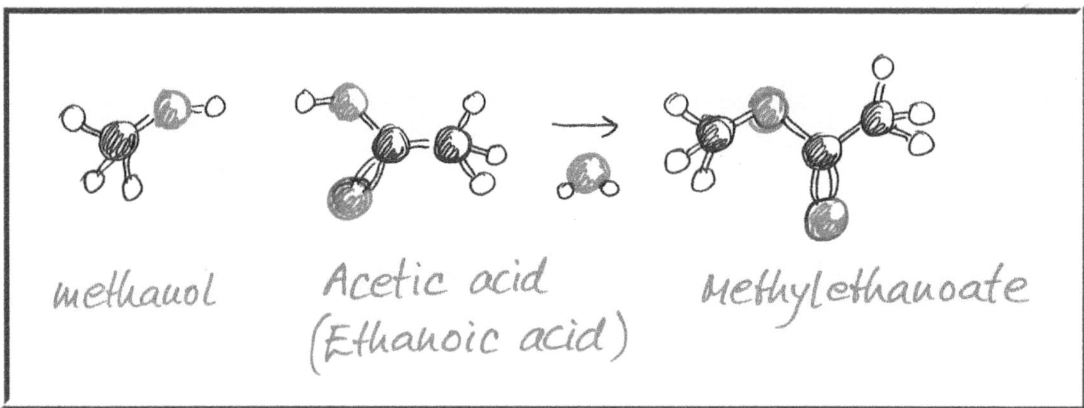


Another important dehydration reaction can occur between two alcohols (like or unlike) leading to the molecular family of **ethers**, characterized by the **oxygen bridge** between the two R groups (**$R_1$-O-$R_2$**).

**Fig. 5.25**

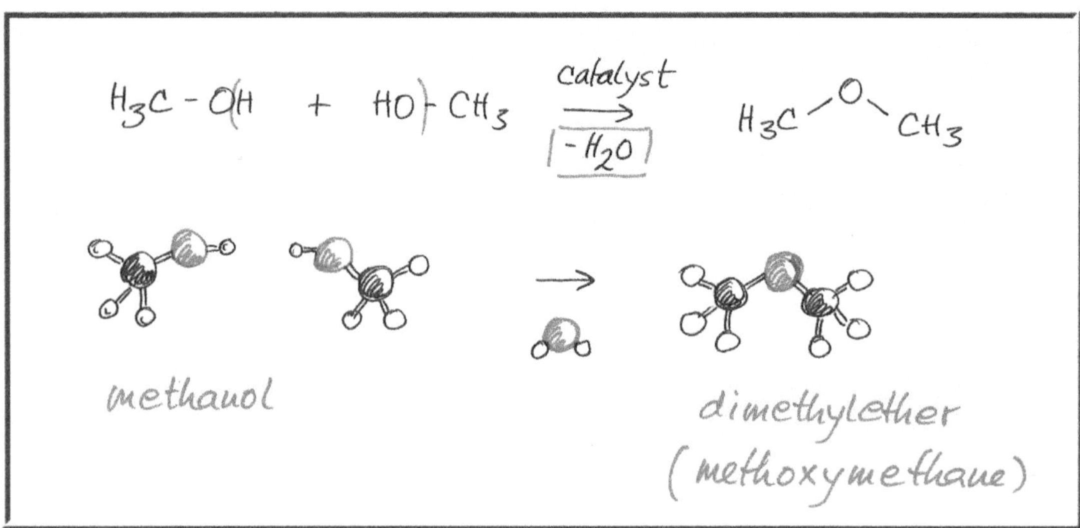

**Example Problem 5.10**
**Complete the following reactions:**

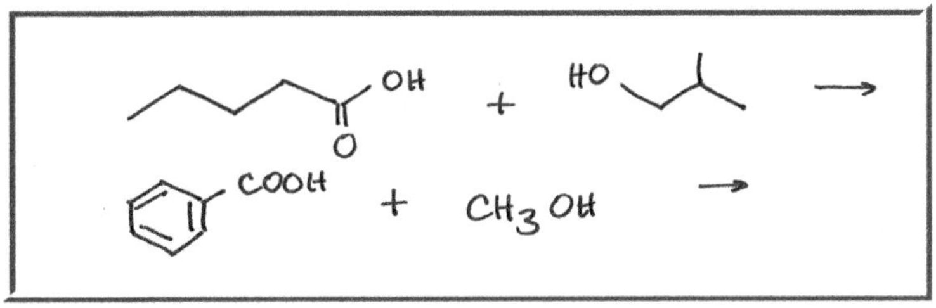


Ethers are very volatile, flammable liquids, and sometimes dangerous due to their tendency to form explosive mixtures with air. They have a low boiling point and therefore high vapor pressure. The vapors travel quickly and the slightest flame or spark can ignite them, causing a huge fire in seconds.

Ethers can be considered a higher derivative of $H_2O$, with the 2 hydrogen atoms each being substituted by a hydrocarbon R-group. They cannot function as hydrogen bond donors, since the hydrogen atoms attached to the carbons, in the alkyl groups, are not polarized enough. However, the oxygen in their structure can still function as a hydrogen bond acceptor. The ethers are therefore non-polar and non-miscible with water, finding wide use as excellent non-polar solvents. Compared with water and an alcohol the relative boiling points/vapor pressure in regards to the degree of hydrogen bonding capacity is outlined below:

**Fig. 5.26**

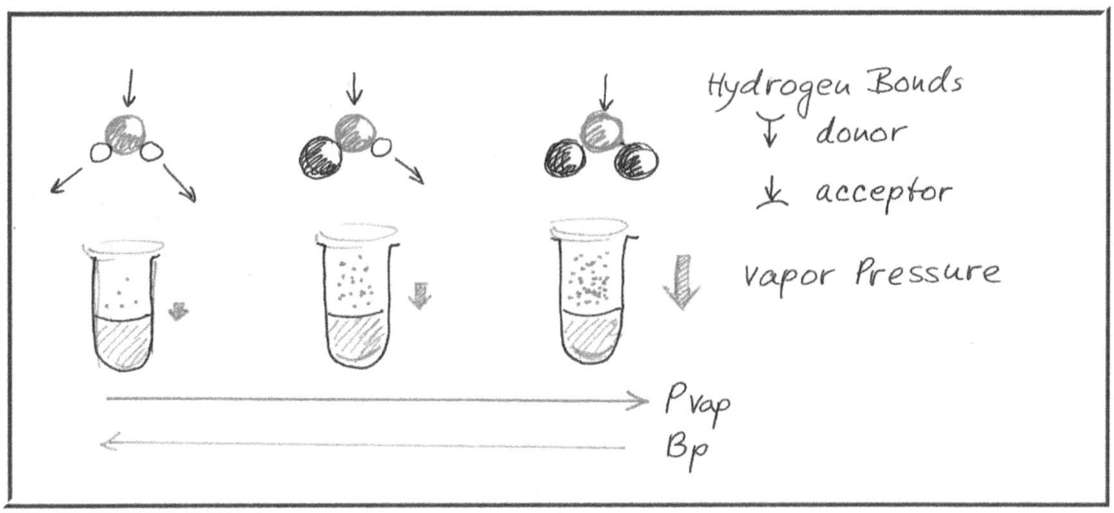


There are two ways of naming ethers:

1. The name "ether" is used as the parent and the two groups named individually ahead of the ether. With two like groups attached to the oxygen bridge, the numerical prefix "di" is used.

2. One of the groups can be used as a normal hydrocarbon parent like alkane, the other

named as “**alkoxy**” (prefix “alk + oxy”) and used as a group.

**Fig. 5.27**

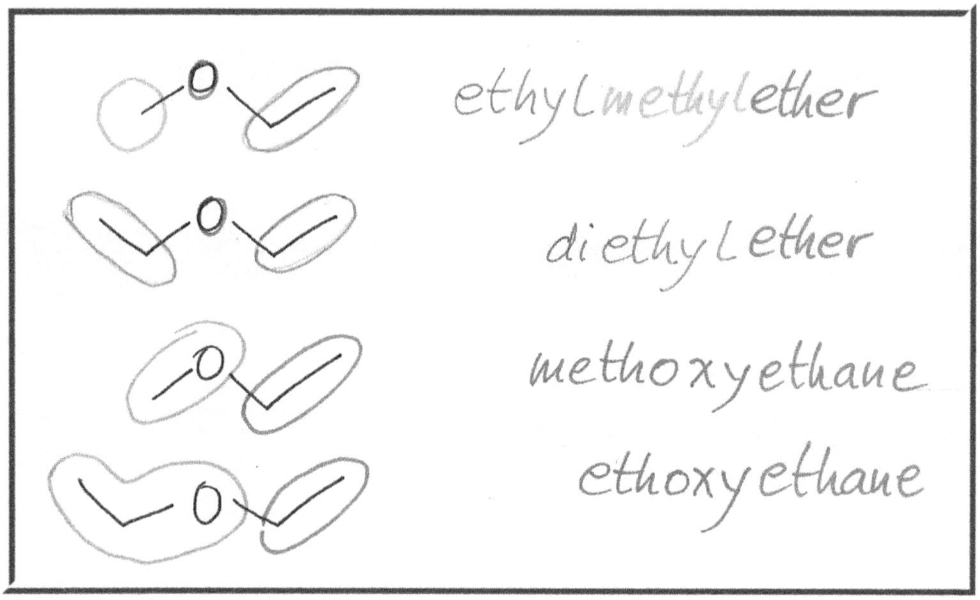


**Example Problem 5.11**
**Name the following Ethers:**

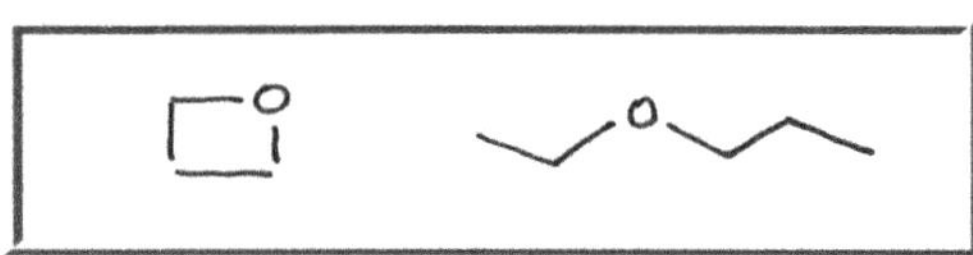

**Example Problem 5.12**
**Write the structures for the following names**
***Methoxypentane***
***Diethyl ether***
***Ethoxybenzene***
***IsopropylMethyl ether***

Ethers are the best and most effective anesthetics known and were, along with chloroform ($CHCl_3$), the first anesthetizing chemicals used in the operating room. Due to the dangers of handling ether, it has been replaced by, often halogenated, higher molecular derivatives. Ether is a chemical that is able to cross the blood-brain barrier, as some other non-polar, lipophilic molecules, thereby causing the individual (human or animal) to lose consciousness.

An important method for the preparation of ethers is known as the **Williamson synthesis,** by which an alkyl halide is reacted with an alkoxide to give ether and an inorganic salt.

**Fig. 5.28**

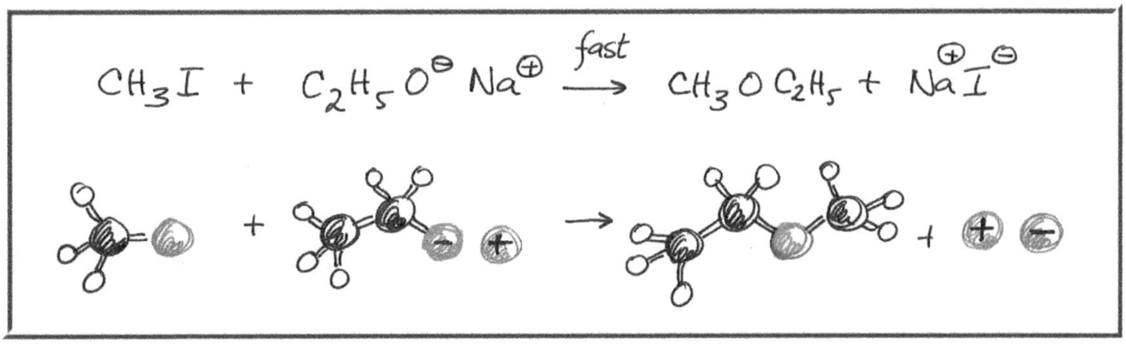


**Cyclic ethers** formed via intra-molecular dehydration reaction between two hydroxyl groups of the same molecule, are widely spread in nature, particularly in the cyclic sugars, the so-called furanoses and pyranoses, based on the hydrogenated rings of furan and pyran.

Certain cyclic ethers, containing many oxygen bridges, often called **"crown ethers"**, are valuable solvating agents for metals, often changing the metal's properties. The lone pairs of the oxygen surround the metal (atom or ion) by coordination, entrapping them in their cage.
An important group of small cyclic ethers is represented by the **epoxydes,** often formed by an oxygen atom adding to a double bond. They are widely found in nature as insect pheromones, unusual fatty acids and often as intermediate metabolites of the body. Technically, they are precursors to the **epoxy** adhesives.

**Fig. 5.29**

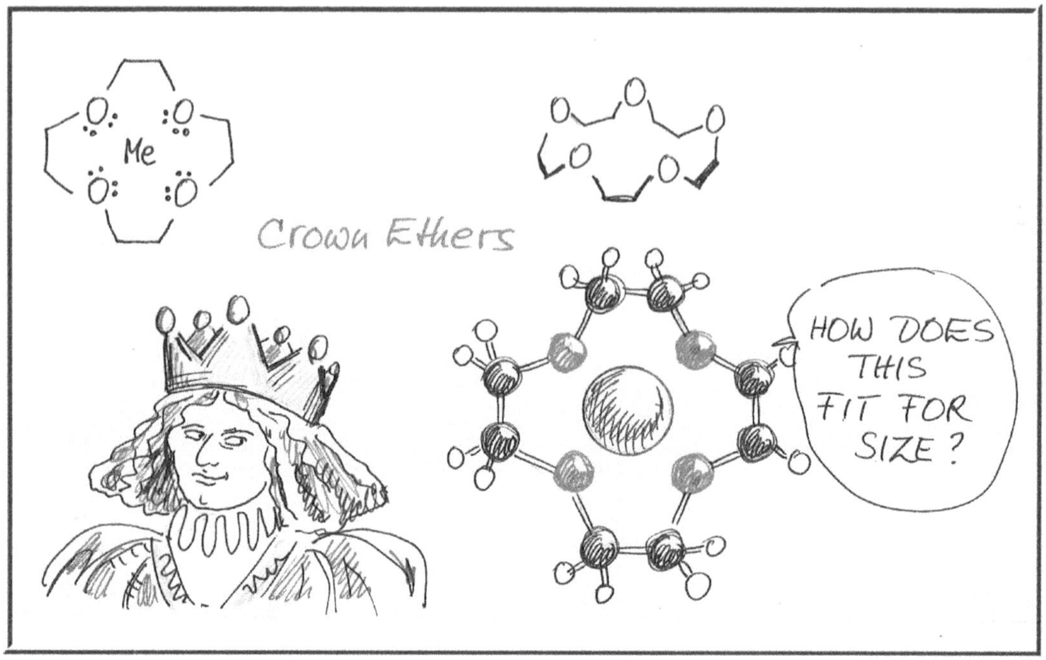


**Fig. 16.30**

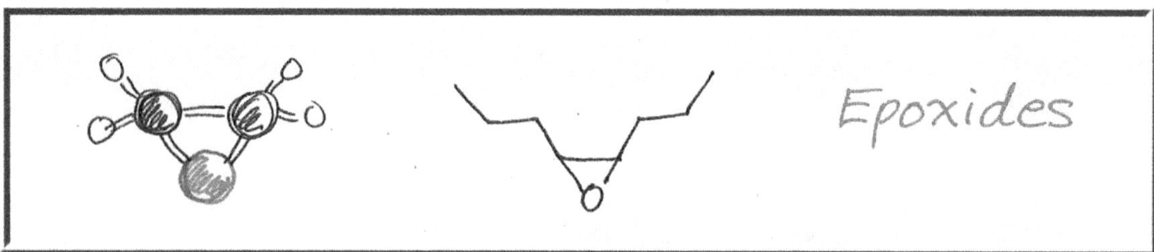


**Bisphenols** have received notoriety as health compromising chemicals. They are used to make a variety of plastics and resins and have become ubiquitous in the environment.

**Fig. 5.31**

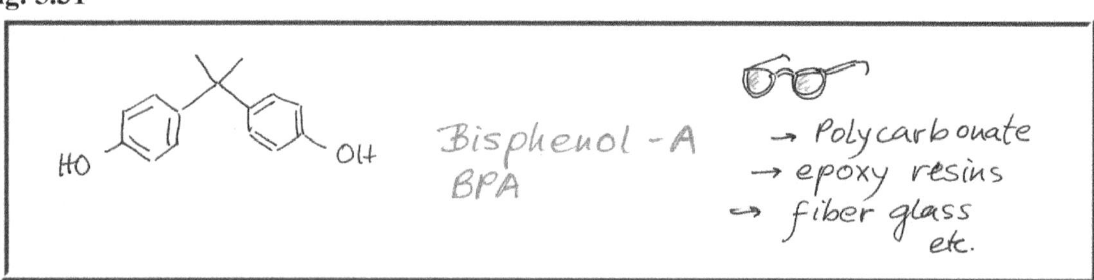


An important group related to the alcohols is the one that has a sulfur atom in place of the oxygen. Molecules of this class are called the **thioalcohols**, **thiols**, or **mercaptans**. The latter name reflects the strong chemical affinity of sulfur for mercury (Hg). The group **-SH** can be called by itself the **sulfhydril** group or **mercapto** group. **"Thio"** is another name for sulfur.

**Fig. 5.32**

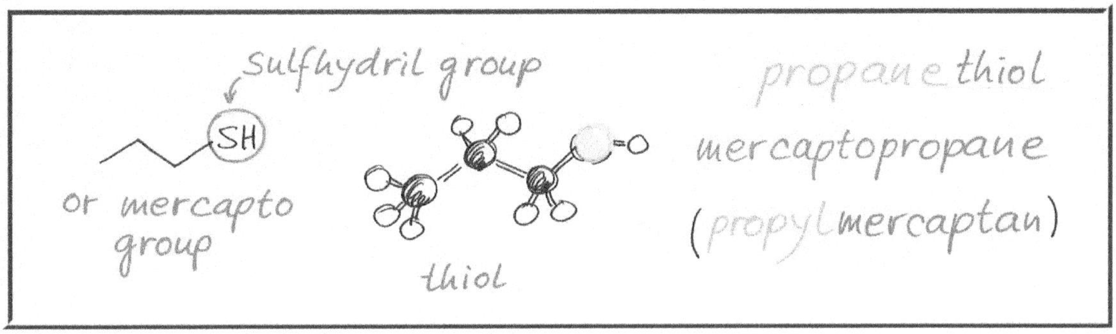


There are three ways for naming thiols:

1. Using the name thiol as parent and naming the hydrocarbon part as full alkane in the front.

2. Using the hydrocarbon as parent and naming the -SH group as "mercapto" in the front as a normal substituent.

3. Using "**mercaptan**" as parent and placing the hydrocarbon group as alkyl in the front as normal substituent.

**Example Problem 5.12**
**Name the following thiols using two or three different methods:**

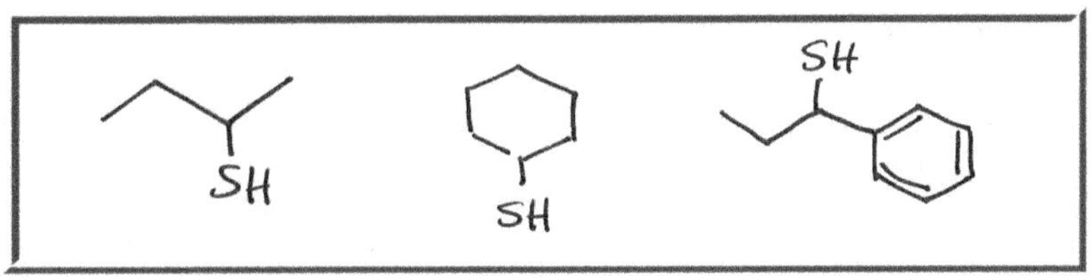


**Example Problem 5.13**
**Write the structures for the following names:**
***Ethenethiol***
***Methyl mercaptan***
***2-Mercaptoethanol***
***1,2-Propanedithiol***

All thiols have a pungent odor similar to rotten eggs. They are very smelly molecules, often volatile and used as territorial markers or as defensive stink bombs by various animals (skunks, weasels, "spraying" male cats, etc.).

**Fig. 5.33**

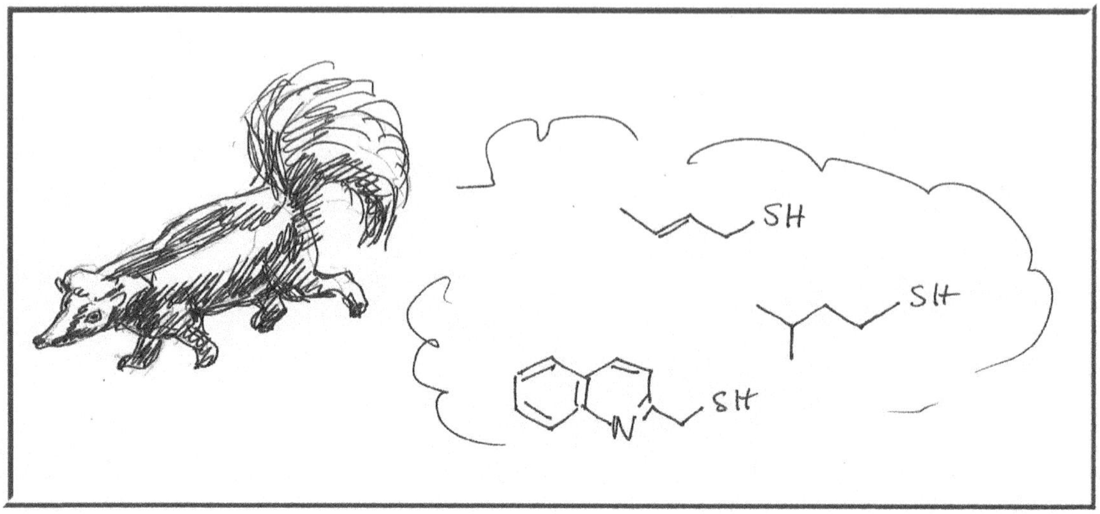


Fig. 5.34

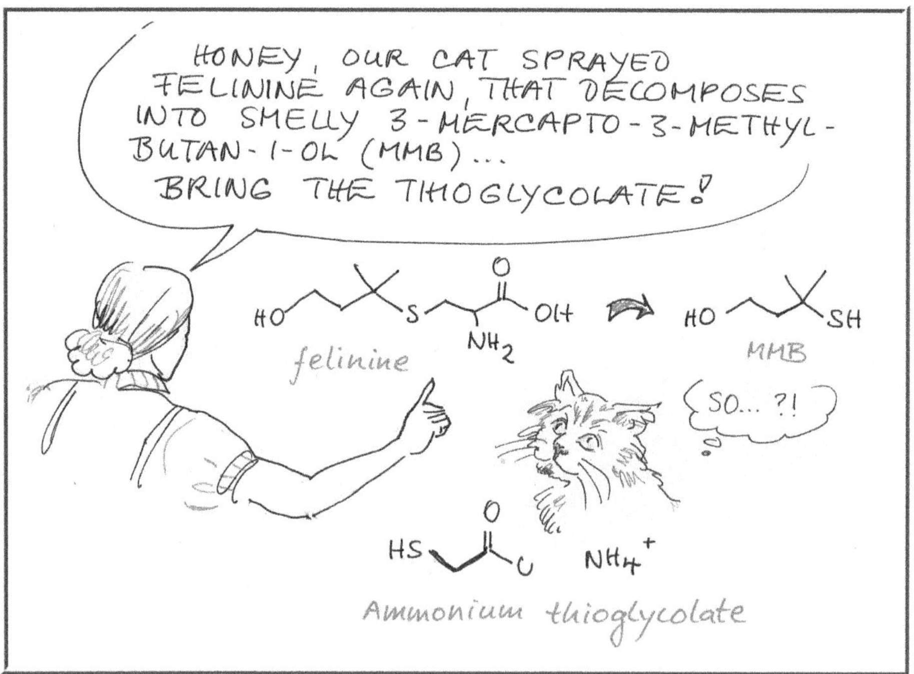


Thus they are a class of the most powerful odorous substances known. The distinctive odor property has proven to be a very effective detector for gas leaks. Natural gas used for heating in stoves and furnaces, etc., has no odor, which makes gas leaks very dangerous, exposing unknowing consumers to possible explosions (see New London, Texas, 1937). Nowadays, mercaptans, such as t-butylmercaptan, and sulfides are added, so that gas leaks can be detected.

**Fig. 5.35**

When oxidized, thiols are converted to dimers, known as **disulfides**, the connecting element being referred to as a **disulfide (-$S_2$-) bridge**.

**Fig. 5.36**

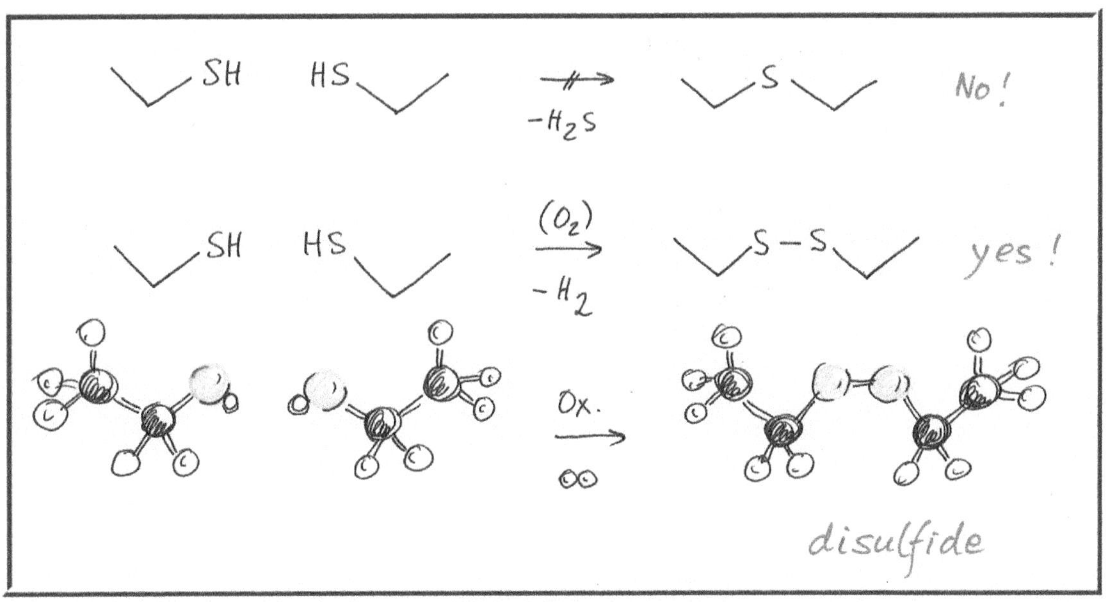


These products can be reverted back to thiols by reduction. This makes them powerful subjects for biochemical redox processes. Many proteins contain sulfhydryl groups that, upon oxidation, create covalent links between strands, literally tying certain areas together as disulfide groups. Copper catalysts help destroy the "nasty" odors of many thiols.

**Fig. 5.37**

The redox system of sulfhydryl and disulfide groups allows for powerful and effective control as well as regulatory influence on protein activity based on changes in their structure. This will be discussed in detail in later chapters.

Thiols do not undergo a reaction similar to the dehydration of alcohols, resulting in compounds known as **thioethers** or **sulfides**. This has to be done by alkylation of thiols with alkyl bromides in the presence of the Lewis acid $AlBr_3$.

**Fig. 5.38**

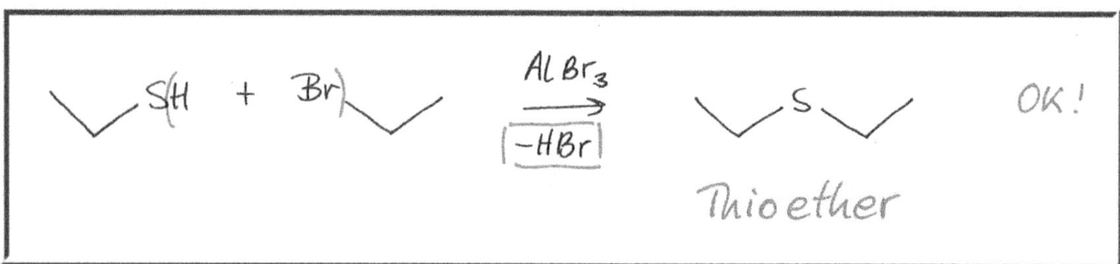


An important aromatic thiol is **Thiophenol**, equivalent to Phenol – a most malodorous molecule, which could evacuate a large hotel simply by putting a little bit of the liquid into the ventilation system.

As mentioned before, thiols are easily converted into mercaptides by action with mercury (Hg)

**Fig. 5.39**

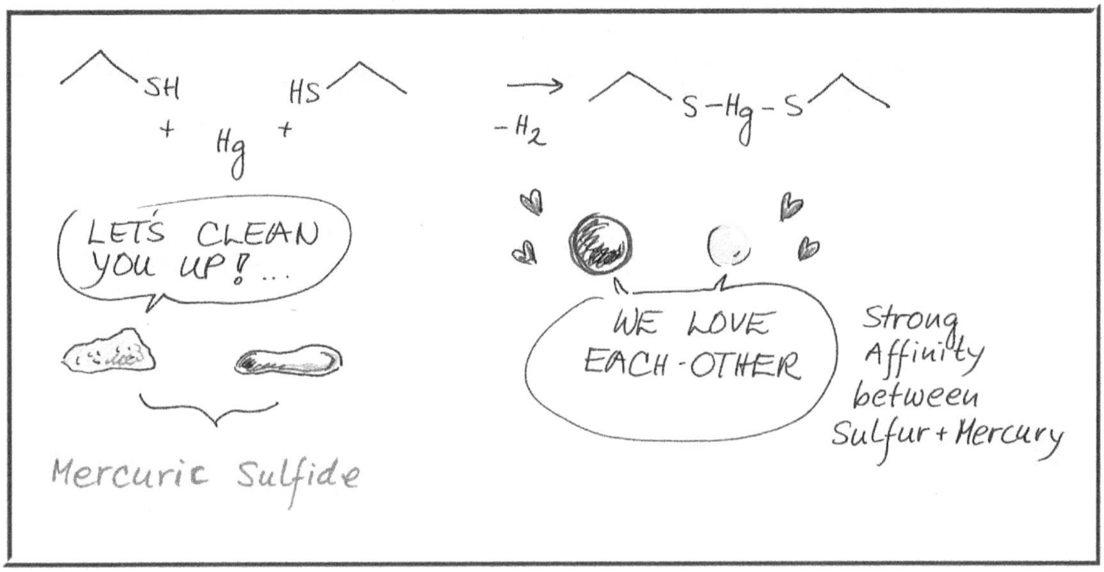


The thiol group has a bonding angle of about 90°. Since S is not as electronegative as O, the chemistry is different, which is reflected especially in the much lower polarity and H-bonding characteristics, meaning lower Bp, lower solubility in $H_2O$ and other polar solvents
The sulfhydryl group can also react with other heavy metals like silver (Ag) and lead (Pb). These can lead to heavy metal poisoning when ingested and, through the conversion of the free sulfydryl groups of the proteins into mercaptides, impair important metabolic functions. The special redox behavior is especially important for biochemical reactions involving the amino acid **cysteine(cys).** An oxidizing agent, itself a disulfide, will by its hydrogen removing capacity, join 2 sulfhydryl groups ($-2H^+$, $2e^-$) via the disulfide bridge together making a **cysteine ligament,** an actual **dimer ($cys_2$).** With a heavy metal covalently attached this dimerization via oxidation is blocked.

However, the denaturation of the protein involved is reversible when drugs, containing sulfhydryl groups themselves that compete for the heavy metals, are administered. Milk proteins can compete for the heavy metals by virtue of their own cysteine groups.

The aliphatic thiols have a higher relative acidity than the corresponding alcohols. By the same token, Thiophenols are more acidic than phenols.

**Fig. 5.40**

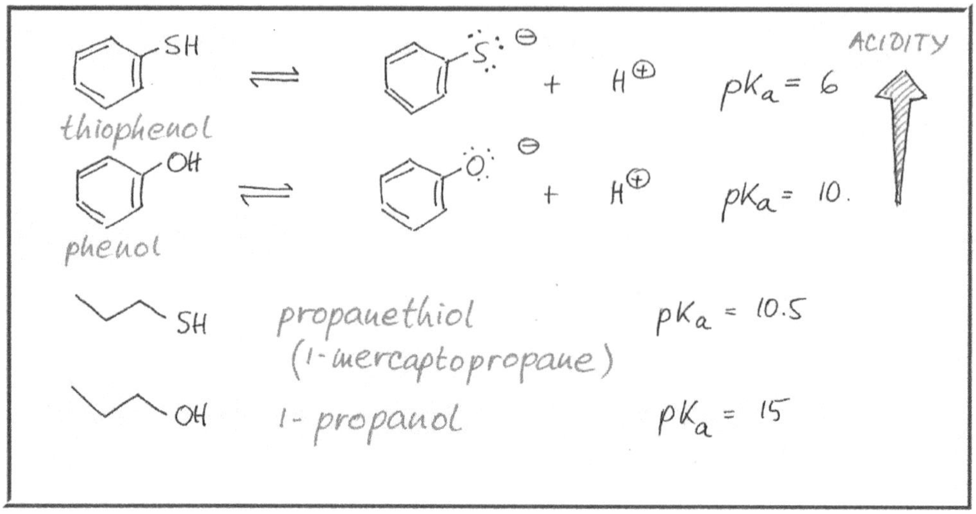


Thiols can be prepared by nucleophilic (electron-rich reagents loving electron-deficiency areas) substitution of alcohols with hydrogen sulfide ($H_2S$) or by treatment of alkylbromides with sodium sulfhydride (NaSH).

All sulfur containing molecules or ions can serve as excellent nucleophiles for various reactions even in the realm of biochemistry.

**Example Problem 5.13**

**Complete the following Redox reactions involving the following thiols and disulfides:**

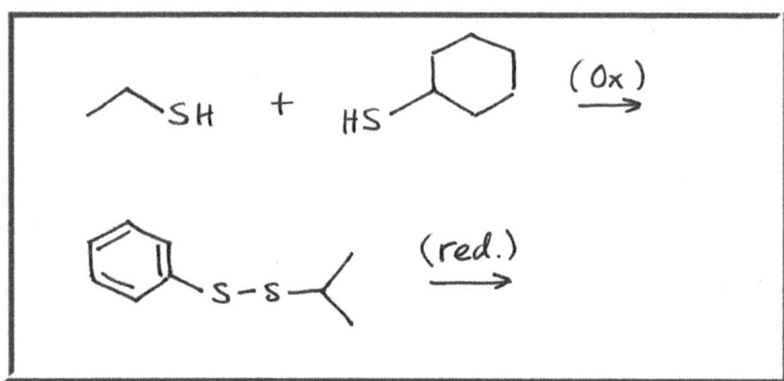

# EXPERIMENTS – Ch. 5

## *Alcohols, Ethers and Thiols*

**Key Terms:** Hydroxyl group, hydrogen bonding, primary (1°), secondary (2°), tertiary (3°) alcohols, nomenclature (common and IUPAC), polyhydric alcohols, valence of alcohols (mono-valent, di-valent etc.), aromatic alcohols (phenol) physical and chemical properties, reactions (oxidation, dehydration), enols and ferric chloride test, ethers and nomenclature, ether synthesis and properties, thiols and properties, redox reactions, disulfide bonds and natural protein fibers.

## I Molecular Modeling of Primary, Secondary and Tertiary Alcohols, Di- and Trivalent Alcohols and Model Reactions for Dehydration and Oxidation Reactions

### Chemicals:

*None*

### Equipment and Materials:

*Organic chemistry modeling kits (Molymod)*
*C (black), H (white), O (red)*
*Single and double bond pieces*

### Procedure:

Build all molecules according to the directions and sketch the results in the table shown below.

1. Build 1-propanol as model for a primary alcohol. .
2. Move the hydroxyl group to the center carbon to produce 2-propanol as example for a secondary alcohol.
3. Build a methyl group and replace the remaining hydrogen atom at C-2 to obtain a tertiary alcohol.
4. Build a new molecule representing methanol and connect to the molecule built in #3 via dehydration reaction (elimination of a water molecule to create a new product).
5. Re-build a primary alcohol and insert another oxygen atom between C-1 and the attached hydrogen. Eliminate a water molecule and connect the remaining oxygen atom via a double bond to C-1 (first oxidation step). Insert another oxygen atom between C-1 and the remaining hydrogen to illustrate the second oxidation step.
6. Rebuild a secondary alcohol and oxidize at C-2. Attempt a second oxidation and evaluate oxidation possibilities.
7. Rebuild a tertiary alcohol and evaluate oxidation possibilities.
8. Build a 1,2-dihydroxyethane molecule ("ethylene glycol", 1,2-ethanediol)
9. Extend the molecule in #8 to form 1,2,3-trihydroxypropane ("glycerol or glycerine", 1,2,3-propanetriol).

## Observations and Results:

| Molecular Model: | | Oxidation Product(s) | Dehydration Product(s): |
|---|---|---|---|
| Name: | Structure: | | |
| | | | |

## Questions and Problems:

1. Draw the chemical equations for all the reactions modeled above.
2. Why can't tertiary alcohols be oxidized?
3. Why does the presence of more hydroxyl groups within a molecule increase the water solubility of alcohols?
4. Why is ethylene glycol (or any other polyhydric alcohol) such an effective antifreeze?

## II Physical Characteristics of Alcohols, Phenols, Esters and Thiols (Color, Odor, Acidity, Solubility and Miscibility)

### Chemicals:

*Ethanol*
*Ethylene Glycol*
*Glycerol*
*Isopropanol*
*Cholesterol*$_{(s)}$
*Phenol*$_{(s)}$
*Solvents:*
*Water (demineralized,* $H_2O$*)*
*Hexane (*$C_6H_{14}$*)*
*Acetone (*$CH_3COCH_3$*)*
*Dichloromethane (*$CH_2Cl_2$*)*

### Equipment and Materials:

*Test tubes and rack*
*Test tube holder, droppers with bulb*
*pH paper (universal), glass rod*

## Procedure:

1. Set up six test tubes in a rack. Fill each separately 1-2 cm high with one of the liquid alcohol samples or a spatula-tip with one of the solid alcohol samples.

2. Add 1-2 mL of the first solvent, mix the contents and watch for solubility or miscibility
3. Empty the contents (waste container), rinse and dry, refill as above and test the second, third and fourth solvent accordingly.
4. Record all observations in the grid below.

## Observations and Results:

**Relative Solubility and Miscibility**

| **Sample:** | Water ($H_2O$) | Hexane ($C_6H_{14}$) | Acetone ($CH_3COCH_3$) | Dichloromethane ($CH_2Cl_2$) |
|---|---|---|---|---|
| Ethanol | | | | |
| Ethylene glycol | | | | |
| Glycerol | | | | |
| Cholesterol | | | | |
| Phenol | | | | |

## Questions and Problems:

1. Does water solubility depend on the number of hydroxyl groups present in the molecule? Explain.
2. What makes an alcohol less soluble in water and more soluble in non-polar solvents?
3. Did the experiment prove the concept "like dissolves like"?

# III Chemical Identification Tests for Alcohols and Phenols

## Chemicals:

*Ethanol*
*Ethylene Glycol*
*Glycerol*
*Isopropanol (2-propanol)*
*Tert-butanol*
*Cyclohexanol*
*Phenol (20% solution)*
*Jones Reagent ($CrO_3$ in dil. $H_2SO_4$ and acetone)*
*Ferric chloride or Iron (III) chloride ($FeCl_3$, 1%)*

## Equipment and Materials:

*Test tubes and rack*
*Test tube holder*

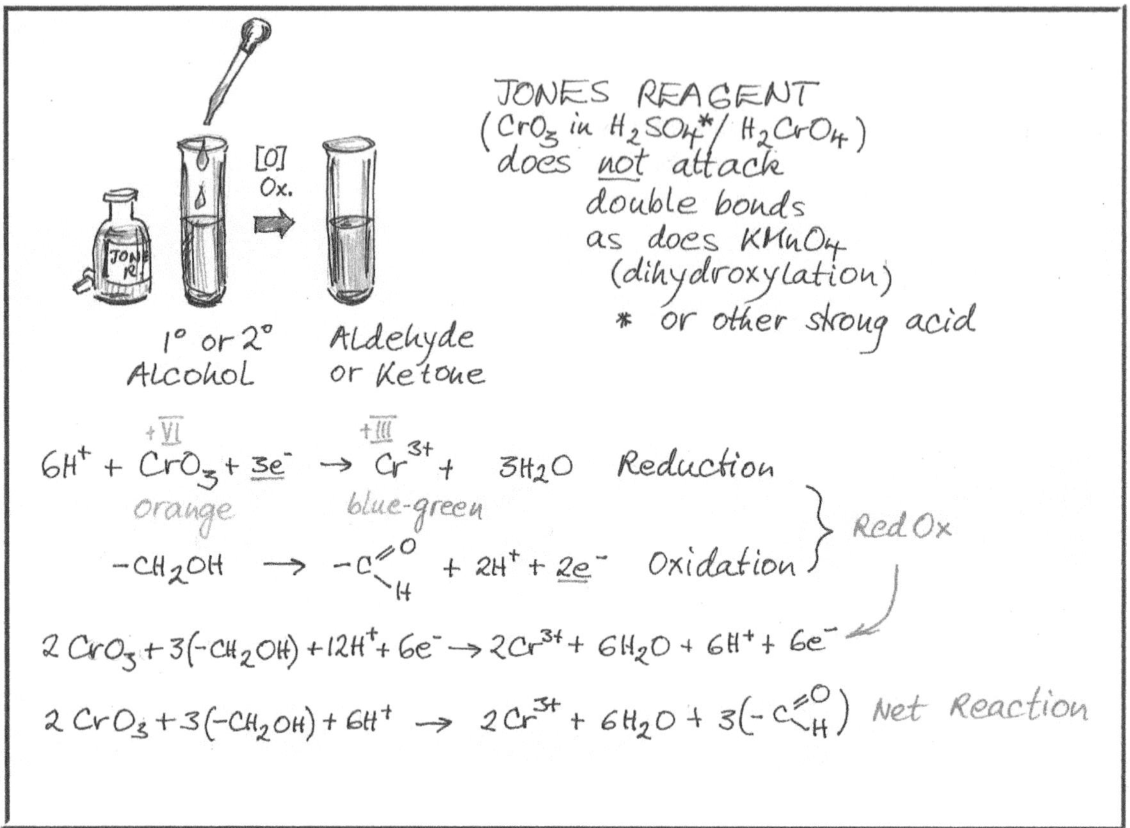


## Procedure:

1. Prepare seven medium sized test tubes in a rack.
2. In each test tube fill one of the five alcohol samples to a level of 1-2 cm. Add 3-5 drops of the yellow to orange colored Jones reagent. Shake the tube back and forth and watch for any color change. Change of the color to green (actually blue) indicates the presence of a primary or secondary alcohol.
3. Fill a clean dry test tube with 1-2 mL of phenol solution and add 3-5 drops of $FeCl_3$ - solution, swirl the contents and watch for any change in color. The appearance of a deep purple color indicates the presence of an **enol**, a molecule with a double bond and directly attached hydroxyl group, as is the case for phenols.

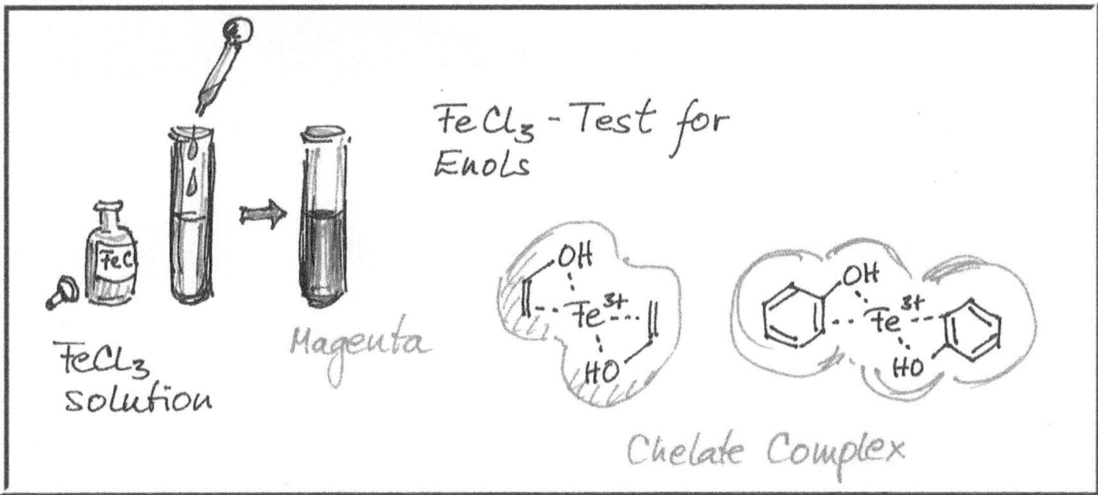


## Observations and Results:

| Alcohol: | Reaction with Jones Reagent: | Reaction of Phenol with $FeCl_3$ solution: |
|---|---|---|
| • **Ethanol** | | |
| • **Ethylene Glycol** | | |
| • **Glycerol** | | |
| • **Isopropanol** | | *N.A.* |
| • ***Tert* – butanol** | | |
| • **Cyclohexanol** | | |
| • **Phenol** | | ------------ |

*Third column: only fill out for phenol!*

## Questions and Problems:

1. What type of general reaction was involved?
2. For the Jones reagent Cr (VI) is reduced to Cr (III). How many electrons were transferred?
3. Salicylic acid is a phenol derivative with an adjacently attached carboxylic group. Draw structural formula and explain why it would give a positive ferric chloride reaction.

# III Ether Synthesis via Acid-Catalyzed Condensation (Intermolecular Dehydration) of Alcohols (Fume Hood, Instructor Guidance)

## Chemicals:

*1-propanol*

*1-butyl alcohol*

*Sulfuric acid (conc. $H_2SO_4$, 15%)*
*Iodine solution ($I_2$ in KI)*

## Equipment and Materials:

*Erlenmeyer flask (wide neck) with two-holed rubber stopper and glass tube*
*Metal stand with clamps*

*Hot Plate and stir bar*
*Thermometer (mercury)*

*Round-bottom flask (2 or 3-neck)*
*Distilling column with rubber hose*
*Thermometer adapter*
*Heating mantle with control box*
*Small funnel*

## Procedure:

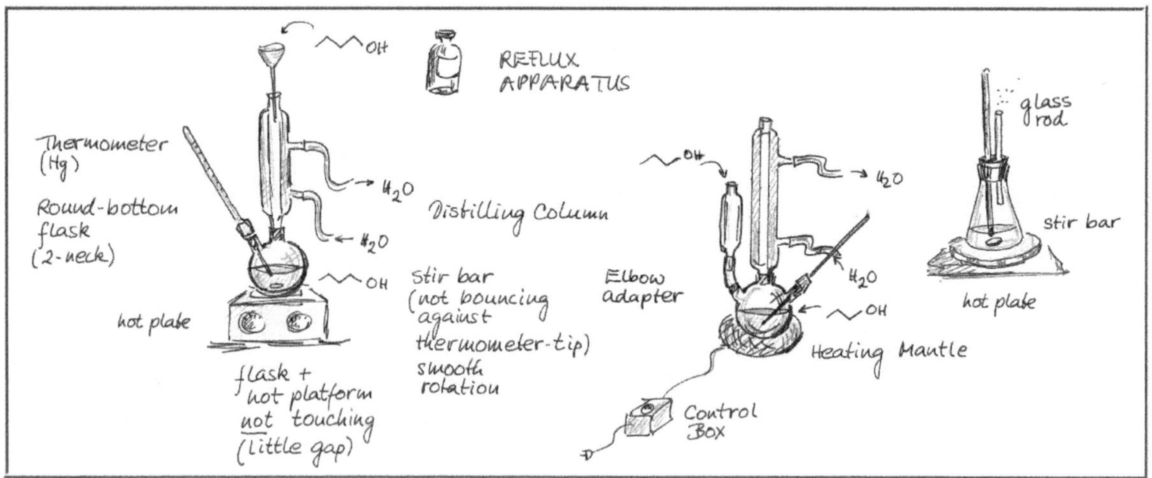


1. Place about 2-3 mL of the alcohol(s) in the Erlenmeyer flask. Add 3-5 drops of $H_2SO_4$, swirl, add the stir bar and place the flask on the hot plate in the fume hood. Attach the rubber stopper with the inserted glass tube in one of the opening, which will serve as air vent, and the thermometer in the other opening to the flask,
2. Heat contents to about 140 °C for about 5-10 minutes.
3. If an ether is formed, two phases should develop, since ethers are not soluble in water. Add a bit of demineralized water if phase separation is not apparent. Add a few drops of Iodine solution, mix thoroughly and see if the organic phase turns to a red - magenta color.

4. An alternative and safer method is to use the reflux procedure, which apparatus can be set-up according to the images shown above with the help of the instructor - (fume hood!).

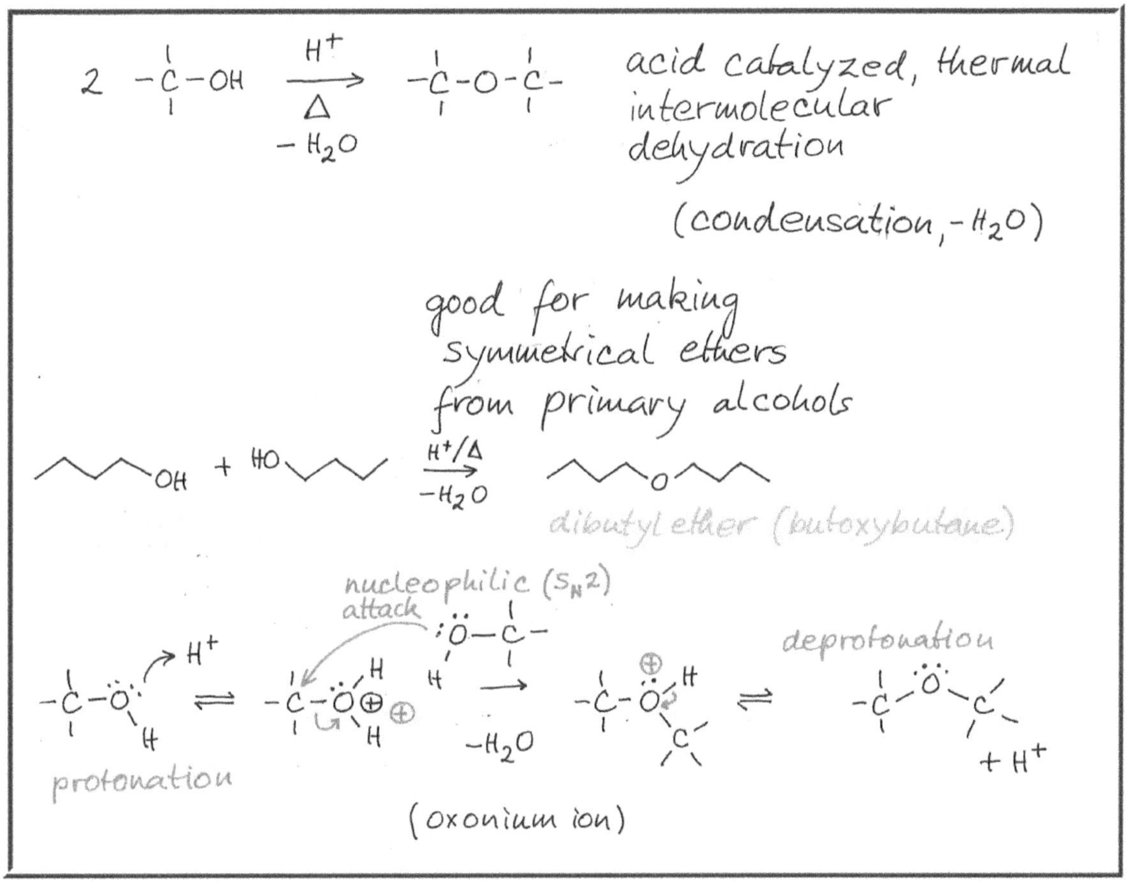


## Observations and Results:

Tabulate the observed results and add the other information required according to the following table:

| **Alcohol(s) reacted:** | **Expected product (ether):** | **Phase separation observed? (yes or no):** | **Color of organic phase with $I_2$:** | **Position of organic phase (top or bottom):** |
|---|---|---|---|---|
| 1-propanol (alone) | | | | |

| | | | | |
|---|---|---|---|---|
| | | | | |
| 1-butanol with 1-propanol | | | | |

## Questions and Problems

1. Write the equations for the dehydration reactions.
2. Why are ethers not soluble in or miscible with water?
3. What are the densities of the ethers formed in the dehydration reactions? (internet research)
4. Would the ethers be found below or above the water layer?
5. Why is the formation of a phase difference proof that an ether was formed?

# IV Redox Reactions of Thiols

## Chemicals:

*Thioglycolate (as $Na^+$, $NH_4^+$ or $Ca^{2+}$ salts)*
*Commercial thioglycolate (hair products)*
*Hydrogen peroxide ($H_2O_2$, 3% and 6%)*
*Wool (10 cm strand)*

## Equipment and Materials:

*Test tubes*
*Droppers with bulb*
*Glass stirring rod*

## Procedure:

1. Place a few mL of thioglycolate solution from the stockroom into one of the test tubes. In another tube place an equivalent amount of the commercial thioglycolate product. By waving the hand over the opening towards the nose, record evidence of any odor and describe its characteristics.
2. To each tube add a few drops of 3% $H_2O_2$ and test for odor. If the odor still persists add a few drops of the more concentrated $H_2O_2$ (6%).
3. Saturate the wool strand sample with thioglyoxalate solution for a few seconds. Rinse with some water and test its consistency. Wrap this strand around the glass stirring rod. Submerge in $H_2O_2$ solution for a while, then remove and dry with a hair dryer.
4. Remove the dried strand of wool from the rod and study its consistency and shape.

## Observations and Results:

Thioglycolate Oxidation:

| **Thiol:** | **Odor characterization:** | **Addition of $H_2O_2$ and change in odor:** |
|---|---|---|
| **Thioglycolate (stockroom sample)** | | |
| **Ammonium Thioglycolate (commercial sample)** | | |

Wool treatment with Thioglycolate:

| | **Characterization before treatment with thioglycolate:** | **Characterization after treatment with thioglycolate:** | **Characterization after treatment with $H_2O_2$:** |
|---|---|---|---|
| **Wool Sample:** | | | |

## Questions and Problems:

1. Were any changes observed during treatment of the samples with thioglycolate and subsequent hydrogen peroxide? Explain!
2. What products are formed upon oxidation of thiols? Write an equation for this reaction.
3. How many electrons are transferred?
4. What type of substance is wool? How does it relate to hair?
5. The fibers of wool and hair are held together by covalent disulfide bonds ("bridges"). What groups are formed upon treatment with thioglycolate? How are they reverted back to disulfide bonds?
6. The skunk smell is mainly composed of 3-methyl-1-butanethiol and a combination of *cis/trans*-2-butene-1-thiol. What are the formulas for each of these three compounds?
7. What would be a good way to "neutralize" the unpleasant skunk odors?
8. Freshly cut onions release propanethiol. What is the formula of this compound?

# Chapter 6 Aldehydes and Ketones

## Ch. 6 Aldehydes and Ketones (Carbonyl Compounds)

Aldehydes are oxidation products of primary alcohols and ketones of secondary alcohols. Both are characterized by the **Carbonyl** (-CO-) functional group, having a terminal position in the former and an internalized position in the latter.

**Fig. 6.1**

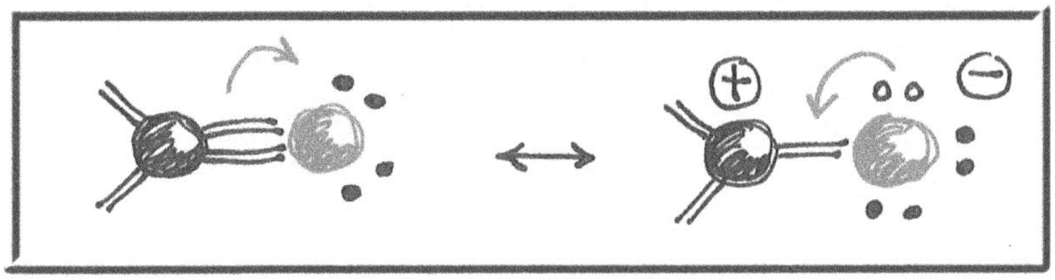

When terminal, it is also known as the **aldehyde group** or the **aldo function**. When located internally it is referred to as the **ketone** group or **keto function**.

**Fig. 6.2**

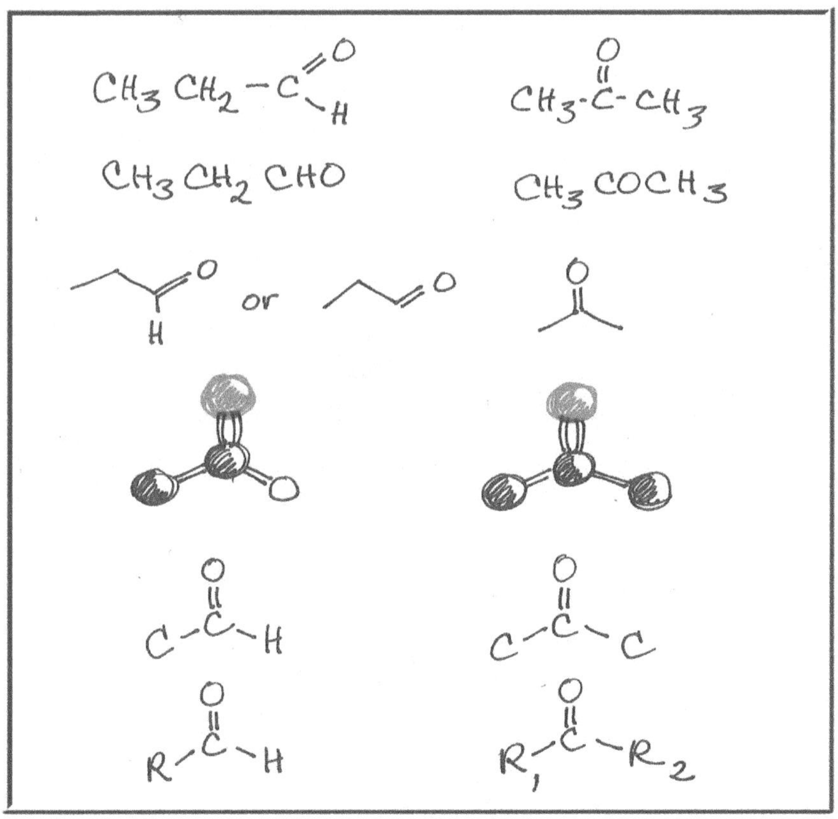


The carbonyl group is highly polarized due to the strong electron affinity and electronegativity of the oxygen atom.

**Fig. 6.3**

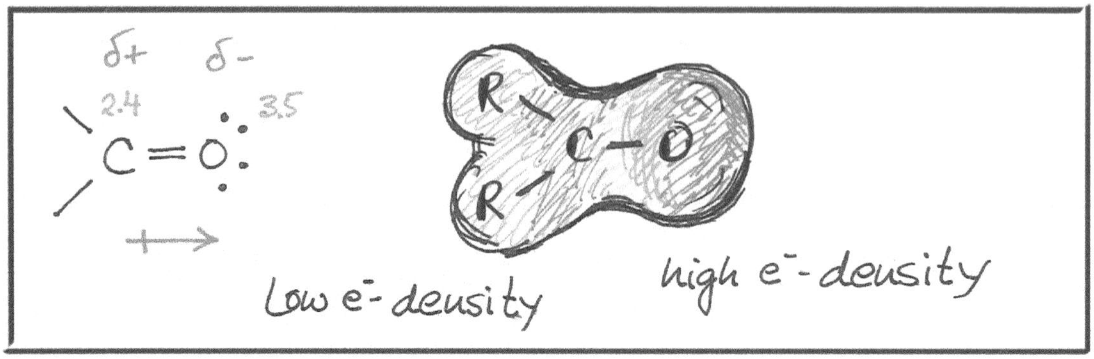


The electron density, or electron "cloud," is shifted towards the oxygen, making the carbon atom more positive and thus a target to nucleophilic ("loving" low electron density areas) attack. As a result, the carbonyl is a highly reactive group. It is resonance stabilized, having two oscillating resonance states.

**Fig. 6.4**

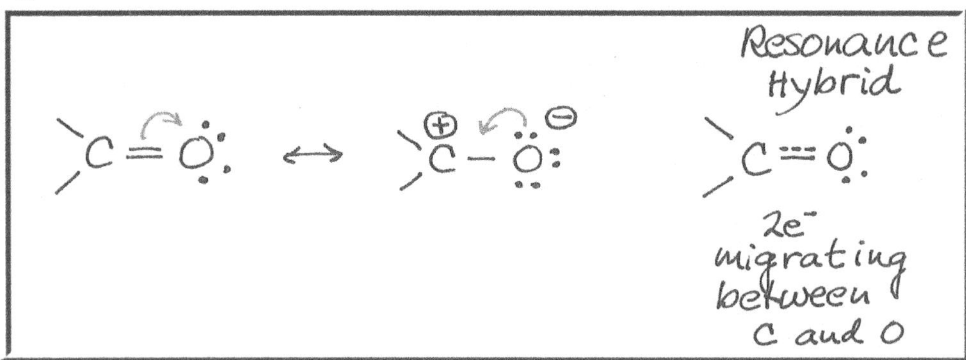


The hybrid between the two extremes shown above is the actual "true" state of the electron distribution, indicated by the dotted line (2 electrons spread between the carbon and oxygen as shared pair and the oxygen atom alone as lone pair).

Formaldehyde and Acetaldehyde are gases and both open-chained (acyclic), aliphatic (non-aromatic) molecules. Of course, aldehydes and ketones can run the whole gamut of different structural types, such as carbocyclic, heterocyclic aliphatic and/or aromatic systems. They can be involved with other functional groups in the same molecule. Other lower aldehydes and ketones are liquids, relatively polar due to the carbonyl group, which can receive H-bonds and thus contribute to their miscibility with water. There is also some miscibility with non-polar solvents, especially if the hydrocarbon section of the molecule gets is more represented than the polar carbonyl group within a molecule (carbon to oxygen ratio being larger). Many aldehydes and ketones are known by their common names, like acetone, benzaldehyde, testosterone, estrone, progesterone etc.

Fig. 6.5

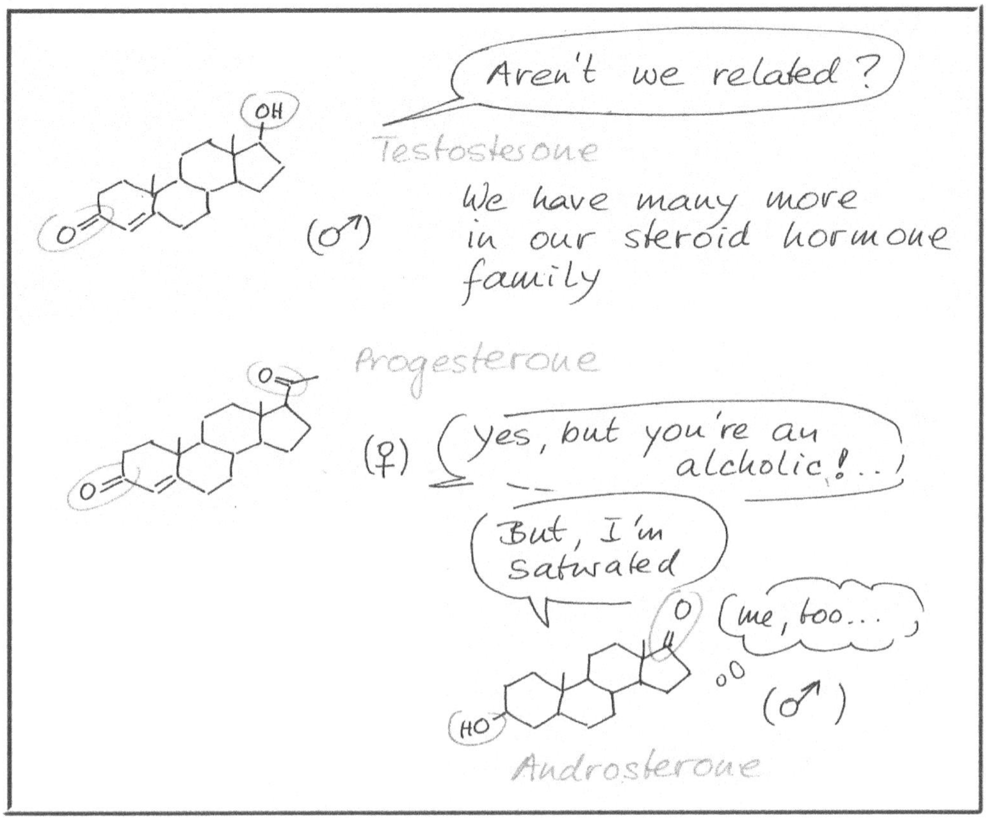


Fig. 6.6

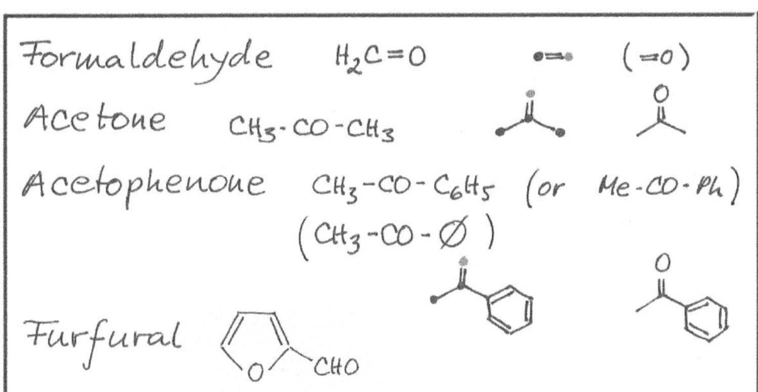


According to the systematic, standardized naming rules (IUPAC), all aldehydes should end with the suffix **-al**, so the names are similar to the alcohols.

methanol → **methanal**
ethanol → **ethanol**
**etc.**

The **"aldehyde"** term as such can also be used as a parent and the alkyl group named up front.

**Fig. 6.7**

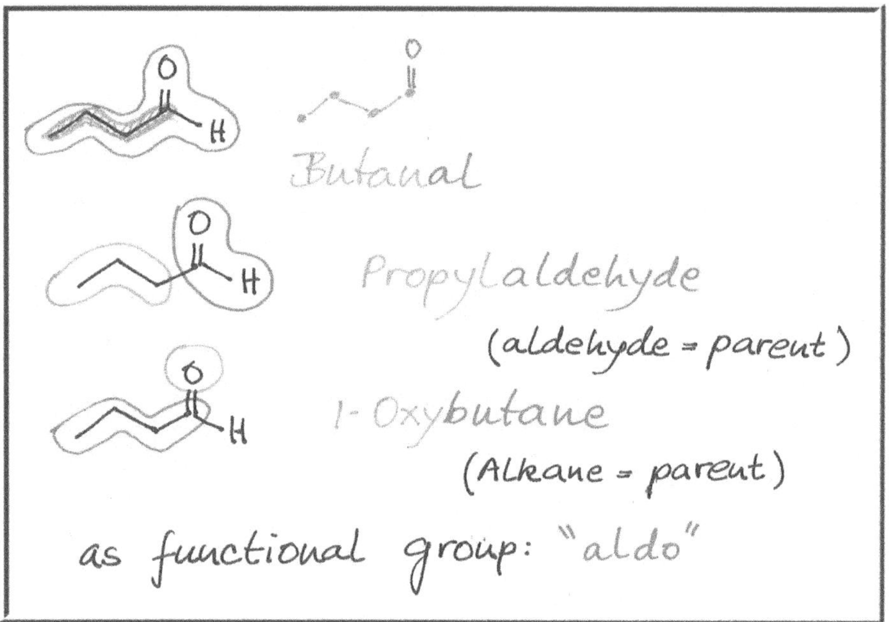


The group is called **"oxo"** when the alkane is used as a parent. An oxy group refers to any group containing any double bonded oxygen (=O) to a carbon or any other element like sulfur. The carbon attached via a double bond to an oxygen counts as part of the main chain.

**Fig. 6.8**

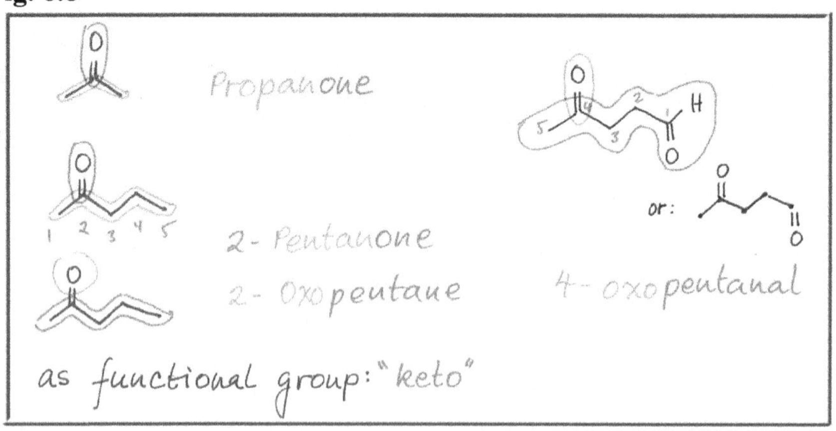

In a condensed form the formula designating an aldehyde is written as **-CHO** and the functional group separately referred to as the **aldo** group. The functional group **–CO-** for ketones is referred to as the **keto** group.

The ketones are all ending in **-one**, where the carbonyl group has to be identified by position in the parent framework if more than one internal positions are possible (structural isomers).

Another naming possibility is using **"ketone"** as the parent and naming the alkyl groups up front, similar to the ethers.

**Fig. 6.9**

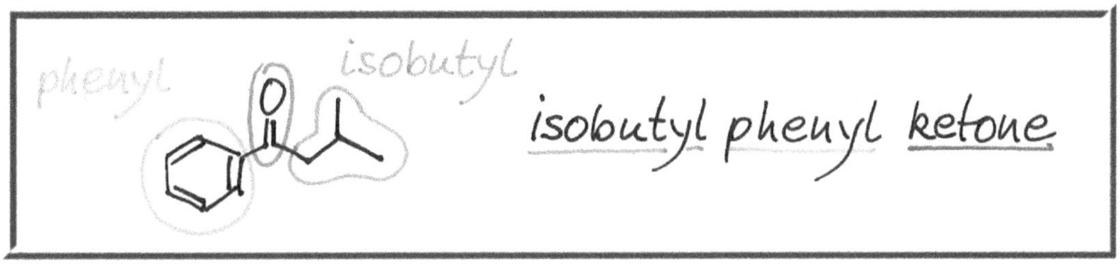


**Example Problem 6.1**
**Name the following aldehydes and ketones using two different methods:**

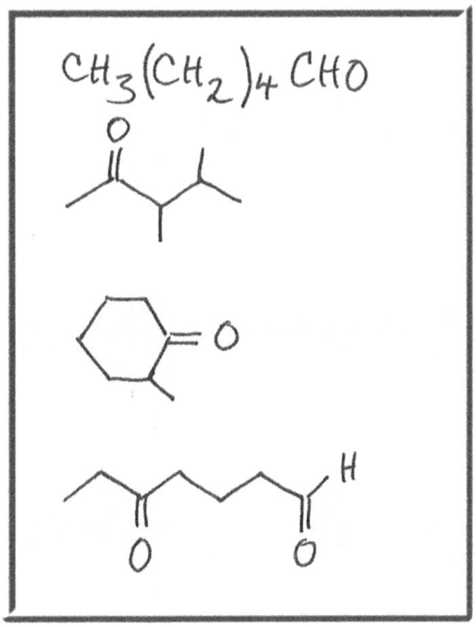


**Example Problem 6.2**
**Draw the structural formulas for the following names (SF and ASF/LAF)**
***3-pentanone***
***4-methylhexanal***
***Propyl aldehyde***
***Ethyl methyl ketone***
***3-Buten-2-one***
***1-Cyclobutylethanone***

When more than one functional group occurs, then the suffix **–dial** is added for a di-aldehyde and **–dione, trione** etc. added as suffix for a poly-valent ketone, all the carbons, including the carbonyl carbons, contributing to the over-all hydrocarbon chain, named up front. If a carbonyl group is involved with another function that also is a terminal group, then it has second rank and is named as **"oxo"** (Fig. 6.8).

**Fig. 6.10**

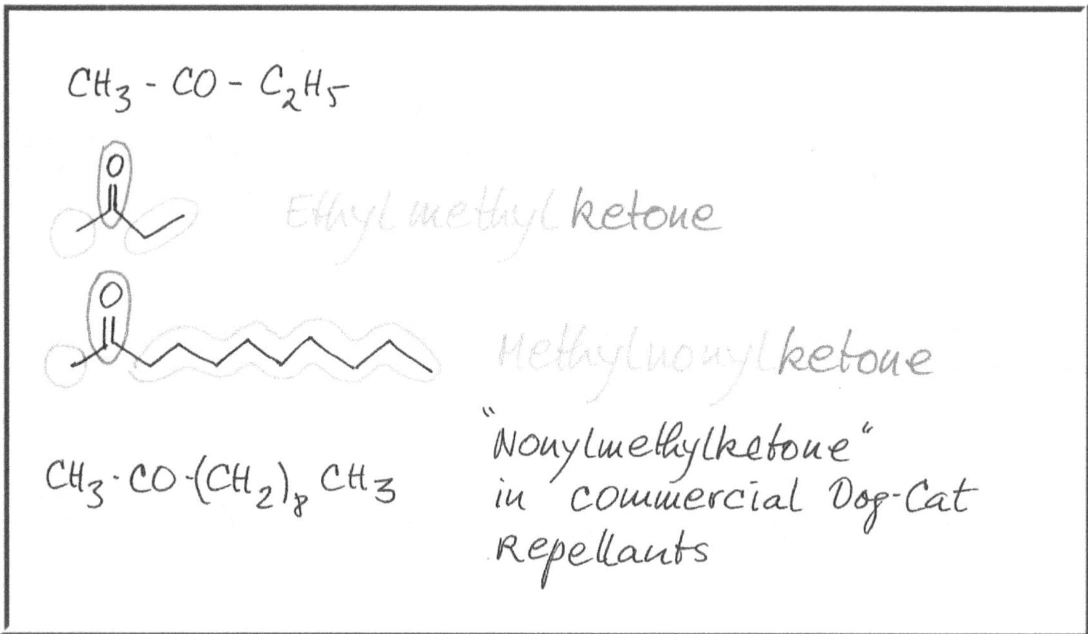


**Example Problem 6.3**
**Give the structural formula for the following names**
***2,4-Pentanedione***
***3-Oxopentanal***
***Pentanedial***
***Ethanedial***

In general, any carbonyl group, be it as a ketone or an aldehyde, can be indicated by "**oxo**", if another functional group is of primary rank.

Another important group is the combination of the carbonyl group attached to an R group, usually alkyl, alkenyl, or aryl. This is called the **acyl (-CO-R) group.** If the R-group is a methyl group then it is the well-known and common function **acetyl (-CO-CH$_3$)**, derived from acetic acid (ethanoic acid).

**Fig. 6.11**

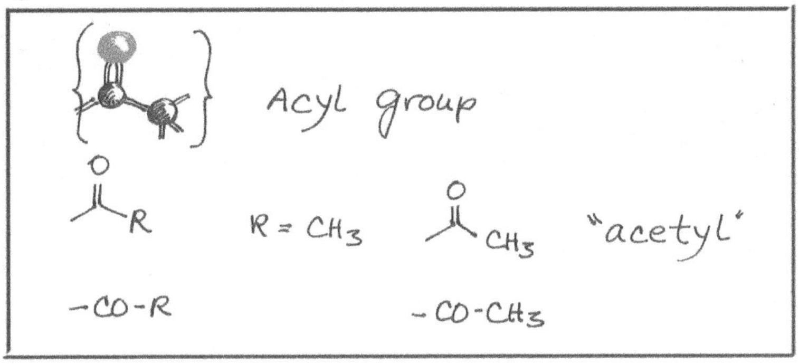


Acyl halides are the chemicals that are often used to introduce the acyl group into another (substrate) molecule. The reaction is referred to as **acylation**. Alkylhalides lead, by the same mechanism and conditions (Lewis Acid needed as catalyst, such as $FeCl_3$ or $AlCl_3$), to **alkylation** of compounds. This valuable synthetic route is known as the **Friedel-Crafts-Reaction**.

**Fig. 6.12**

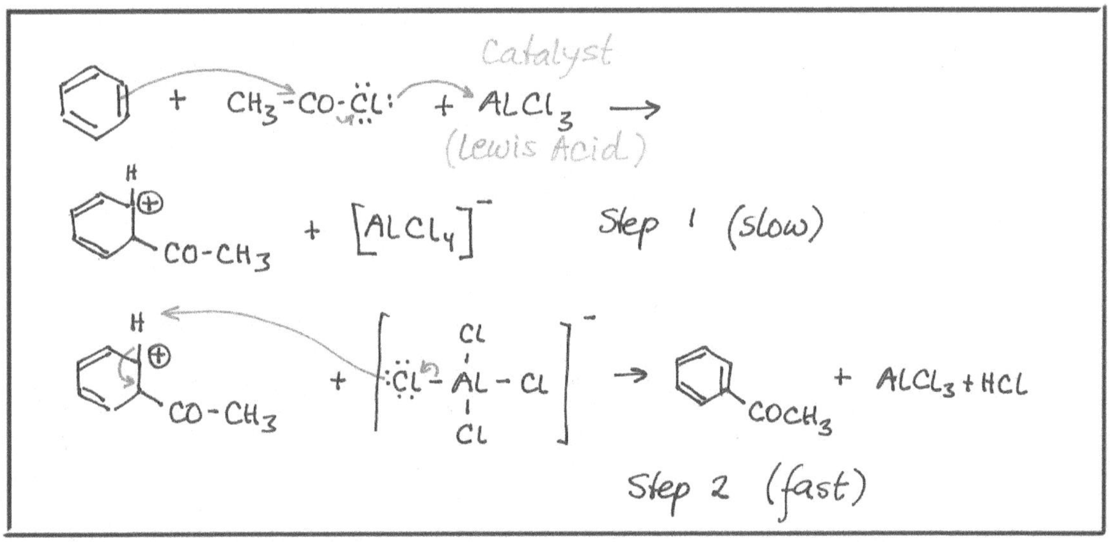


**Example Problem 6.3**
**Complete and balance the following reactions:**

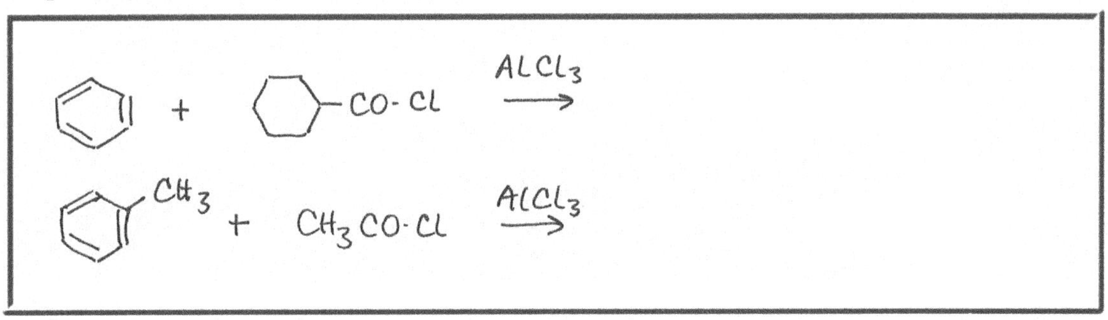

Aldehydes and ketones are widely spread in nature. They are components of many **essential oils** and the volatile principles (essences) of aromatic plants, such as **vanillin** from vanilla, **menthone** from a variety of mint plants, **thujone** as ingredient of the drink absynth, made from anise, **zingerone** from ginger, and the **ionones**, the rose ketones extracted as rose oil and many more Often they are associated with other groups such as hydroxyl or ether functions. Ketones, such as **carvone**, are found in spearmint oil, caraway and clove seeds.

.

**Fig. 6.13**

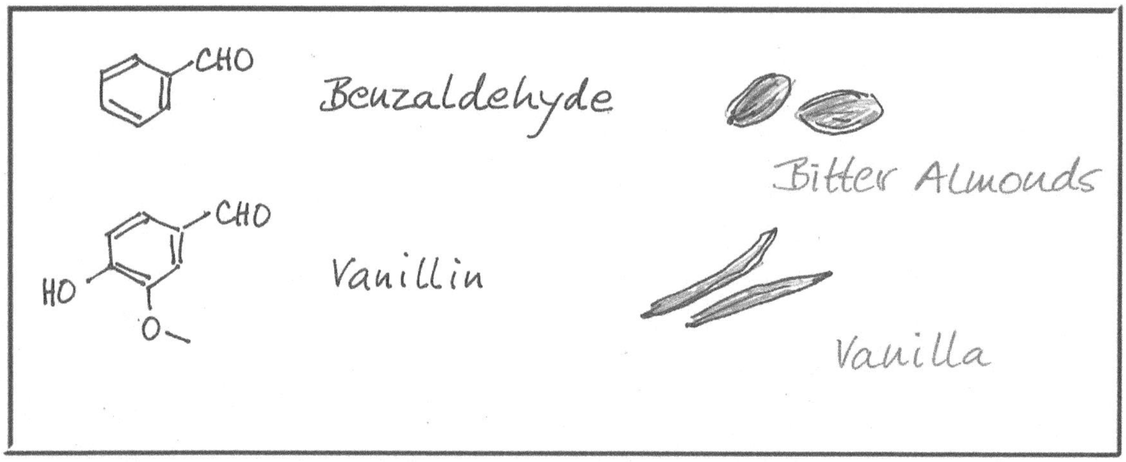


Many animals also produce aldehydes and ketones, especially insects and invertebrate marine species also utilize allopathic ketones as repellants. In mammals the ketones are more of the macrocyclic sort, released by specific glands as signal molecules. **Muscone**, a macrocyclic ketone from the glands of the musk deer (used as a pheromone to attract females), is a costly additive for perfumes. **Civetone,** another macrocyclic ketone, is the sexual attractant for the African Civet cat. The odor is pungent and musky in concentrated form but very pleasant in proper dilutions. Synthetic musks have now replaced the natural sources and civetone is synthesized in a few steps from ethyl oleate found in palm oil.

**Fig. 6.14**

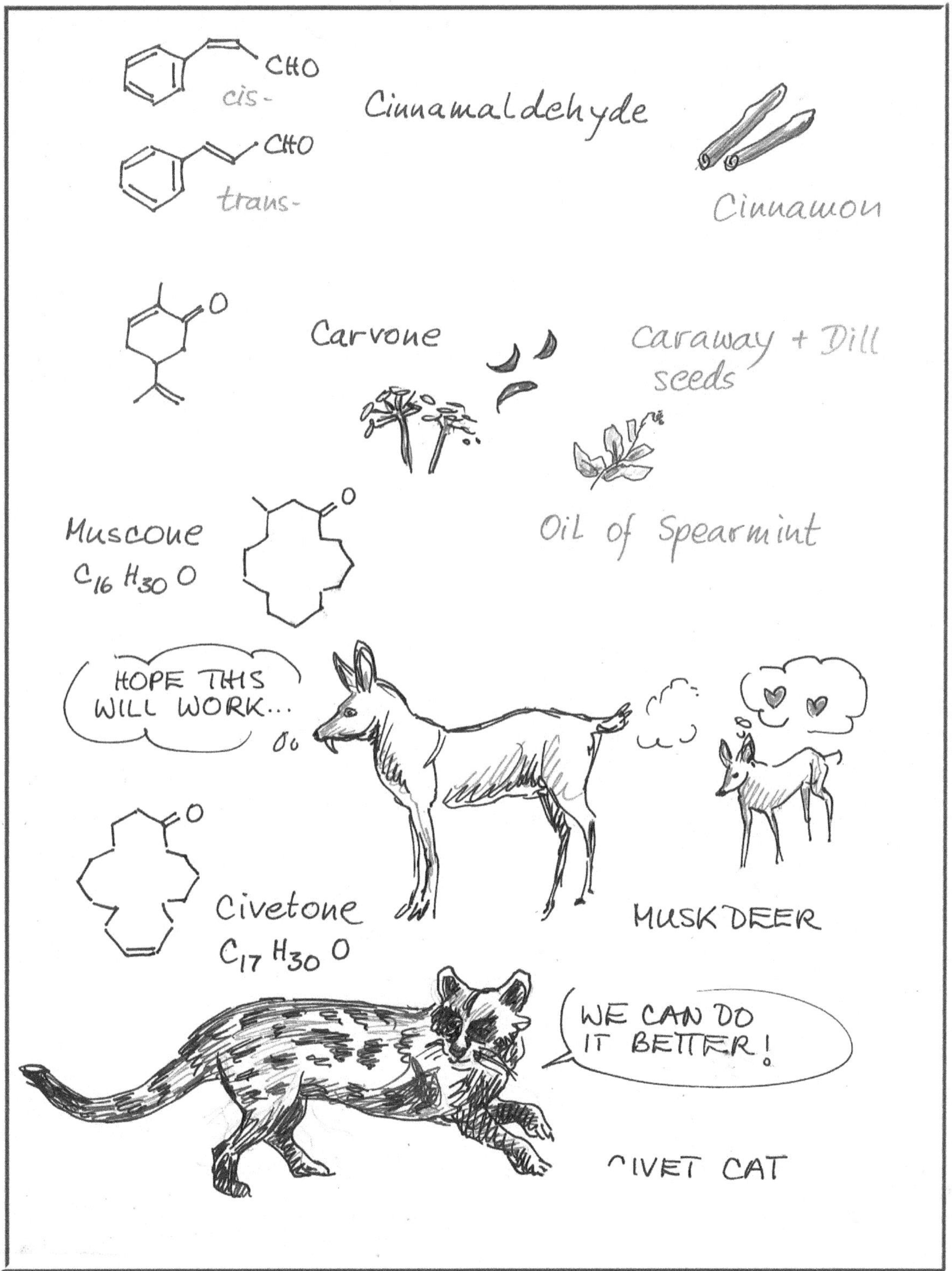
CHO
cis-
Cinnamaldehyde
CHO
trans-
Cinnamon
O
Carvone
Caraway + Dill seeds
Muscone
$C_{16}H_{30}O$
O
Oil of Spearmint
HOPE THIS WILL WORK...
O
Civetone
$C_{17}H_{30}O$
MUSK DEER
WE CAN DO IT BETTER!
IVET CAT

As already indicated, ketones are often used as natural repellents for other species but often also as special attractants for their own species in the animal world. An interesting coincidence is the occurrence of **acetophenone** in the sweat of Italian wine workers as well as in the territorial markers of the Japanese weasel.

**Fig. 6.15**

Ants produce a variety of aldehydes and ketones as repellents in territorial "warfare."

**Fig. 6.16**

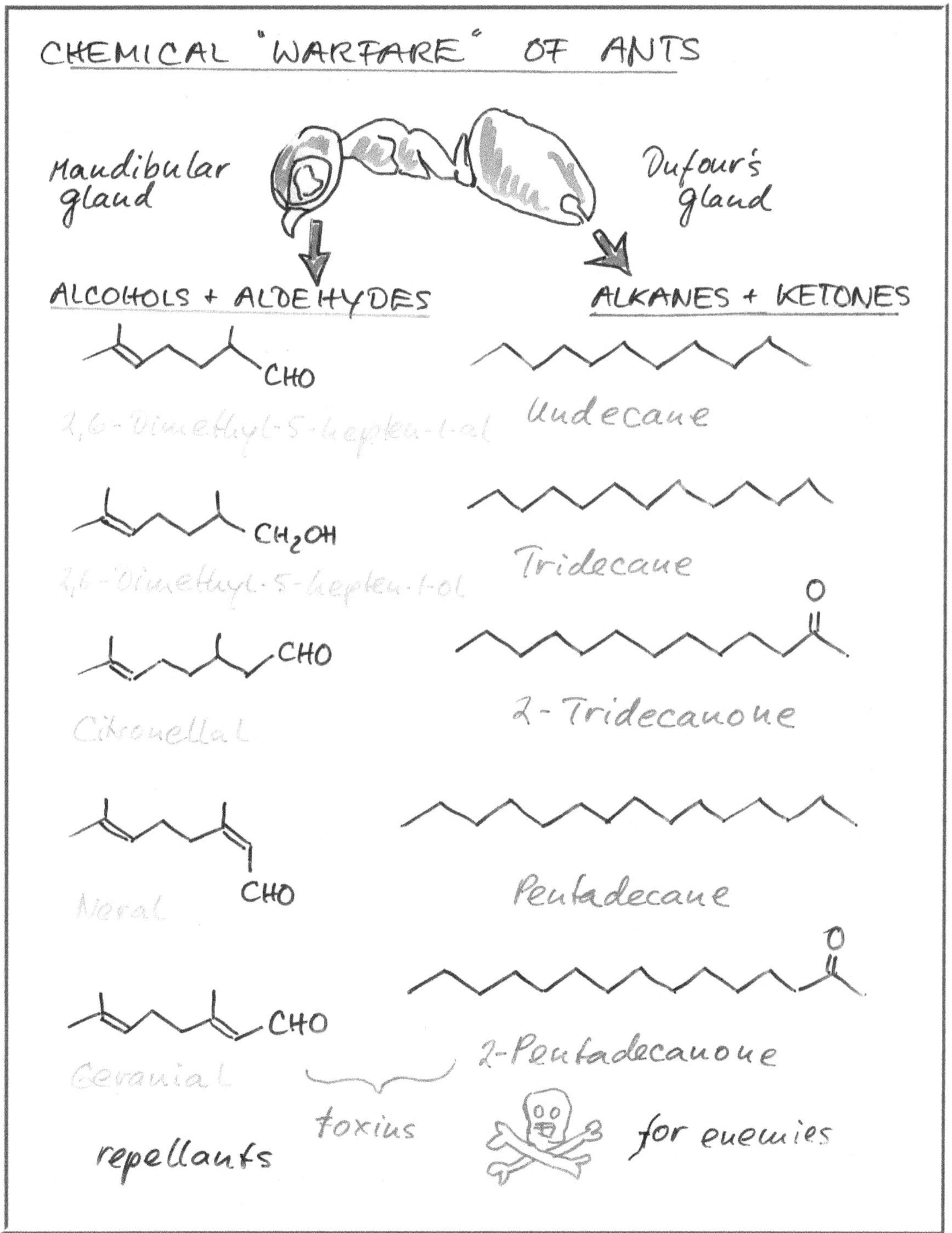

The dog and cat repellent was the outcome of research related to natural "keep-off" molecule 2-Undecanone or methylnonylketone (listed as nonylmethylketone on the label).

**Fig. 6.17**

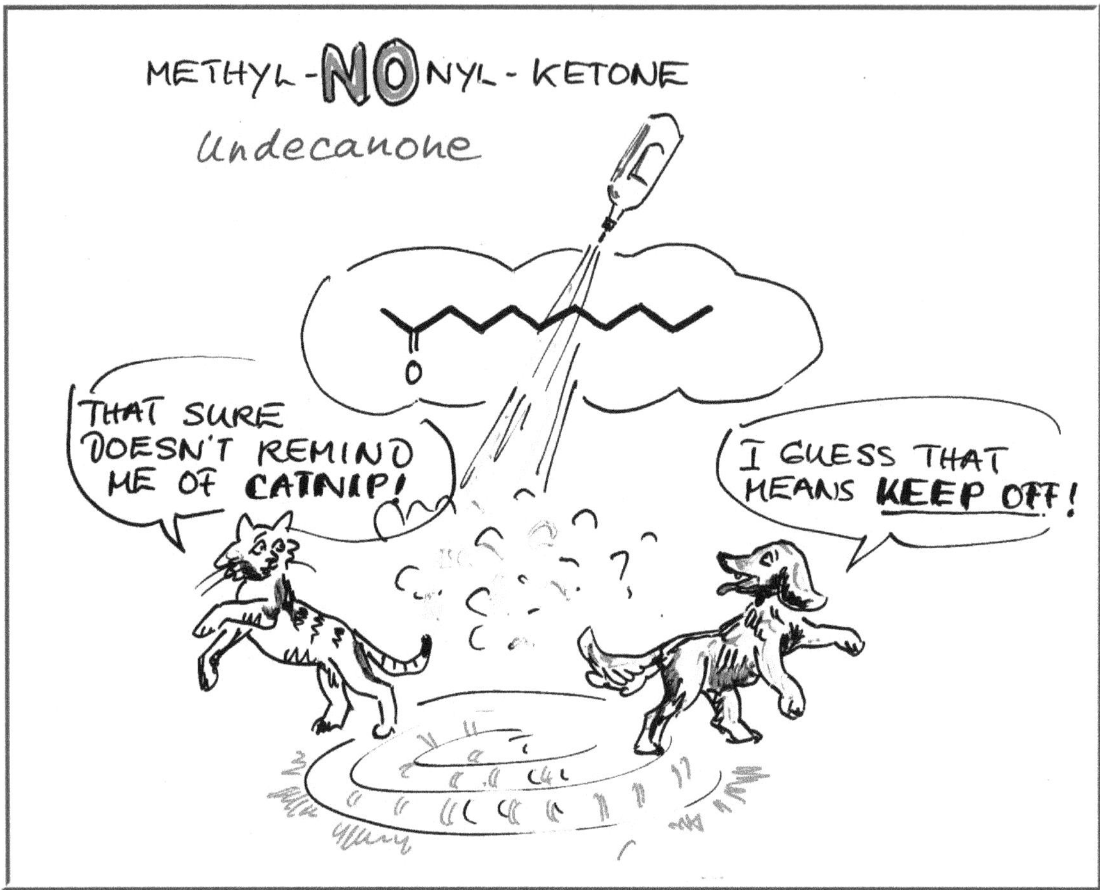


An interesting territorial marker and sexual pheromone "dianeackerone" (3,7-Diethyl-9-phenyl-2-nonanone) is one used by the African dwarf crocodiles

**Fig. 6.18**

SCENT OF A CROCODILE
I hope this will make me more attractive today:
O
1 Volatile nonanone
10 Steroid
Pheromones from skin glands
DIANEACKERNONES

The following table shows a selection of naturally occurring aldehydes and ketones.

**Table 6.2**

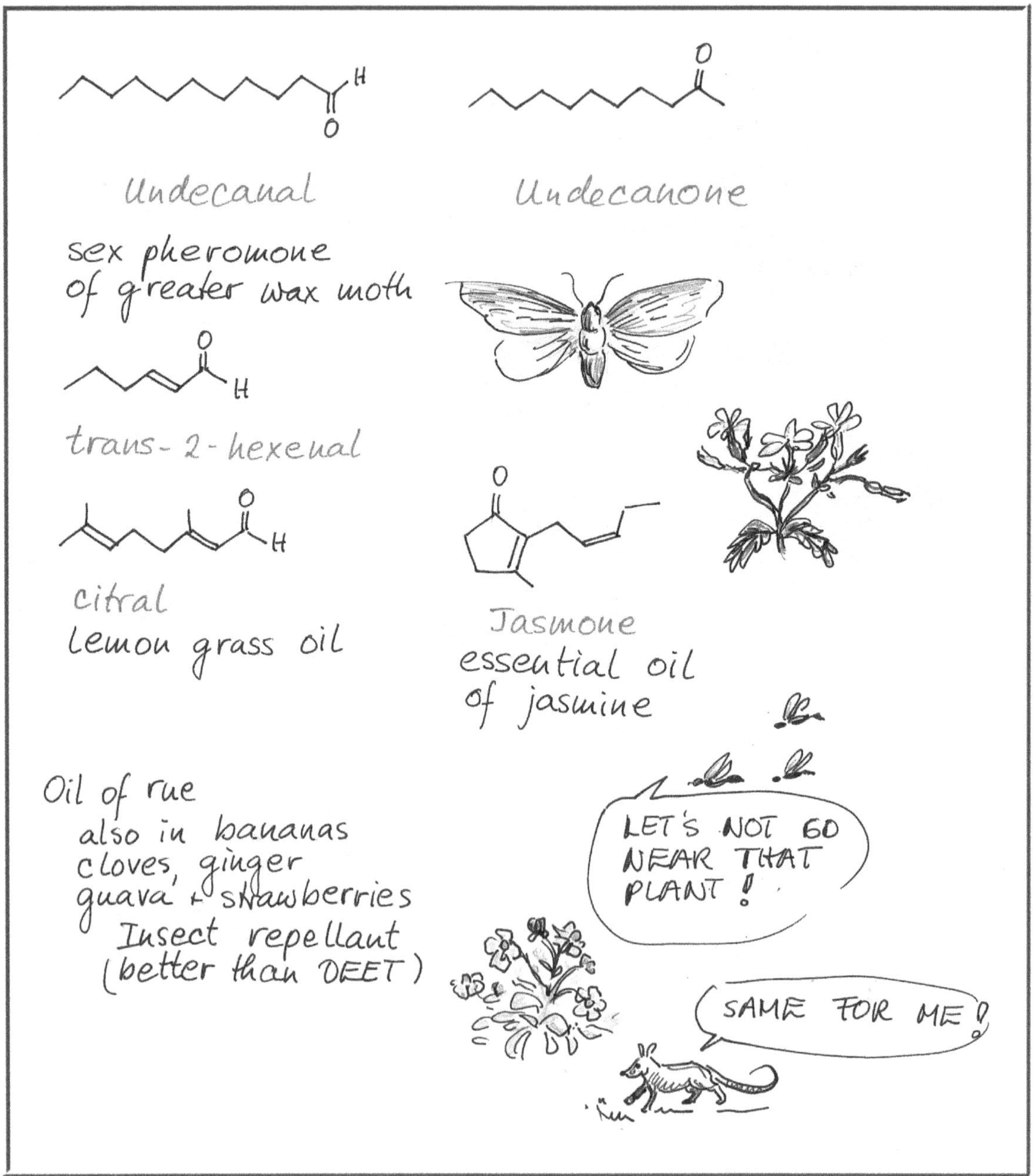


An important interaction between a carbonyl group and an alpha (α)-hydrogen – a hydrogen atom attached to the adjacent carbon – the **α-carbon** (located next to the functional group) – is **tautomerism**. The carbonyl group has an acidifying effect on a hydrogen atom attached to an α-carbon.

**Fig. 6.19**

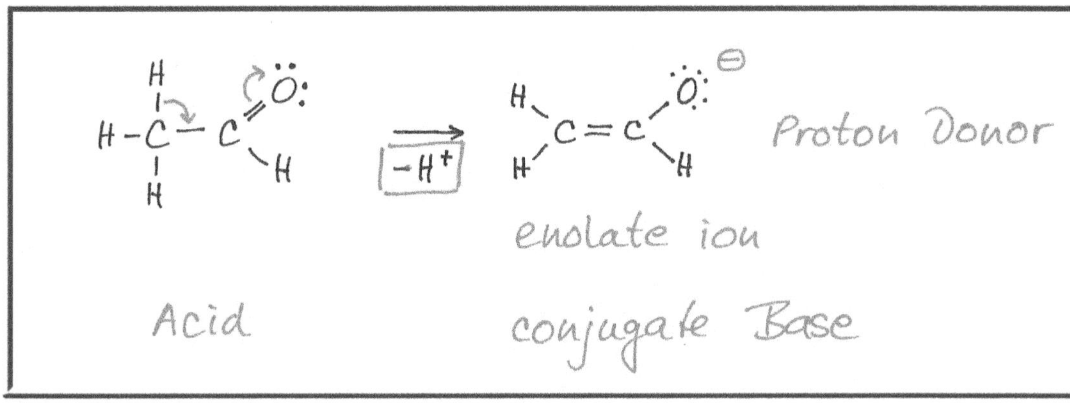


This intra-molecular proton migration between the α-carbon and the oxygen atom, of the -CO- group, leads to **keto-enol tautomers** – a type of interchangeable, constitutional isomer. This **enolization** phenomenon is typical for aldehydes and ketones. An enol (alkenol) is slightly more acidic than the corresponding saturated alcohol (alkanol), being another example for pi-systems (multiple bonds) acting as electron-withdrawing groups. Protons can be abstracted by a strong base resulting in salt formation. After the discharge of a proton, an **enolate ion** is formed (double bond and negative oxygen).

**Fig. 6.20**

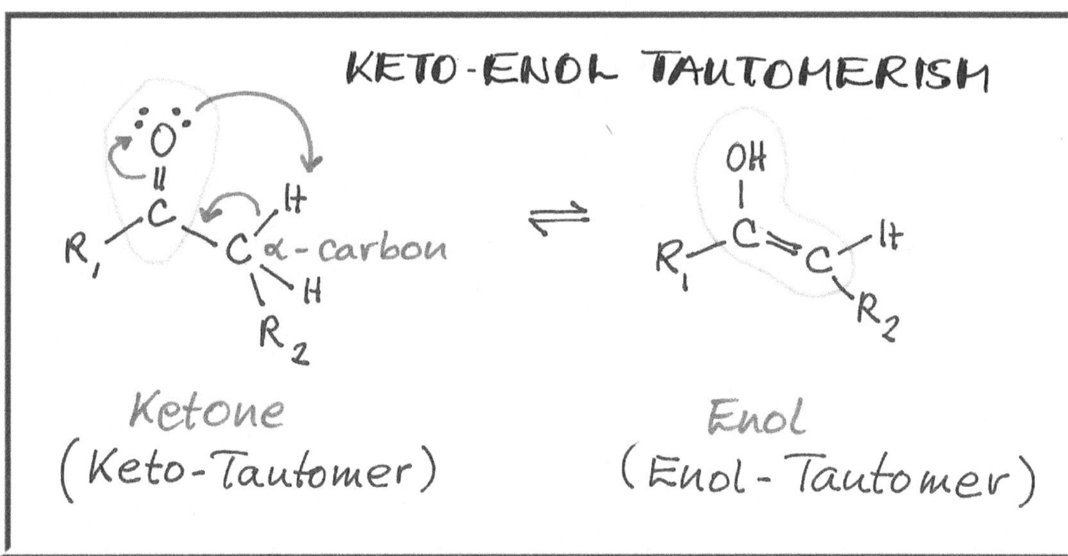


Any **enol** gives a distinct red-purple color reaction when treated with a **Ferric Chloride** ($FeCl_3$) solution, which itself is yellow. The metal coordinates of the enol to a colored **chelate complex** (scissor-like lobes) (see Ch. 5).

**Fig.6.21**

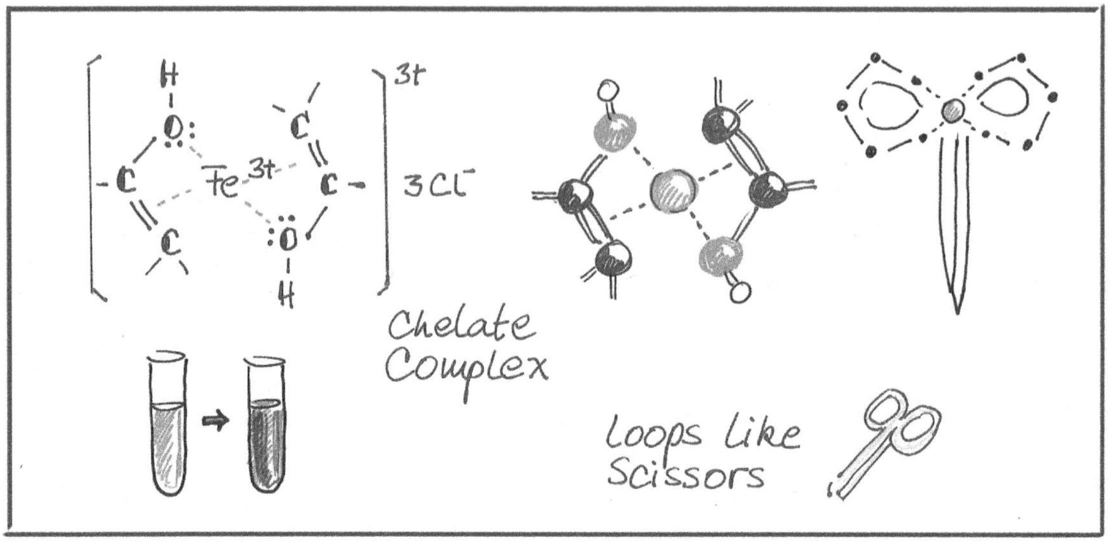


**Example problem 6.4**
**Draw the corresponding enol structures for the following ketones:**

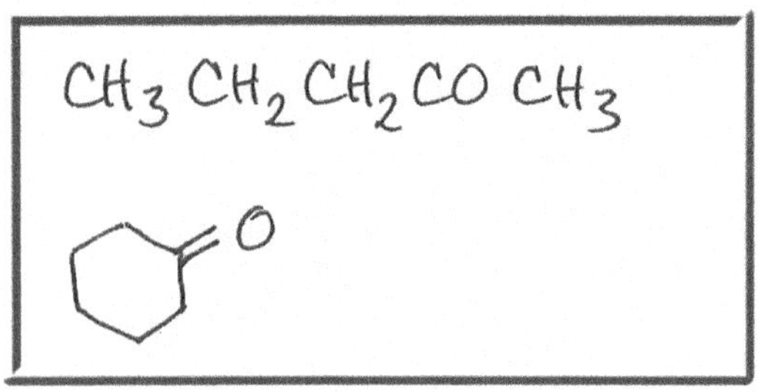


**Example Problem 6.5**
**Which of the following molecules would give a positive Ferric Chloride test?**

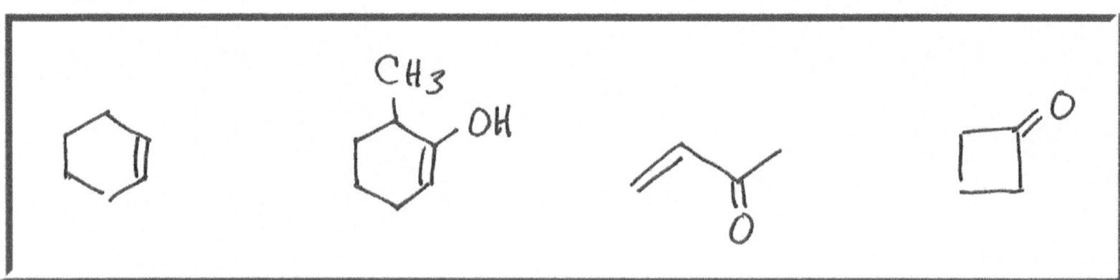

**Example Problem 6.6**
**What is the outcome for the following molecules treated with strong base?**

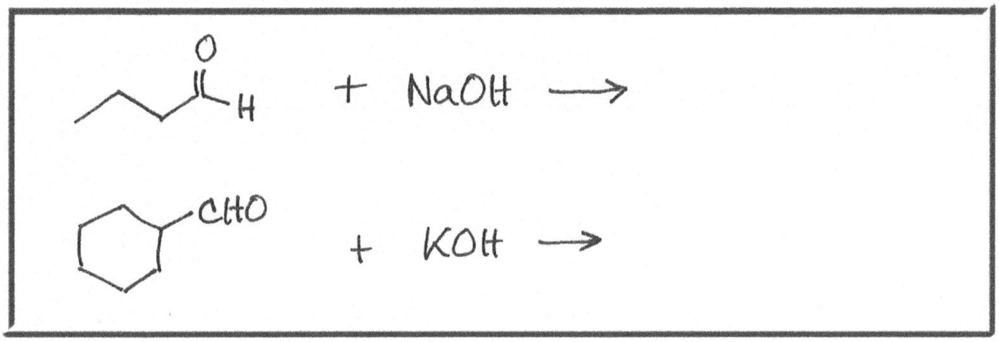


Depending on the molecule and the position of the equilibrium, one form might be preferred over the other.

Since Aldehydes can be oxidized to the corresponding carboxylic acid or reduced to give the related alcohol, it is easy to identify their presence. Two important reagents are used for this purpose, both based on redox-reactions.

1. **Tollens Reagent** consists of a $AgNO_3$ solution in ammonia. The $Ag^+$ ions are the electron acceptors for the oxidation of the aldehyde, reducing them to elementary silver according to the following reaction:

**Fig.6.22**

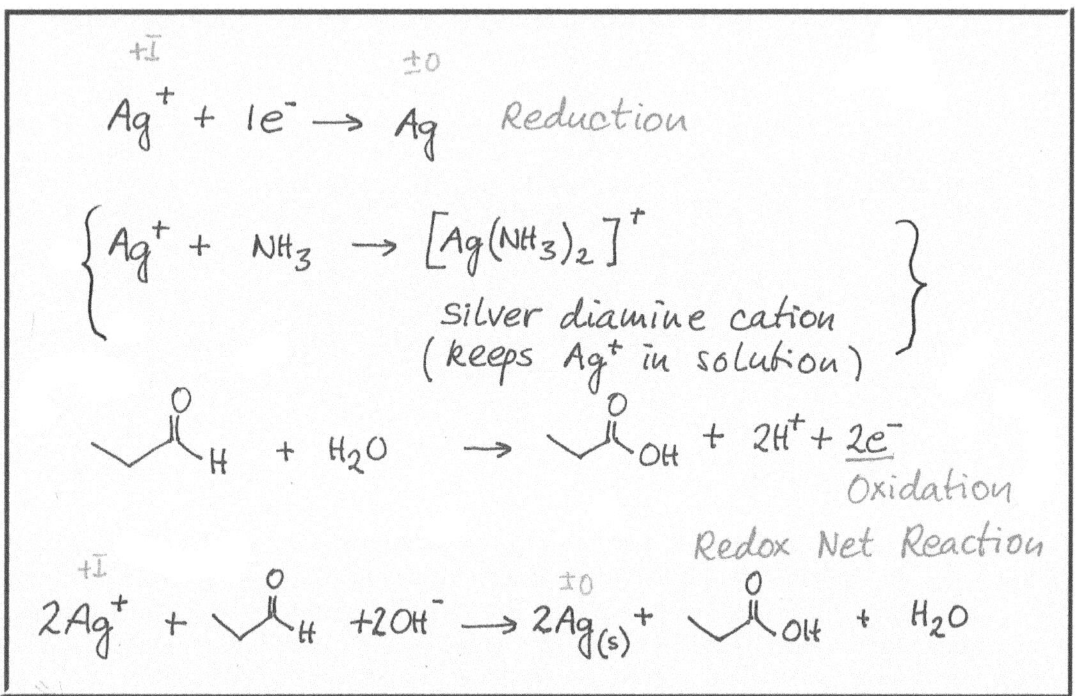

The silver, in its elementary form, coats the container with a silver mirror. Commercially, this reaction is applied for the actual manufacture of mirrors. A fine Ag coating is achieved by spraying a mist of $AgNO_3$ and ammonia onto the surface with a spray gun.

**Fig. 6.23**

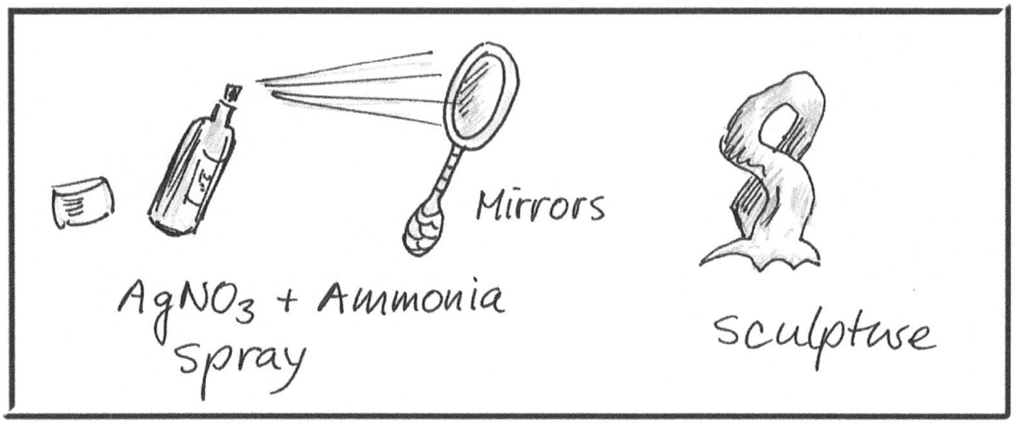


The method is also used by glass artists who produce silvered glass sculptures. $AgNO_3$ stains everything brownish black, including hands and anything it comes into contact with.

1. The other test is the **Benedict Reaction**, which relies on the use of $Cu^{2+}$ ions as oxidizing agent. Here, the aldehyde must have an adjacent hydroxyl group on the α-carbon to give a positive result, shown by the color change of the blue aqueous $CuSO_4$ solution to the brick-red colored precipitate of $Cu_2O$ (Cuprous oxide).

**Fig.6.24**

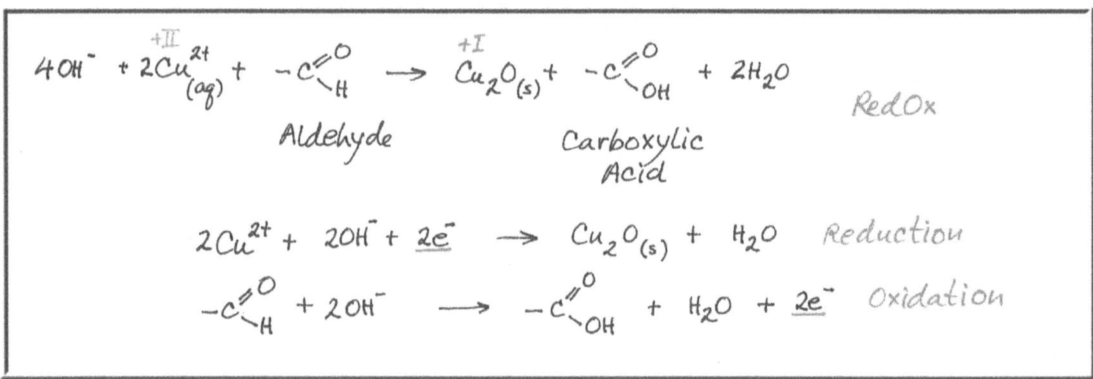


Another, similar reagent is **Fehling's solution,** also based on the reduction of blue $CuSO_4$ to the red insoluble $Cu_2O$. Instead of formulation with Sodium Citrate and Sodium Carbonate, Sodium Potassium Tartrate and Sodium Hydroxide are used. Both can distinguish between aldehydes and ketones (exception: alpha hydroxyketones). However, Benedict's and Fehling's reagents fail for the identification of aromatic aldehydes. Here, the Tollens reagent gives better results.

**Fig. 6.25**

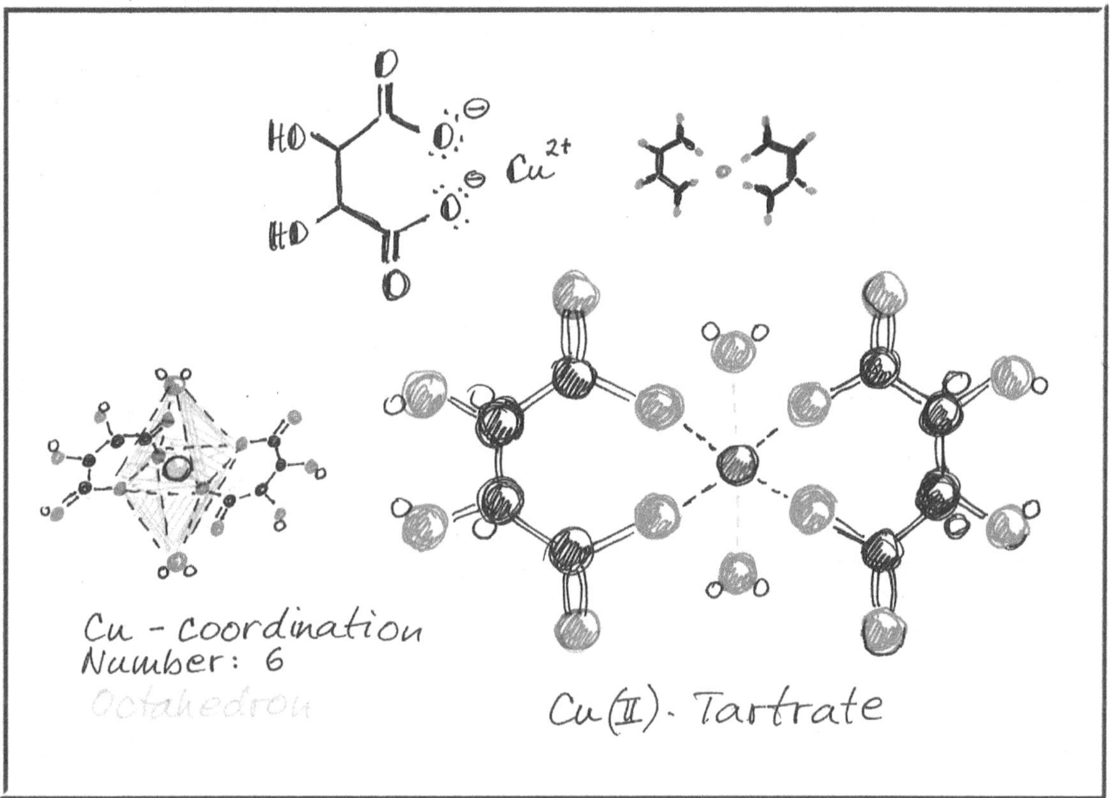


The $Cu^{2+}/Cu^{+}$ redox reaction is especially valuable for sugar molecules, such as aldoses, is covered in the biochemistry text Ch.5 (B. Heinz "Essentials of Biochemistry", Kendall-Hunt, 2013).

Ketones resist oxidation, so conventional reactions cannot be used for their identification. Hydroxy ketones, however, being the case for many sugars, can convert slowly into aldehydes (see Ch. 5, Biochemistry)
However, methyl ketones, such as acetone and methylethylketone, can be identified by the **Iodoform Reaction** (haloform reaction). This is a type of redox reaction that leads to the yellow, solid iodoform, a common name for triiodomethan. Although the reaction works with any halogen, iodine is best for analytical purposes, since it produces a well visible yellow precipitate. Chloroform ($CHCl_3$) – a well-known anesthetic is a homolog of this haloform class. Iodoform is found in many antiseptic tinctures and has a typical "medicinal" odor.

**Fig. 6.26**

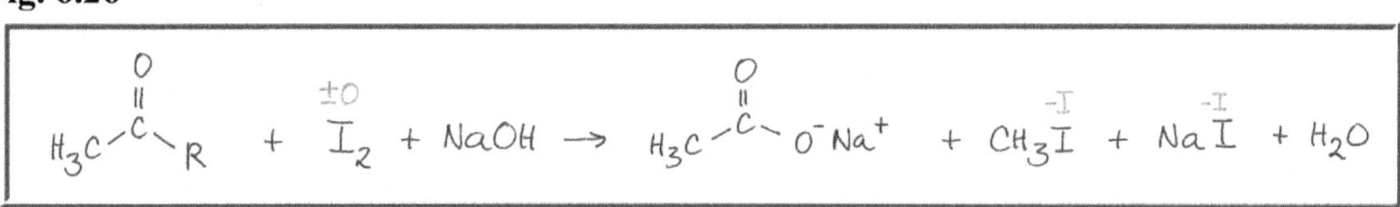

It is worthwhile to have a closer look at the mechanism of this reaction, since it involves the formation of an enolate ion by the action of a strong base. Enolate ions are a key initiator for many reactions involving aldehydes and ketones due to the tautomeric effects described above. The mechanism of the haloform reaction involves three successive halogenation steps of the α-carbon of the methyl group. The iodine ($I_2$, oxidation number ± 0) is reduced to iodide ($I^-$, oxidation number -1) in each step by the action of hydroxide ions. The final step leads to the cleavage of the original molecule into the haloform/iodoform and a carboxylic acid salt containing the R-group of the original ketone.

**Fig. 6.27**

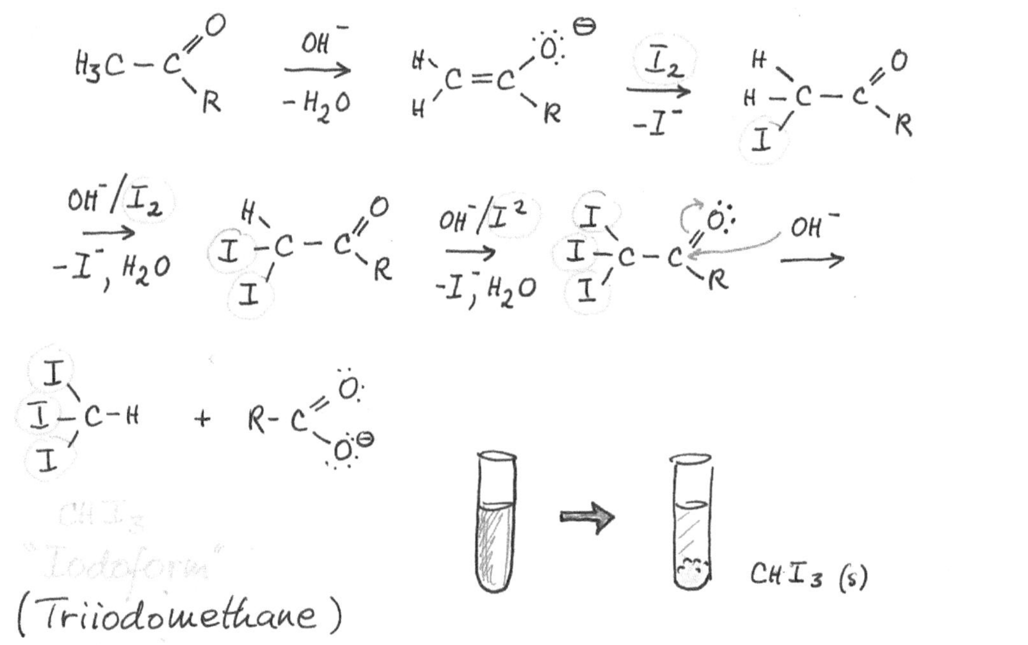


Net Reaction:

$$CH_3\text{-}CO\text{-}R + \overset{\pm 0}{3I_2} + 4\,NaOH \rightarrow \overset{-I}{CHI_3} + \overset{-I}{3NaI} + R\text{-}COONa$$

**Example Problem 6.7**
**Which of the following molecules would give a positive Tollens reaction?**

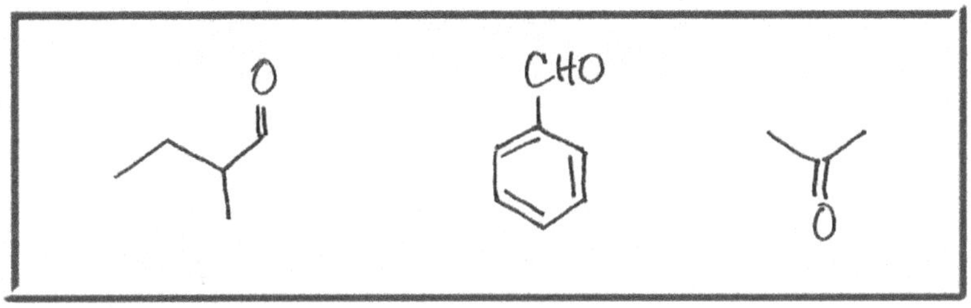

**Example Problem 6.8**
**Which of the following molecules would produce a red orange precipitate when treated with Benedict or Fehling reagent?**

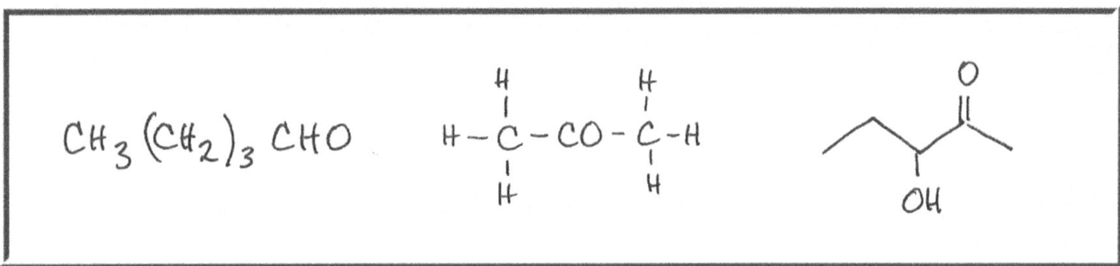

**Example Problem 6.9**
**Which of the following molecules would give a positive Iodoform reaction?**

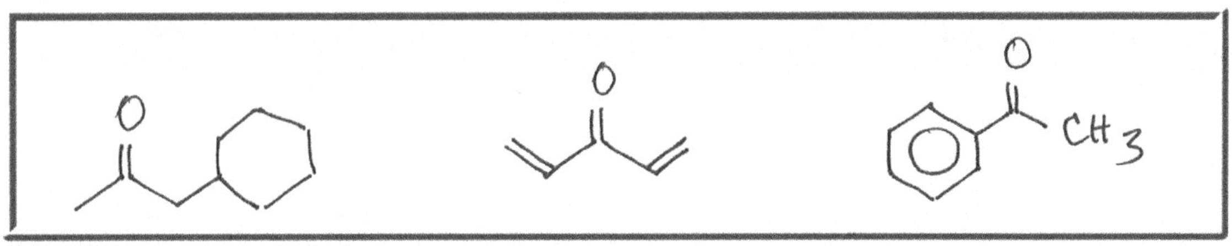

Aldehydes and ketones can also be reduced to the 1° and 2° alcohol, respectively.

The boiling points of aldehydes and ketones are between the ethers and the alcohols (alkanols). Aldehydes such as formaldehyde and acetaldehyde are colorless gases at room temperature (~20° C). The higher homologs are colorless liquids with faint acrid odors. The aromatic aldehydes have a pleasant, spice-like odor (benzaldehyde, cinnamaldehyde, etc.) and are usually solid.

The following table shows some physical properties such as melting point (Mp.), boiling point (Bp.) and water solubility of selected aldehydes and ketones:

**Table 6.3**

| Aldehyde/Ketone | Mp. (°C ): | Bp. (°C): | Water Solubility: |
|---|---|---|---|
| **Formaldehyde (methanal) $CH_2O$** | **-92** | **-21** | + |
| **Acetaldehyde (ethanal) $CH_3CHO$** | **-121** | **21** | + |
| **Propylaldehyde (propanal) $CH_3CH_2CHO$** | **-81** | **49** | + |
| **Propenylaldehyde (propenal) $CH_2$=CHCHO** | **-87** | **52** | + |
| **Butylaldehyde (butanal) $CH_3CH_2CH_2CHO$** | **-99** | **76** | (+) |
| **Benzaldehyde (benzenecarbaldehyde) $C_6H_5CHO$** | **-26** | **179** | (± ) |
| **Acetone (propanone) $CH_3COCH_3$** | **-94** | **56** | + |
| **Ethylmethy ketone (butanone) $CH_3COCH_2CH_3$** | **-86** | **80** | + |
| **Cyclobutanone $C_4H_6O$** | | **99** | + |
| **Cyclopentanone $C_5H_8O$** | **-58.2** | **130** | + |
| **Cyclohexanone $C_6H_{10}O$** | **-45** | **155** | + |
| **Acetophenone (methylphenylketone) $C_6H_5COCH_3$** | **20.5** | **202** | (± ) |
| **Diacetyl (butanedione) $CH_3COCOCH_3$** | | **88** | + |
| **2,4-pentanedione $CH_3COCH_2COCH_3$** | **-23** | **139** | + |

*+ very soluble, (+) slightly soluble, (± ) very slightly soluble*

Phenyl groups and longer hydrocarbon moieties tend to lower the water solubility.
Aldehydes and ketones serve as hydrogen bond acceptors, thereby allowing good solubility and miscibility with water.

**Fig. 6.28**

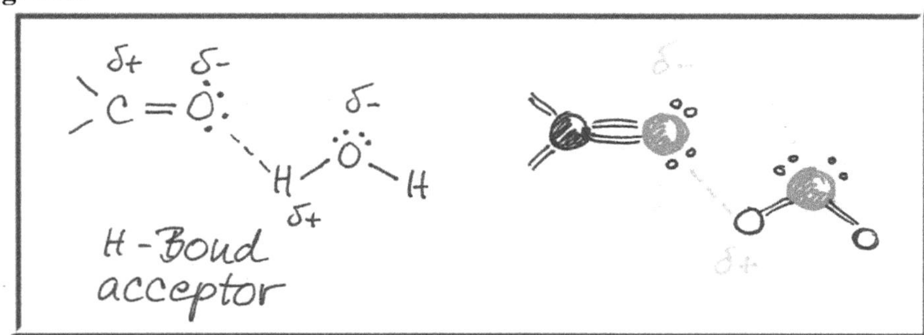


Due to their high chemical reactivity, aldehydes have many applications for chemical synthesis in laboratory and industry. They play an important role as intermediates in the manufacture of resins, plasticizers, solvents, dyes, and pharmaceuticals. The unsaturated aldehydes, like acrolein and crotonaldehyde, are tear-producing chemicals known as **lacrimators.**

**Fig. 6.29**

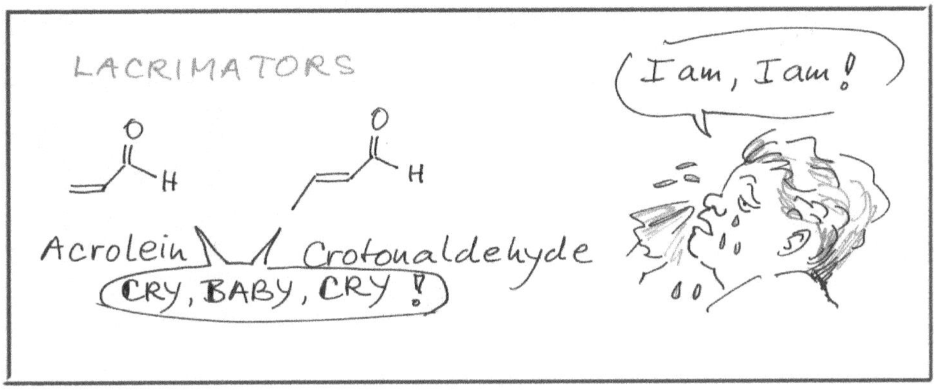


Chloral is trichloroethanal, which can add water to a relatively stable form called **chloralhydrate.** Mixed with alcohol, it becomes a powerful "knock-out" drug called a "Mickey-Finn"

**Fig 6.30**

Normally, **geminal hydroxyl groups** attached to the same carbon are not very stable, instantly separating out as water and returning to the aldehyde. The strong electron-withdrawing power of the three chlorine groups (EN = 3.0) keeps the hydrate stabilized.

**Fig. 6.31**

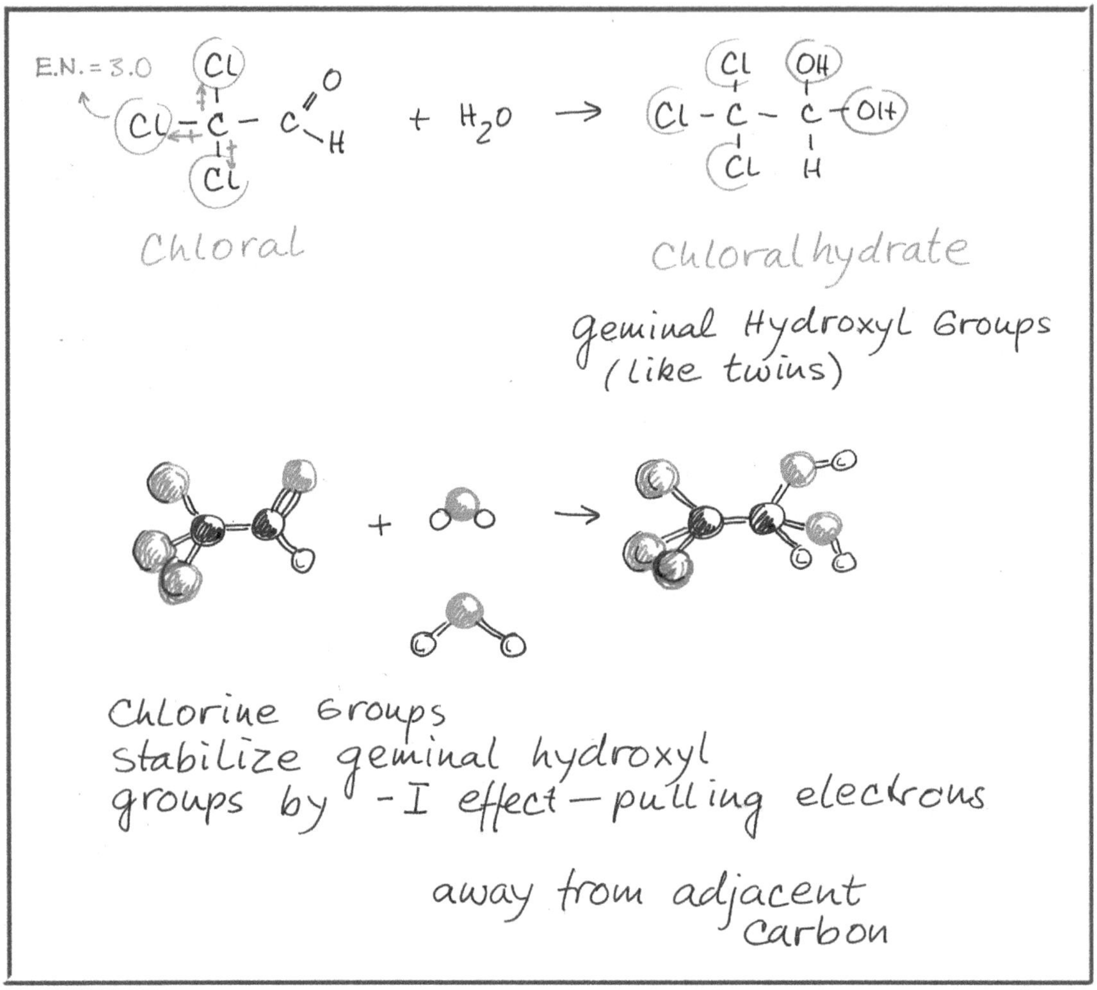


Besides oxidation and reduction reactions, aldehydes can undergo many preparatory valuable reactions, such as the **aldol addition,** producing **"ald**ehyde-alcoh**ols",** that can continue via condensation to more derivatives. This provides a valuable method for creating new carbon-carbon bonds.

**Fig. 6.32**

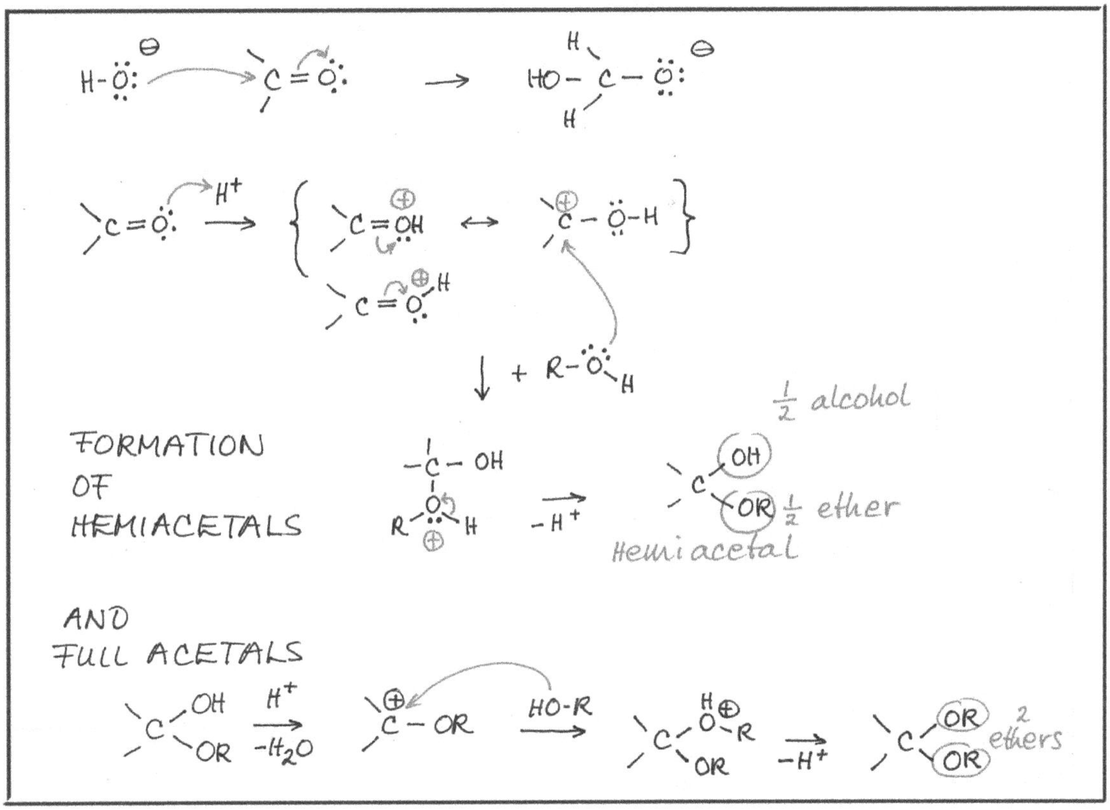


Of special importance are the **hemiacetal-** and **acetal**-forming nucleophilic additions, which will receive special coverage in Ch.25 (Carbohydrates ).

The **Canizzaro reaction** produces via redox reactions both a carboxylic acid and an alcohol (disproportionation), when treating an aldehyde with a strong base such as KOH.
Most addition reactions concerning the carbonyl compounds – aldehydes and ketones - proceed via **nucleophilic addition ($A_N$)**, which is a fundamental reaction type (see also Ch. 11).

**Fig 6.33**

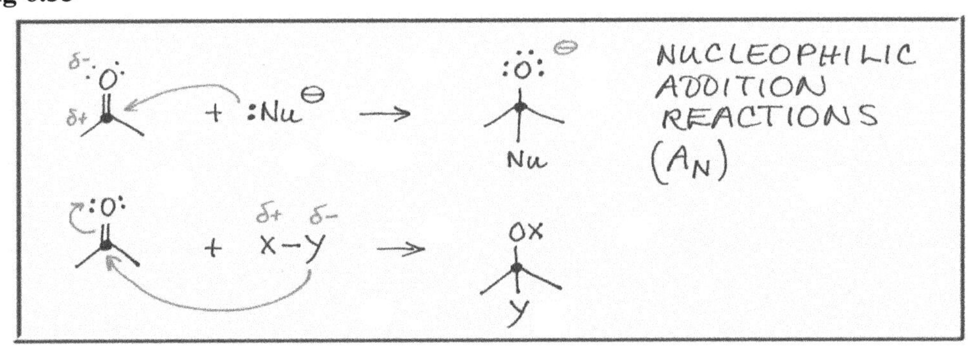

**Example Problem 6.10**
**Complete the following reactions and indicate what type of reaction is involved:**

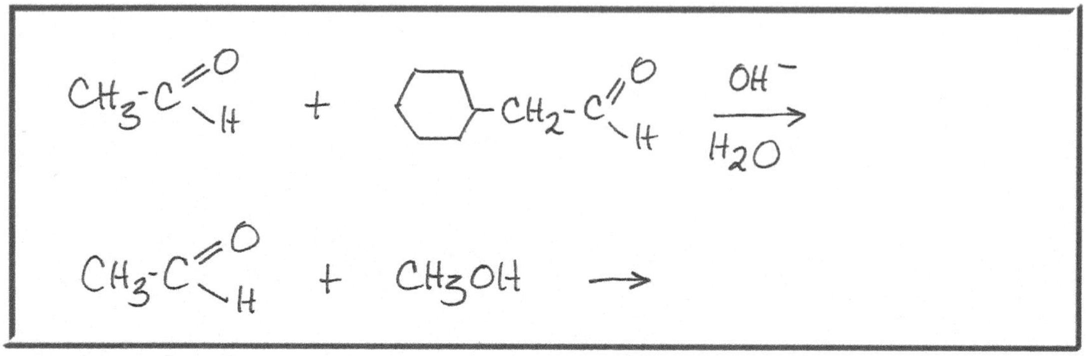

**Preparatory methods** for aldehydes and ketones can consist of:

1. **Catalytic dehydrogenation of 1° alcohols or 2° alcohols**

**Fig. 6.34**

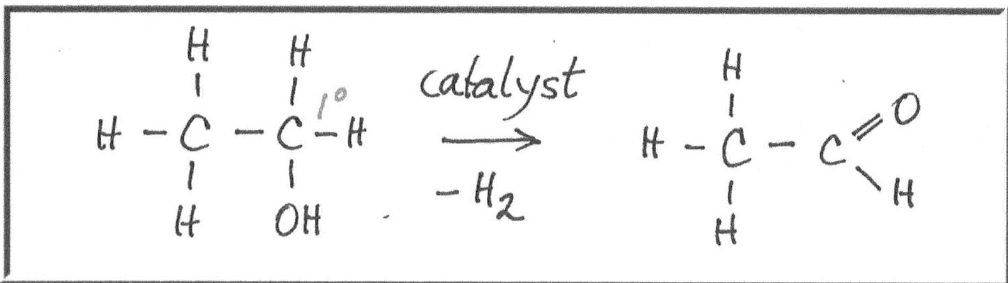

2. **Oxidation of primary alcohols (aldehydes) or secondary alcohols (ketones)**

**Fig. 6.35**

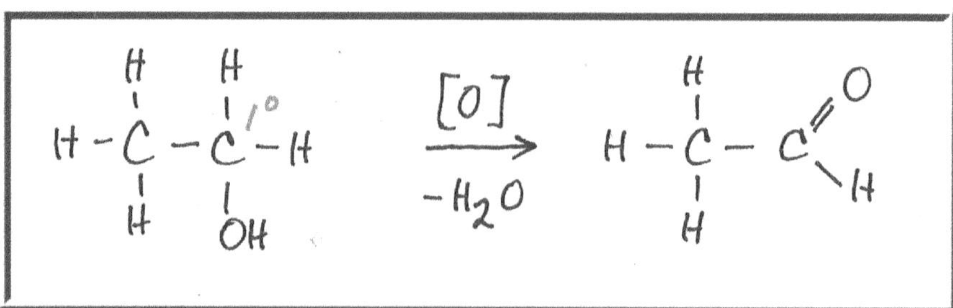

3. **Oxidation of alkenes ("olefins")**

Fig. 6.36

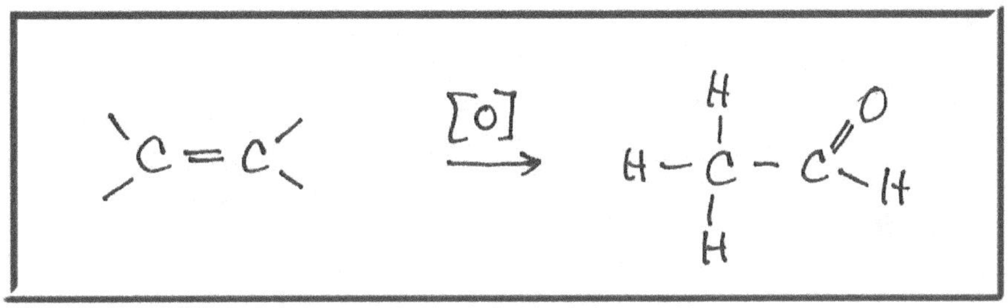


**4. Hydroformylation (catalytic addition of formaldehyde) of alkenes (olefins) ($A_E$)**

Fig. 6.37

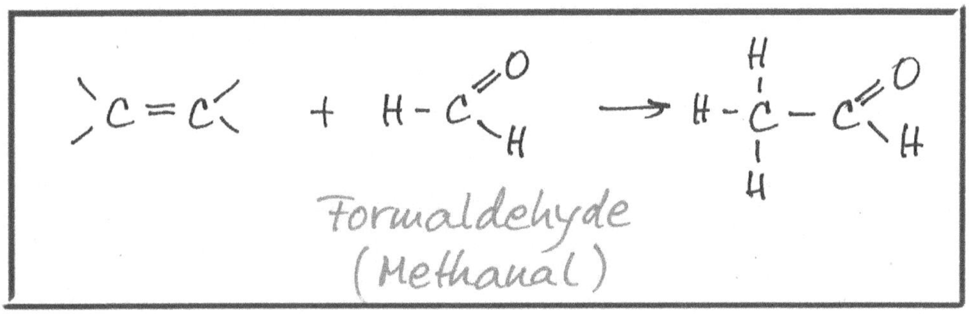


**Example Problem 6.11**
**Complete the following preparatory reactions leading to a ketone or aldehyde using the indicated methods:**

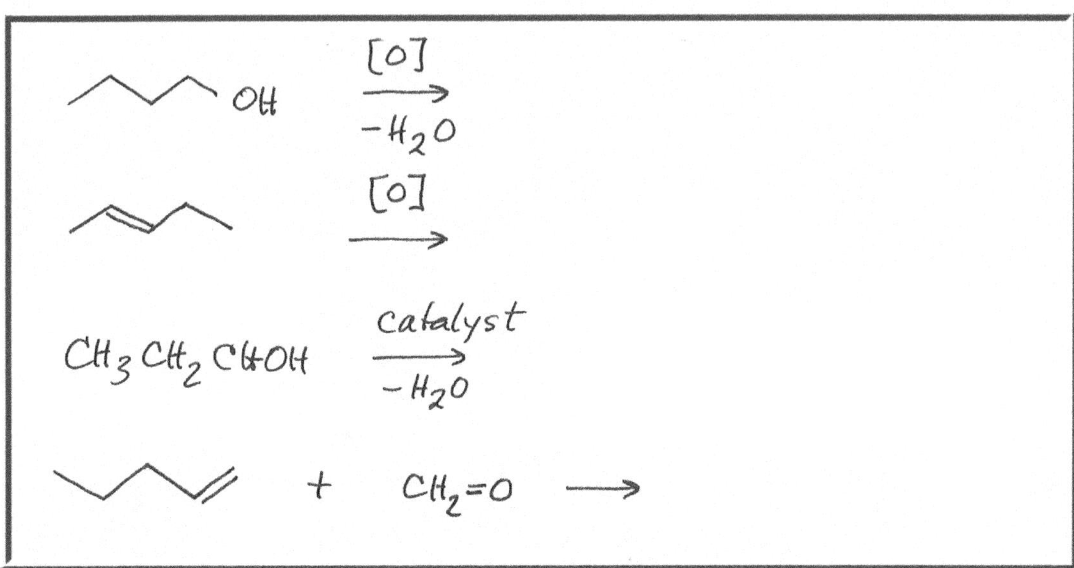


Of the ketones, acetone and methylethylketone are universal solvents and water miscible. They have the pungent, acetone-like odor typical of nail-polish remover.

**Fig. 6.38**

They are not as reactive as the aldehydes, but can still undergo many addition reactions typical of the carbonyl group. Besides the liquid types used mainly as solvents, ketones are often used as flavoring agents, as well as aroma essences in the perfume industry (muscone). **Butanedione** is the chemical that gives butter its characteristic flavor. It is artificially added to popcorn to give it a buttery flavor but without the associated fat. It has been held responsible for some adverse reactions such as asthma-like attacks that are initiated after the consumption of artificially buttered popcorn.

**Fig. 6.39**

Associated with diabetes, the so-called **ketone-bodies** are often released into the blood stream including high levels of acetone. Increased amount of acetone, for instance, can also be associated with fasting and high-protein diets.

**Uric acid**, a metabolite often associated with these conditions, is a cyclic ketone responsible for the symptoms of gout. Strictly speaking it belongs to the class of the cyclic amides, molecules containing the –CONH- group (see chapter 19).

**Fig. 6.40**

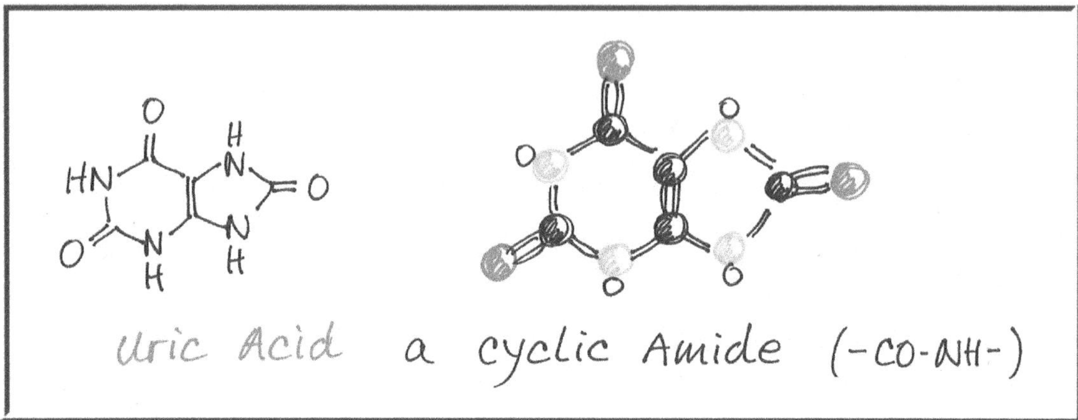


Ketone bodies occur in the plasma after three days of starvation. They are acetoacetate and its reduction product 3-hydroxybutyrate.

They substitute for glucose as fuels for the muscles and brain and are produced in the liver by fatty acid metabolism. Decarboxylation (-$CO_2$) converts acetoacetate into acetone.

**Fig. 6.41**

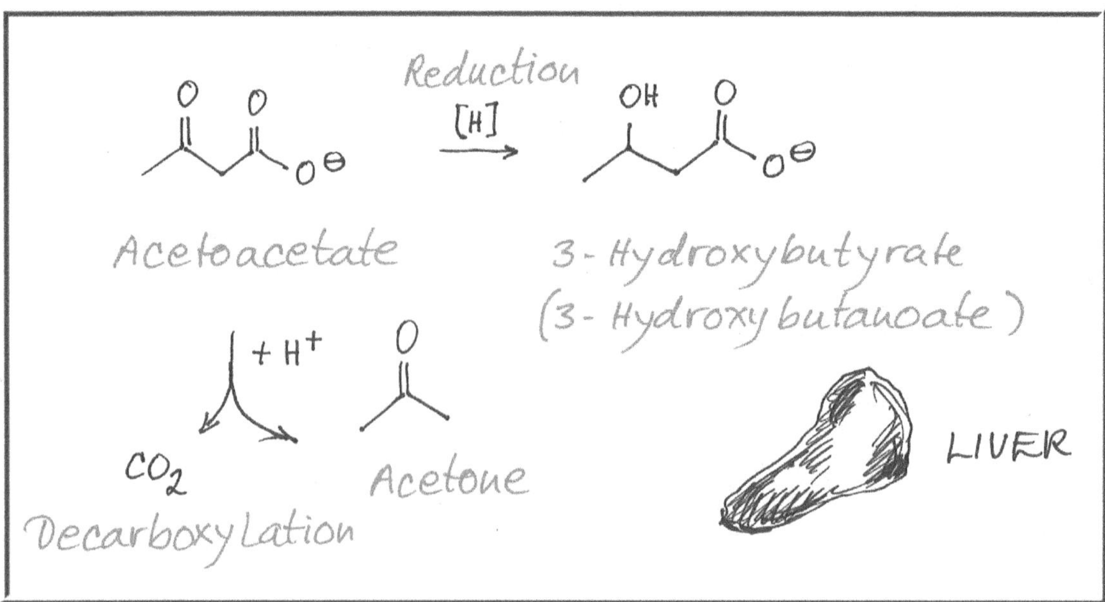

# EXPERIMENTS – Ch. 6

## *Aldehydes and Ketones*

**Key Terms:** The carbonyl group, resonance and polarity of the group, keto-enol tautomerism (intra-molecular proton migration), oxidation reactions, benedict reagent, Iodoform test, methylketones, nomenclature, hydrogen bond acceptor properties.

## I Molecular Modeling of Aldehydes and Ketones

### Chemicals:

*None*

### Equipment and Materials:

*Molecular model kit (Molymod)*

### Procedure:

1. Link up three carbon atoms and attach a double bonded oxygen atom to an end carbon atom (terminal carbon) and saturate the rest of the bonds with hydrogen atoms.
2. Remove the double bonded Oxygen and situate it on the middle carbon atom, replacing and moving the hydrogen atoms to their new positions.

### Observations and Results:

Record names and the different formulas in the table below (MF molecular formula, ESF expanded structural formula, CSF condensed structural formula, ASF/LAF abbreviated structural formula or line angle formula):

**Aldehyde/Ketone:**

| Common Name: | IUPAC Name: | MF | ESF | CSF | ASF/LAF |
|---|---|---|---|---|---|
| | | | | | |
| | | | | | |

## Questions and Problems:

1. Are the two constructed molecules constitutional (structural) isomers? Explain.
2. Draw a resonance structure to each molecule and insert the proper formal charges.

# II Physical properties (color, odor), Relative Solubility and Miscibility of Aldehydes and Ketones

## Chemicals:

*Sample aldehydes and ketones*
*Acetaldehyde, acetone, acetophenone,*
*Undecanone, benzaldehyde, citral*

*Water ($H_2O$, demineralized)*
*Ethanol*
*Hexane*
*Diethyl ether*
*Dichloromethane*

## Equipment and Materials:

*Test tubes and rack*
*Droppers with bulb*

## Procedure:

1. Observe the given sample of aldehydes and ketones and record color and odor as well the physical state.
2. Prepare separate test tubes for each sample aldehyde and ketone. Fill with 1-2 ml of each sample solution or a spatula tip if a solid is involved, and add the first solvent (1-2 mL) to each sample, swirl and watch for solubility or miscibility.
3. Repeat the procedure above with fresh solutions in cleaned out test tubes and add the second solvent to each sample. Repeat for the other two solvent and record all results in the table below.

## Observations and Results:

Physical characteristics and relative solubility of aldehydes and ketones:

| **Sample aldehyde or ketone:** | Color | Odor | Physical state | Ethanol solubility | Hexane solubility | Diethyl ether solubility | Dichloro-methane solubility |
|---|---|---|---|---|---|---|---|
| **Acetaldehyde** (ethanal) | | | | | | | |
| **Acetone** (dimethyl ketone, methoxymethane) | | | | | | | |

| | | | | | | | |
|---|---|---|---|---|---|---|---|
| **Undecanone** (dog/cat repellant) | | | | | | | |
| **Benzaldehyde** | | | | | | | |
| **Citral** (oil of lemon grass) | | | | | | | |
| **Other** | | | | | | | |

## III Chemical Identification Tests for Aldehydes and Ketones and their Distinction

### Chemicals:

*Sample aldehydes and ketones*
*Acetaldehyde, acetone, acetophenone,*
*Undecanone, benzaldehyde, citral*

*Benedict reagent ($CuSO_4$/sodium citrate)*
*Iodoform reagent ($I_2$/NaOH)*
*Tollens reagent ($AgNO_3$ in ammonia $NH_4OH$)*

### Equipment and Materials:

*Test tubes and rack*

*Droppers with bulb*
*Hot water bath (stand, ring clamp, wire gauze)*
*Beaker (large)*

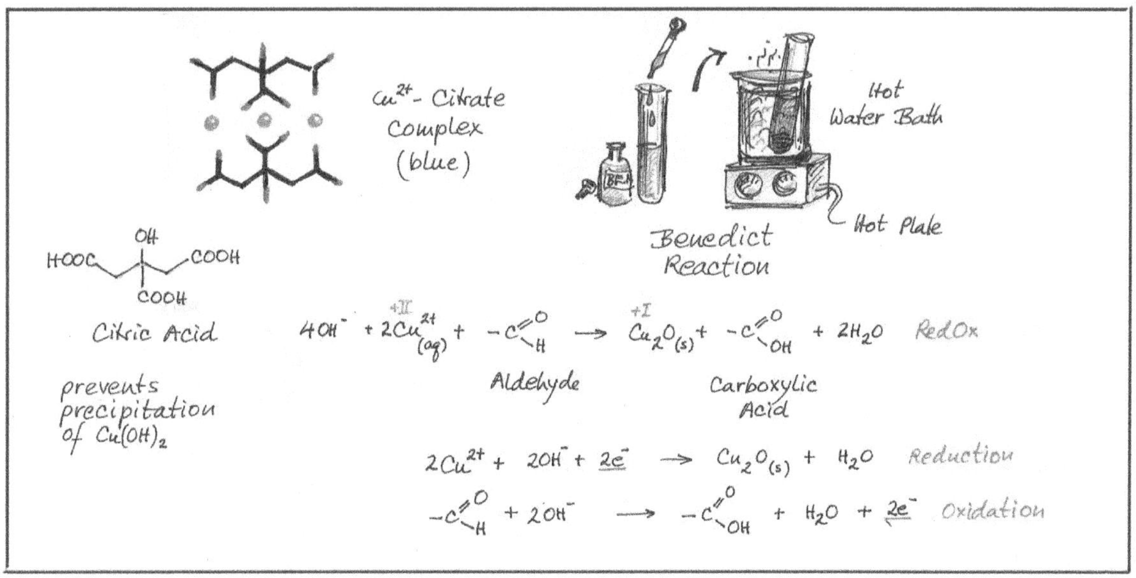


### Procedure:

1. Prepare a hot water bath.
2. Set up test tubes in a rack and fill each with a separate sample of each listed aldehyde and ketone.

3. Add a few drops of Benedict reagent, heat and watch for the formation of a red $Cu_2O$ precipitate
4. Empty out all test tubes, rinse and dry, and refill them with the same sample solutions as above. To each sample add 10 drops of 10% NaOH solution, heat shortly in a warm water bath (50-60 °C) and add about 20 drops of the iodine reagent and watch for the development of a yellow precipitate. This "Iodoform test" is typical for methyl ketones ($CH_3$-CO –R)
5. To a separate clean and dry test tube add 1-2 mL of acetaldehyde (ethanal)., then add a few drops of Tollens reagent and watch for the formation of a silver "mirror" on the inner surface of the test tube. (careful: $AgNO_3$ solution leaves dark stains if spilled!).
   Record the results in the following table:

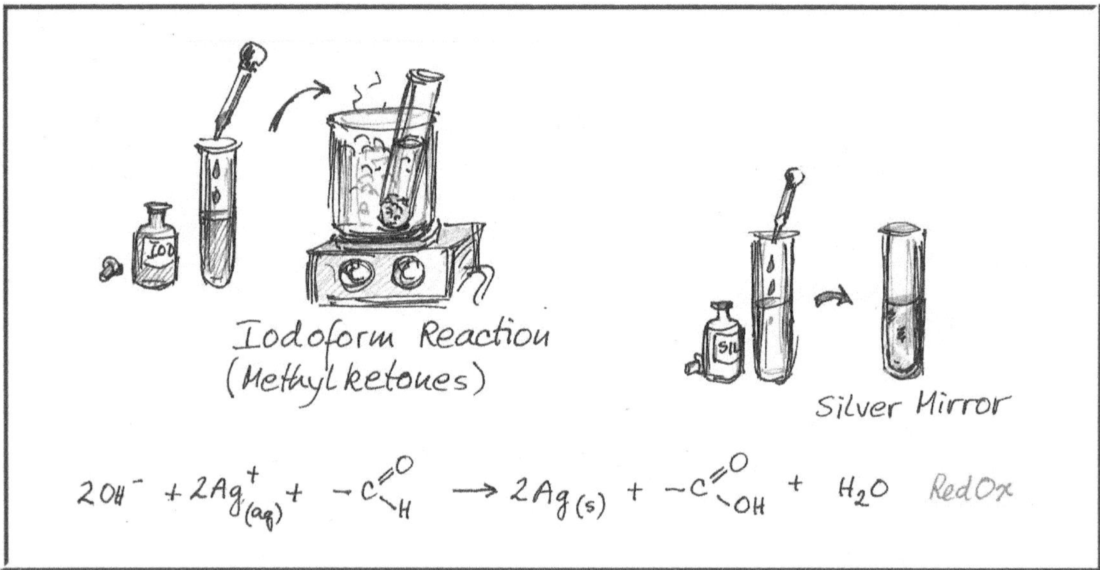


## Observations and Results:

| **Aldehyde/Ketone:** | **Benedict reaction:** | **Iodoform reaction:** | **Tollens reaction:** |
|---|---|---|---|
| **Acetaldehyde** (ethanal) | | | |
| **Acetone** (dimethyl ketone, methoxymethane) | | | |
| **Acetophenone** (ethyl methyl ketone, methoxyethane) | | | |
| **Undecanone** (dog/cat repellant) | | | |

| | | | |
|---|---|---|---|
| **Benzaldehyde** | | | |
| **Citral** (oil of lemon grass) | | | |

*Do not perform for shaded areas.*

## Questions and Problems:

1. Which ketones gave a positive iodoform reaction? Draw their structural formulas.
2. An unknown deterrent molecule was found in an ant. It could be a 2-octanone or a 3-octanone. What chemical test could distinguish between the two possibilities?
3. An artist wants to give a silver mirror to all his glass sculpture pieces and intends to use acetone and $AgNO_3$ solution for this purpose. What would you tell him?

# Chapter 7 Carboxylic Acids and Esters

## Ch. 7 Carboxylic Acids and Esters

Carboxylic Acids are the organic acids that act as proton donors and the amines the organic bases serving as proton acceptors. They are the organic version of acid and base opposites with the ability via proton transfer to form salt and water.
The carboxylic acid functional group consists of a carbonyl group and a hydroxyl group connected to the same carbon.

**Fig. 7.1**

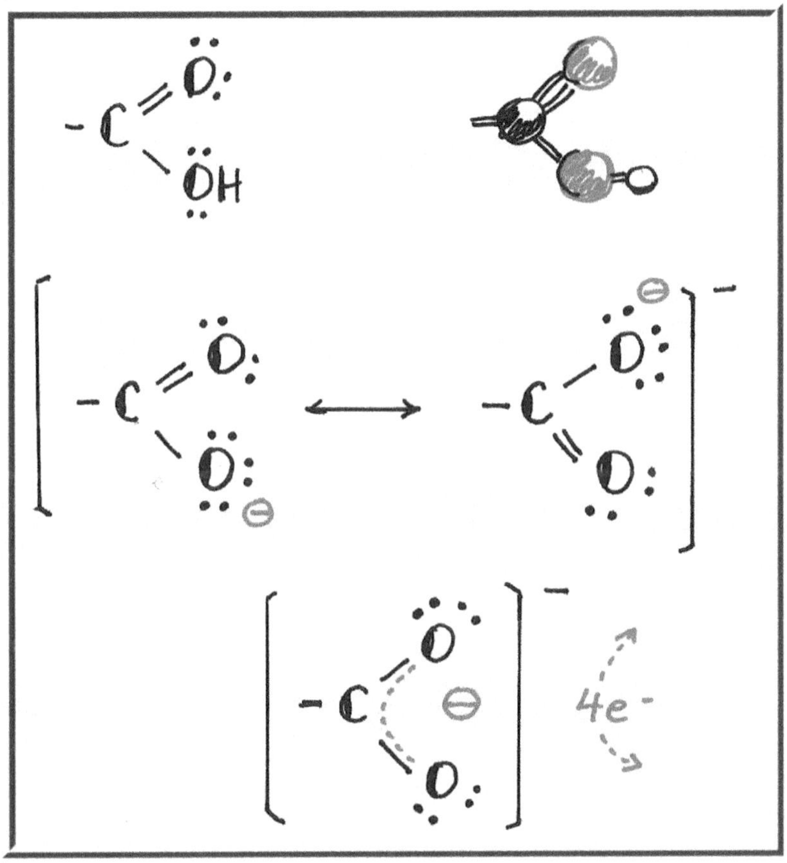


By releasing the proton, the carboxylic group becomes resonance stabilized, consisting of two resonance contributors.

The resonance hybrid for the carboxylate ion is shown by a dashed line between the three atoms O-C-O, and represents four electrons spread between them. The one negative charge is also distributed between the three atoms. This delocalization of the electrons gives the ion its special stability.

The negative charge and four electrons can also be considered to oscillate between the two oxygen atoms, the actual meaning of resonance. This desired energy-lowering state facilitates the release of the proton, making the carboxylic acids relatively good, although weak acids in comparison to the strong inorganic representatives ($HNO_3$, $H_2SO_4$). The ionization constants of simple, monovalent carboxylic acids ($K_a$) range between 3 and 5, acetic acid having a $pK_a$–value of 4.75. The acidity constant and pKa value for acetic acid are calculated as follows:

$$K_a = [H^+][CH_3COO^-]/[CH_3COOH] = 1.8 \times 10^{-5}$$
$$-\log(1.8 \times 10^{-5}) = 4.75$$

Acidity constants ($K_a$) and pKa values of various acids are listed in table 18.1.

The strong electron withdrawing effect of the carbonyl group leads to a de-shielded carbon atom (δ+) that in turn attracts electrons from the other negatively charged O-atom (+I inductive effect). The resulting "push and pull" culminates in the delocalization of the π-electrons. As already pointed out above, the equal distribution of the electron density between the "forked" O-C-O bond arrangement provides the extra stability for the carboxylate ion, a resonance energy similar to that of the aromatic compounds. The electron or **electrostatic potential maps** shown below, illustrate the even electron distribution between the two oxygen atoms of the carboxylate ion in contrast to the neutral, un-dissociated species.

**Fig. 7.2**

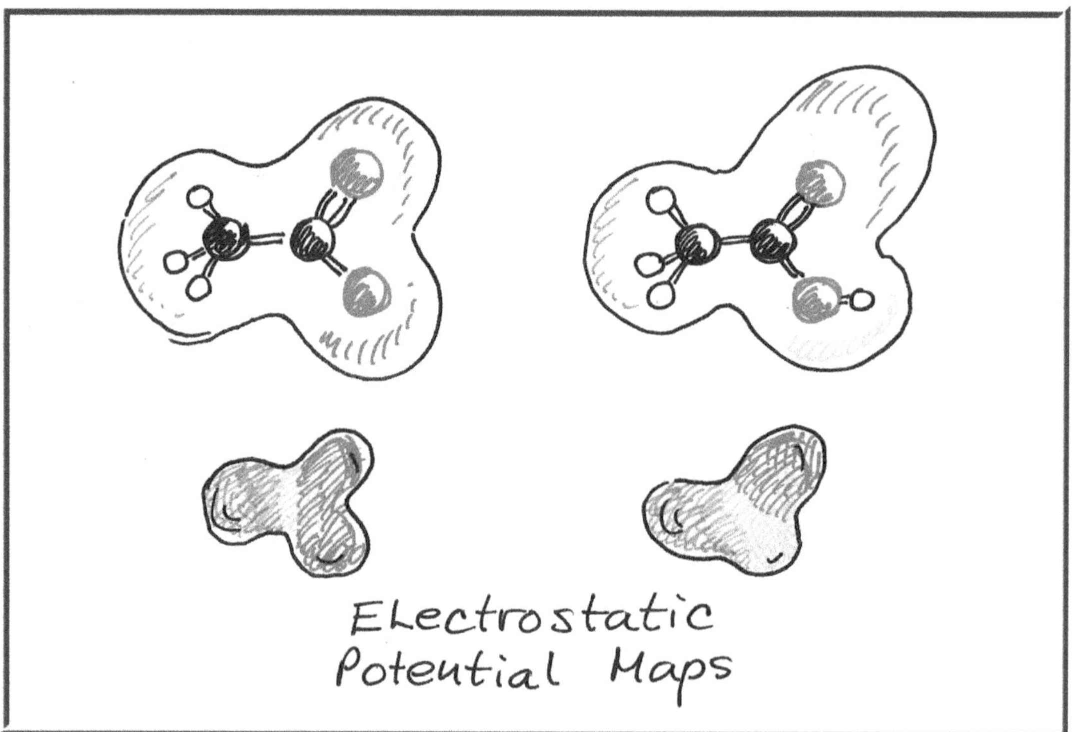


Carboxylic acids represent a homologous series with general formula R-COOH. Liquids are those with one to five carbon atoms, getting viscous with higher number of carbons such 6C hexanoic acid, better known as caproic acid giving goats their particular smell. Later they become colorless solids. They have characteristic, acrid or pungent odors. Formic acid is the repellant chemical of ants ("formica"), an "ant" acid, not to be confused with an antacid, an "anti"-acid drug preventing excess stomach acid (HCl) that can lead to heartburn.

Fig. 7.3

Many carboxylic acids have an acrid "biting", pungent smell. Acetic acid is found in vinegar – the actual oxidation product of alcohol in nature –, propionic acid (3C) is a colorless pungent smelling liquid found in degraded sugars, butyric acid (4C) smells like rancid butter and is the typical smell of unwashed underarms as well as dirty socks. Valeric acid (5C) also has a very unpleasant smell. Caproic acid (6C) along with caprylic (8C) and capric acid (10C) are all found in and on goats giving them that particular "funky" aroma. Dogs are able to trace human beings as well as other creatures by their special spectrum of organic molecules exuded from their bodies, many of which belong to the carboxylic acid group.

**Fig. 7.4**

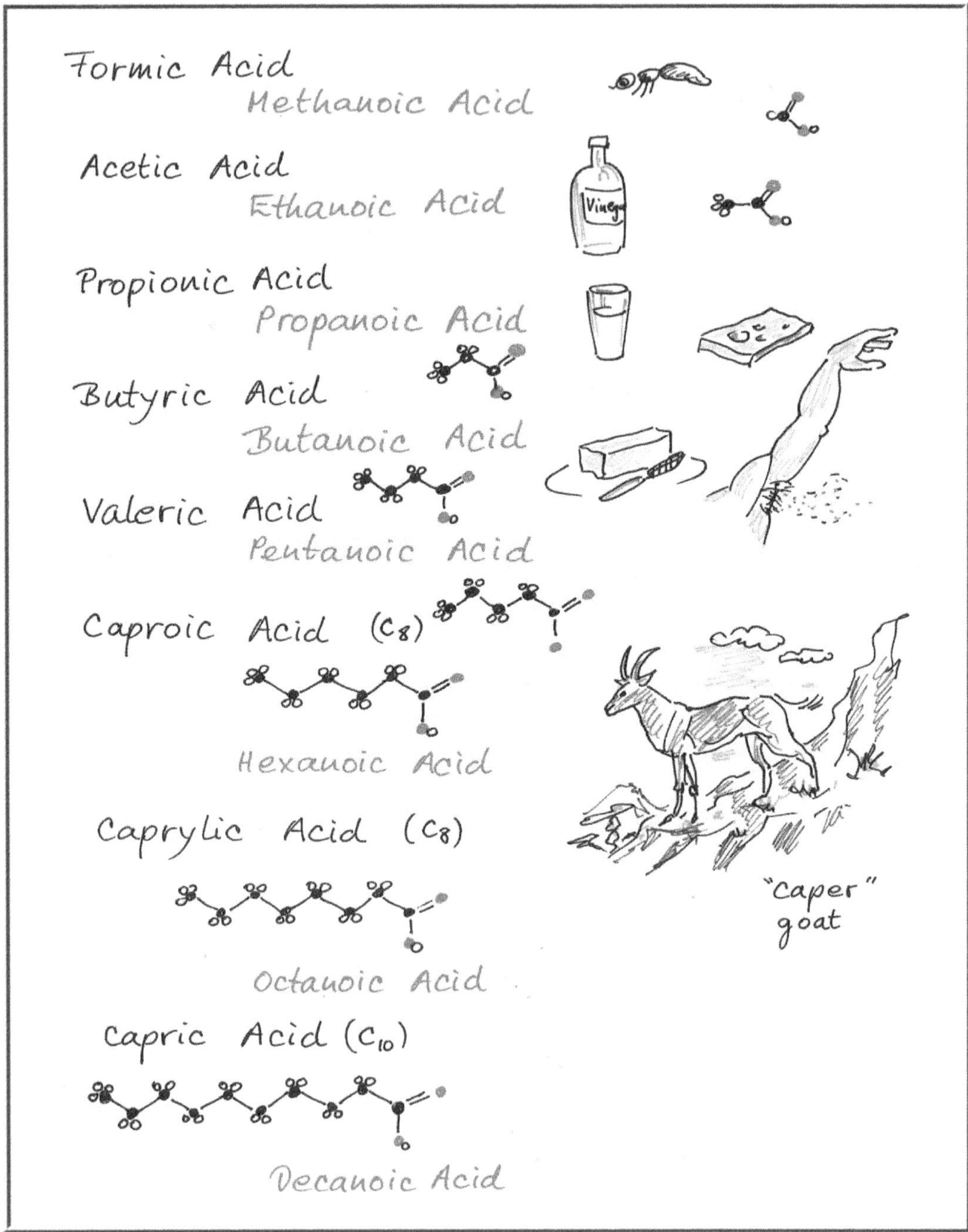


The lower molecular weight representatives are soluble in water being hydrophilic. The longer the hydrocarbon chain becomes, the more hydrophobic the molecules react, being less and less soluble in water. Some aromatic carboxylic acids, even though they are polar, show no signs of being water-soluble. The best remedy for making them water-soluble is to convert them into the

corresponding salt by adding a strong, inorganic base. Most organic salts are easily dissolved in water.
Carboxylic acids have relatively high conversion points (Bp, Mp), due to their tendency to form intra or intermolecular hydrogen bonds.

**Fig.7.5**

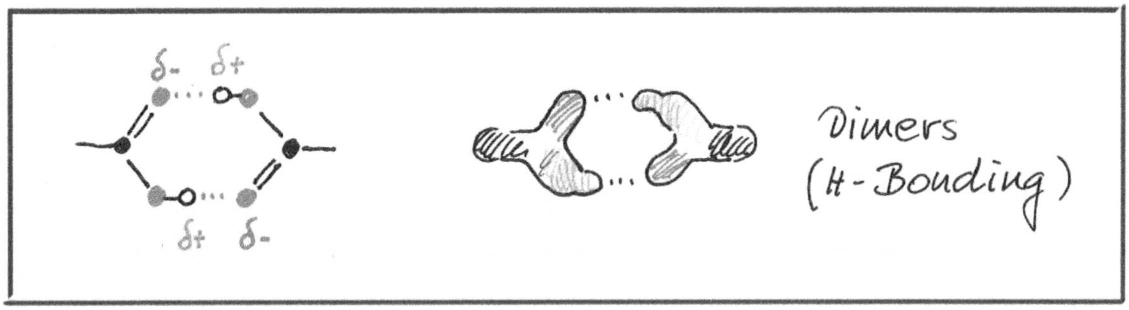


The acidity can be increased by electron withdrawing groups attached to the α-carbon (Carbon atom adjacent to the functional group) such as halogen, alkoxy, or nitro groups. Trifluoroacetic acid (TFA) has a much higher acidity than normal acetic acid ($pK_a = 0.23$). It is actually 100,000 times more acidic. Groups containing multiple bonds (π-systems), such as alkenyl or phenyl groups, have a positive influence on acidity, but to a weaker extent.
Introduction of electron-withdrawing groups such as halogens – fluorine being the most powerful – increases the acidity, due to the strong negative induction effect (-I) on the carboxylic group. Other groups of this sort that enhance acidity is the nitro group ($-NO_2$), any carbonyl-containing group (ketone, aldehyde, carboxylic, amido, ester groups) as well as the cyano or nitrile group (-CN). The longer the hydrocarbon chain of the carboxylic acid is, the less water soluble the acid becomes. "Fatty acids", which will be discussed in more detail in the Lipid chapter (Ch.26) are completely insoluble in water becoming more and more hydrophobic or lipophilic.

**Fig. 7.6**

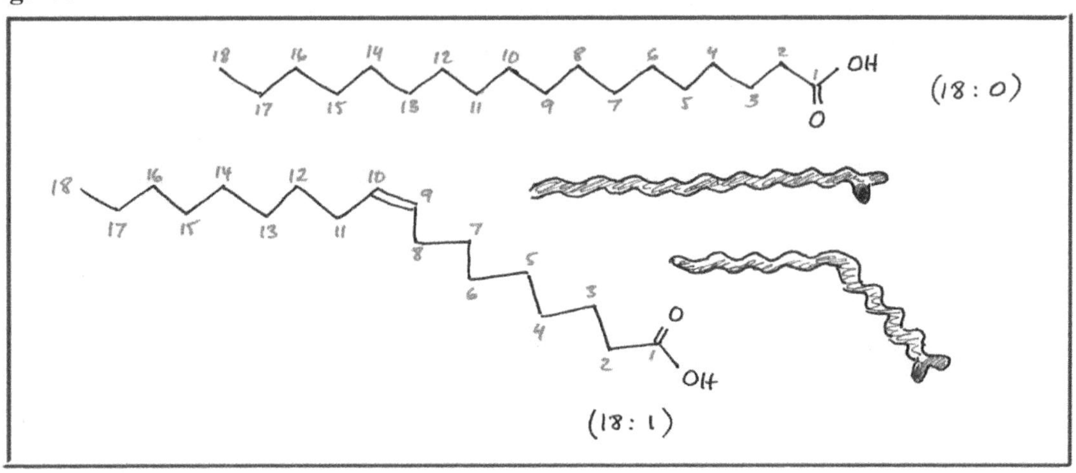

The following table shows some of the physical properties of mono-valent carboxylic acids and their derivatives:

**Table 7.1**

| **Acid IUPAC name; common name ( )** | **Structure (CSF)** | **Mp.** | **Bp.** | **Water solubility** | **$K_a$** | **$pK_a$** |
|---|---|---|---|---|---|---|
| Mathanoic (formic) | $HCOOH$ | 8.4 | 100.7 | +++ | $1.77\times 10^{-4}$ | 3.77 |
| Ethanoic (acetic) | $CH_3COOH$ | 16.6 | 118.1 | +++ | $1.77\times 10^{-5}$ | 4.75 |
| Propanoic (propionic) | $CH_3CH_2COOH$ | -22 | 141.1 | +++ | $1.3\times 10^{-5}$ | 4.89 |
| Butanoic (butyric) | $CH_3CH_2CH_2COOH$ | -8 | 163.5 | +++ | $1.5\times 10^{-5}$ | 4.82 |
| Methylpropanoic (isobutyric) | $CH_3CH(CH_3)COOH$ | -47 | 154.5 | + | $1.4\times 10^{-5}$ | 4.85 |
| Pentanoic (valeric) | $CH_3(CH_2)_3COOH$ | - 34.5 | 187 | (+) | $1.6\times 10^{-5}$ | 4.80 |
| Trichloroethanoic (trichloroacetic) | $CCl_3COOH$ | 58 | 195.5 | + | $0.7\times 10^{-1}$ | 0.70 |
| Trifluoroethanoic (trifluoroacetic acid | $CF_3COOH$ | -15 | 72.4 | +++ | strong | -0.25 |
| Benzenecarboxylic (benzoic) | $C_6H_5COOH$ | 122 | 249 | (+) | $6.5\times 10^{-5}$ | 4.19 |

Many carboxylic acids are still known by their common names (formic acid, acetic acid, butyric acid, caproic acid, etc.). **IUPAC nomenclature** is based on the parent alkane, alkene, or alkyne substituting the ending **–e** with the ending **-oic acid**. Acetic acid thus becomes **ethan**(e)**oic acid.** A six carbon, mono-valent acid becomes **hexanoic acid.** A four-carbon acid with a double bond originating at carbon C-2 would be called **2-butenoic acid.**

**Fig.7.7**

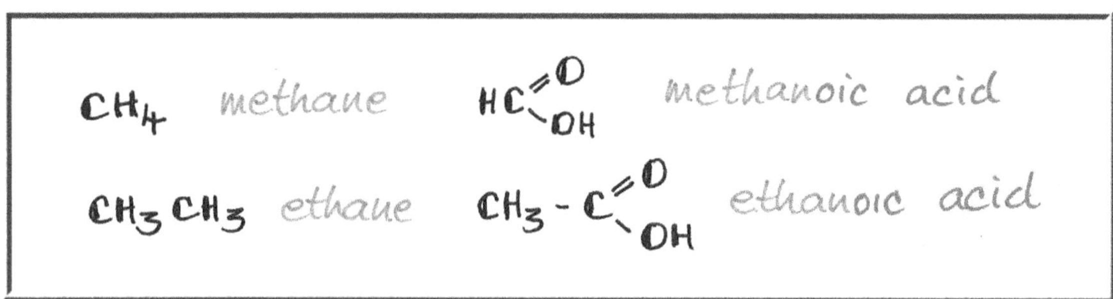

**Example problem 7.1**
**Which of the following pairs represents the stronger acid?**

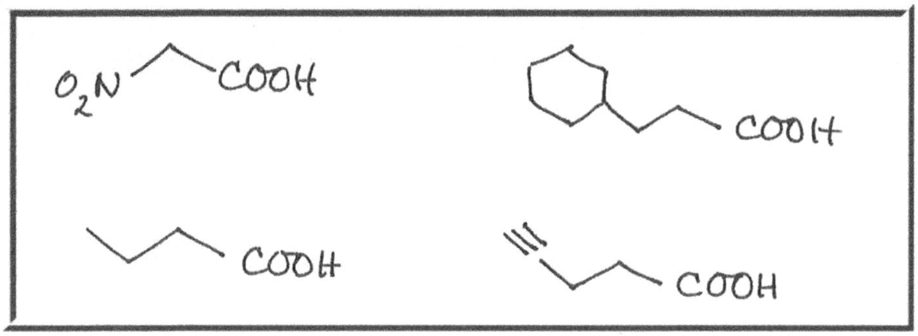


**Example problem 7.2**
**Name the following molecules using the IUPAC system.**

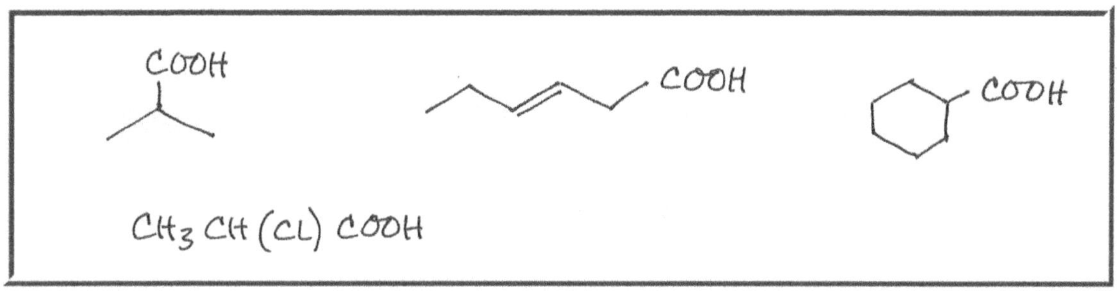


The carboxylic carbon automatically counts as number one, so branched carboxylic acids would be named accordingly, for instance **3-methylbutanoic acid** would have the following structure:

**Fig.7.8**

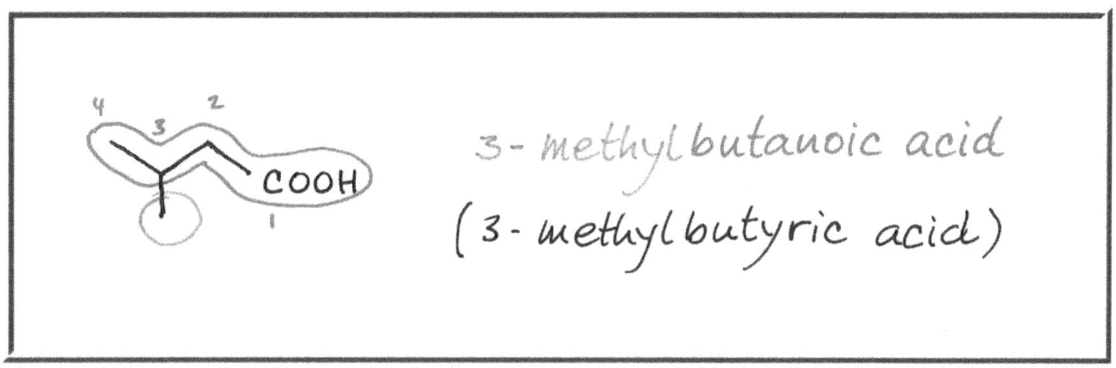


**Example Problem 7.3**
**Give the chemical structures to the following names:**
***3-butenoic acid***
***5-Chlorodecanoic acid***

***Cyclohexylacetic acid***
***2-Carboxybutane***

Since carboxylic acids can form salts with inorganic and organic bases, the resulting anions have their own names.

**Fig.7.9**

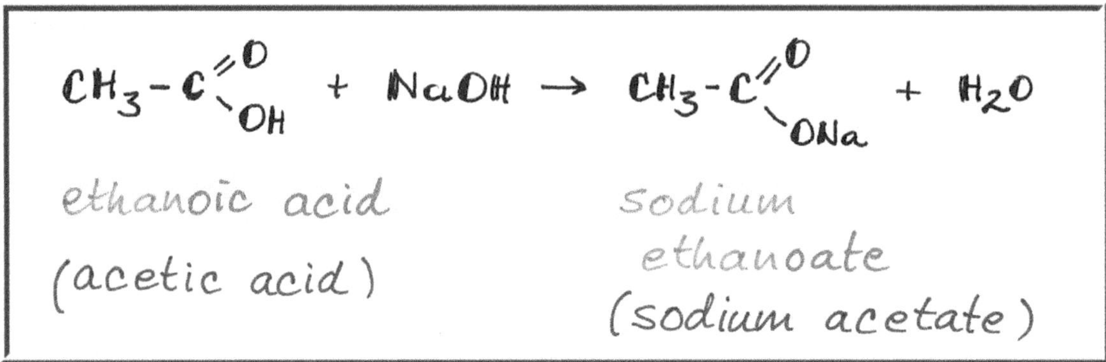


Here again the hydrocarbon is named in full, having the "e" dropped from the end and substituted by the ending **-oate**.

**alkanoates**
**alkenoates**
**alkynoates**

A salt produced by the reaction of ethanoic acid with NaOH would be named **sodium ethanoate.** A salt with an amine such as ethyl amine and the same acid would be called **ethylammonium ethanoate.**

**Fig.7.10**

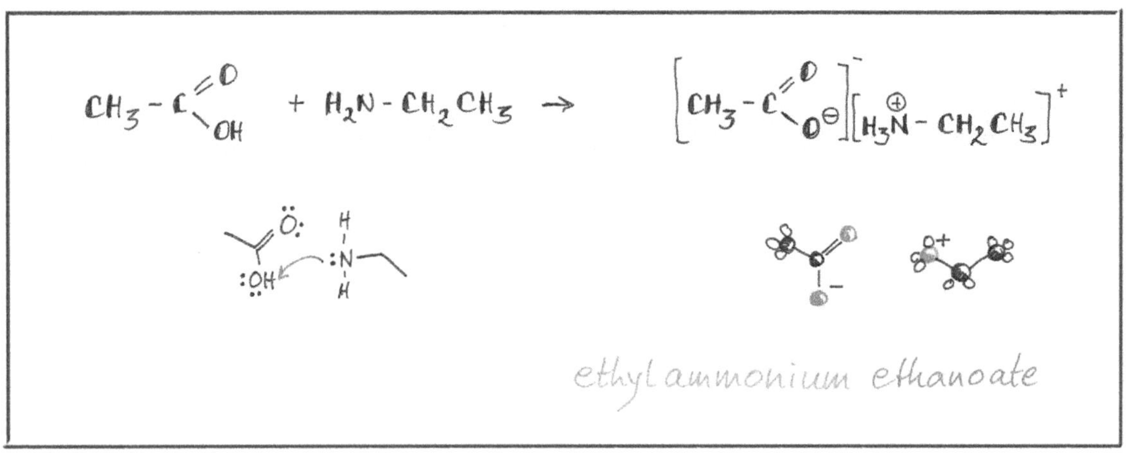

**Example Problem 7.4**
**Give the structures of the following salts:**
***Sodium acetate***
***Magnesium methanoate***
***Potassium isobutyrate***
***Lithium octanoate***

**Example problem 7.5**
**Name the following salts:**

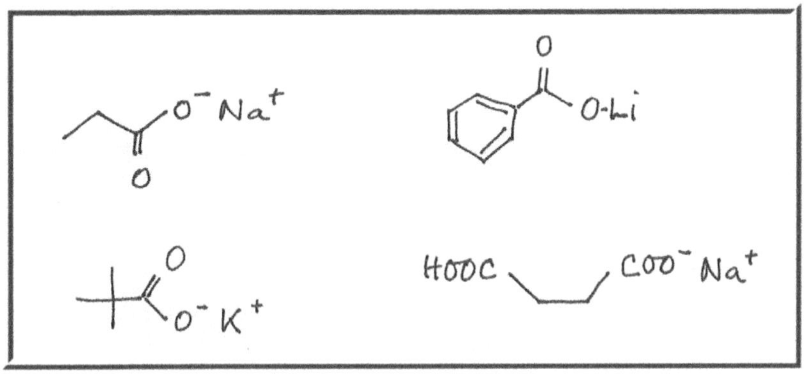

Carboxylic acids can be divided into mono-, di-, or tri-valent acids, depending on the number of -COOH groups involved in the molecule. The terms monoprotic, diprotic, triprotic all the way up to polyprotic can also be used. Here the number of carboxylic functional groups directly relates to the number of protons that can be released per molecule.
Important divalent acids are oxalic acid, terephtalic acid and malonic acid for instance. One important trivalent acid is citric acid.

**Fig. 7.11**

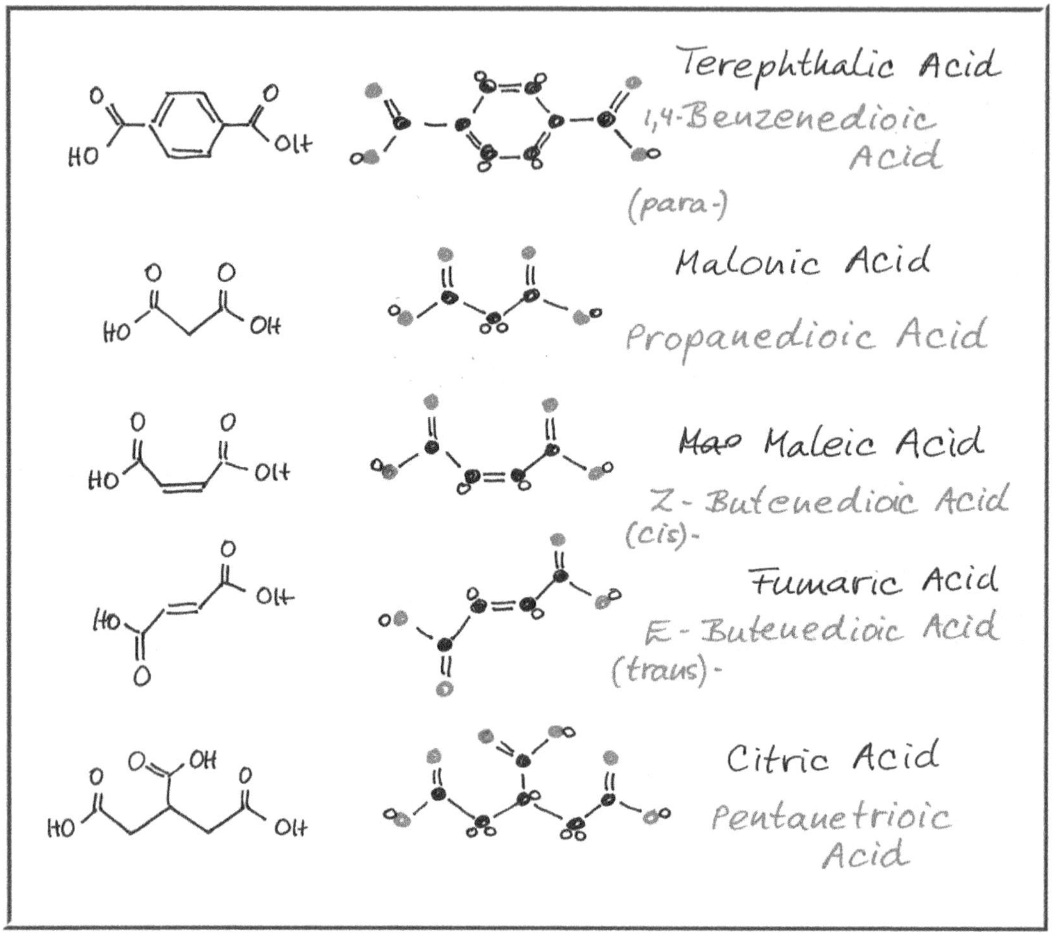


The closer the carboxylic functions are located within the molecule, like on adjacent carbons, the higher the compounded inductive effect becomes, ultimately leading to a lower $pK_a$ and thus higher acidity of the first carboxylic group.

**Fig. 7.12**

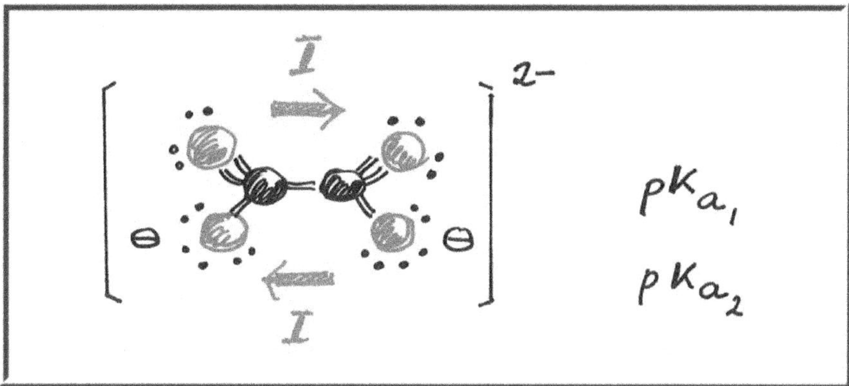


The others are slightly more basic. Each carboxylic function of multifunctional molecules is characterized by its own pK-value.

Some physical data of some important di-carboxylic acids are shown in the following table:

**Table 7.2**

| **Acid IUPAC name; common name ( )** | **Structure (CSF)** | **Mp.** | **$K_a$ (25℃ ) $K_{a1}\times 10^{-5}$ $K_{a2}\times 10^{-5}$** | **$pK_a$-values** |
|---|---|---|---|---|
| Ethanedioic (oxalic acid) | HOOC-COOH | 189 | 3500<br>5.3 | 1.23<br>4.18 |
| Propanedioic (malonic) | HOOC-$CH_2$-COOH | 136 | 171<br>0.22 | 3.77<br>5.66 |
| *Cis*-butenedioic (maleic) | HOOC-CH = CH -COOH | 130 | 1170<br>0.026 | 3.10<br>6.59 |
| 1,2-Benzenedicarboxylic (phthalic) | HOOC-$C_6H_4$-COOH | 231 | 130<br>0.39 | 3.89<br>5.41 |
| 1,4-Benzenedicarboxylic (terephthalic) | HOOC-$C_6H_4$-COOH | > 300 | 31<br>1.5 | 3.51<br>4.82 |

Acids with more than one functional carboxyl group are usually well soluble in water since more opportunities to form hydrogen bonds arise. They have an inductive effect on each other giving rise to the acidity expressed by lower pK-values. The closer they are to each other the more pronounced this effect will be. If far apart they tend to form cyclic anhydrides upon heating, such as maleic (acid) anhydride and phthalic (acid) anhydride. Otherwise they simply loose carbon dioxide (decarboxylation).

**Example Problem 7.6**
**Draw the structure of the product for**

**(1) Decarboxylation of oxalic acid**
**(2) Salt formation with NaOH and oxalic acid (molar ratio 1:1)**
**(3) Anhydride formation by heating hexanedicarboxylic acid**

Carboxylic acids are widely spread in nature. Many fruits contain acids – ascorbic acid (vitamin C), malic acid (apples), oxalic acid (rhubarb and spinach), citric acid (citrus fruits).

**Fig.7.13**

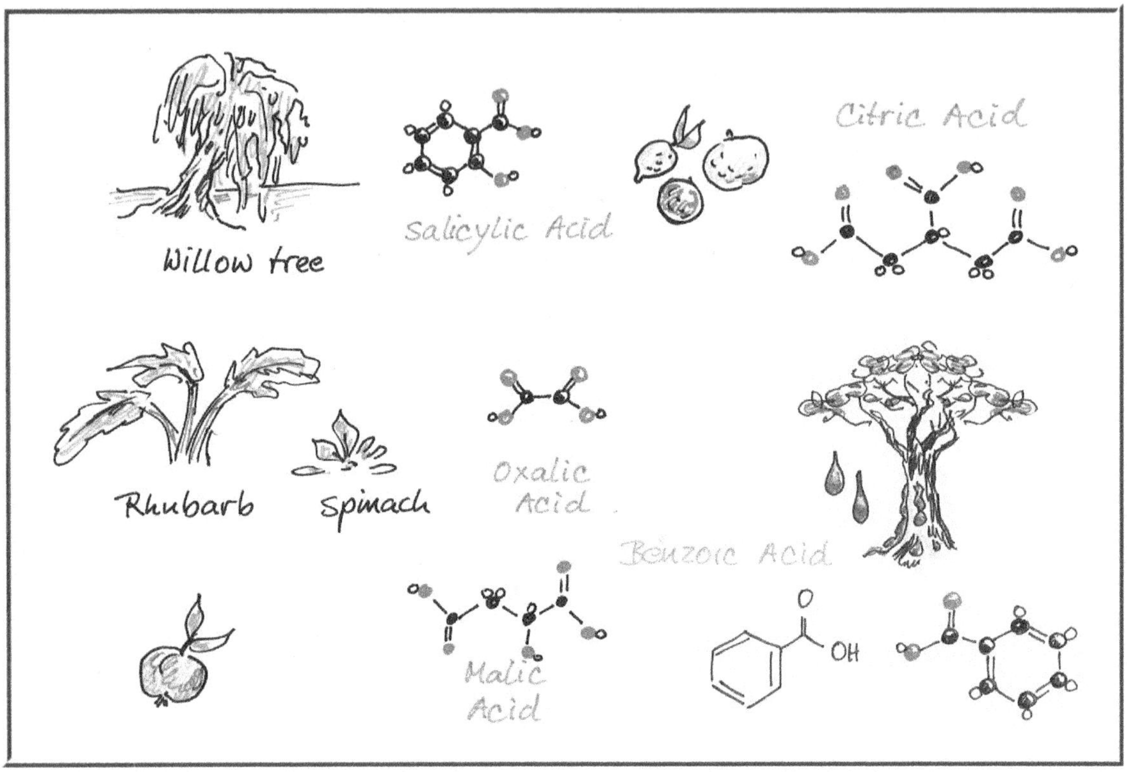


They are also found in other parts of the plant such as leaves, stems and roots. Benzoic acid - the food preservative per se - is found as "Gum Benzoin" in the resins of various trees ( *styrax)*. Salicylic acid (2-hydroxybenzoic acid) is found in all parts of the willow trees (salix*)*. Commercially it is used as a corn and wart remover and as the precursor molecule for the manufacture of the analgesic drug Aspirin (Acetylsalicylic acid). The trees use salicylic acid mainly as an allopathic, chemical warfare agent to prevent other plants from rooting in their vicinity.

Retinoic acid is an important carboxylic acid found in vertebrate animals that is derived from retinol (vitamin A), being essential for embryonal development.

Fatty acids are found in all fats and oils, waxes, and in the membrane molecules phosphoglycerides, (triacylglycerols). Arachidonic acid, a highly unsaturated fatty acid, converts into prostaglandins - molecules that control many functions in the body.

There are many "unnatural" carboxylic acids that play an important role in many industries. Terephthalic acid is used to make polymers, such as PETE plastics (polyethyleneterephthalates). There is some controversy concerning their role as endocrine disrupters leading to the "feminization" of society.

**Fig. 7.14**

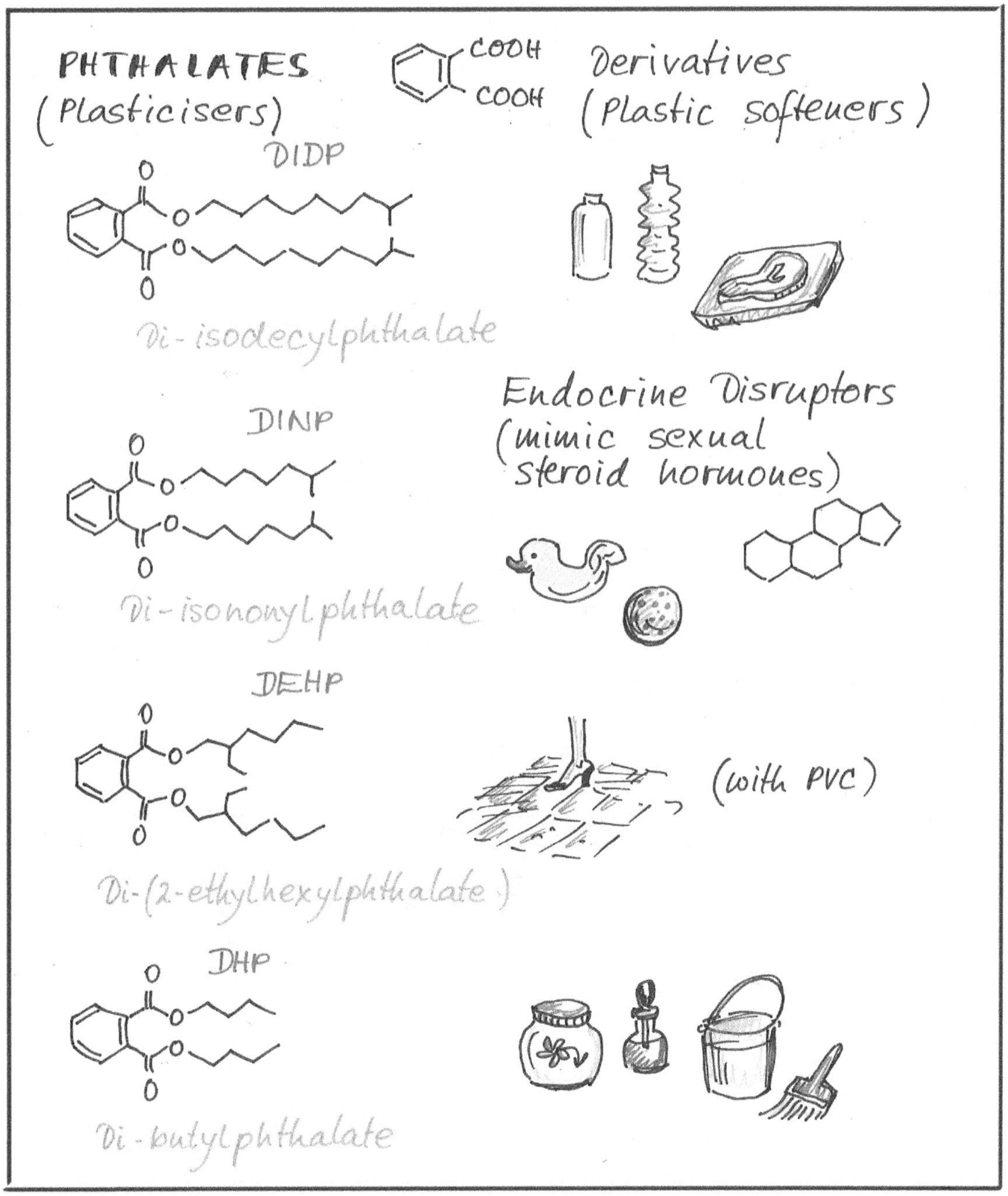


Acrylic acid is used for making various acrylates that can be utilized for paints, varnishes, resins, etc.

**Fig. 7.15**

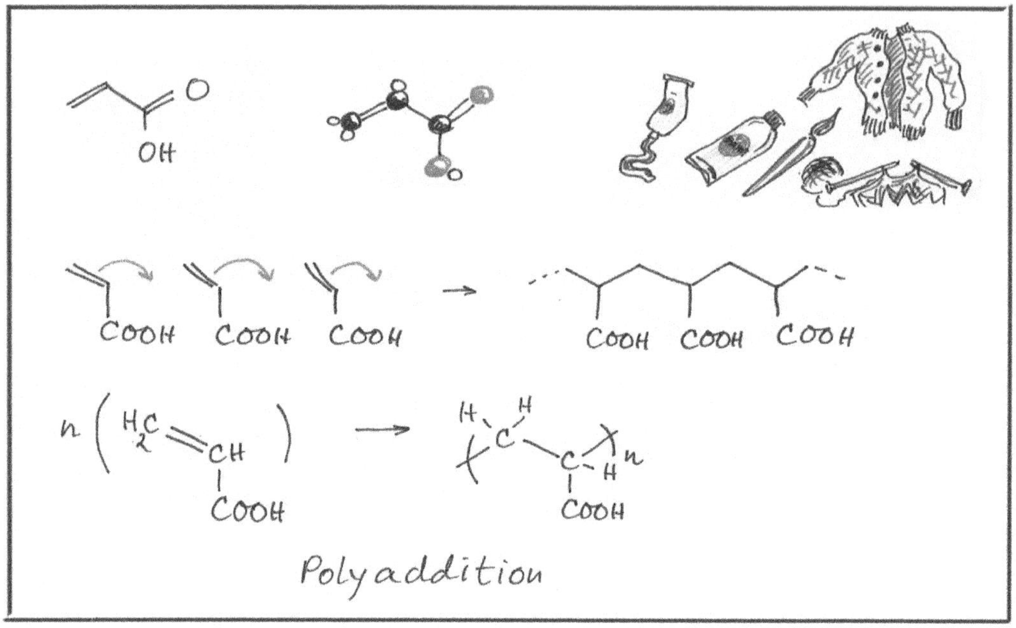


Important acids for cosmetics are the **α-hydroxycarboxylic acids** (AHA's). Although they are abundant in the natural world, they can also be synthesized to give various unnatural representatives. Examples are lactic acid from sour milk, glycolic acid from sugar cane, the already mentioned Citric acid from citrus fruits, malic acid from apples, tartaric acid from grapevine, mandelic acid from bitter almonds and many more.

The acidity for alpha-hydroxy acids is enhanced due to the ability of the neighboring hydroxyl group to form a hydrogen bond with the oxygen atoms of the carboxylic group, thereby facilitating the release of the proton.

**Fig. 7.16**

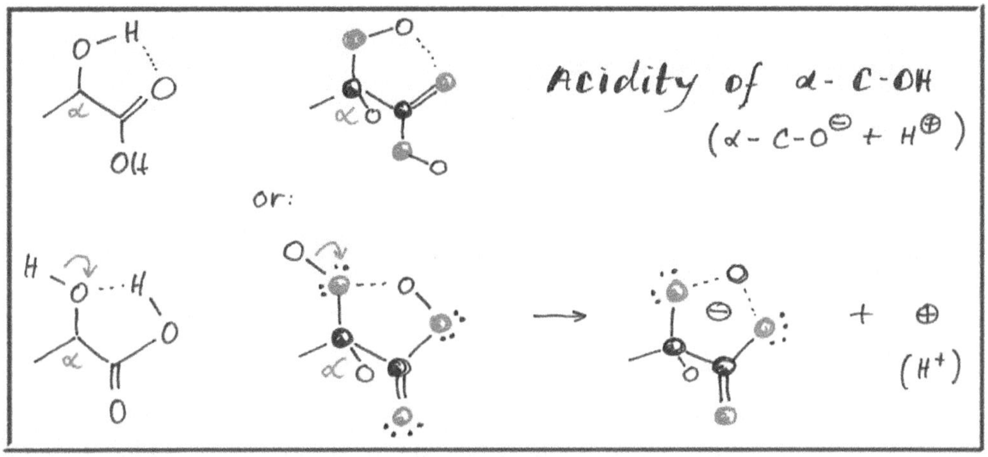


These acids are considered anti-wrinkling and beautifying for the skin. Ancient Egyptians were fond of bathing in milk.

**Fig. 7.17**

Carboxylic acids undergo various reactions, such as:

1. acid base reactions with inorganic or organic bases to form **salts**

**Fig.7.18**

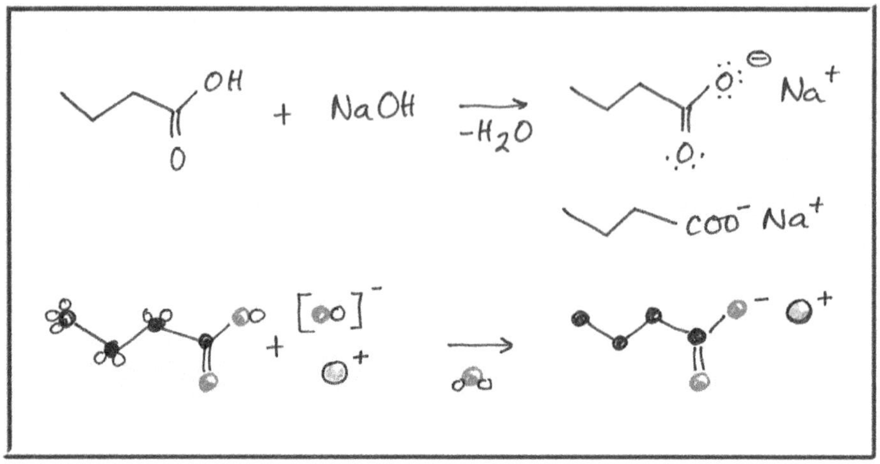


2. Dehydration reactions between two carboxylic acid molecules (same or different) to form **acidic anhydrides**

**Fig.7.19**

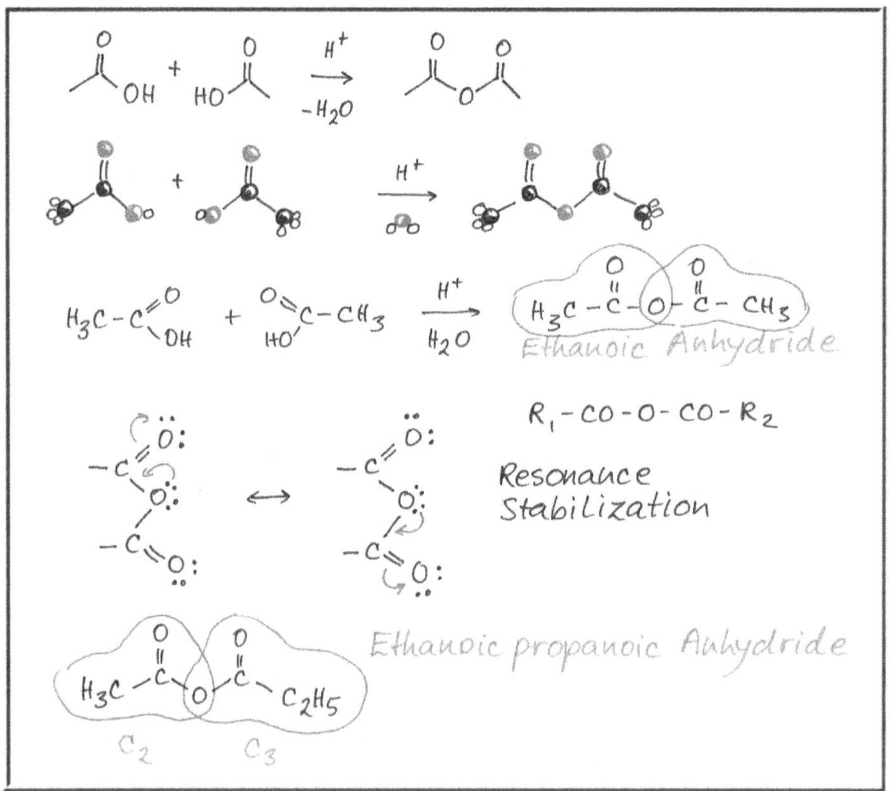


Figuratively 2 acyl groups (R-CO-) are linked via an ether bridge (-O-)
There is some resonance stabilization, but both carbonyl groups compete for the oxygen bridge lone-pairs.

3. Dehydration Reactions between carboxylic acid and alcohol groups, which leads to the formation of **esters**.

**Fig. 7.20**

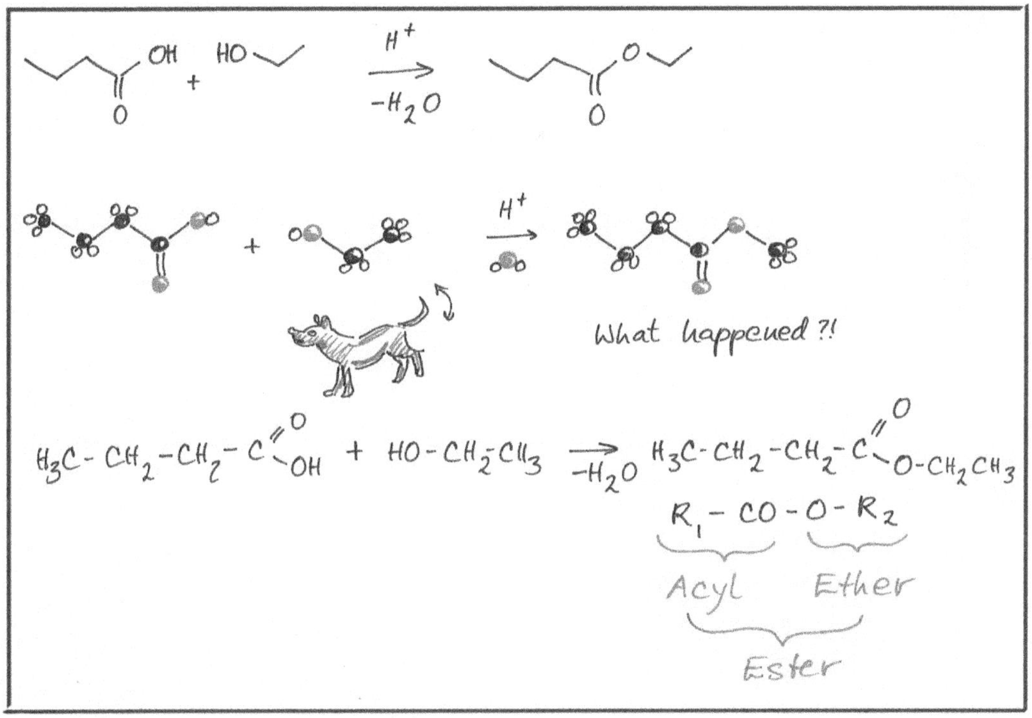


The general structural formula of the ester group is $R_1$-CO-O-$R_2$, the R-groups representing different organic rests or "radicals".
It is figuratively an acyl group linked to an ether group, also known as an **oxo group**.

4. Dehydration reactions with ammonia or amines to **amides**

**Fig. 7.21**

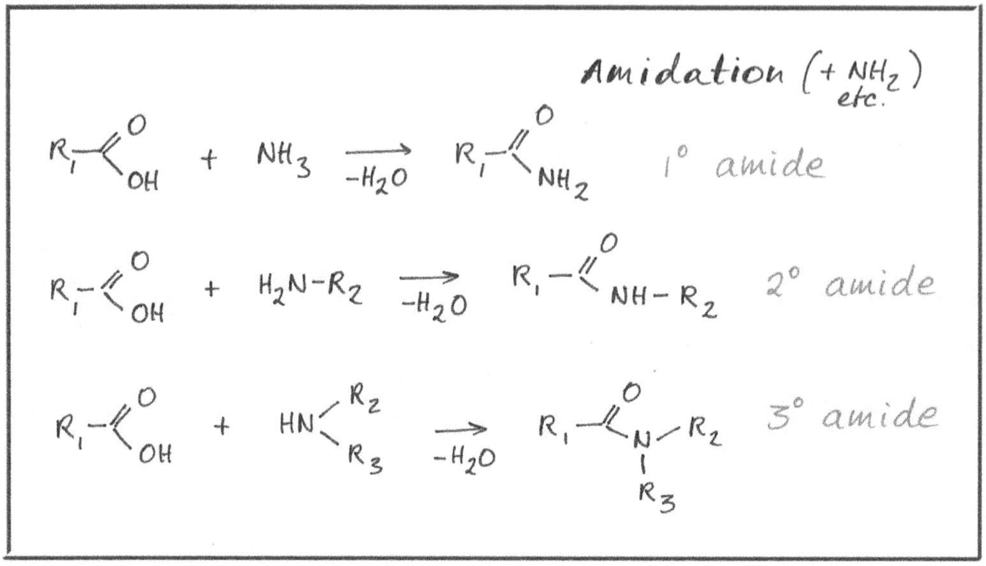


5. Halogenation to form **acyl halides.**

**Fig.7.22**

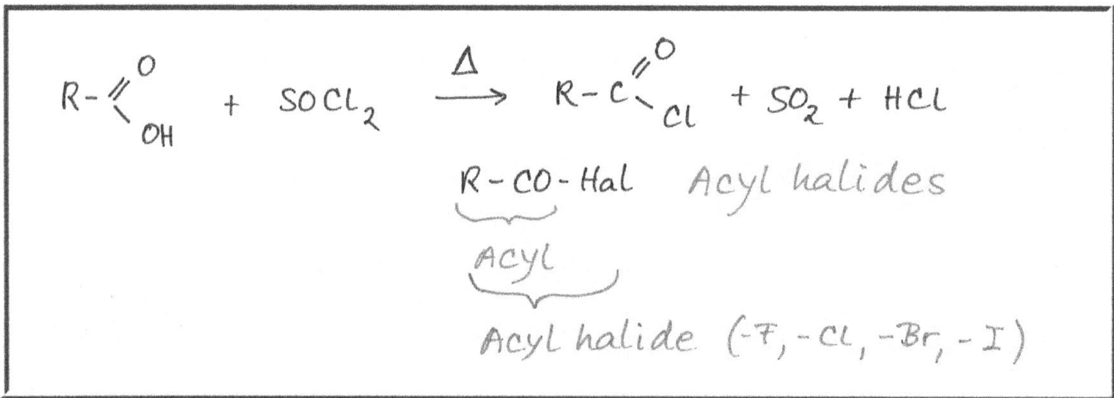


Acyl halides are valuable **acylating reagents** (Friedel-Crafts-Reaction), discussed in chapter 6 and 10.

**Example Problem 7.7**
**Give the names and structures for the esters produced when the following sets of two molecules react:**
***2,2-Dimethylpentanoic acid and methanol***
***Tert-butanol and propanoic acid***

Carboxylic acids are widely used in a variety of ways. Vinegar is one of the most well-known commercial products containing carboxylic acid. Benzoic acid and its salts, propionic acid, and citric acid are important food preservatives. Monosodium glutamate (MSG) is a meat tenderizer and flavor enhancer. They also function as a starting material for a variety of reactions leading to other material (flavoring agents, plastics, etc.).

The **nomenclature for esters** is based on the groups coming from the alcohols, expressed as a regular alkyl group and the acid expressed as the anionic form similar to the salts of carboxylic acids ending in –**oate**.

**Fig. 7.23**

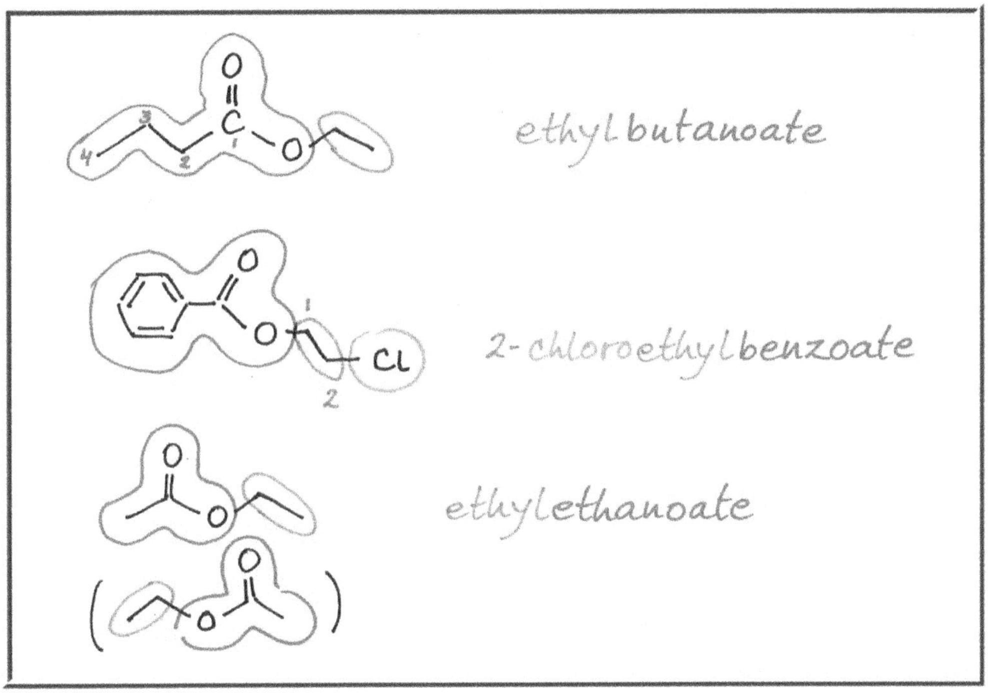


Esters mostly have a very pleasant aroma. Depending on the R-groups, the nuances of various aromas can be expressed. Many distinct aromas of fruits are based on particular esters. However, sometimes more than one ester can be present in the same plant, giving rise to a whole variable spectrum of various aromas and tastes.

**Fig.7.24**

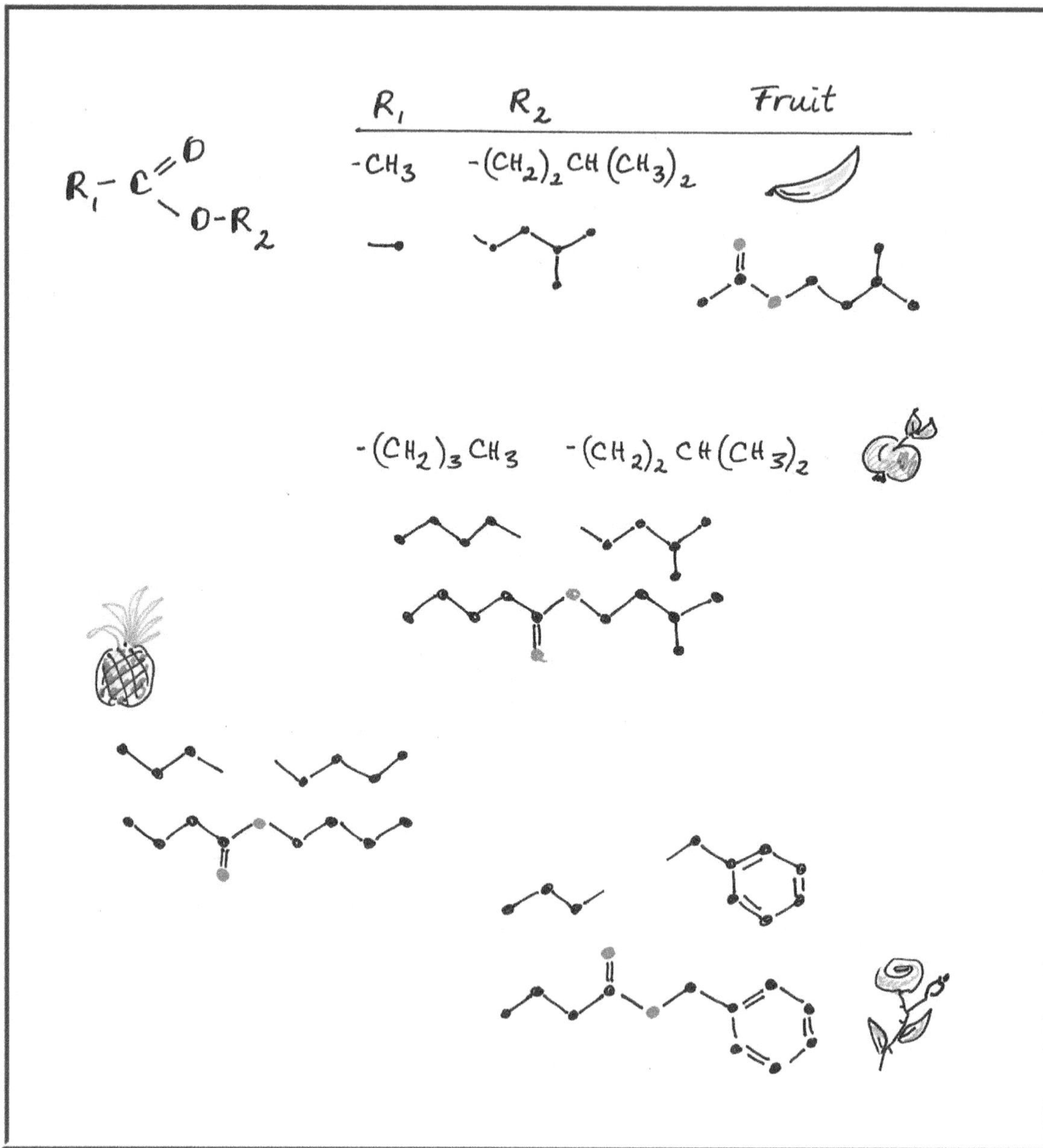


Some esters found in plants also serve as pheromones for certain insects such as ethyl cinnamate being the sexual attractant for the male oriental fruit moth. As a result of coevolution processes this connection most likely makes the moth serve as a pollinator for the plant.

**Fig. 7.25**

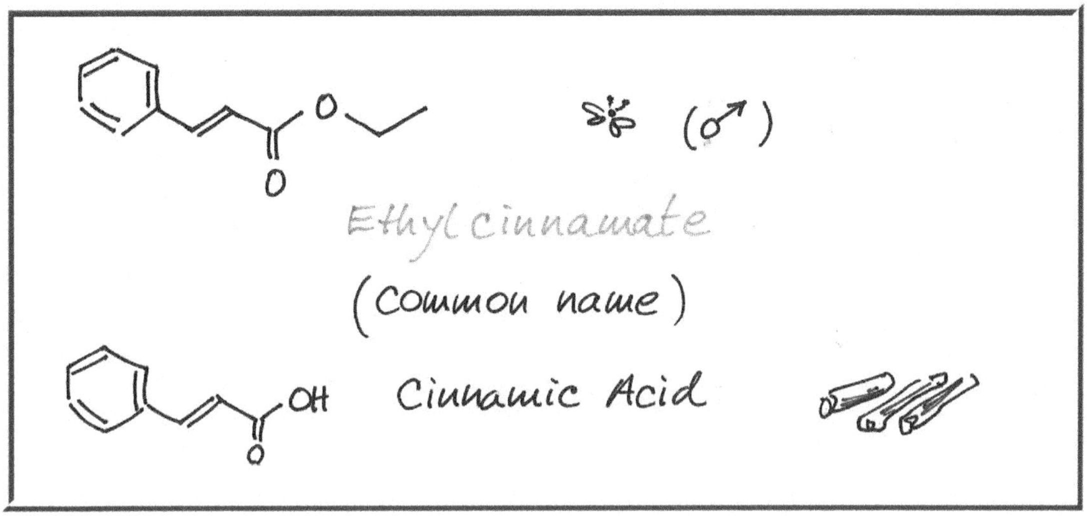


In order to have aroma, the esters must be relatively volatile. Different esters derived from the same molecule can have completely different properties. As alpha hydroxy acid, salicylic acid can form two ester derivatives, one going off the carboxylic group leading to "Oil of Wintergreen", found in many evergreens with mint-like aroma and taste (methylsalicylate), the other going off its hydroxyl group leading to a molecule like aspirin (acetylsalicylic acid), a powdery, bitter tasting substance without any aroma.

**Fig. 7.26**

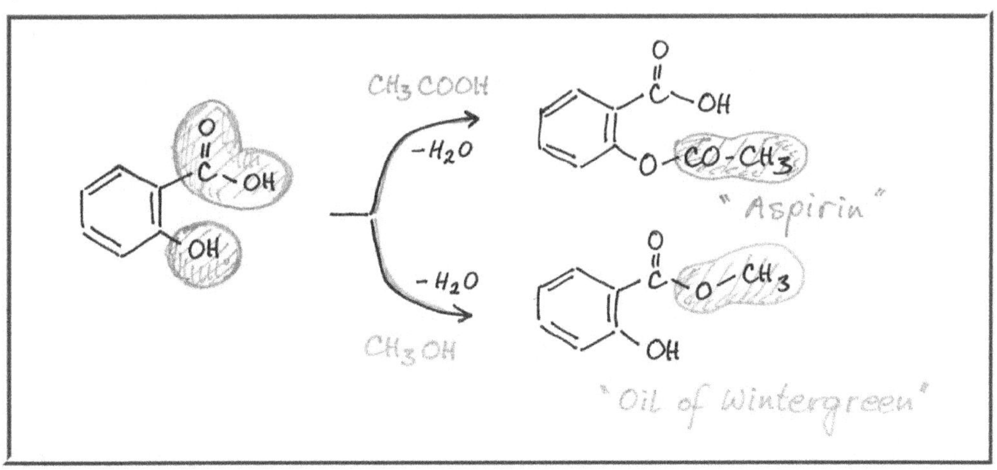


Of importance are non-volatile solid esters that make up the molecular family of the **waxes,** lipophilic long-chained molecules, discussed in more detail in chapter 26, dealing with lipids. A well-known wax is found in the foreheads of the sperm whale, cetyl palmitate, that is an ester between cetyl alcohol and palmitic acid.

**Fig. 7.27**

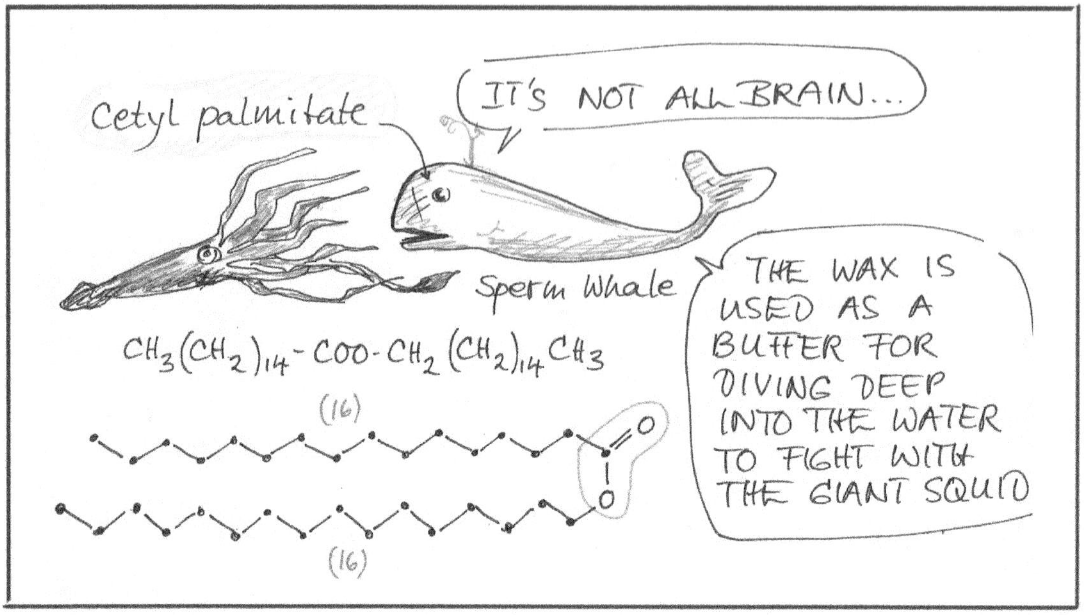


Cyclic esters are called **lactones**. Here the carboxylic group and alcoholic group must be within the same molecule, separated by 2 or more carbon atoms, so that intra-molecular dehydration to the cyclic product without much ring strain can occur.

**Fig.7.28**

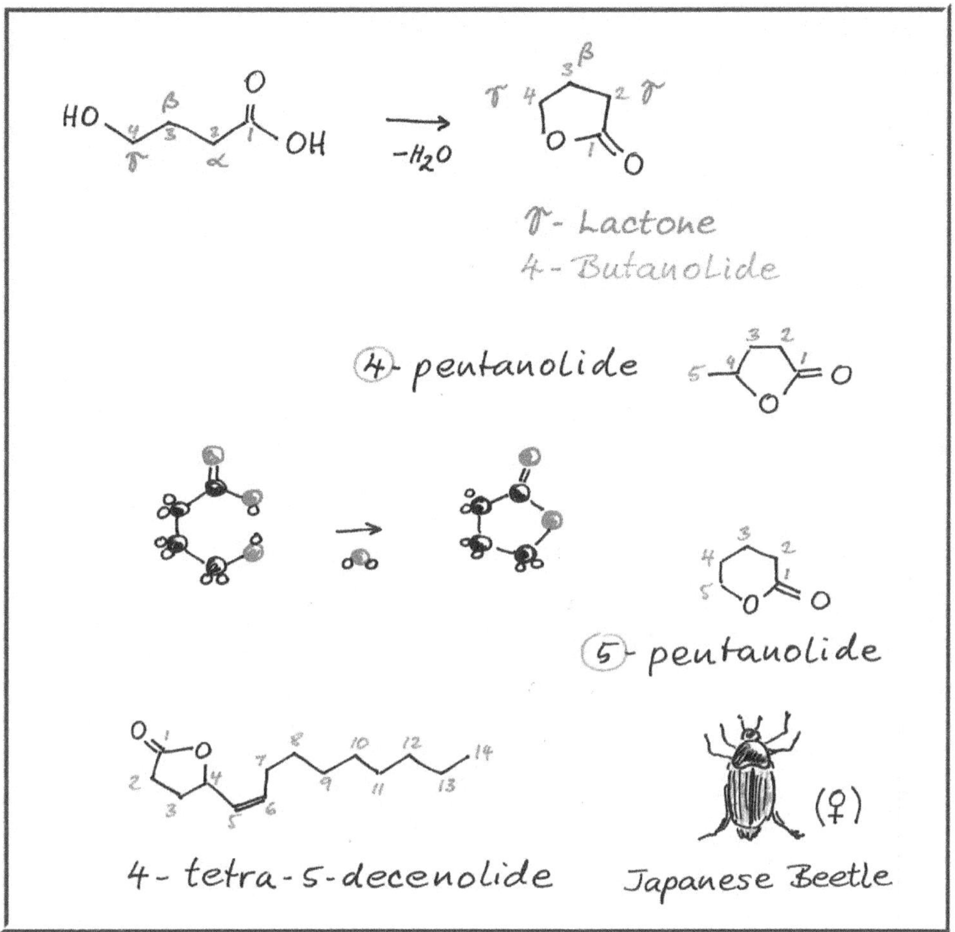


They are common in nature as pheromones (inter-individual hormones that work as sex-attractants), **macrolide antibiotics** (Erythromycin), **tumor inhibitors** (Vernolepin), and much more.

Fig. 7.29

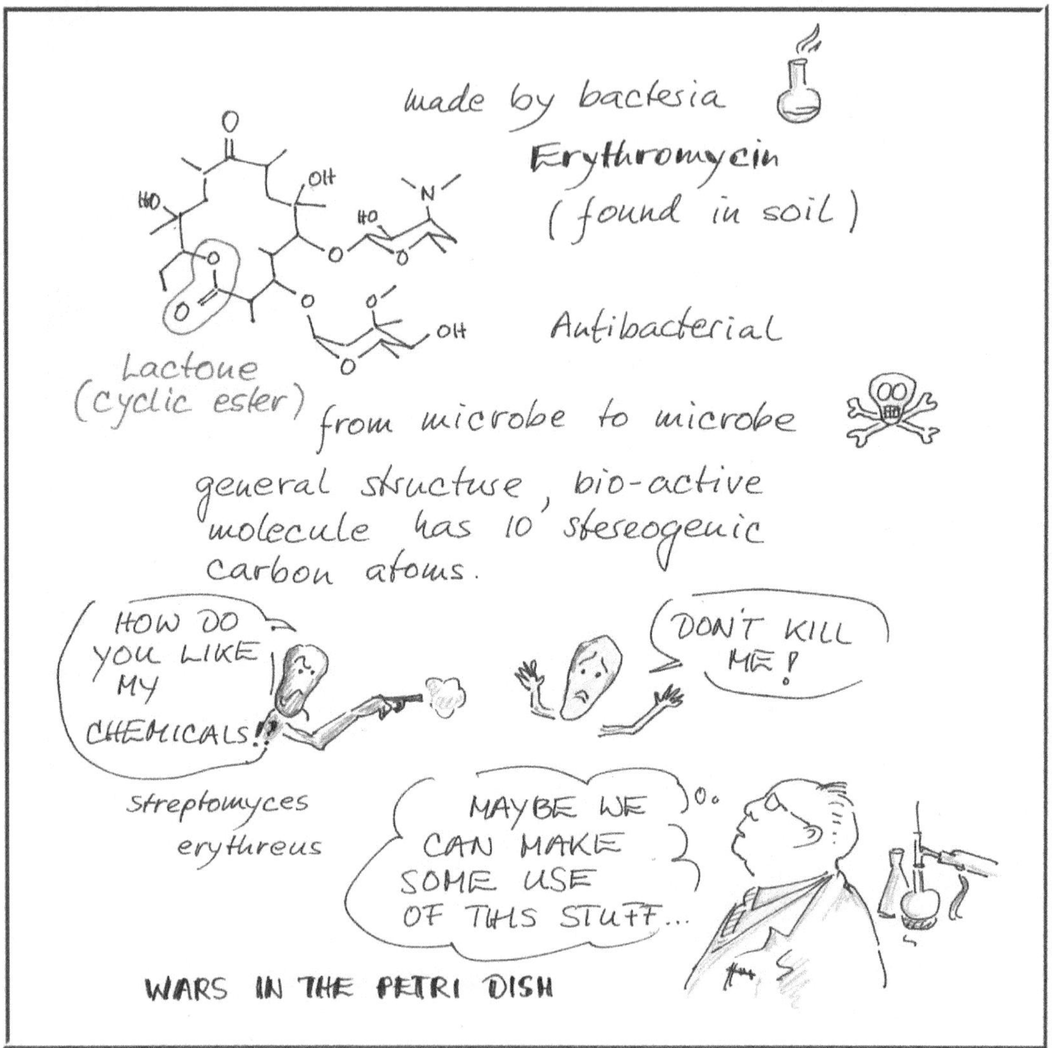


Fig. 7.30

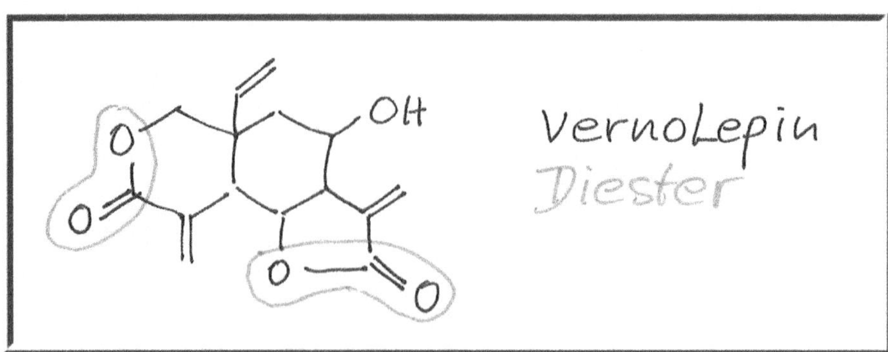

An essential nutrient for humans is **ascorbic acid** or vitamin C, which would cause the debilitating and deadly collagen disease **scurvy**, if missing from the diet. Structurally it is a lactone and the acidic hydrogen atoms are actually coming from the carbons of the double bond via induction effect of the cyclic ester oxygen atoms. It can be thought of as an intramolecular dehydration product of the corresponding unsaturated "sugar" acid.

**Fig. 7.31**

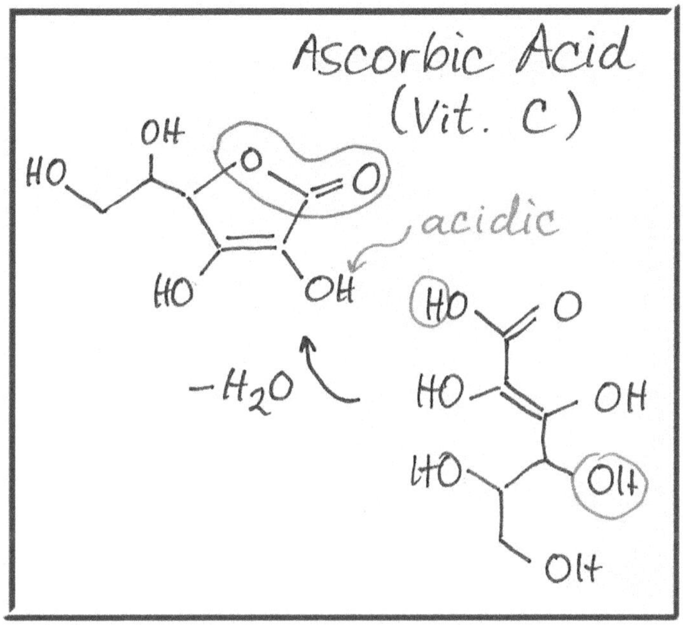


**Example Problem 7.8**
**What lactones are formed via intra-molecular esterification (dehydration) of the following molecules?**
***5-hydroxyhexanoic acid***
***4-hydroxy-2-pentenoic acid***
Acid anhydrides are usually not found in nature, an exception being **cantharidin** or "spanish fly," which is an aphrodisiac for humans and is derived from a beetle.

**Fig.7.32**

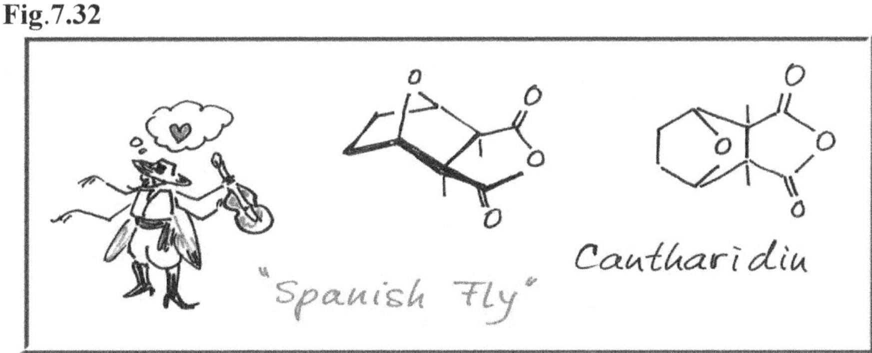

Esters can be hydrolyzed (splitting with water) back to the carboxylic acid and alcohol compounds, which is the reversal of **esterification**. The process is called **saponification** if it occurs in the presence of a base. The name is derived from soap, since the original soap-making process is hydrolyzing fats and oils with **lye** (caustic soda / NaOH or potash / KOH).

**Fig.7.33**

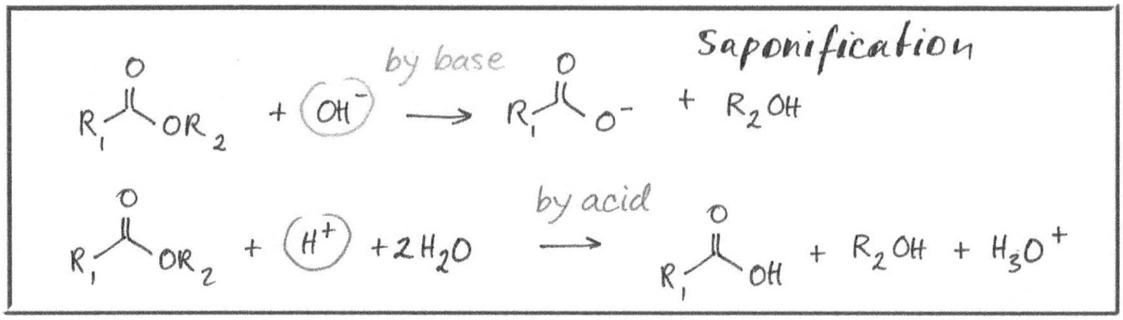


As will be discussed in the chapter dealing with the biochemistry of lipids, fats and oils are very important esters between fatty acids and glycerol (1,2,3-trihydroxypropane), Phospholipids are similar only that one ester group is established with phosphoric acid in the form of phosphatidyl groups.

**Fig.7.34**

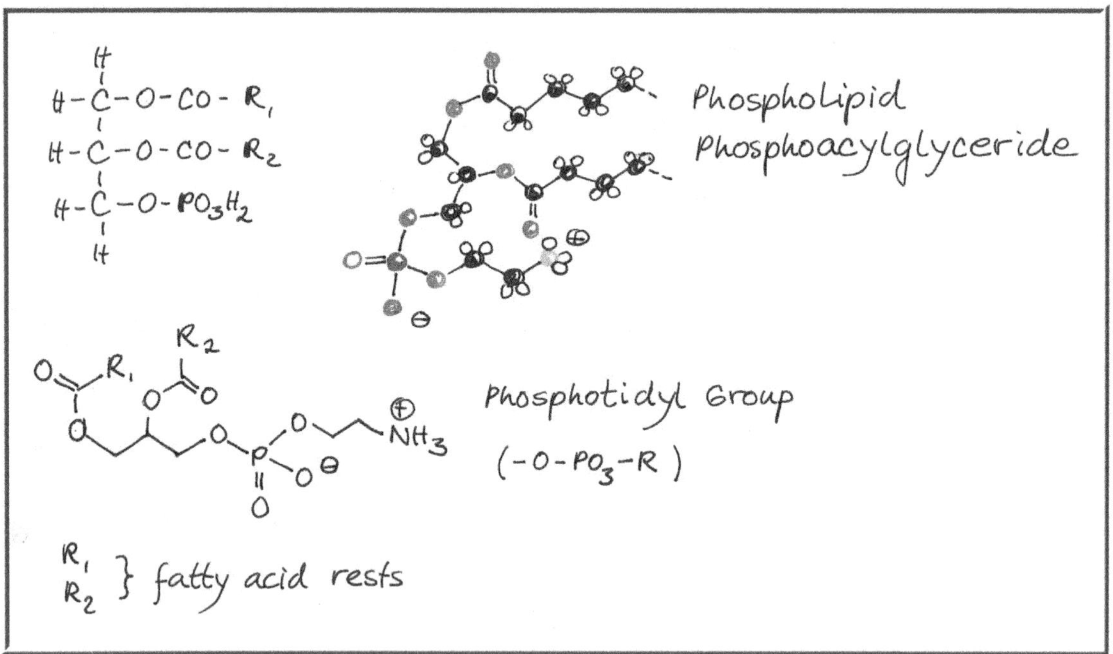


Amides will be discussed as derivatives of the **amines** in the next chapter (Ch. 19)

A most interesting ester found in nature and commonly used as an additive for many commercial products is 4-hydroxybenzoic acid methyl ester or methyl-4-hydroxy benzoate, better known as

**Methylparaben**. It is found in the vaginal secretions of female dogs in estrus (in "heat") and as preservative in food, beverages and cosmetics. It is almost always present in lotions and creams along with other parabens and listed on the labels as such.

**Fig. 7.35**

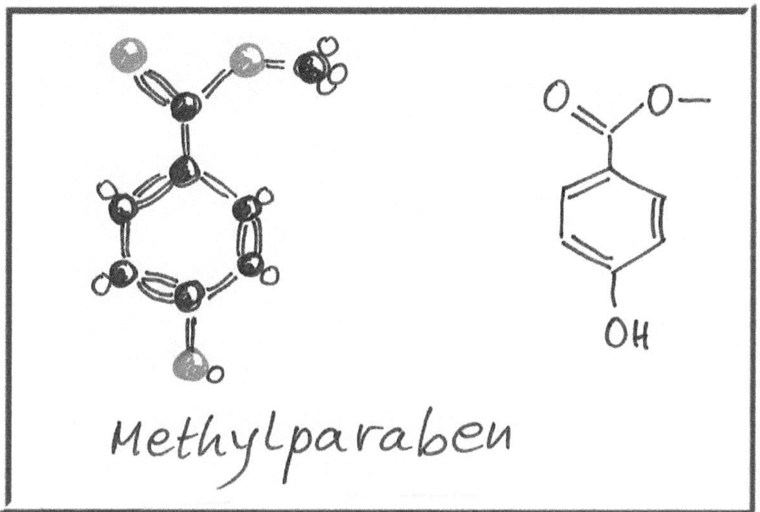


**Fig. 7.36**

Carboxylic acids as well as their esters often serve as **pheromones** or as inter-individual hormones. Acrylic acids that are commercially so valuable in producing many types of plastics, fibers, resins and paints etc. are also involved in many species as pheromones. Underarm effervescent chemicals from humans have an uncanny resemblance with sexual attractants utilized by bees.

**Fig. 7.37**

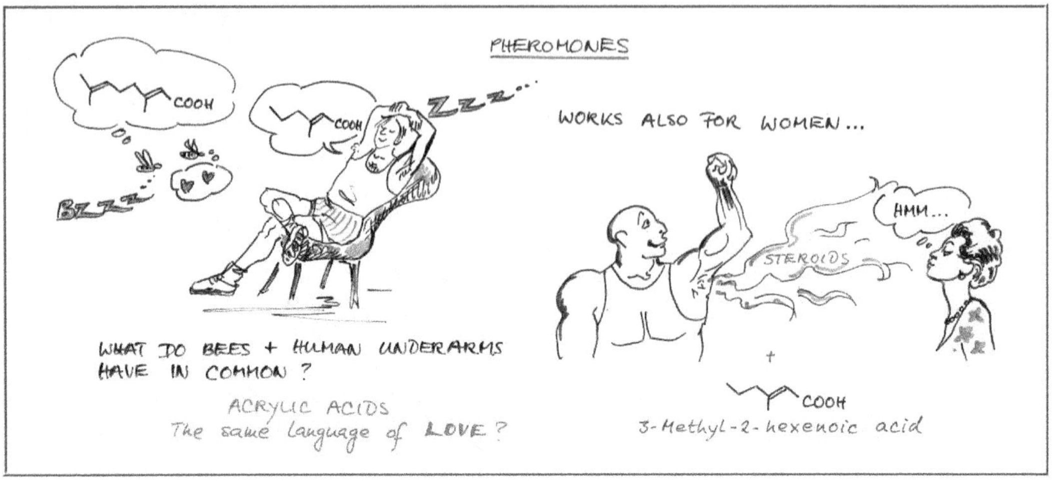


Although most males of the human species release a cocktail of volatile steroids and unsaturated carboxylic acids such as 3-methyl-2-hexenoic acid as pheromones, females are more inclined to release various saturated carboxylic acids for their "chemical communications".

Elephants communicate romantically via esters used as pheromones, such as *cis*-dodecen-1-yl acetate, released by females as signal for their readiness to mate. Strangely enough, this pheromone is also used by some insects such as moths.

**Fig. 7.38**

Some notoriety was received by LSD (lysergic acid diethylamide), a derivative of lysergic acid produced by a fungus, which grows on rye and wheat as well as some other grains. The molecule was first fully synthesized by the Swiss chemist Hofmann, who tested the substance on himself and wrote about his "surrealistic" experiences. Bread made from these grains caused "strange" hallucinogenic behavior in people which consumed it, a condition known as St. Anthony's Fire during the middle ages in Europe, medicinally classified as "ergotism". Lysergic acid derivatives are also found in the "Morning Glory" varieties (Ipomea species)), extensively used for ritualistic, visionary purposes in pre-Columbian times in Central America and somewhat to the present day.

**Fig. 7.39**

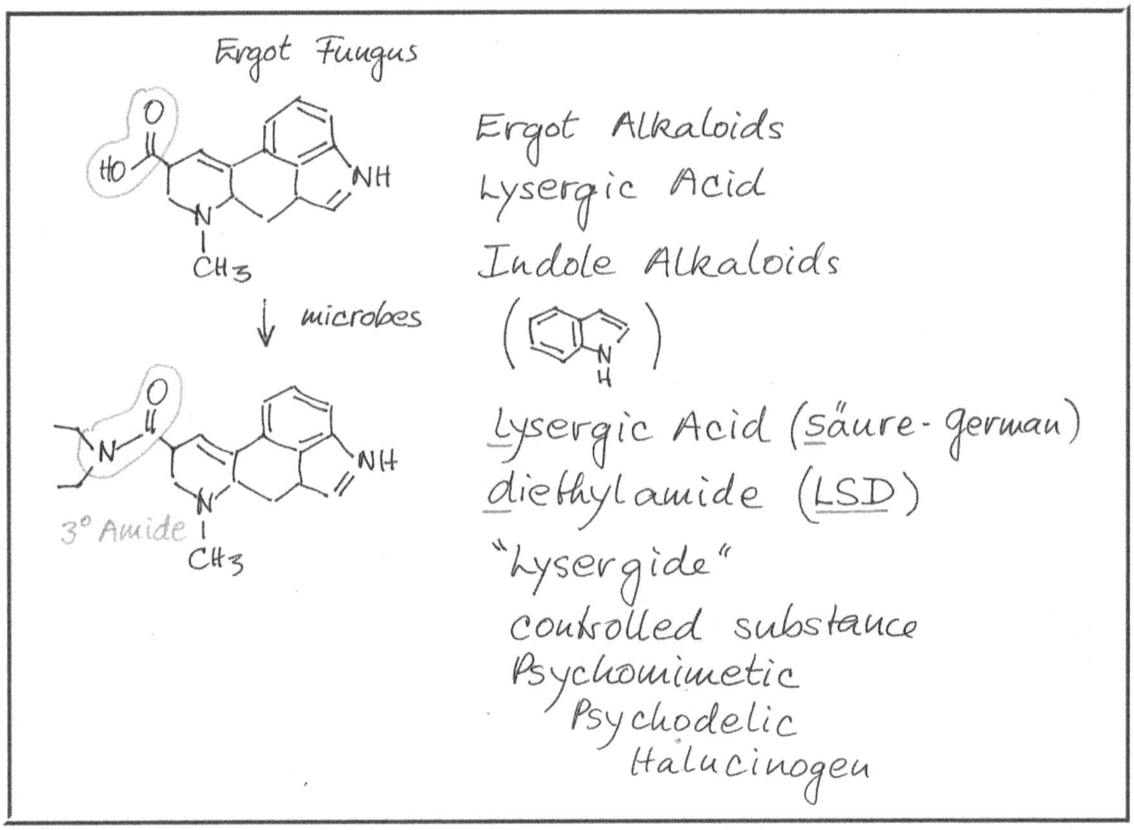


The American psychologist and research scientist Timothy Leary made LSD well known to the general public, leading to abuse in many cases and thus making the chemical legally a controlled substance today.

**Fig. 7.40**

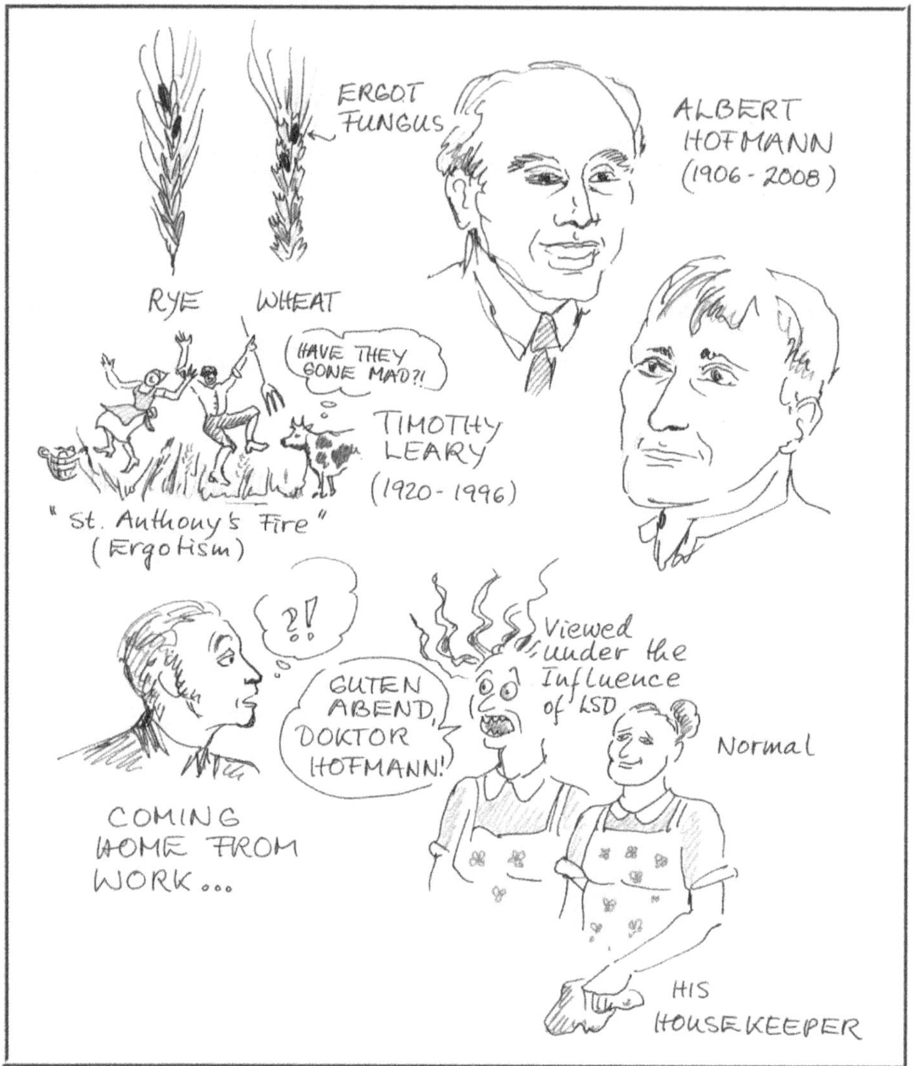


Another acid with neurological effects known as valproic acid (4-carboxyheptane) is commonly prescribed as an anti-seizure medicine.

**Fig. 7.41**

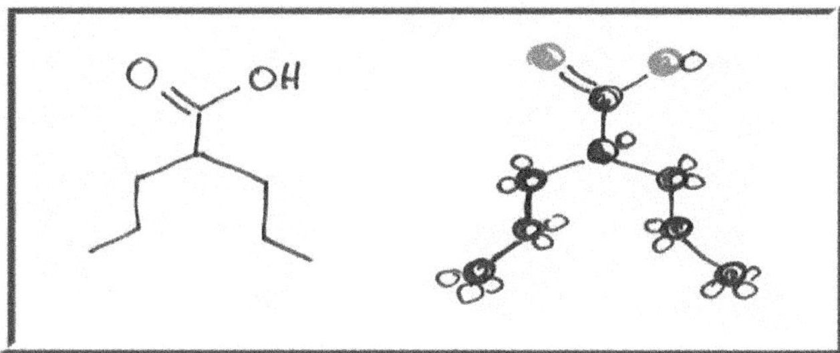


Another well-known drug of the most prescribed pharmaceuticals, **the statins** as cholesterol-lowering drugs, is **lovastatin,** a chemical naturally produced by the mold-like fungus *Aspergillus terreus,* found predominantly in East Asian soils but also in the United States. This fungus is widely researched since a produces a variety of other potentially valuable pharmaceuticals.

**Fig. 7.42**

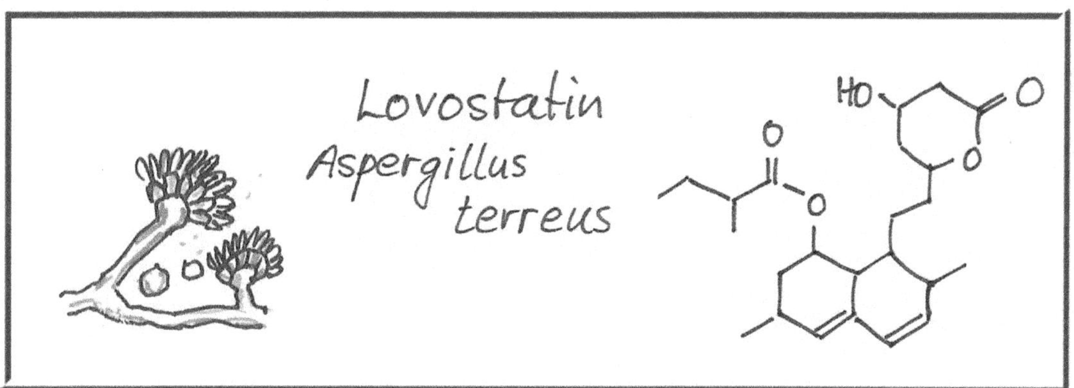


The question always remains when seeing signs of convergent "chemical evolution", as to what purpose medicinal chemicals actually have for the plant itself. This holds also true when dealing with various amines and amides including the family of the alkaloids, the latter so widely produced by species of the plant kingdom, to be discussed in the next chapter.

# EXPERIMENTS –Ch.7

## *Carboxylic Acids and Esters*

**Key Terms:** The carboxylic group, the carboxylate ion and its resonance structures, acidity and $pK_a$ values, buffering capacity, salt formation, conversion points and hydrogen bonding, alpha-hydroxy acids, esters, esterification, saponification, amides

## I Molecular modeling of Carboxylic acids, Carboxylates (as Salts), Anhydrides and Esters

### Chemicals:

*none*

### Equipment and Materials:

*Molecular modeling kit*

### Procedure:

1. Build an ethane molecule
2. Build a carboxyl group, attaching a carbonyl group (carbon with a double bonded oxygen, C=O) and a hydroxyl group (-OH) to the same carbon atom of the carbonyl group and substitute it for one of the methyl groups of the ethane molecule prepared in #1, making it **ethanoic acid** (IUPAC name for acetic acid) derived from ethane.
3. Temporarily release a proton (hydrogen atom leaving the bond link behind) to create an **ethanoate ion** (acetate ion). Look at the carboxylate ion and inspect the possibility of each oxygen atom taking turns in having the double bond. This phenomenon is called **resonance** and stabilizes the ion (resonance stabilization).
4. With another carboxylic acid model (from another student group) inspect the possibility of making hydrogen bonds between the two carboxylic groups when positioned across from each other.
5. Eliminate a water molecule inter-molecularly (between two separate molecules) to construct an anhydride with the typical bond element –CO-O-CO- to form ethanoic acid anhydride, commonly known as **acetic anhydride**.
6. Reverse the reaction and condense (eliminating a water molecule by dehydration) the free acid with and alcohol molecule such as methanol to make an **ether.**

### Observations and Results:

Draw all the modeled structures in the following table:

| Molecule/Molecular ion: | Structural formula ESF and (ASF/LAF): | Resonance structures: | H-Bonding |
|---|---|---|---|
| **Ethane** | | | |
| **Ethanoic acid** | | | |
| **Ethanoate ion** | | | |
| **Ethanoic acid anhydride (acetic anhydride)** | | | |
| **Methyl ethanoate (ester)** | | | |

*Shaded areas need not be performed (n.a.)*

## Questions and Problems:

1. What influence does hydrogen bonding have on some physical properties of carboxylic acids?
2. Consider the hydroxyl- acid: 5-hydroxypentanoic acid. This could form a cyclic ester by intra-molecular dehydration. What would be its formula?
3. Are acid anhydrides and the esters still acidic? Explain.
4. Draw the structure of the product that forms between propanoic acid (propionic acid) and sodium hydroxide (NaOH).

## II Physical Characteristics of Carboxylic Acids and their Derivatives

### Chemicals:

*Selection of different carboxylic acids*
*Salts, anhydrides and esters*
*Glacial acetic acid* ($CH_3COOH$)
*Propionic acid or butyric acid*
*Sodium acetate* ($CH_3COONa$)
*Benzoic acid* ($C_6H_5COOH$)
*Acetic anhydride* ($CH_3CO\text{-}O\text{-}COCH_3$)
*Oxalic acid (HOOC-COOH)*
*Salicylic acid*
*Acetyl salicylic acid*
*Water (demineralized / deionized)*

### Equipment and Materials:

*Test tubes and rack, test tube holder or clamp*
*Droppers with bulb*
*Glass stirring rod*

*pH paper or blue lithmus paper*
*Hot water bath*

### Procedure:

1. Prepare 8 test tubes in a rack
2. In each separate tube place a spatula tip of the solid samples or about 1-2 cm high if the sample is a liquid. Be very careful with **glacial acetic acid** which is a **very caustic liquid** and make sure to wear **protective goggles**!
3. Add a few mL of water to each tube and test for solubility, if difficult add some more solvent, shake rigorously and as last resort place the test tube in a hot to boiling water bath.
4. If solubility is not affected by increase in temperature, add a few drops of 10% NaOH until the solution becomes clear.
5. Prepare two test tubes each containing a spatula tip of acetyl salicylic acid (Aspirin). Add 3 mL of water and place in the hot water bath. To one give 20 drops of 10% NaOH the other leave alone. Boil for about 5 min. Remove them from the water bath and test both with 1-2 drops of ferric chloride solution ($FeCl_3$).

### Observations and Results:

Record the results in the following table:

| **Sample:** | **Physical state:** | **Color** | **Odor/Aroma** | **Solubility in water:** | **pH of liquid or aqueous solution** |
|---|---|---|---|---|---|
| **Glacial acetic acid** | | | | | |
| **Propionic acid or butyric acid** | | | | | |
| **Sodium acetate** | | | | | |
| **Benzoic acid** | | | | | |
| **Acetic anhydride** | | | | | |

| | | | | | |
|---|---|---|---|---|---|
| **Oxalic acid** | | | | | |
| **Salicylic acid** | | | | | |
| **Acetyl salicylic acid** | | | | | |

In the following table record observations made before and after hydrolysis with water:

| **Sample:** | **Color with $FeCl_3$** | **Odor:** |
|---|---|---|
| **Un-hydrolyzed ester: Acetyl salicylic acid (Aspirin)** | | |
| **Base hydrolyzed ester: Acetic acid and salicylic acid** | | |

## Questions and Problems:

1. Which chemicals were the most soluble in water? Explain
2. Which chemicals needed to be heated to enhance solubility?
3. Why does addition of dilute sodium hydroxide solution (NaOH) to an insoluble carboxylic acid increase its solubility in water?
4. Why does hydrolyzed acetylsalicylic acid (Aspirin) give a positive ferric chloride test?

## III Condensation of Carboxylic acids with Alcohols to form Esters

### Chemicals:

*Glacial acetic acid*
*Salicylic acid*
*Methanol*
*1-Propanol*
*1-Pentanol*
*1-Octanol*
*Benzyl alcohol*
*Phosphoric acid ($H_3PO_4$, 85%, conc.)*
*Ferric chloride solution ($FeCl_3$, 1%)*

### Equipment and Materials:

*Test tubes and rack*
*Graduated cylinder (10 mL)*
*Droppers with bulb*
*Hot water bath*

### Procedure:

1. Place about 3 mL of the different alcohols in 5 separate test tubes.
2. Carefully add 2 mL of the glacial acetic acid (caustic, wear goggles!) into four of the test tubes as indicated in the grid below. To the methanol sample add a spatula tip of salicylic acid.

3. To each of the 5 mixtures add 15 drops of the concentrated phosphoric acid ($H_3PO_4$), gently swirl holding the opening of the test tube away from anybody's face and transfer to the hot to boiling water bath. Boil for about 15 minutes.
4. Fanning the vapors gently with the hand towards the nose, record and describe the smell or aroma of the resulting esters.

| **Acid:** | **Methanol** | **1-Propanol** | **1-Pentanol** | **1-Octanol** | **Benzyl alcohol** |
|---|---|---|---|---|---|
| Glacial acetic acid | | *Mix 1* | *Mix 2* | *Mix 3* | *Mix 4* |
| Salicylic acid | *Mix 5* | | | | |

## Observations and Results:

Tabulate the observed smells/aromas and give the names for the esters formed in *Mix 2-5* corresponding to example shown for *Mix 1* in the following table:

| **Ester:** | **Smell/aroma:** | **Equivalent fruit or general plant aroma:** |
|---|---|---|
| *Mix 1:* Propyl ethanoate | | |
| *Mix 2:* | | |
| *Mix 3:* | | |
| *Mix 4:* | | |
| *Mix 5:* | | |

## Questions and Problems

1. What is the chemical equation for the formation of an ester between propanoic acid and propanol? What is the name of the ester formed?
2. What happens if the ester is treated with HCl?
3. What happens if the ester is treated with NaOH?
4. The term saponification comes from the old practice of making fatty acid salts as soaps by treating various fats and oils with lye (calcium hydroxide).

# IV Synthesis of Aspirin (Acetyl Salicylic Acid)

## Chemicals:

*Salicylic acid*
*Acetic anhydride*
*Phosphoric acid ($H_3PO_4$, 85%)*
*Ferric chloride solution ($FeCl_3$, 1%)*

## Equipment and Materials:

*Erlenmeyer flask (125 mL)*
*Beaker (400 mL or larger) for hot water bath*
*Hot plate with stir facility, stir bar (spin vane)*
*Büchner filtration apparatus, fitting round*

*filter paper*
*Metal stand with clamp*
*Melting point device*

## Procedure:

1. Prepare a boiling water bath on a hot plate in the fume hood.
2. In a tarred out Erlenmeyer flask weigh out 2.00g of salicylic acid.
3. Carefully add 5mL acetic anhydride and gently swirl the flask (fume hood).
4. Slowly add 10 drops of phosphoric acid ($H_3PO_4$), secure the flask with a clamp to the metal stand and submerge it half-way into the boiling water bath.
5. After having added the stir bar to the solution, stir and heat until the salicylic acid is fully dissolved.
6. Remove and let cool.
7. Add 20 drops of water to the reaction mixture, remove stir bar with forceps, and place the flask in an ice bath until aspirin crystals appear.
8. Suction off the solvent from the crystals by vacuum filtration in the fume hood. Rinse with ice water and weigh the crude, "dried" crystals on the scale.
9. Test a few crystals in a bit of water with the ferric chloride reagent ($FeCl_3$). If a pink to purple color remains, recrystallize the crystals by re-dissolving them in a minimal amount of hot to boiling water and let stand until the next class period. Test the melting point and chromatic behavior (TLC) of the purified, recrystallized aspirin.

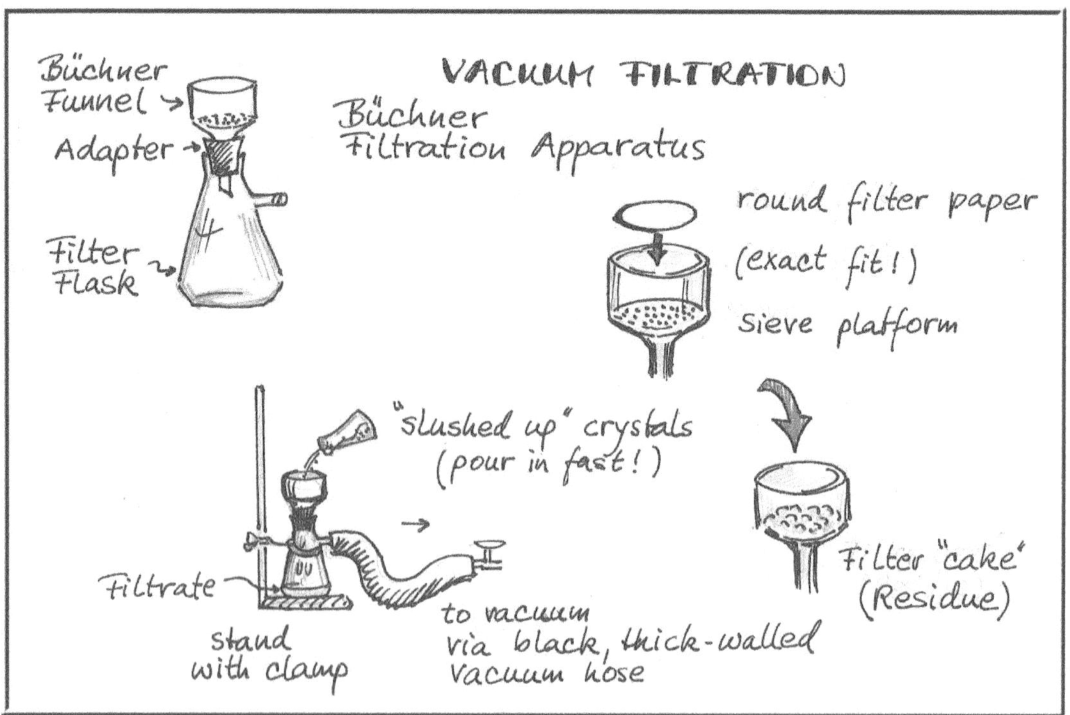

## Observations and Results:

| Chemicals (reactants and products): | Mass (g): | Molar mass (g/mol): | Yield (g) (theor.) | Yield (g) Practical (crude) | Yield (g) Practical (recryst.) | $FeCl_3$ Test: | Mp: (theor.) (°C) | Mp: (pract.) (°C) |
|---|---|---|---|---|---|---|---|---|
| Salicylic acid | 2.00 | 138 | | | | | 157-159 | |
| Acetic anhydride | | | | | | | | |
| Phosphoric acid | | | | | | | | |
| Acetyl salicylic acid (crude) | | 180 | 2.61 | | | | 135 | |
| Acetyl salicylic acid (recryst.) | | 180 | 2.61 | | | | | |

## Questions and Problems:

1. Verify, that 2.61g is the theoretical yield of aspirin in the performed synthesis reaction? Use dimensional analysis to work out this problem.

$$\text{---g}_{SA}\left(\frac{1\ mol\ SA}{---gSA}\right)\left(\frac{1\ mol\ A}{1\ mol\ SA}\right)\left(\frac{---g\ A}{1\ mol\ A}\right) = \text{---g}_{A}$$

2. What is the practical or experimental yield of the actual product obtained before and after recrystallization?

$$\frac{practical\ yield\ (g)}{theoretical\ yield\ (g)}\ or\ \frac{have}{should\ have} \times 100\% = \%\ \text{yield}$$

# Chapter 8 Amines, Amides and Alkaloids

8.1 The amino functional group

8.1.1 Proton acceptor function, basicity and pK-values

8.1.2 Degree of substitution (primary, secondary, tertiary or quaternary)

8.1.3 Hydrogen bonding

8.2 Nomenclature

8.3 Physical and chemical properties

8.4 Reactions

8.4.1 Salt formation with inorganic acids

8.4.2 Amide formation with carboxylic acids (the amido group0

8.5 Natural distribution and uses

8.6 Relative toxicity, physiology and medicinal uses

8.7 amides

8.7.1 Structure and properties

8.7.2 Distribution in nature and industry

8.7.3 Formation and synthetic routes

8.7.4 Nomenclature

8.7.5 Toxicity

8.8 Alkaloids

8.8.1 Heterocyclic natural compounds with diverse structures

8.8.2 Basicity and pK-values

8.8.3 Non-aromatic and aromatic ring structures

8.8.4 Physiological functions and properties

8.8.5 Relative toxicity and medicinal uses

<u>**Chapter 8  Amines and Amides**</u>

The basic or alkaline principle of organic molecules is expressed by the amines and characterized by the **amino group -$NH_2$.**

**Fig. 8.1**

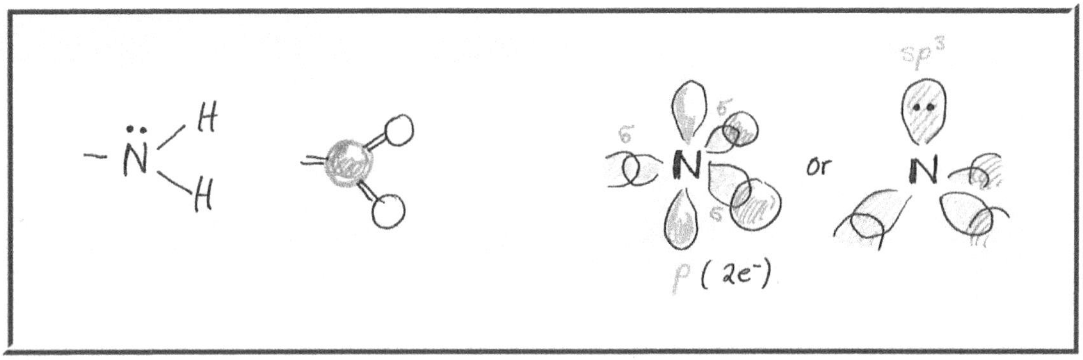


The lone pair acts as the actual proton acceptor.

**Fig. 8.2**

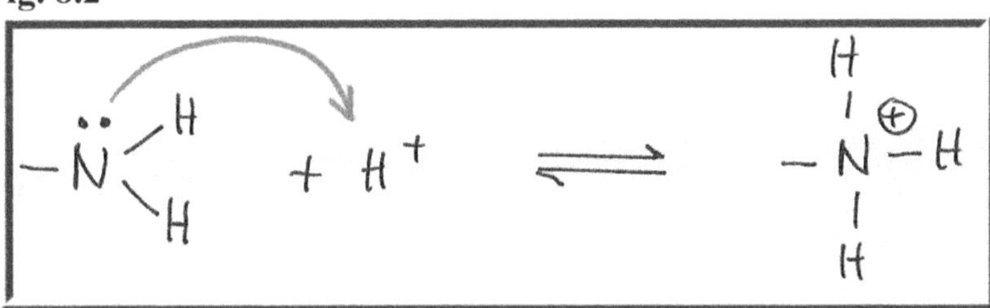

The conjugate acid/base pair being:

**Fig. 8.3**

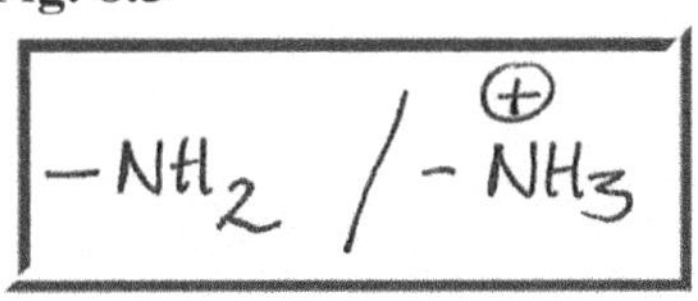


Three types of neutral amines are possible depending on the **degree of substitution** (H-atoms replaced by R-groups, radicals or organic rests).

**Fig. 18.4**

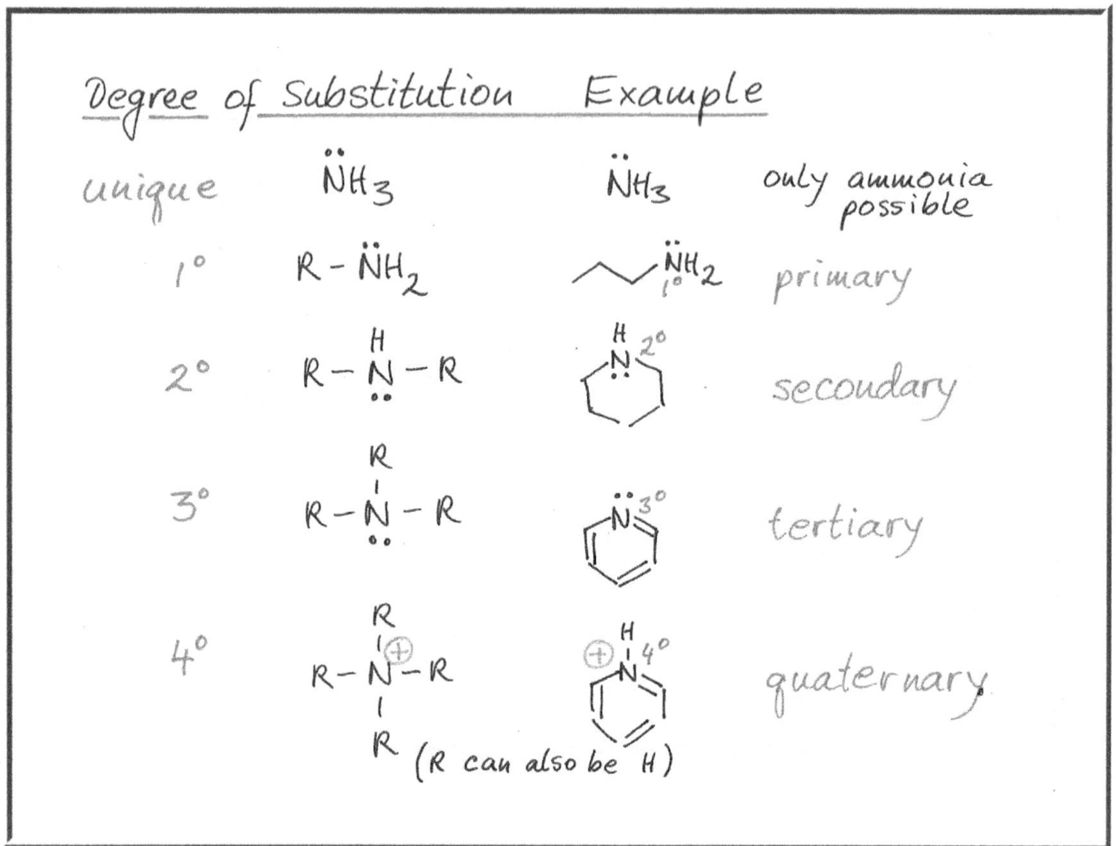


A fourth type, but not neutral, species is called a **quaternary amine**. This can involve any degree of substitution including the unique form – the ammonium ion ($NH_4^+$).

It is actually any four-bonded amine carrying a positive formal charge.

Amines can be considered derivatives of ammonia, where the one to three H-atoms are replaced by R-groups.

**Fig. 8.5**

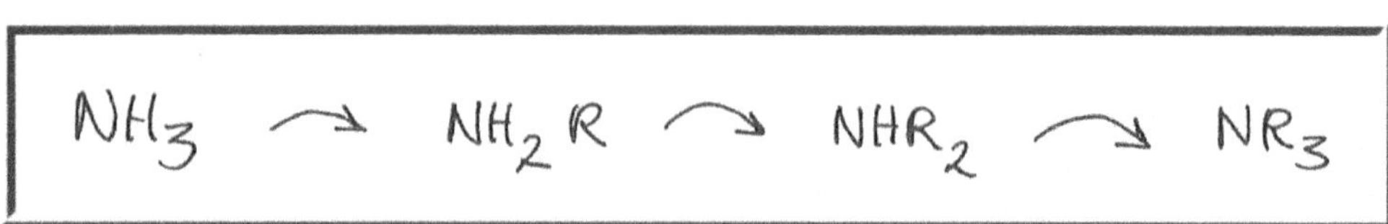


Another distinguishing factor is the molecular type of R groups involved. This divides the group of primary amines into **alkylamines** (alkanamines), **cycloalkylamines** (cycloalkanamines), and **arylamines** (arenamines). The number of amino groups present within a molecule determines the valency or valence of the amine, distinguishing them as a monoamine, **diamine**, **triamine** or even a **polyamine**. Important diamines, that occur during the process of decay and putrefaction of cadavers, are **putrescine** and **cadaverine**. Polyamines in general are toxic to living creatures since they can bind via their positive ammonium groups to the negative phosphate groups of DNA molecules, compromising their biological function.

**Nomenclature for primary amines** can be based on:

1. The "amine" being the parent and the attached group named as alkyl, aryl or other up front, using prefixes such as mono, di, and tri if needed.
2. The amino group named as substituent to the parent alkane or arene

**Fig. 8.6**

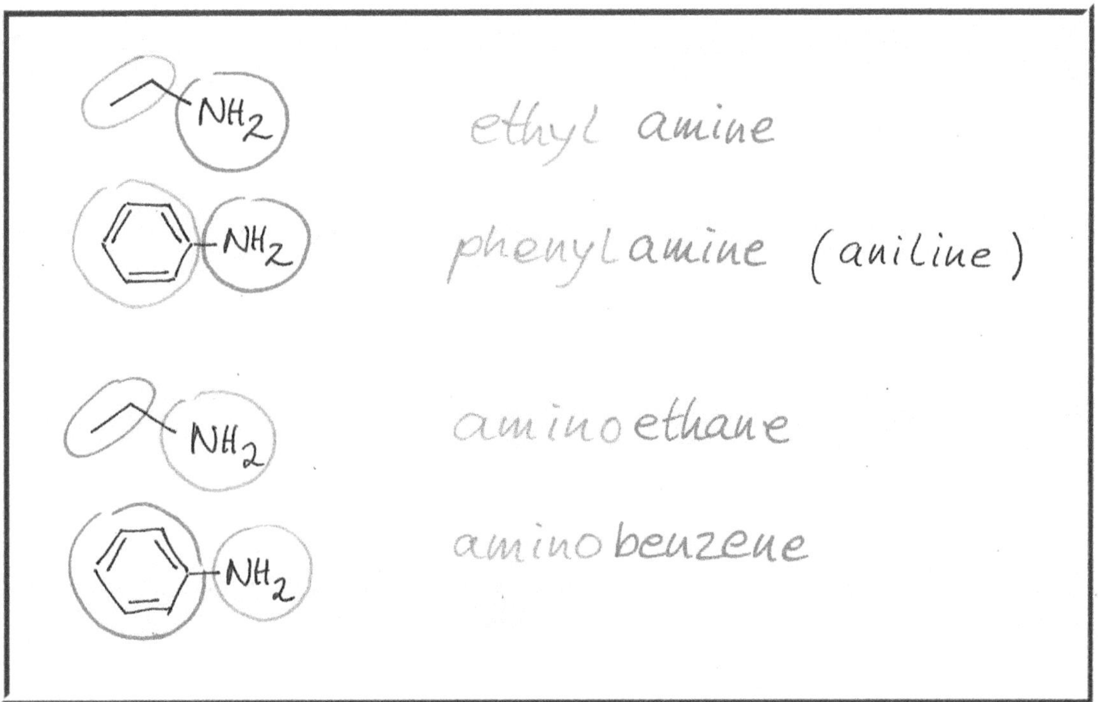


**Diamines** are named by using the basic alkane or arene to which the amino groups are attached, hereby <u>not</u> dropping the "e," followed by the name "diamine" as parent.

**Fig. 8.7**

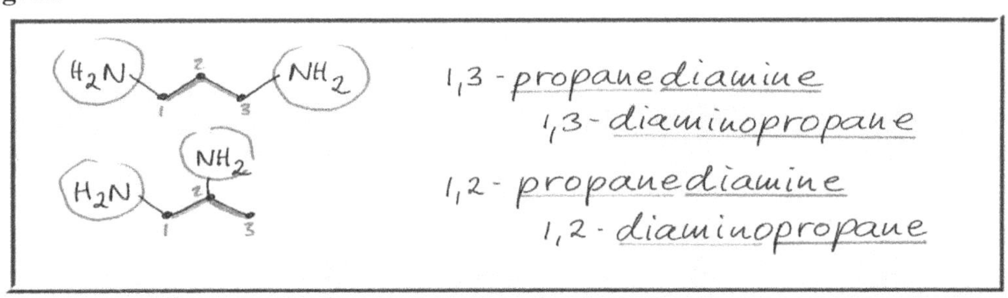

Secondary and teriary amines are a bit more difficult to name. When naming **secondary amines**, the longer chain attached to the 2° amino group will be named just like a 1° amine. The lower priority group attached to the Nitrogen of the amino group will be named by the normal alkyl or aryl name, only with a capital N attached (via hyphen/dash) to it as a prefix.

**Fig. 8.8**

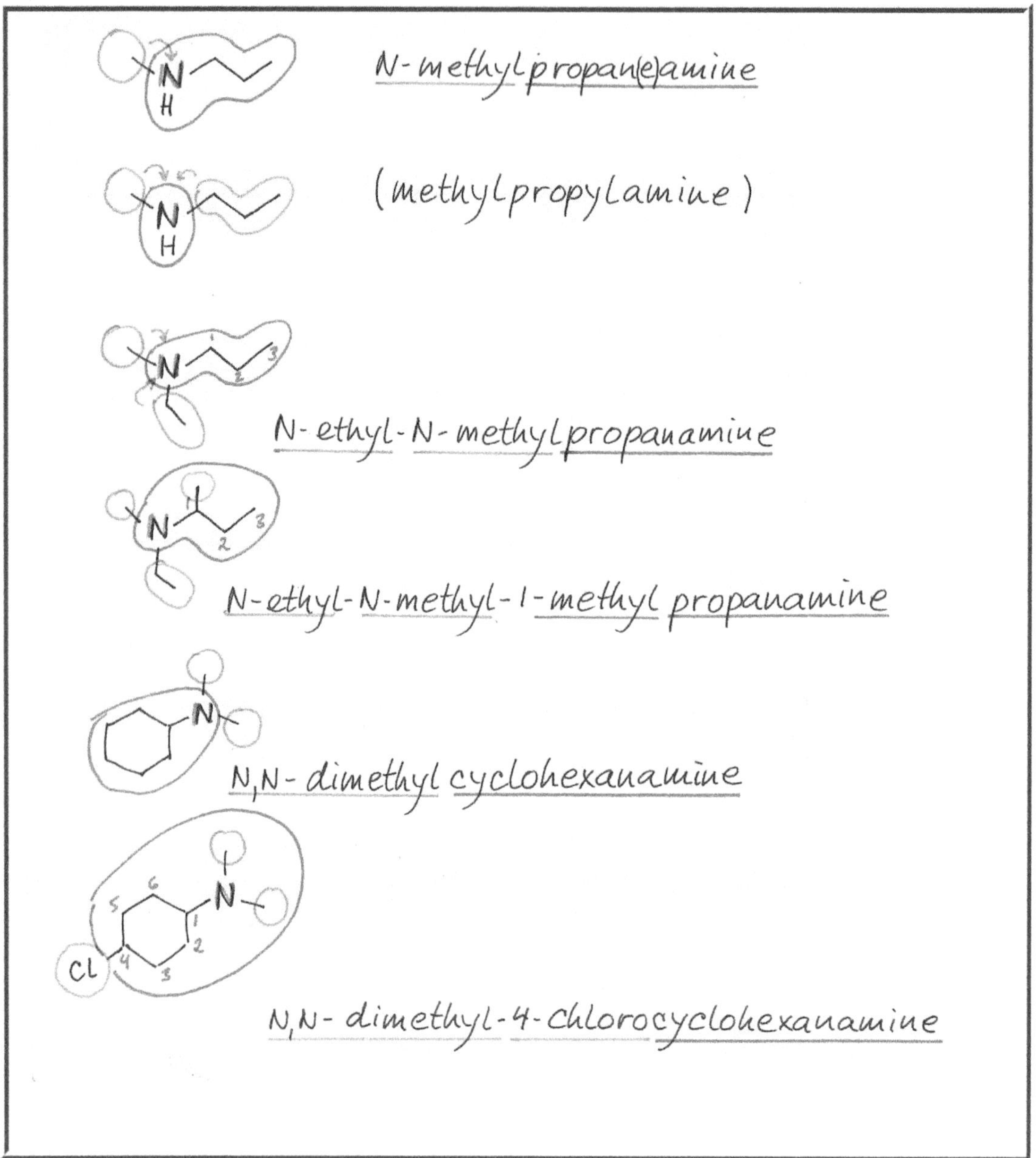

The N takes the place of a numbered location.

**Tertiary amines** are treated the same, only that the two separate groups will have to be indicated, each with an N- in front. So the groups attached to the nitrogen must have their own N- prefix. Alphabetical order counts here as well.

**Quaternary amines** are named as substituted ammonium salts.

**Fig. 8.9**

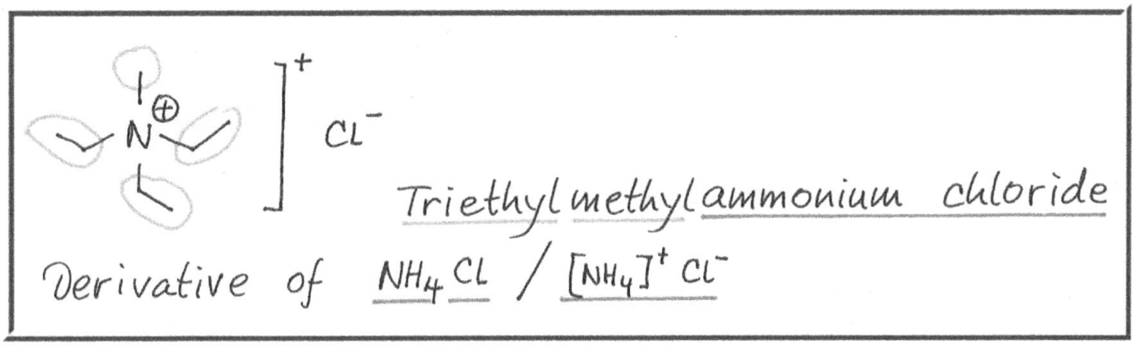


**Example Problem 8.1**
**Assign a name to each of the following molecules using two methods:**

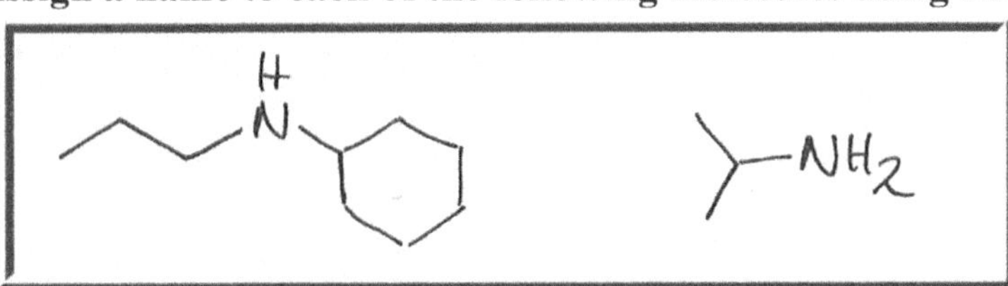

**Example Problem 8.2**
**Give the structures represented by the following names:**
***Ethylmethyl amine***
***2-Aminotoluene***
***N,N-Dimethylpropanamine***
***N-Isoproylcyclohexanamine***

Amines are not as polar as alcohols, having lower dipole moments as well as dielectric constants and therefore lower boiling points. Their intermolecular bonding forces (via H-bonding) between themselves are present but not as pronounced as is the case for alcohols since nitrogen has a lower electronegativity than oxygen (EN 3.0 vs. 3.5). However water-solubility is often better when compared to the corresponding alcohols of similar molecular weights. The amino groups act as stronger hydrogen bond acceptors than hydroxyl groups with the result of stronger hydrogen bonds with the solvent water molecules.
The boiling point decreases with the degree of substitution, the 1° amine being the highest and the 3° amine being the lowest, the latter having no polarized H-atoms left to allow H-bonding. However, like ethers they can act as H-bond acceptors.

**Fig. 8.10**

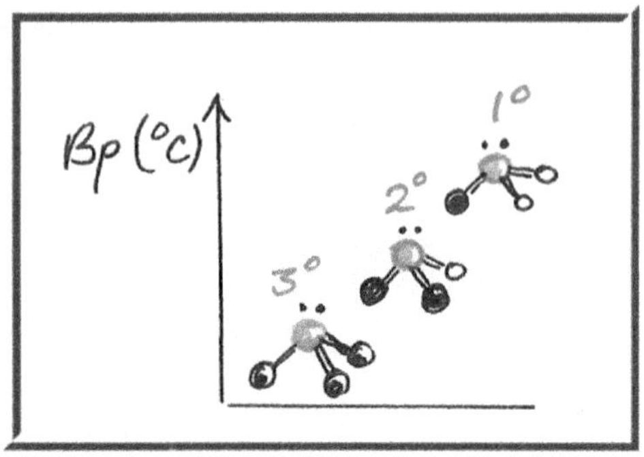


Ethylamine (Bp = 17° C) and Trimethylamine (Bp = 3° C) are gases at room temperature. Amines with less than 6-7 C-atoms are soluble in water. The arylamines all have high boiling points and, except for aniline, have a low solubility in water, becoming more and more hydrophobic.

The **basicity constant $K_b$**, which measures the strength of the amine as a base, depends on the degree of substitution, the base strength increasing from the primary, via the secondary to the tertiary amine. For instance, methylamine would therefore be a stronger base than $NH_3$.

This is due to the electron pushing quality of alkyl groups, enriching the electron density around the nitrogen atom, which thus becomes more enticing for a proton to land.

**Fig. 8.11**

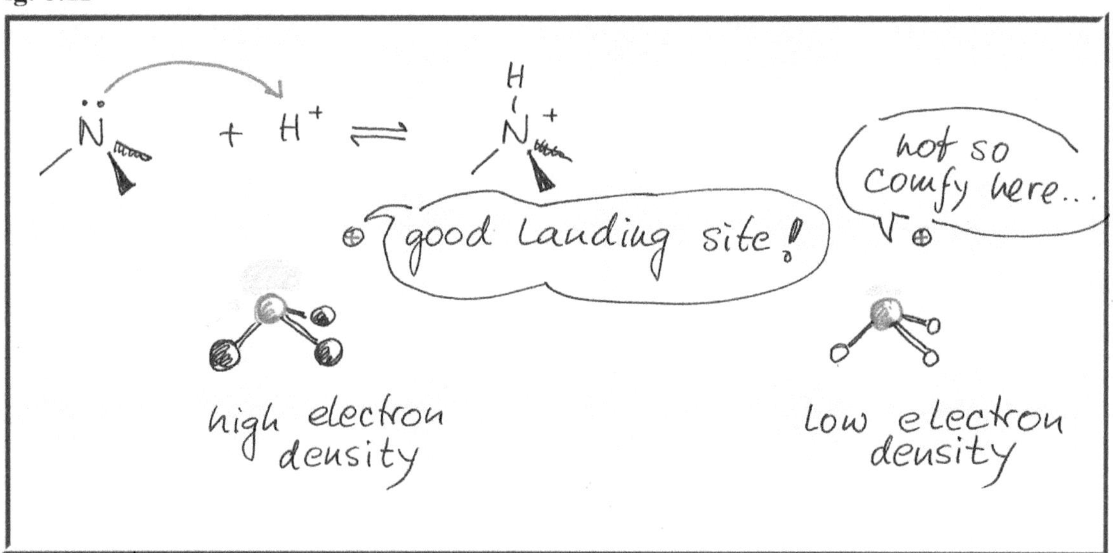


This also stabilizes the conjugate acid form ($R_3NH^+$), making it a very weak acid. So the trend of increasing acidity runs in the opposite direction.

**Fig. 8.12**

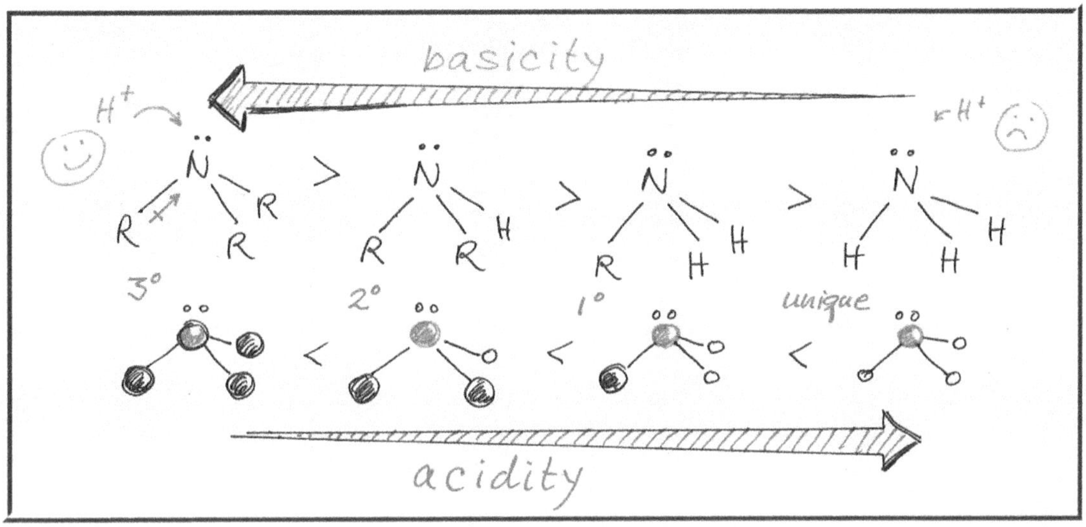


The average $K_b$ values for the alkanamines and cycloaalkanamines are around $10^{-4}$, which means the more usable $K_a$ values are around $10^{-10}$, derived from the water constant $K_aK_b = 10^{-14.}$

The corresponding $pK_a$ can be calculated from $pK_a = -\log K_a$.

**Fig 8.13**

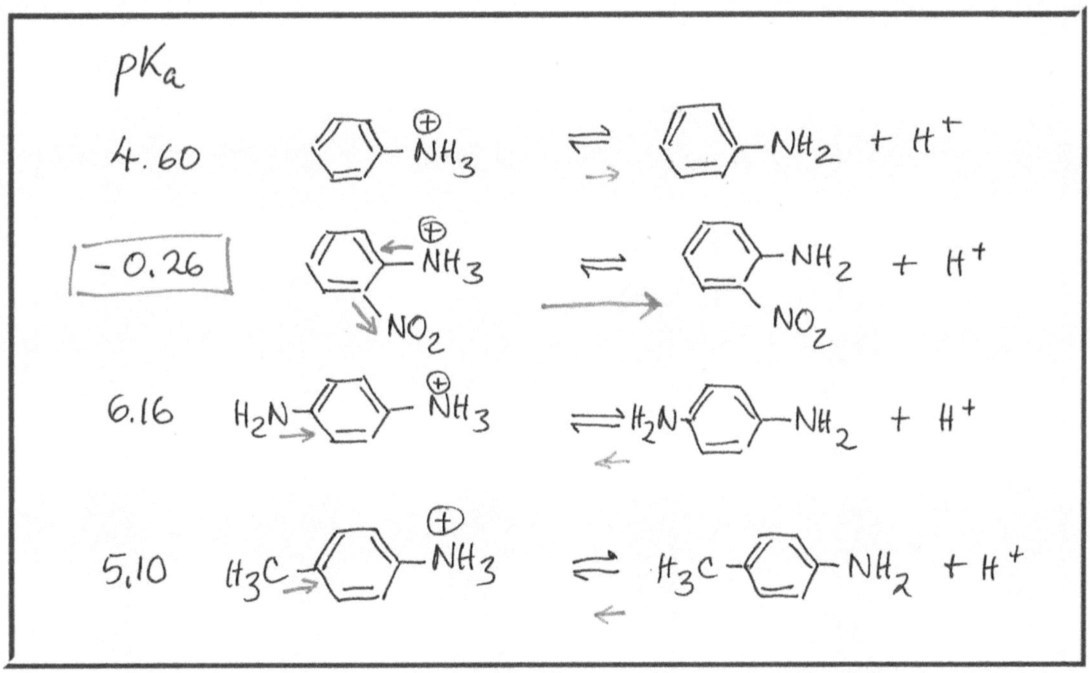

Arylamines are weaker bases than the alkylamines or ammonia, the reason for this being that the aromatic system has an electron withdrawing effect on the amino group, making it less attractive for the electron seeking proton. Induction and resonance effects lower the electron density around the nitrogen and thus less attractive. As a result aniline is much less basic than aminocyclohexane.

**Fig. 8.14**

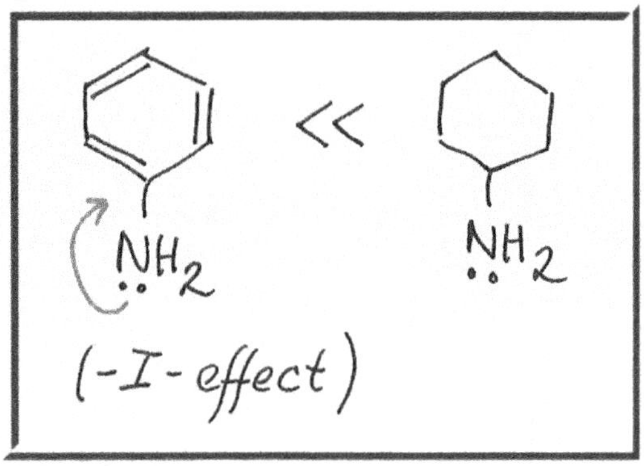


The strongest, electrically neutral, organic base ($pK_b = 0.4$) appears to be the molecule **guanidine**, a relative of the molecule **urea**. When picking up a proton, the guanidinium ion becomes an extremely well stabilized conjugate acid, provided by four possible resonance structures.

**Fig. 8.15**

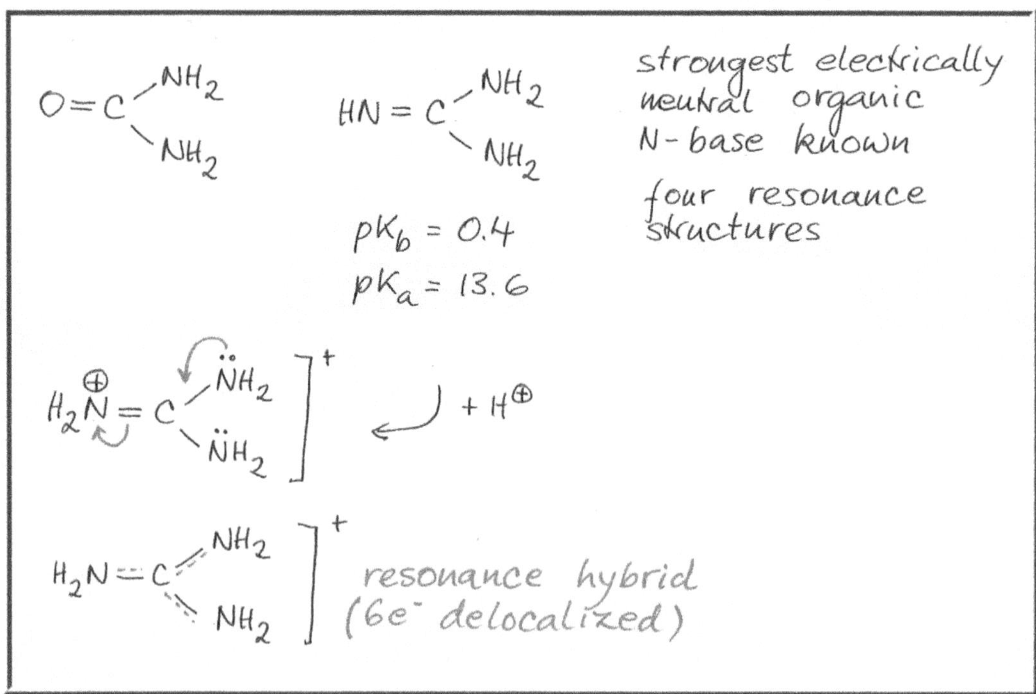

Amines can also act as acids, being able to donate a proton. However, they are extremely weak acids

**Example Problem 8.3**
**Compare each of the example pairs of amines as to their basicity or strength as a base and explaining your reasoning:**

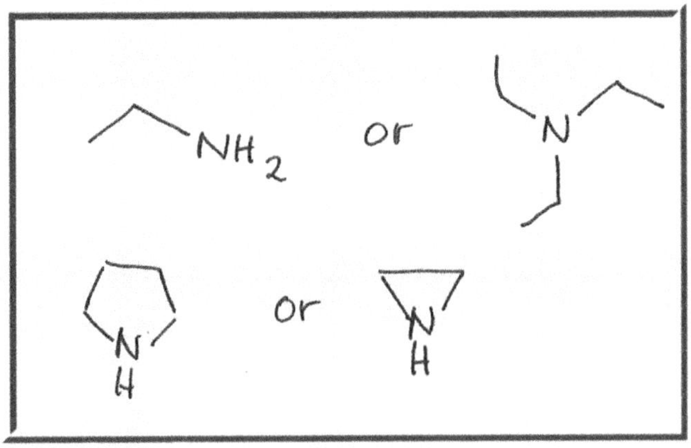

**Example Problem 8.4**
**Which of the following functional groups would help increase the basicity of aniline? Give an explanation for your choices.**

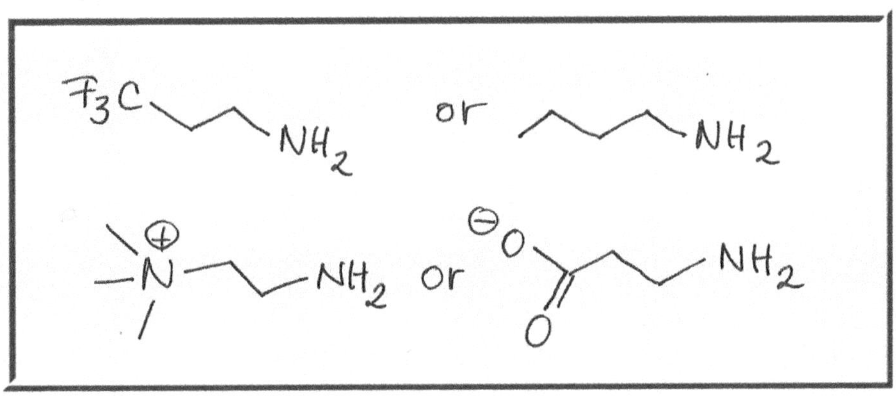

Alkylamines can be prepared by direct alkylation of ammonia, which leads to primary amines. Nucleophilic substitution ($S_N2$) of an alkyl halide with a primary amine, being an excellent nucleophile leads to secondary amine. In the same manner, a tertiary amine could be prepared by using a secondary amine as nucleophile. They can also be prepared by catalytic reduction with hydrogen of nitro-compounds, that carry the nitro ($-NO_2$) functional group.

**Fig. 8.16**

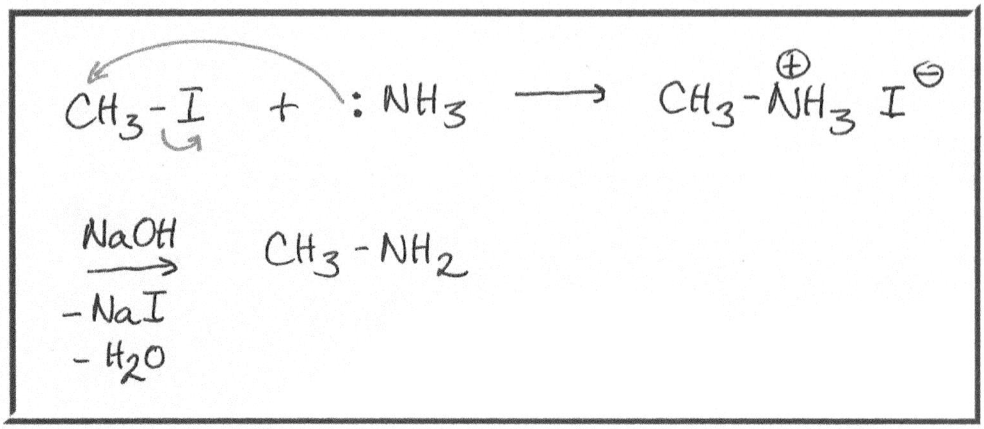

Conversely, amines can be oxidized to nitrogen compounds with higher oxidation states such as hydroxylamines (azanols), nitroso- (-NO) and nitro- (-$NO_2$) compounds.

**Fig 8.17**

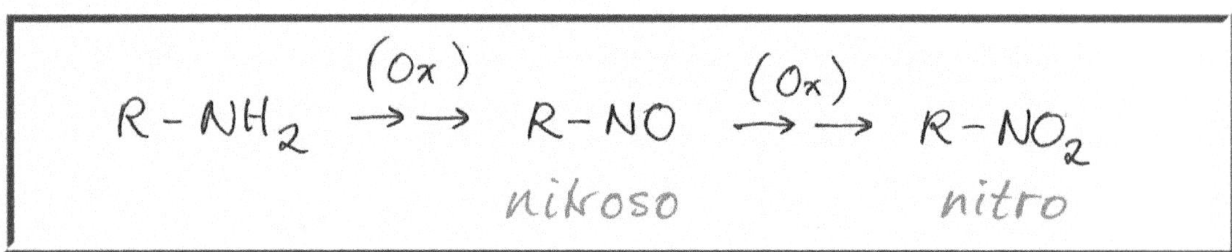

Aromatic amines are the key precursor molecules for the synthesis of **diazonium salts**, which via **diazo coupling reactions** lead to many **dyes and pigments.**
The initial step involves treatment of the arenamine with sodium nitrite in the presence of a strong acid such as HCl at 0°C. The nitrous acid ($HNO_2$), produced in situ (freshly, "on site") will lead via dehydration to the diazonium salt. The diazonium ion will then as the electrophilic reagent couple with another aromatic system to form the diazo compound via electrophilic substitution ($S_E$). Depending on the type of substituents, whether "pushers" or "pullers", and the type of aromatic ring (benzene or naphthalene systems), the resulting long drawn-out delocalized pi-systems can lead to various colors of the visible spectrum. The "stretching" of the pi-system throughout the length of the molecule has a **bathochromic effect**, (shift to longer wavelengths and lower frequencies of the electromagnetic spectrum) meaning, like a longer harp string having a darker note, the more extended electron system can resonate with longer wavelengths, giving the molecules visible colorations.

**Fig. 8.18**

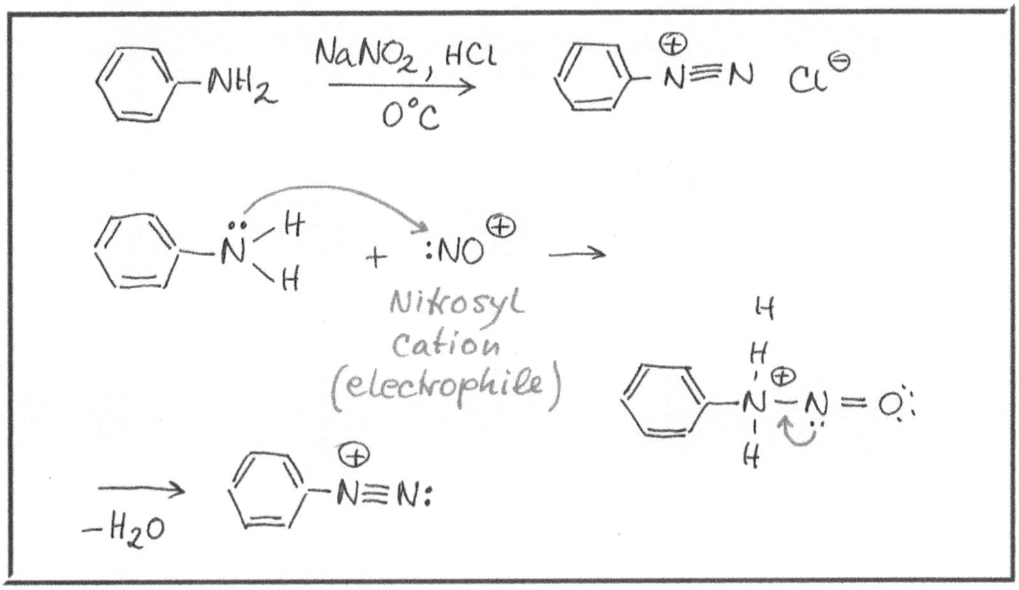


The following image shows a representative for an **azo-dye:**

**Fig. 8.19**

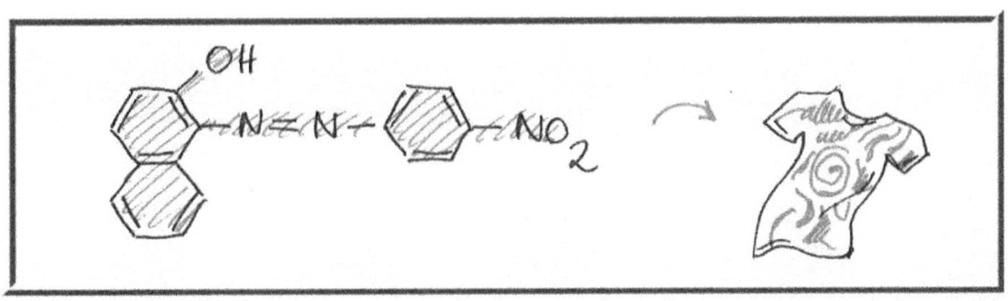


Due to their pronounced nucleophilic properties, amines can react with a variety of carbonyl compounds such as acyl halides, esters, and acidic anhydrides to give **amides**, compounds with the functional group called the **amido** –group (-CO-$NH_2$). These can also be prepared by dehydration reactions between a carboxylic acid and an amine.

**Fig 8.20**

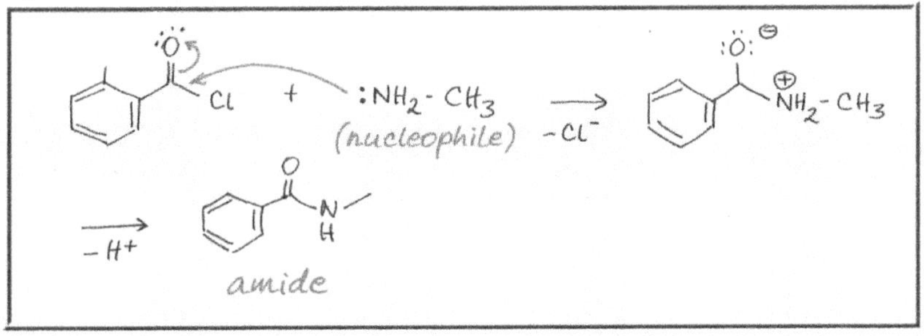

Like the regular amines, amides also have the distinction of either being primary, secondary, or tertiary types depending on the degree of substitution of the oxo-amino group. The nomenclature for the alkan(e)amides is as follows:

1) The amide group is named as suffix at the end of the molecule and the full alkane, including the carbon of the carbonyl group, named up front as prefix. For secondary and tertiary amides, the groups attached to the nitrogen are named as normal alkyl groups preceded by N- .

**Fig. 8.21**

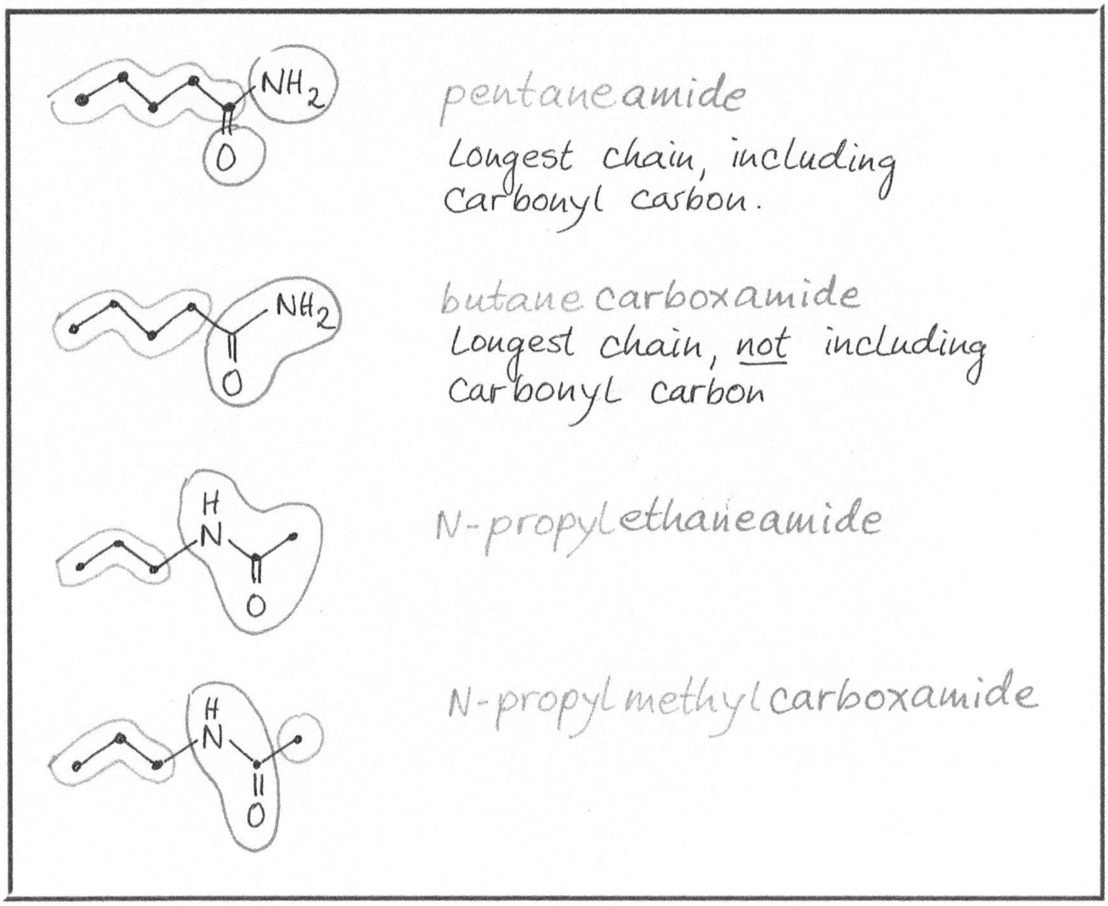


2) If the carbon of the carbonyl group is not included in the alkane chain, then the ending becomes **carbox(y)amide** with otherwise the same naming rules applied as above.

**Example Problem 8.5**
**Give the names for the following molecules**

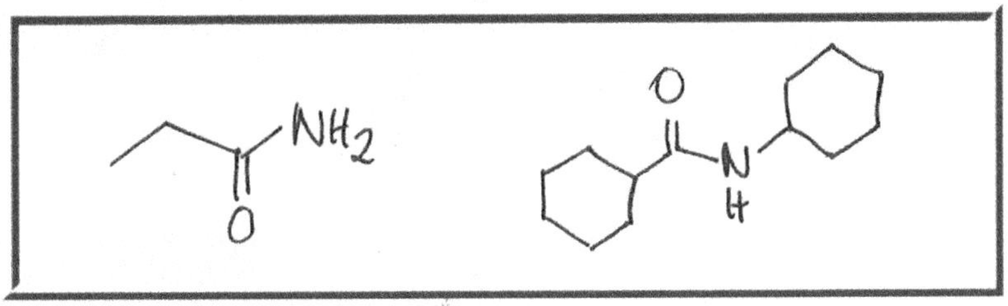


**Example Problem 8.6**
**Give the structures for the following names:**
***Pentaneamide***
***N-Methylethaneamide***
***N-Butyl, N-ethylpentanecarboxamide***
***Hexanediamide***

Hydrolysis of amides reverts back to the corresponding carboxylic acids and amine.
The amido group does not engage in proton exchanges, so it does not act as a base or acid. However, it has excellent hydrogen bonding properties. This often leads to dimerization and polymerization via hydrogen bonds.

**Fig. 8.22**

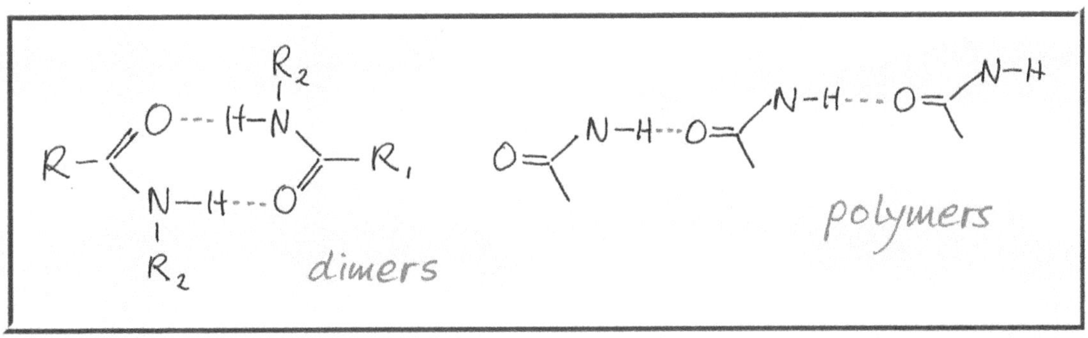


**Acetaminophen** is an important analgesic, over-the-counter (OTC) drug that can easily be prepared in the laboratory by reacting *para*-Aminophenol with Acetic anhydride according to the following reaction:

**Fig 8.23**

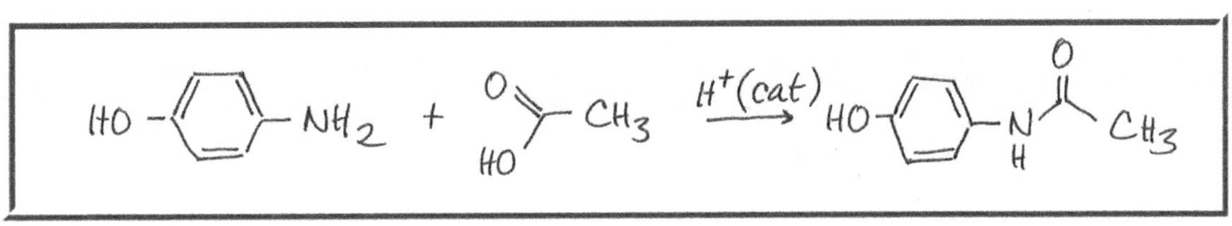

**Example Problem 8.7**
**Complete the following reactions and predict the type of reaction involved**

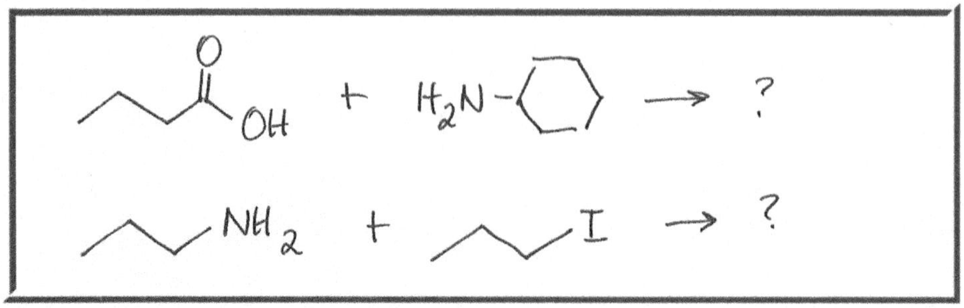


**Sulfonamides** represent another important family of drugs, used as the major infection fighting agents before the arrival of penicillins and other antibiotics. They are also amides insofar that they came about by aminating an acid, only here the acid is a sulfonic acid. They are still in use today in many countries and are commonly applied in veterinary medicine. The drugs can be prepared by nucleophilic attack by an amine on an aromatic sulfonylchloride molecule in the presence of base. As an antibacterial agent it interferes with the synthesis of folic acid in bacteria, mimicking the part of para-Aminobenzoic acid in the molecule.

**Fig. 8.24**

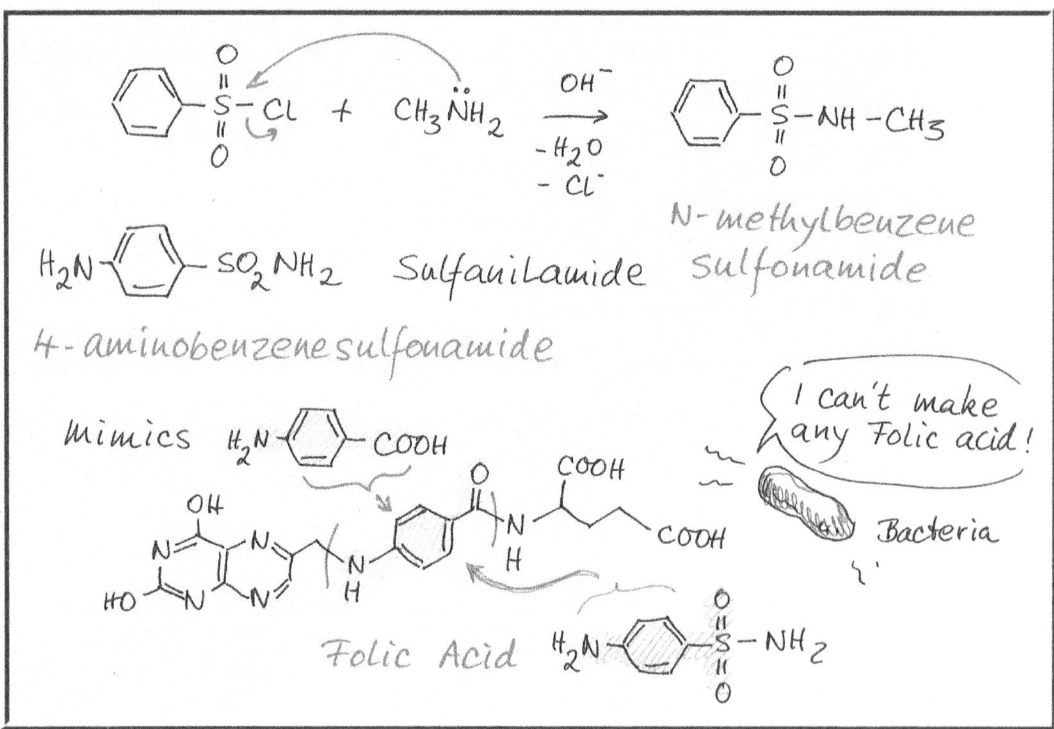


**Penicillins** are also important amides, consisting of two amido groups, one secondary, one tertiary (also known as a beta-lactam). They act by interfering with the formation of bacterial cell walls

**Fig. 8.25**

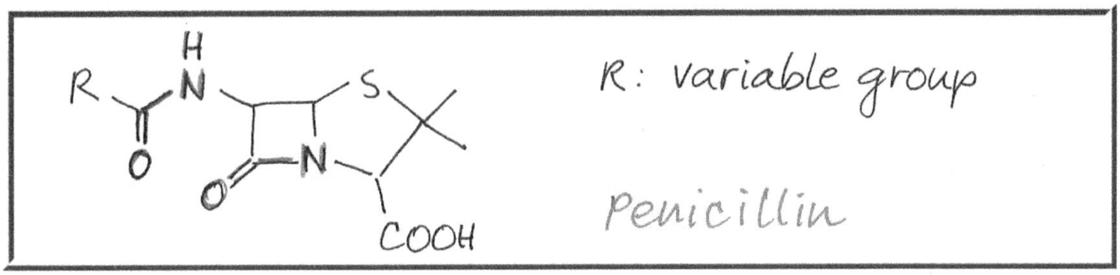


A major problem with amines, including some diazo compounds, is that it comprises many molecules that have shown carcinogenic effects in humans and animals. One of the first diazo dyes "Butter Yellow", used to colorize margarine, was withdrawn due to indications of being a potent liver carcinogen.

The following table shows some important carcinogens:

**Table 8.1**

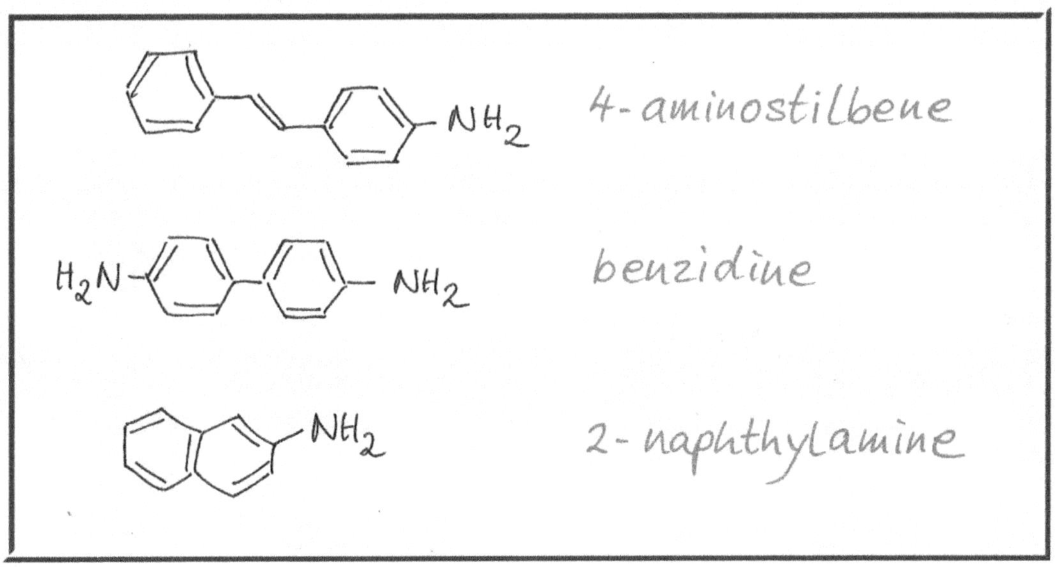

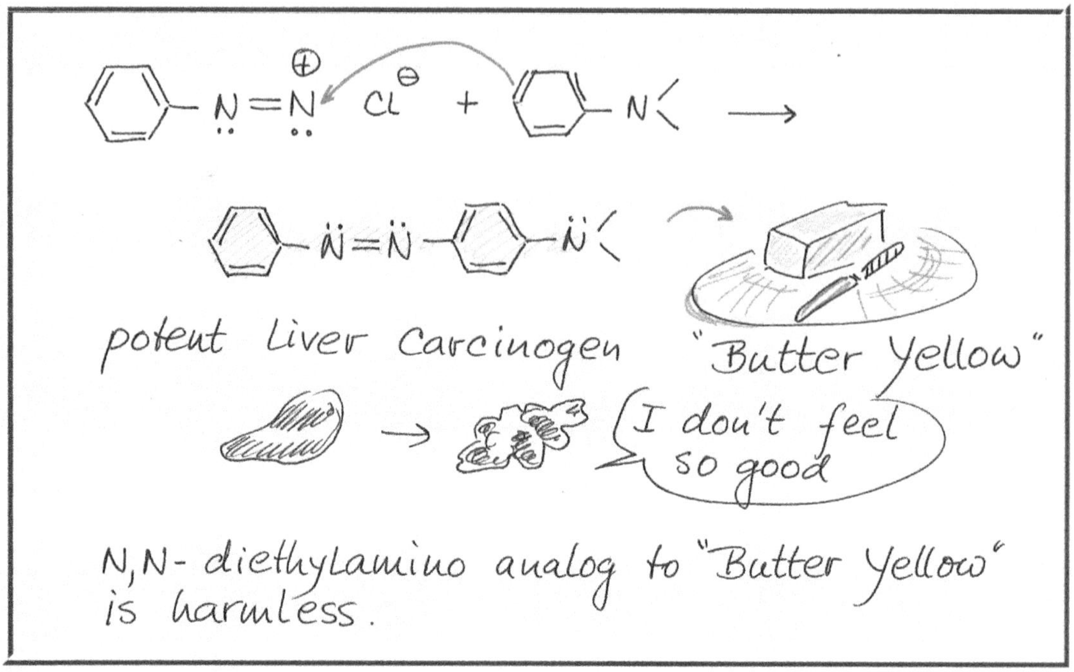


Next to the thiols, amines are the other group of organic malodorous (smelly) molecules. The smell of thiols can be subdued by oxidation to disulfides, the smell of amines neutralized by addition of an acid. When cadavers are found, the putrid smell of the decaying process is controlled by pouring lemon juice or vinegar over it. The malodorous amines are converted to ammonium salts, which do not smell. The detestable odor from decaying organic animal matter has also the purpose to keep living animals away, since toxic components (ptomaines) causing many diseases are part of the mix. However cadaver search dogs are trained on commercially available death formulation I and II solutions, since there olfactory organs are exceptionally well developed, that they can even detect trace amounts. Polyamines carry positive charges under physiological conditions and are therefore dangerous molecules since as wrap around DNA molecules which are negatively charged, thereby blocking its function as a code for protein synthesis.

**Fig. 8.26**

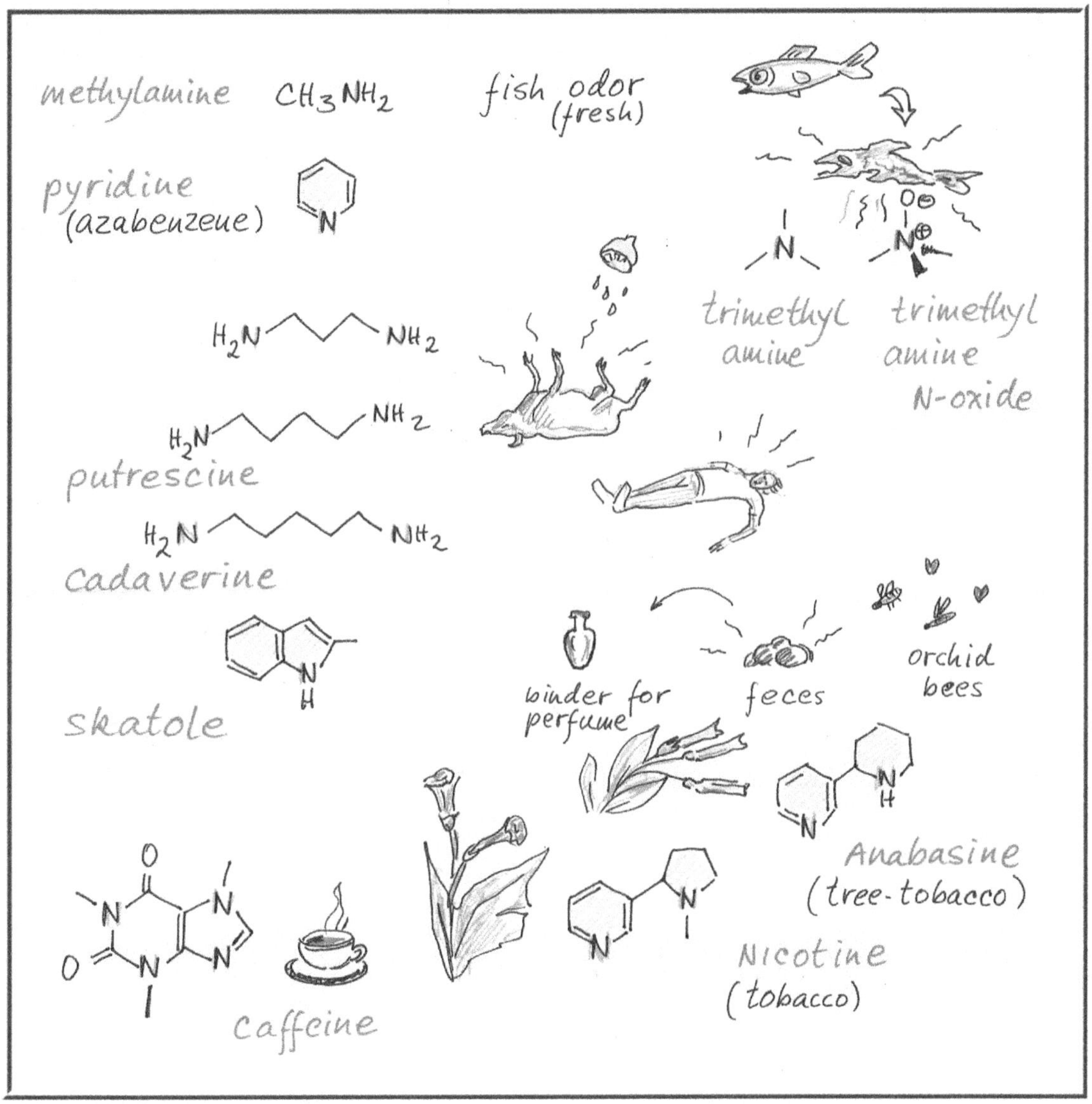


**Alkaloids** are mostly cyclic amines with alkali-like behavior or basic properties. They are nitrogen containing **heterocycles** – aromatic (hetarenes) or non-aromatic. They can be simple structures such as nicotine or highly complex molecules like Strychnine. They are extracted from a variety of plant materials with the help of organic solvents such as hexane, treated with inorganic acid (HCl) to make a salt, which improves their water solubility, separated from the rest of any present molecules, and finally reproduced as "free" base by treatment with inorganic base such as NaOH, which destroys the salt by removing the chloride ions. This type of procedure is often represented by a **flowchart**.

**Fig. 8.27**

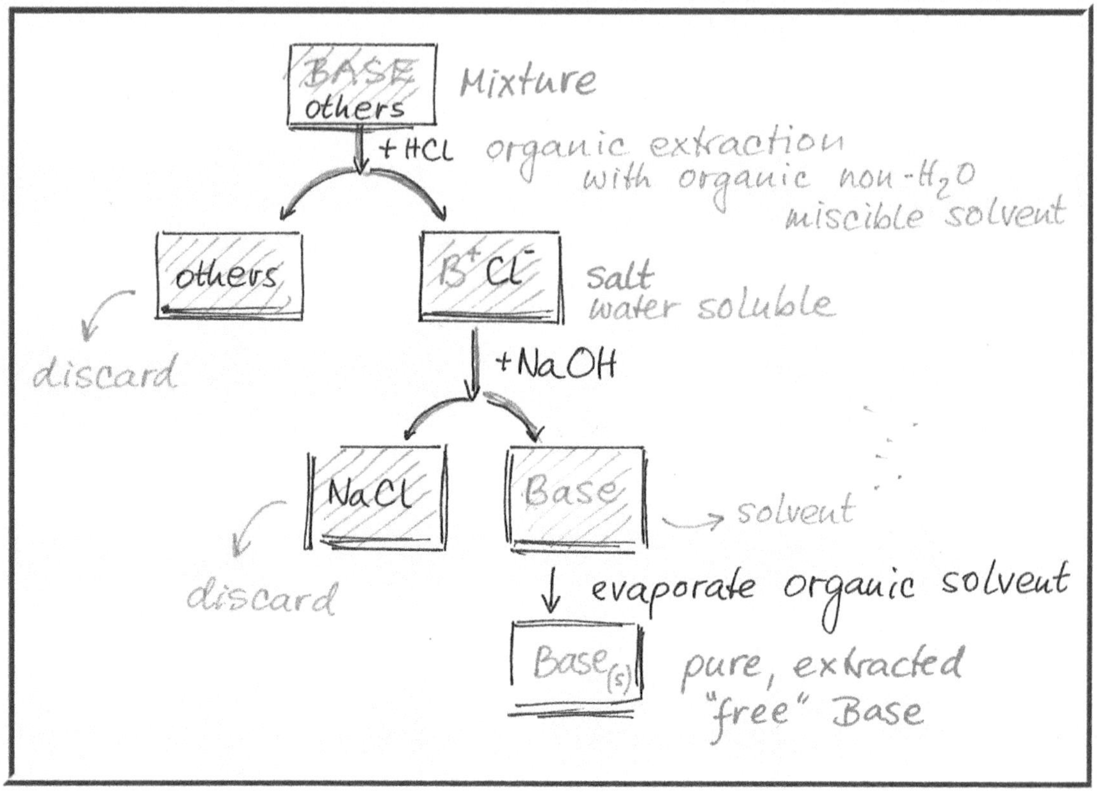


Any amine when treated with acid makes a salt, thereby becoming water-soluble.

**Fig. 8.28**

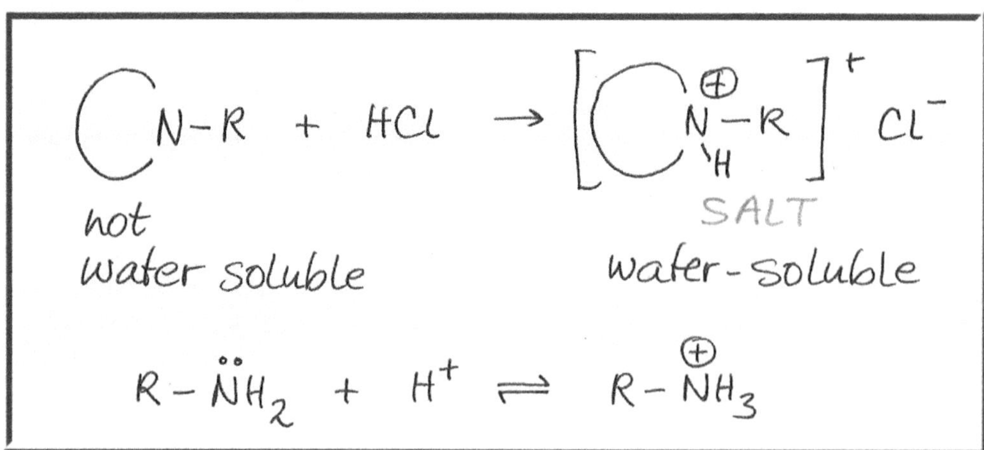


Caffeine and Nicotine are aromatic heterocycles. Atropine and cocaine are non-aromatic heterocycles. The same is true for morphine, a non-aromatic alkaloid found in poppy plants.

Coniine, a poisonous alkaloid found in the hemlock plant and used to poison Socrates (an ancient Greek philosopher who was sentenced to death), is a non-aromatic substance as well.

**Fig. 8.29**

Some have both types of rings such as Lysergic acid diethylamide (LSD). Some are not heterocyclic at all, having external amino groups such as mescaline. Many of them have psychotropic or psychoactive effects, either being a stimulant, hallucinogen or a combination.

**Fig. 8.30**

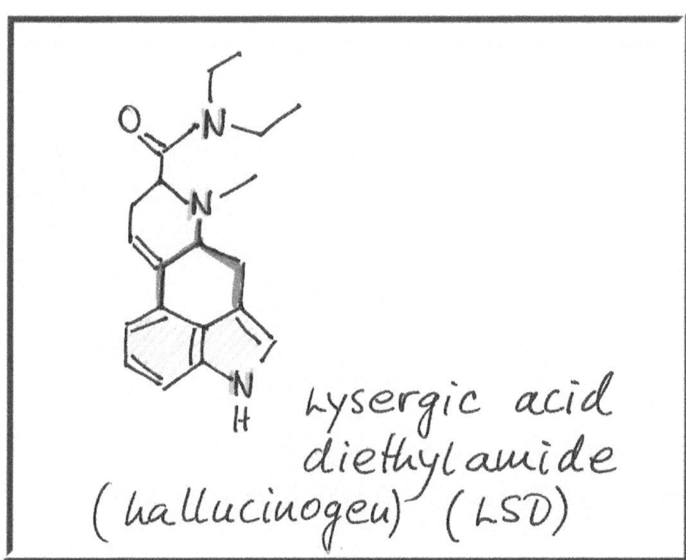

The following table shows a selection of alkaloids:

**Table 8.3**

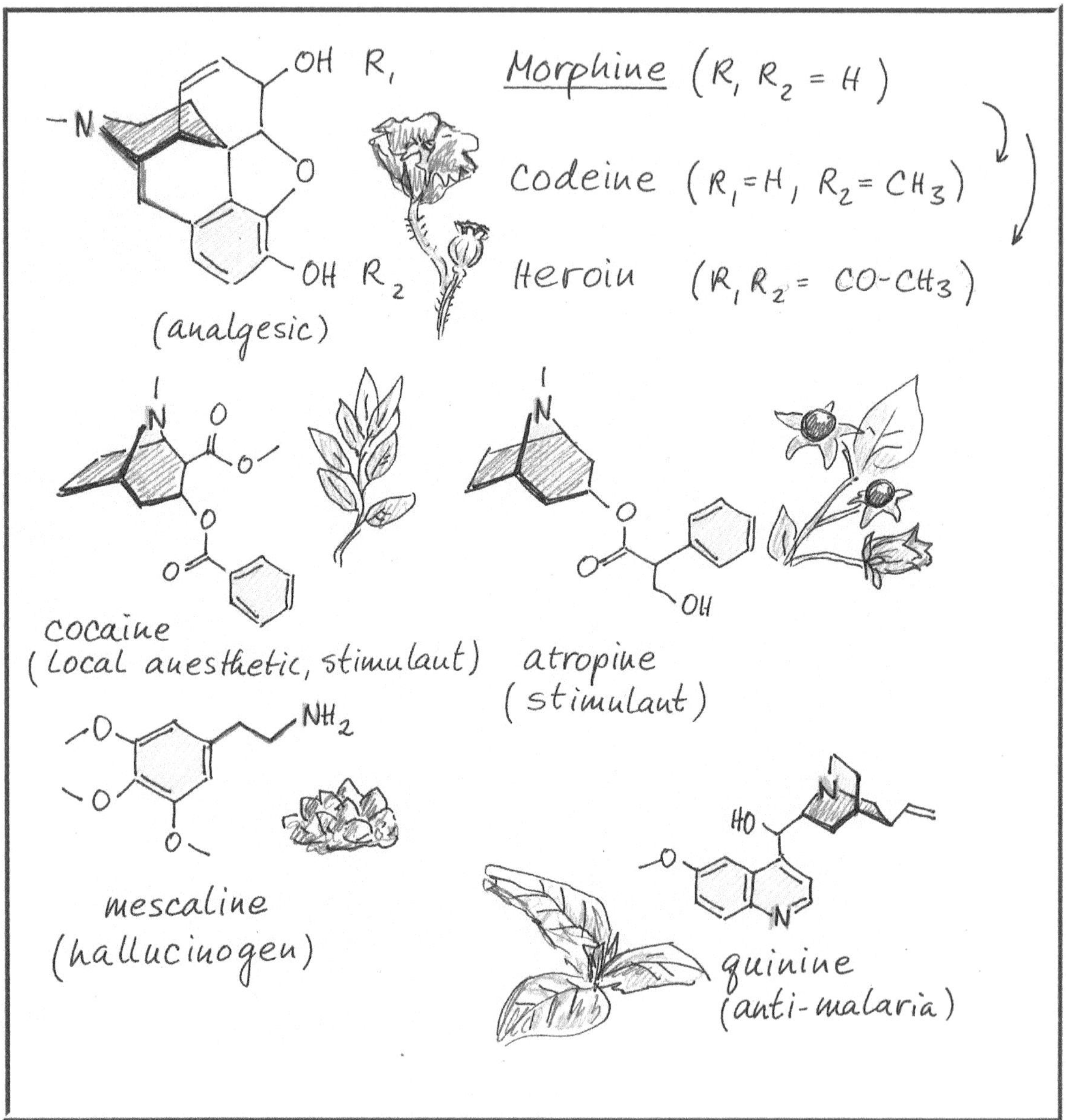


Amines have their basis in many vitamins (vital amines), especially the water-soluble ones, that act as cofactors for enzymes and are also found in other natural compounds, such as the aforementioned alkaloids, hormones, and metabolic and neurological regulators (dopamine, tryptamine, epinephrine, etc.).

**Fig. 8.31**

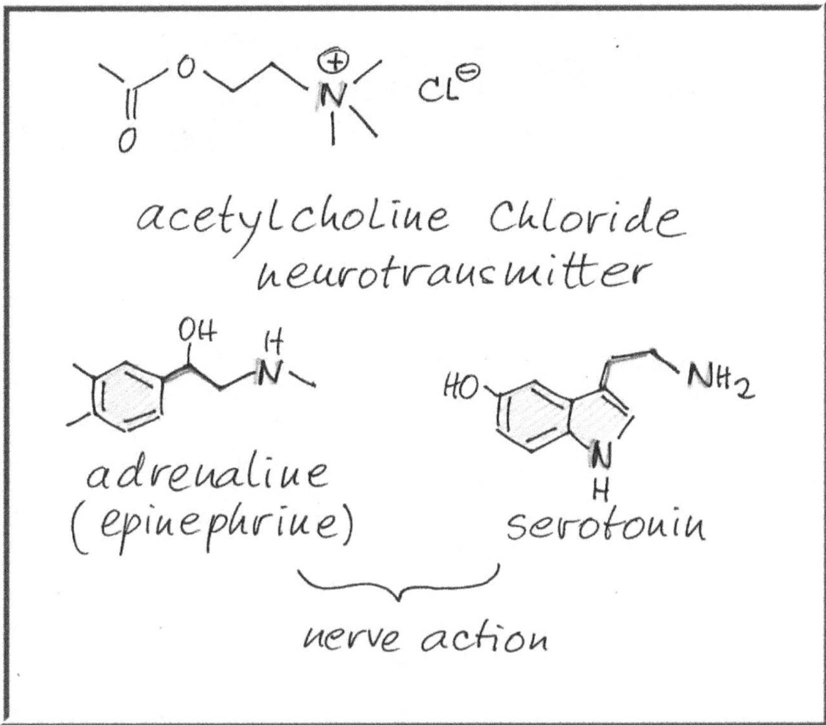


Nitrogen containing aromatic heterocycles are found in the B-group vitamins, which function as important co-factors in enzymes, especially the ones involved in oxidation reduction (oxidoreductases) reactions, by transferring electrons.

**Fig. 8.32**

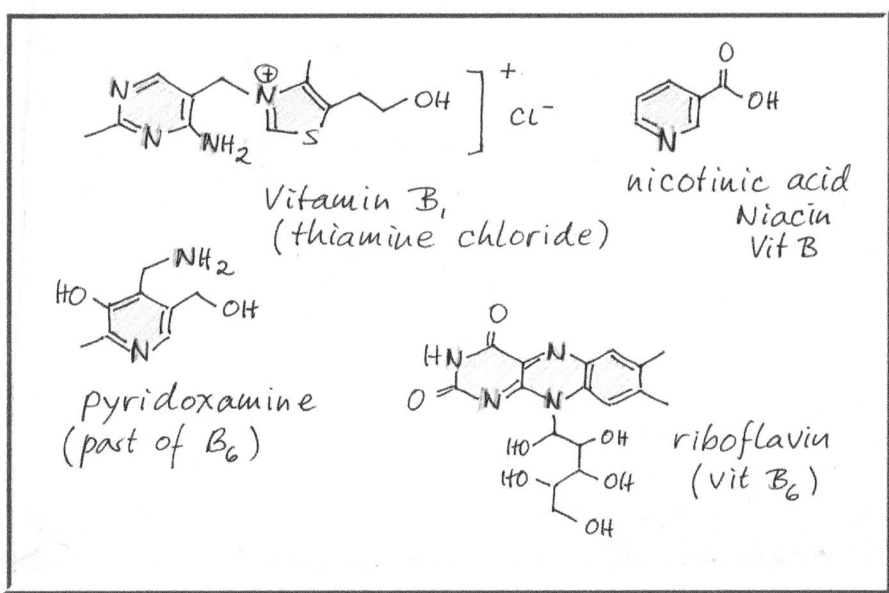

As amino acids they are the building blocks of peptides and proteins. As hetarenes they make up the four (adenine, guanine, cytosine, thymine/uracil) nitrogenous bases of nucleic acids (RNA, DNA) and are part of ATP (adenosine triphosphate), the energy molecule per se for living systems.

**Fig. 8.33**

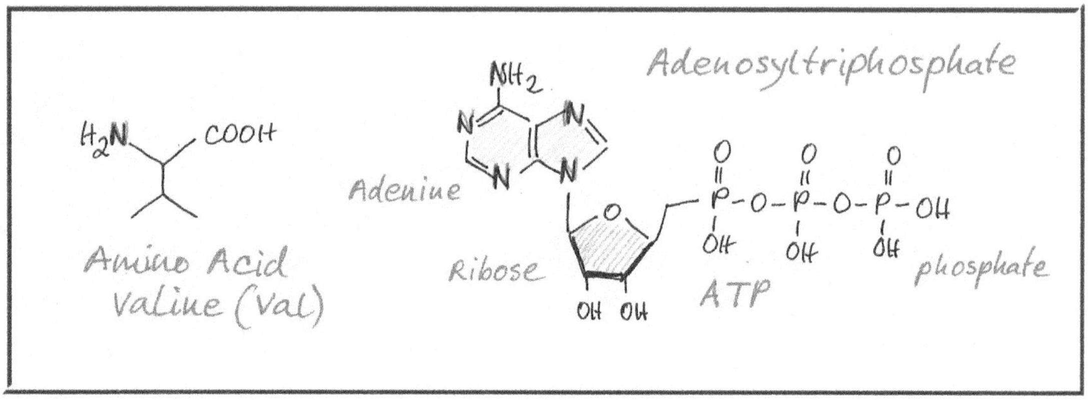


An interesting naturally occurring amide is the endogenous cannabinoid neurotransmitter Anandamide, found in minor amounts in all chocolate products.This chemical has a soothing effect on pigs.

**Fig. 8.34**

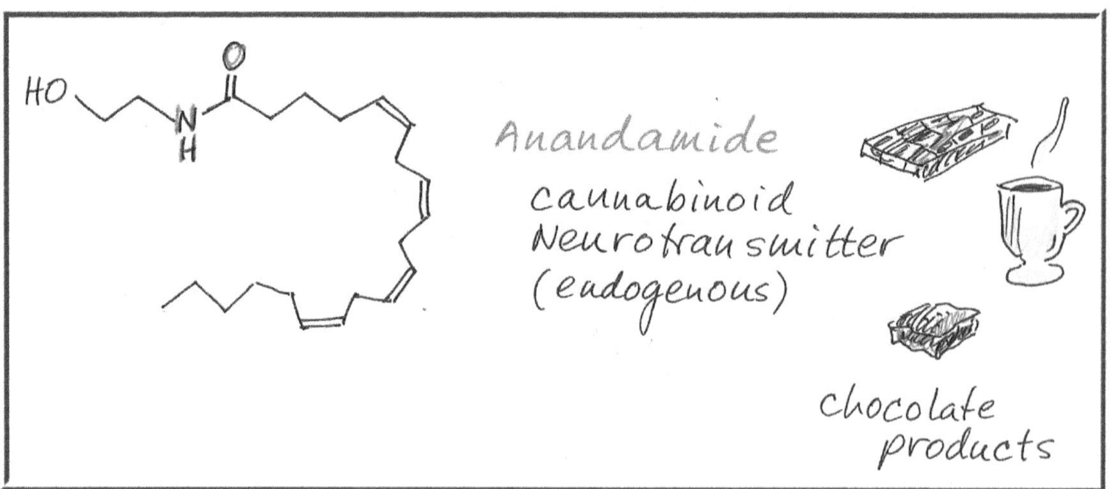


They are also involved in artificial, man-made products such as nylon, a synthetic fiber produced by condensation of diamines with dicarboxylic acids, namely 1,6-diaminohexane and hexanedioic acid for Nylon 6,6. The result is a polyamide fiber that is strengthened by many hydrogen bonds between the individual strands.

**Fig. 8.35**

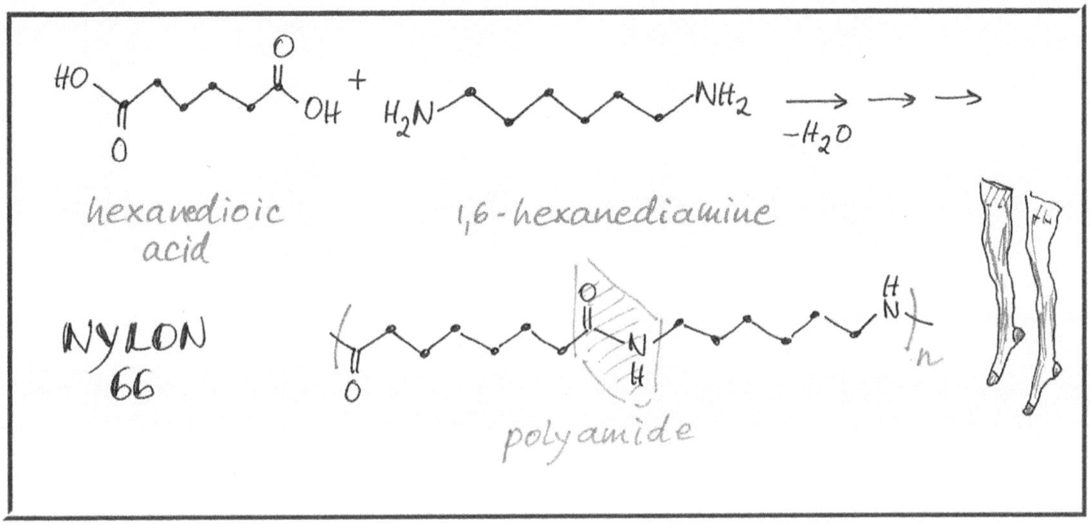


When nitrogen atoms are part of the ring the resulting compounds are known as N-heterocycles or N-hetarenes. Most representatives have common names. However, according to IUPAC, there are certain naming rules. Ring nitrogens are referred to as **aza** –groups. A systematic name for pyrimidine for example would be 1,3-diazabenzene. So the carbocylic ring will be the parent and the ring nitrogens treated like substituents, numbered and named by aza, diaza, triaza etc.

The following table shows a list of common N-containing heterocycles with their common and systematic names:

**Table 8.4**

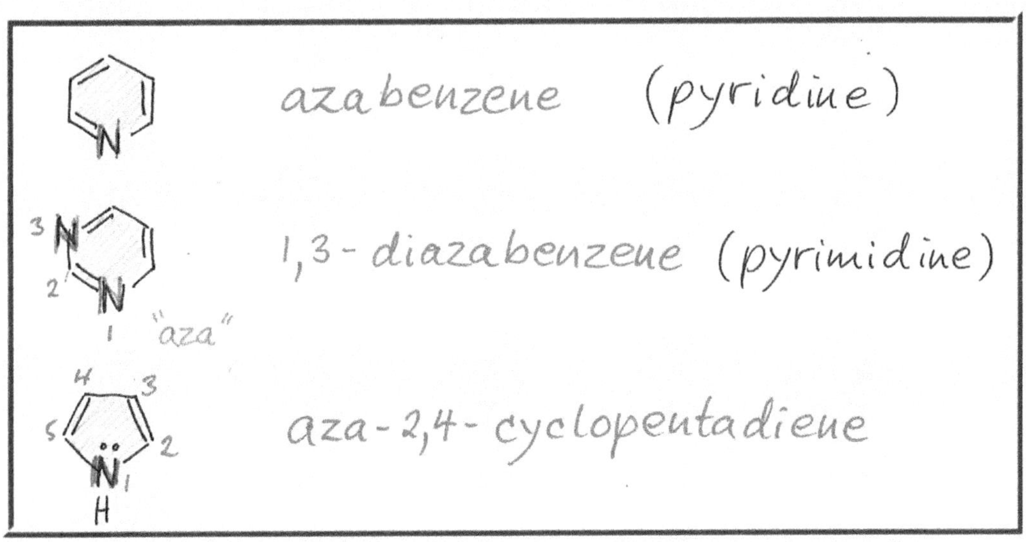

**Example Problem 8.8**
**For the following common names attributed to the shown heterocyclic structures give the systematic (IUPAC) names.**

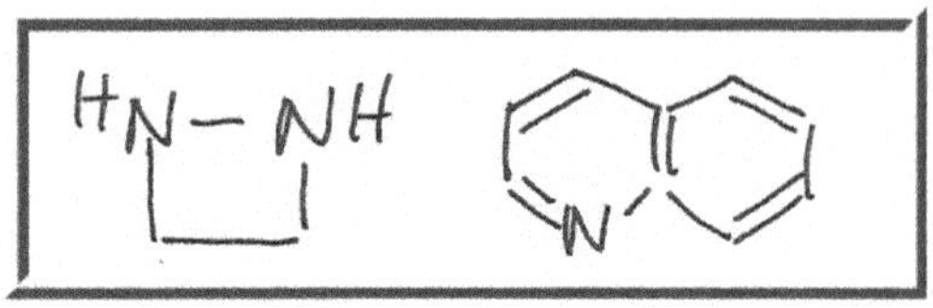

**Example Problem 8.9**
**Give structures and common names for the following systematic names:**
***1,3,4-Triazabenzene***
***Aza-2,4-cyclopentadiene***

**Example Problem 8.10**
**Classify the following molecules according to functional groups, degree of substitution, and molecular class.**

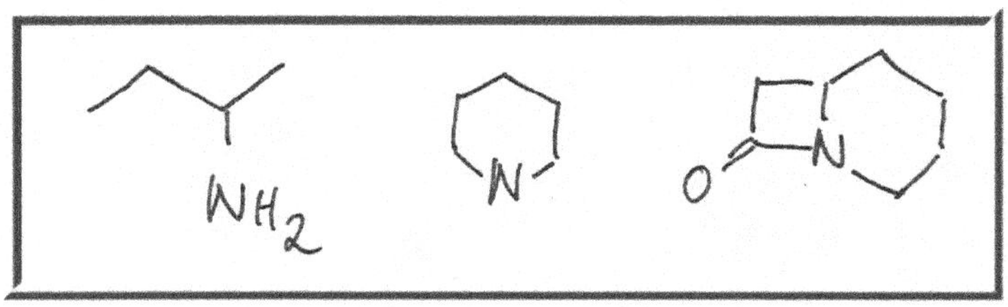

# EXPERIMENTS –Ch. 8

## *Amines and Amides*

**Key Terms:** The amino group, classification by degree of substitution, primary ($-NH_2$), secondary (-NHR), and tertiary amines ($-NR_2$) - ($1^o$, $2^o$, $3^o$), quaternary ammonium compounds ($NR_4^+$), Amides ($-CONH_2$, -CONHR, $-CONR_2$), basic properties (proton acceptor), $pK_a$ values, resonance structures, reactions (hydrolysis, salt formation, diazotization) azo dyes, diazonium coupling reaction

## I Molecular Modeling of Amines and Amides

### Chemicals:

*none*

### Equipment and Materials:

*Molecular modeling kit (Molymod)*

### Procedure:

1. Build a primary, a secondary and tertiary amine with any number of carbons. Name and sketch the molecules in the table below.
2. Build acetic acid (ethanoic acid) and connect via condensation (elimination of a water molecule) to the primary amine and then to the secondary amine. Name and draw the products of the resulting amides in the same table.

### Observations and Results:

| $1^o$ Amine<br>Name and structure: | $2^o$ Amine<br>Name and structure: | $3^o$ Amine<br>Name and structure: |
|---|---|---|
| | | |
| **Amide formation with $CH_3$ COOH<br>Name and Structure:** | **Amide formation with $CH_3$ COOH<br>Name and Structure:** | |
| | | |

*Shaded areas need not be performed (n.a.)*

## Questions and Problems:

1. What are the equations for the amide formation?
2. Why can't a tertiary amine dehydrate to an amide?
3. Why aren't amides bases like the amines? How does the neighboring carbonyl group affect electron density on the nitrogen atom?
4. A very efficient chemical test to distinguish between primary, secondary and tertiary amines is the **Hinsberg test**. Here benzenesulfonyl chloride reacts with free hydrogen groups of the amine to form a derivative upon release of HCl. Primary amines produce a precipitate that is soluble in NaOH, secondary amines produce a precipitate that is not soluble and tertiary amines produce an instable derivative that decomposes back into the original products. How would triethylamine react?

# II Physical Characteristics of Amines and Amides (Color, Odor, Solubility and Miscibility)

## Chemicals:

*Collection of various amines and amides:*
*Triethylamine*
*Aniline*
*N-methylaniline*
*Acetamide*
*Benzamide*
*HCl (10%)*
*NaOH (%)*

## Equipment and Materials:

*Test tubes and rack*
*Droppers and bulb*
*pH paper (red lithmus or universal)*
*Glass stirring rod*
*Hot water bath*

## Procedure:

1. In the fume hood test the vapors of the various amines and amides and record their odor in the table below.
2. Dipping a cleaned stirring rod into each solution, record the pH by tapping the end of the glass rod on pH paper.
3. Add 2 mL of water to each sample to test for water solubility.
4. Add a few drops of 10% HCl until the solution is neutralized (lithmus or pH paper) and investigate any changes in solubility.
5. To the samples containing the two amides add more of the 10% HCl (about 2mL), place them in the hot water bath and watch for signs for hydrolysis (reversal of the amidation reaction to the free amines)
6. Prepare two other samples containing each of the amides in separate tubes, add 2mL of 10% NaOH and also place in the hot water bath for a few minutes.

## Observations and Results:

Untreated amines and amides:

| Sample Solution (amine or amide): | Odor: | color | Solubility in $H_2O$ |
|---|---|---|---|
| Triethylamine | | | |
| Aniline | | | |
| N-methylamine | | | |
| Acetamide | | | |
| Benzamide | | | |

Neutralized amines with HCl:

| Sample Solution (Amine $\cdot$ $HCl$): | Odor: | Color: | Solubility in $H_2O$: |
|---|---|---|---|
| Triethylammonium chloride | | | |
| Anilinium chloride | | | |
| N-Methylammonium chloride | | | |

Hydrolyzed amides with HCl or NaOH:

| Sample Solution (Amide): | Odor: | Color: | Solubility in $H_2O$: |
|---|---|---|---|
| Acetamide | | | |
| Benzamide | | | |

## Questions and Problems:

1. What is the equation for ethylamine treated with HCl?
2. Why are amines treated with HCl or any other acid so much more soluble in water?
3. Personnel working in the forensics department pour lemon juice over cadavers which give off malodorous chemicals such as putrescine and cadaverine in order to reduce the stench. What is the chemistry behind this practice and why is it so effective?
4. Write the equation for the acid and base hydrolysis of acetamide. (Tip: HCl produces ammonium chloride and the free carboxylic acid; NaOH produces the sodium salt of the liberated acid and ammonia vapors.

## III Synthesis of Acetaminophen, an Analgesic Drug

### Chemicals:

*Para-Aminophenol* ***(fresh)*** *or 4-aminophenol*
*Acetic anhydride*
*Phosphoric acid ($H_3PO_4$, 85%)*

### Equipment and Materials:

*Erlenmeyer flask (med.)*
*Hot plate, stir bar (spin vane)*
*Perforated rubber stopper with glass rod (optional)*
*Melting point device*
*Vacuum filtration set-up (Büchner funnel, filter flask, adapters and vacuum hose*
*Metal stand and clamps*
*Ice bath*

### Procedure:

1. Weigh 1.5g of fresh *para*-aminophenol (care must be taken that only light- colored aminophenol is used; if dark or black the chemical has polymerized and is no longer usable, it is sensitive to heat and light) in a tarred out Erlenmeyer flask (125mL).
2. Add 25mL of water and 20 drops of concentrated phosphoric acid ($H_3PO_4$, 85%).
3. Add a stir bar to the mixture and heat on a hot plate to boiling, while setting the magnetic stirring knob to an even rotation. Carefully add 2 mL of acetic anhydride and continue stirring and heating for 1-2 minutes.
4. Transfer the flask to an ice bath and cool for about 30 min. for crystals to develop. Rubbing the sides of the flask with a glass rod or seeding the solution with an actual crystal of the acetaminophen might help accelerate the crystallization process.
5. Vacuum filtrate the crystals on a **Büchner Filtration apparatus** and rinse the residue with ice water.
6. Allow the crystals to dry (might need another class period), weigh the crude product and determine the melting point (theoretical Mp. = 169-171°C for the product and Mp. = 189-190°C for the reactant).
7. Calculate the theoretical and practical (experimental) yield of the product.
8. Recrystallize the crude crystals by re-dissolving them in about 20mL of hot water and allowing them to cool. Impurities will thus be removed. Vacuum filtrate and re-determine the yield and melting point.

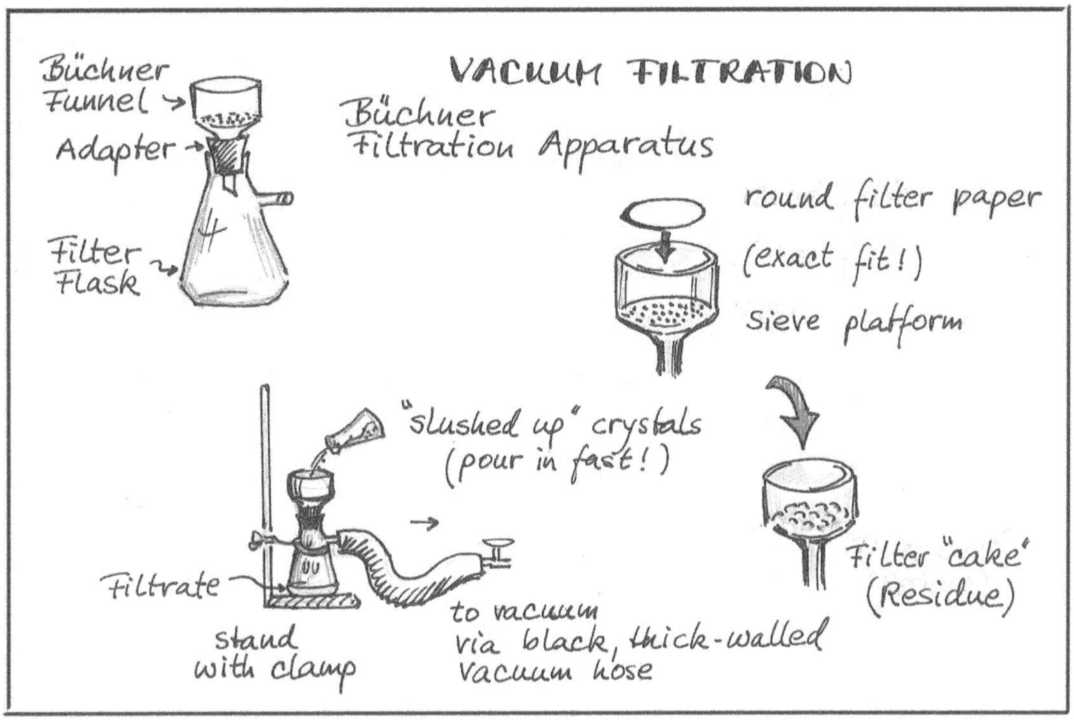


## Observations and Results:

Record the results in the following table:

| Chemical: | Formula | Mass (g) (input/output) | Molar mass (g/mol) | Moles (mol) |
|---|---|---|---|---|
| ***Para*-aminophenol (white or off-white colored, <u>not</u> grey or black)** | | 1.5 | | |
| **Acetaminophen** | | | | |

Calculation of theoretical yield:

$$1.5\text{g}_{\text{ reactant}} \left(\frac{--mol}{--g}\right)\left(\frac{1\ mol}{1\ mol}\right)\left(\frac{--g}{--mol}\right) = \_\_\ \text{g}_{\text{ product}}$$

Calculation of practical yield in %:

$$\frac{practical\ yield\ (g)}{theoretical\ yield\ (g)}\ or\ \frac{(have)}{(should\ have)} \times 100\% = \%\ yield$$

## Questions and Problems:

1. Using the dimensional analysis formula shown above calculate the theoretical yield of the product.
2. What is the practical yield of the synthesized acetaminophen?
3. Write the equation for the reaction.
4. A trade name for acetaminophen is para-acetamol. In which common over the counter drugs (OCD's) is acetaminophen involved?
5. A similar drug is phenacetin, which is produced by condensing 4-ethoxyaniline with acetic anhydride. Write an equation for this reaction)

# IV Diazotization and Synthesis of an Azo Dye

## Chemicals:

*Para-aminophenol (fresh 4-aminophenol)*
*NaOH (10%)*
*2-Naphthol (β-Naphthol)*
*Sodium nitrite ($NaNO_2$)*
*HCl (conc.)*

## Equipment and Materials:

*Digital top-loading balance*
*Erlenmeyer flasks (100-125 mL)*
*Ice bath and beaker with ice water*
*Graduated cylinders*
*Dropper with bulb*
*Glass stirring rod*
*Büchner vacuum filtration apparatus*

## Procedure:

1. Fill a small Erlenmeyer flask with 30mL NaOH (10% solution) and add 1.44g of 2-naphthol (β-naphthol) while cooling the solution in a beaker filled with ice water. Use metal stand and clamp to secure the Erlenmeyer flask in place.
2. In a separate small Erlenmeyer flask, tarred out on the scale, weigh out 0.70g of sodium nitrite ($NaNO_2$) and dissolve it in 5 mL of water. Set aside for use in the next step.
3. In another Erlenmeyer flask (100-125mL) weigh out 1.20g of (light-colored!) 4-aminophenol and add 45 mL of water. Slowly add 12 mL concentrated HCl and stir until fully dissolved. Cool in an ice bath and slowly add with a dropper the $NaNO_2$ solution prepared above (#2), while continuing to stir constantly (formation of the diazonium salt solution).
4. Add the diazonium salt solution, prepared above, slowly, taking about 5 min., to the alkaline 2-naphthol solution, prepared in step #1, while constantly stirring and keeping the solution cooled in an ice-water bath. Stir for another 5-10 min. for the coupling reaction to go to completion.
5. Suction off the brick-red crystals using the Büchner vacuum filtration apparatus, rinsing the residue with ice cold demineralized water. Weigh when dry to determine the practical yield.

6. Test the dying properties of the azo dye as an aqueous solution on a piece of wool, acrylic fiber and cotton. Test its binding quality or color fastness by rinsing with water.

## Observations and Results:

Tabulate the observed results and add the other information required according to the following table:

| **Chemical:** | **Formula** | **Mass (g)** (input/output) | **Molar mass (g/mol)** | **Moles (mol)** |
|---|---|---|---|---|
| **2-Naphthol** | | 1.44 | | |
| **4-aminophenol (→ *diazonium salt*)** | | 1.20 | | |
| **Azo dye (product of coupling reaction)** | | | | |

Dying properties of the azo dye:

| **Wool (protein)** | **Acrylic fibers (synthetic)** | **Cotton (cellulose)** |
|---|---|---|
| | | |

## Questions and Problems

1. Describe the physical characteristics of the synthesized product.
2. Do research on some other azo dyes and draw their structural formulas.
3. What is the reaction leading to the azo dye synthesized using the chemical in the experiment?
4. What fiber best takes on the color of the dye?

5. What makes a good dye? What is the difference between a dye and a pigment?
6. What is the theoretical and practical yield of the dye synthesized? (Show calculation!)

## V Purity Determination of Synthesized Analgesics (Aspirin and Acetaminophen) with Thin Layer Chromatography (TLC)

### Chemicals:

*Acetyl salicylic acid (Exp. 17 IV)*
*Acetaminophen (Exp. 18 )*
*Ethanol*
*Developing solvent :*
*( 75% ethylethanoate/ethylacetate*
*25% hexane or 3:1)*

### Equipment and Materials:

*Thin layer chromatography set-up (TLC)*
*Drawn-out capillary tubes*
*Spot plate*
*Developing chamber (tall beaker with watch glass)*
*TLC plate (with fluorescent silicagel coating)*
*UV lamp (254 nm)*

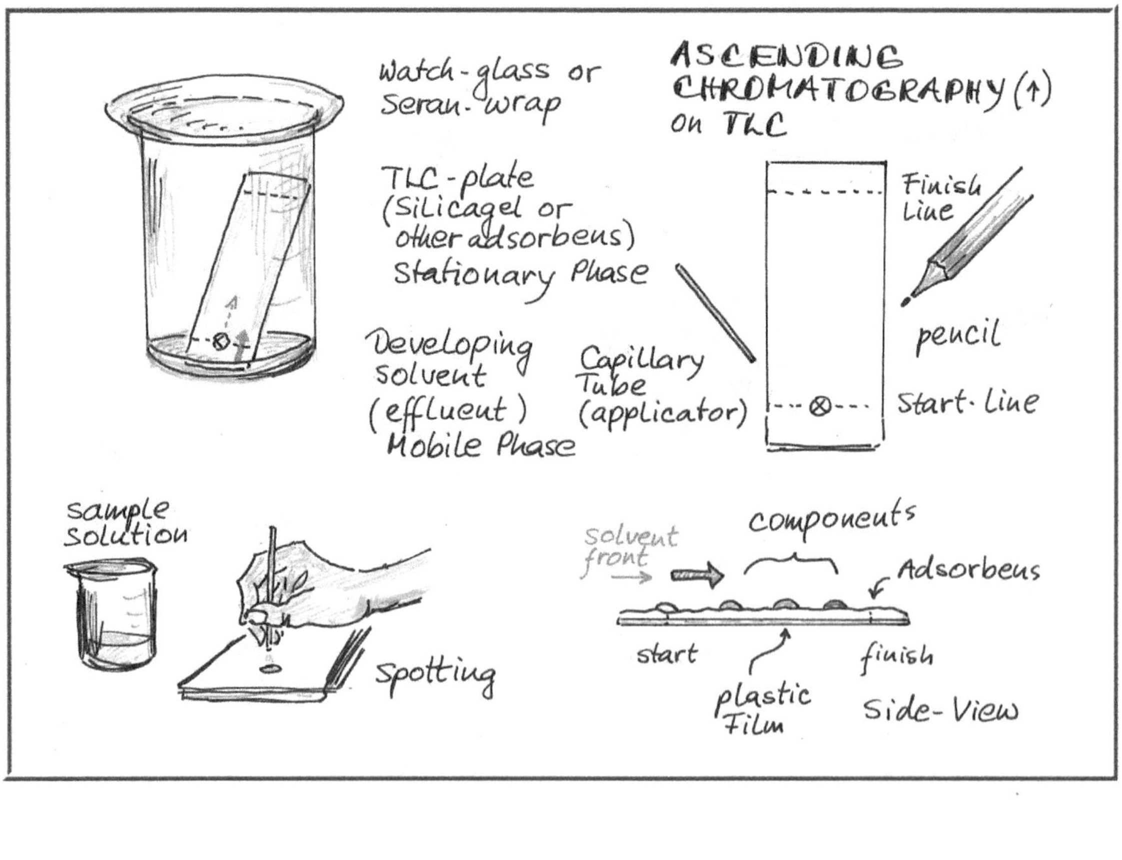

## Procedure:

1. Prepare the developing tank by placing the developing solvent about 0.25 to 0.5 cm high in the tall beaker. Put the watch glass on top, so that the tank can get saturated with the vapors.
2. About 1cm from the bottom edge of the TLC plate (adsorbens side) apply about 0.5 cm from the left edge a spot of alcoholic acetyl salicylic acid and about 0.5 cm to its right another spot of an aspirin standard solution. Then, 0.5 cm further to the right, apply a spot of alcoholic acetaminophen solution and 0.5 cm to the other edge a spot of an acetaminophen standard solution. If needed, dab the spot another time with solution. Allow the 4 spots to dry.
3. Mark the finish line about 1 cm from the top with a pencil or scratch a thin line with the edge of a spatula.
4. Place the TLC plate in the developing tank and let the solvent mixture (mobile phase, effluent, eluent) creep up the silica gel adsorbent phase (stationary phase) via capillary forces and drag the solutes or components up to whatever level at which they have reached their equilibrium with both phases.
5. When the solvent front has reached the finish line, remove the TLC plate and let dry.
6. View the plate under the UV lamp and observe dark spots on the yellow fluorescent background. Circle with pencil and dot the center of the spot(s).
7. Calculate the retention factors ($R_f$) according to the following equation:

$$R_f = \frac{\textit{distance travelled by spot (solute,component)}}{\textit{distance travelled by solvent front}}$$

## Observations and Results:

| **Chemical:** | **Distance of spot(s) (mm)** | **Distance of solvent front (mm)** | **$R_f$ value** |
|---|---|---|---|
| Pure aspirin (standard) | | | |
| Pure acetaminophen (standard) | | | |
| Synthesized aspirin and other components, if present | | | |
| Synthesized acetaminophen and other components, if present: | | | |

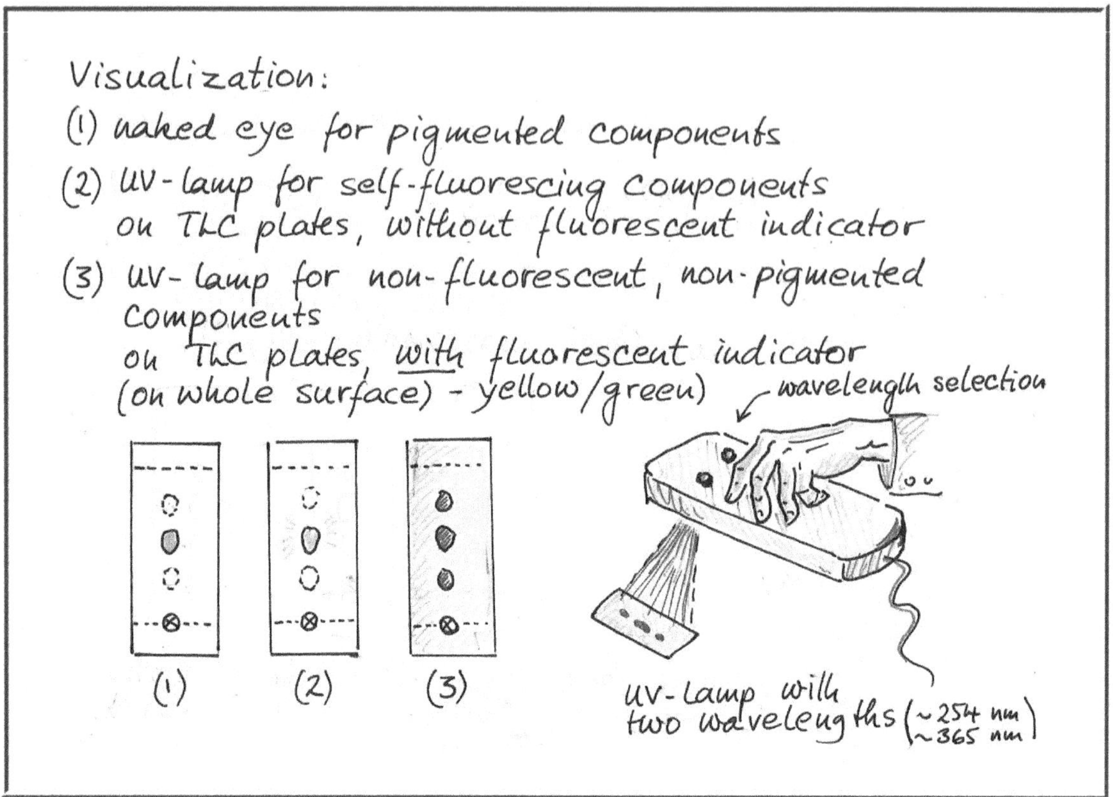


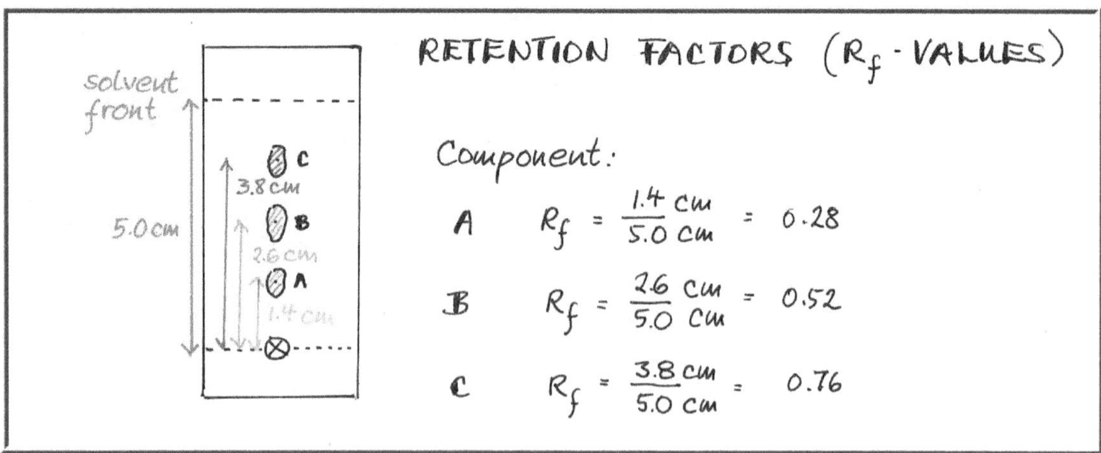


## Questions and Problems:

1. How pure were the synthesized analgesic drugs?
2. What contaminants could still be present and how could they be eliminated?
3. Why do $R_f$ values have no units?

4. How does the solvent system have to be changed if a component travels with the solvent front or remains on the start line?
5. What if a spot produces trailing or a thick elongated spot? How can this problem be remedied or avoided?

# Chapter 9   Stereoisomerism and Chiral compounds

9.1 The stereogenic center or chiral carbon

9.1.1 Stereochemistry and optical isomerism

9.1.2 The number of possible stereo-isomers ($n^2$ rule)

9.1.3 Enantiomers, diastereomers and *meso*-forms

9.1.4 R/S, D/L and +/- classifications

9.2 Polarimetry and specific angle of rotation

9.3 Stereochemical resolution

9.4 Medical implications

9.5 Properties of chiral compounds

9.6 Organic halogen compounds

9.6.1 Nomenclature

9.6.2 Distribution in nature and commercial uses

9.6.3 Physical and chemical properties

9.6.4 Formation and synthetic methods

9.6.5 Relative toxicity

9.6.6 Environmental implications

**Chapter 9 Chiral Compounds and Stereochemistry**

When a tetrahedral, saturated carbon compound has four different groups attached to the central carbon atom, it becomes **asymmetric** or **chiral**. Chiral carbons, often indicated by an asterisk in the structural formula, are thus **stereogenic centers**, which means, that they can produce two stereoisomers or spatial arrangement possibilities around each tetrahedral carbon atom, thus producing two **mirror images**. Even if the molecule has the same molecular formula and general structure, it can be produced in two versions that are not super-imposable, such as left and right handed structures.

**Fig. 9.1**

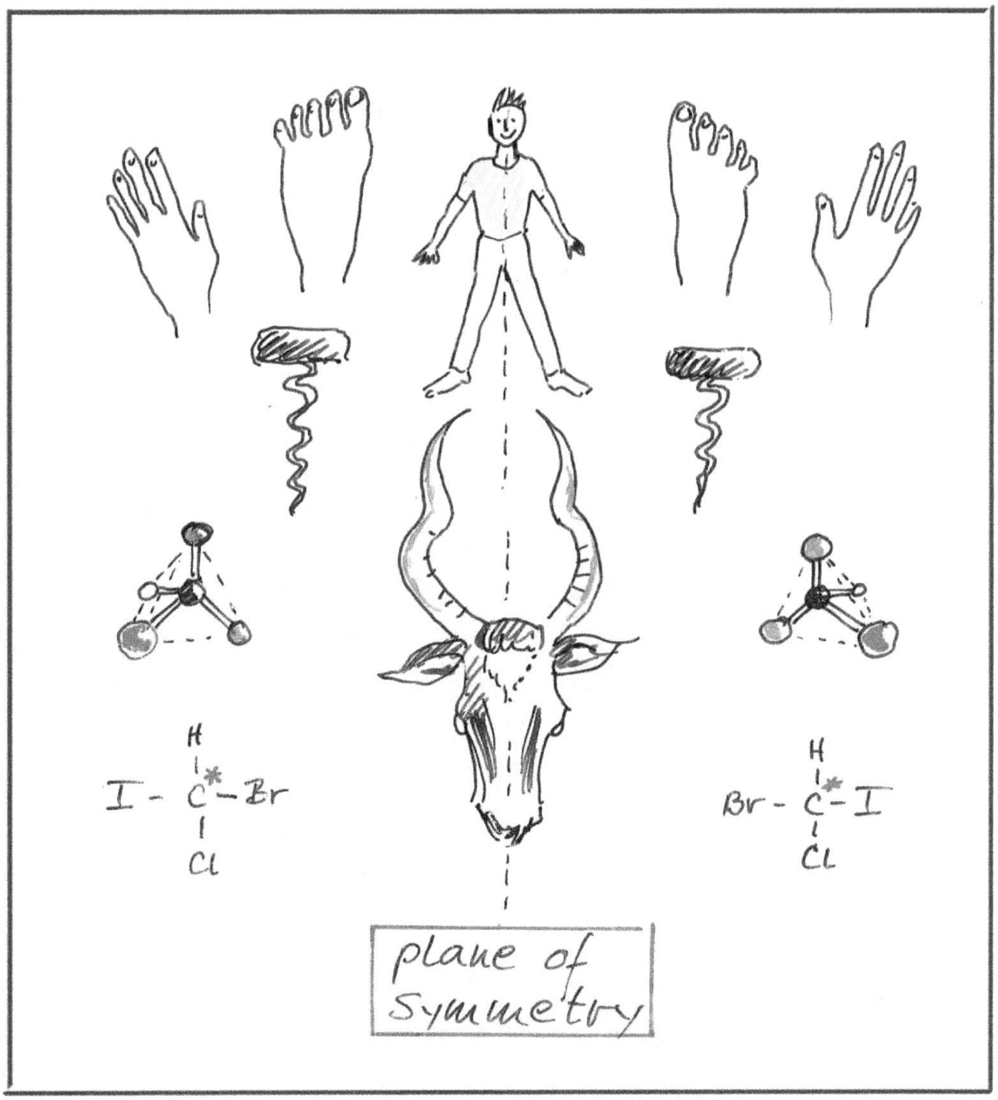

**Chirality** comes from a Greek word that actually means "handedness." "Chiros" is also found in the term chiropractor. In a chemical sense, the carbon located in the center of a **tetrahedron** must have **four different chemical environments**. This means that each corner of the tetrahedron must consist of different atomic arrangements. The two possible isomers have a plane of symmetry between them and it is as if one molecule looks itself in the mirror i.e. everything is the same except reversed. Pure mirror image molecules are called **enantiomers** and always come in twos.

The following two molecules, shown by various representations, are chiral:

**Fig. 9.2**

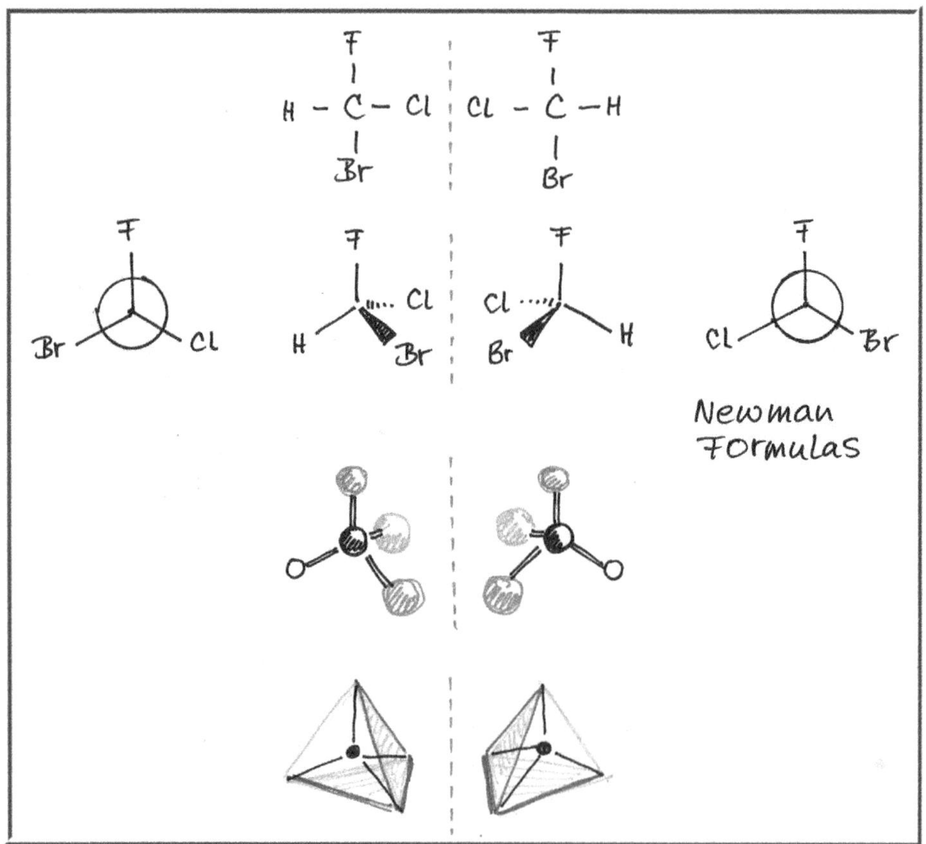


To convert a two-dimensional structure (2D) into the actual three-dimensional (3D) forms, the **dash-wedge representations**, where the wedge comes out of the plane and the dash goes into the plane, are used. Regular bond lines are within the plane. These are **perspective formulas**, which show the tetrahedron as in "real life", the way the eye would see it side-ways, a bit slanted (in "perspective"), as if it would be "sitting" on the table-top. When a molecule is written like a flat cross ( + ), known as the **Fischer formula** or representation, the **Fischer Convention** comes in handy. Fischer was a German chemist (1852-1919), who mainly dealt with sugar chemistry. He suggested that all horizontal bonds should be considered coming out of the plane and that all the vertical bonds going into the plane.

Type equation here.

**Fig. 9.3**

That way the actual 3 -dimensional structure of the tetrahedron surrounding the particular carbon is simulated.

**Example Problem 9.1**

**Convert the following molecules from the shown formula type into the other formula type based on the Fischer or perspective formula types:**

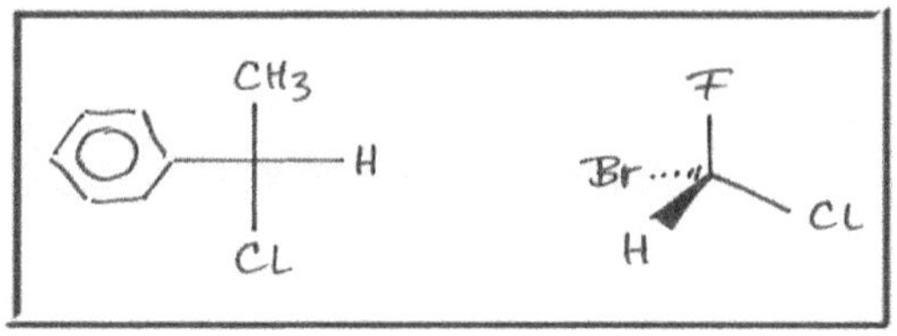


Chiral carbons can be involved in many structures including rings. There can also be more chiral carbons involved in one molecule. The following image shows two natural products, nicotine and carvone, each having one chiral carbon.

**Fig. 9.4**

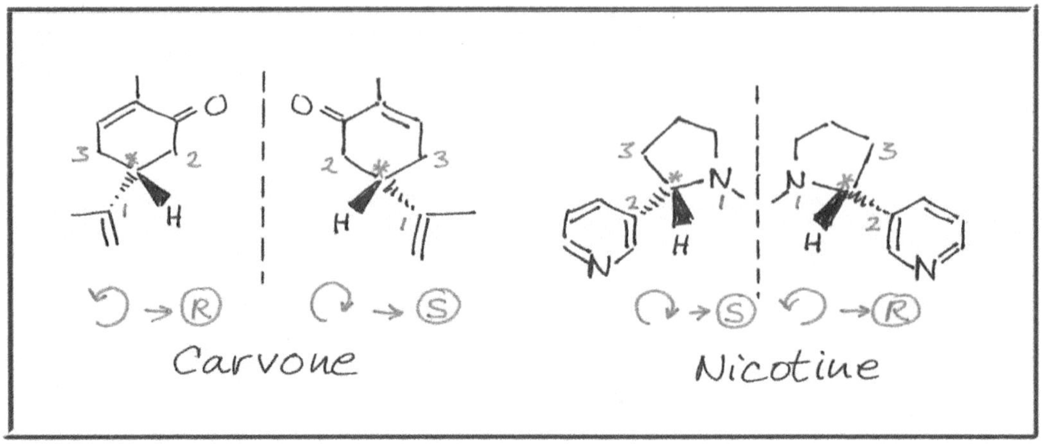


**Example Problem 9.2**
**Indicate the chiral carbon atoms involved in the following molecules by using an asterisk (* ):**

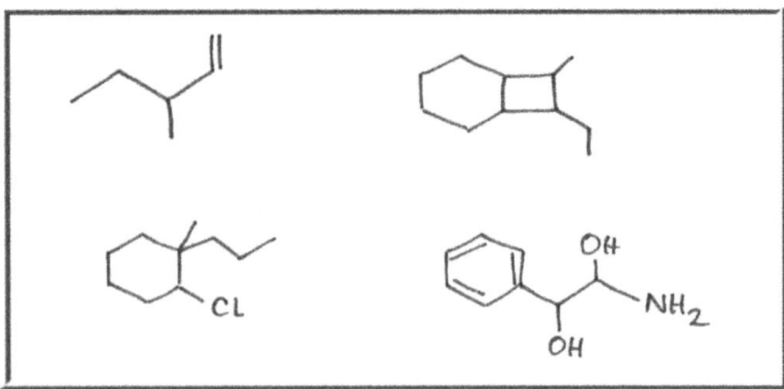

**Example Problem 9.3**
**Which of the following molecules are chiral?**

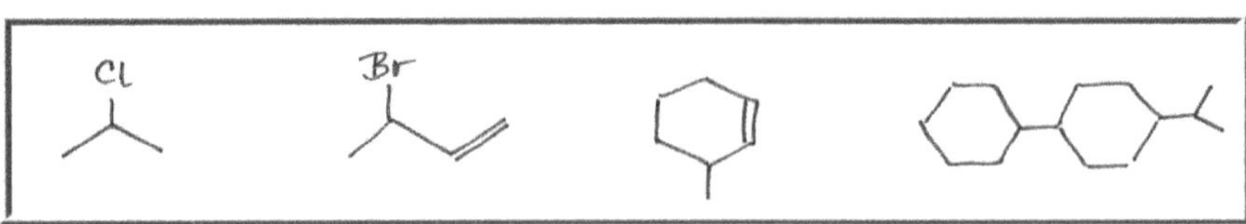

**Example Problem 9.4**
**Draw a chiral structure for a molecule having the molecular formula $C_2H_4ClBr$.**
**Draw an enantiomer to the molecule using different 3D representations.**

Enantiomers always come in pairs, just like hands, feet, ears, and eyes come in pairs. They are built the same but are reflections of each other via a plane of symmetry. One chiral or stereogenic center will lead to a left-handed and a right-handed molecular arrangement. Both of these mirror-image molecules are completely alike in their physical and chemical properties. They have the same density, melting and boiling points, heat capacity, etc., and the same chemical reactivity. There is only one category in which they differ and that is how they interact with **planar polarized light** passing through them as solutes in a liquid solution. This phenomenon is called **optical activity**. One enantiomer shifts the plane of polarized light to the left (-) and the other to the right (+). The former is a **levorotatory** enantiomer and the latter a **dextrorotatory** enantiomer. A pair of mirror image molecules can sometimes also be referred to as **optical antipodes**. If both forms are in a 50:50 % mixture, they cancel each other's optical rotation and become optically inactive. This is known as a **racemic** mixture, the components called **racemates** and the process of producing equal amounts of antipodes **racemization**.

**Fig. 9.5**

The optical rotation of an optically active substance has to be experimentally determined with a measuring device called a **polarimeter**, that allows the determination of the rotation angle in degrees, plus (+) or minus (-).

**Fig. 9.6**

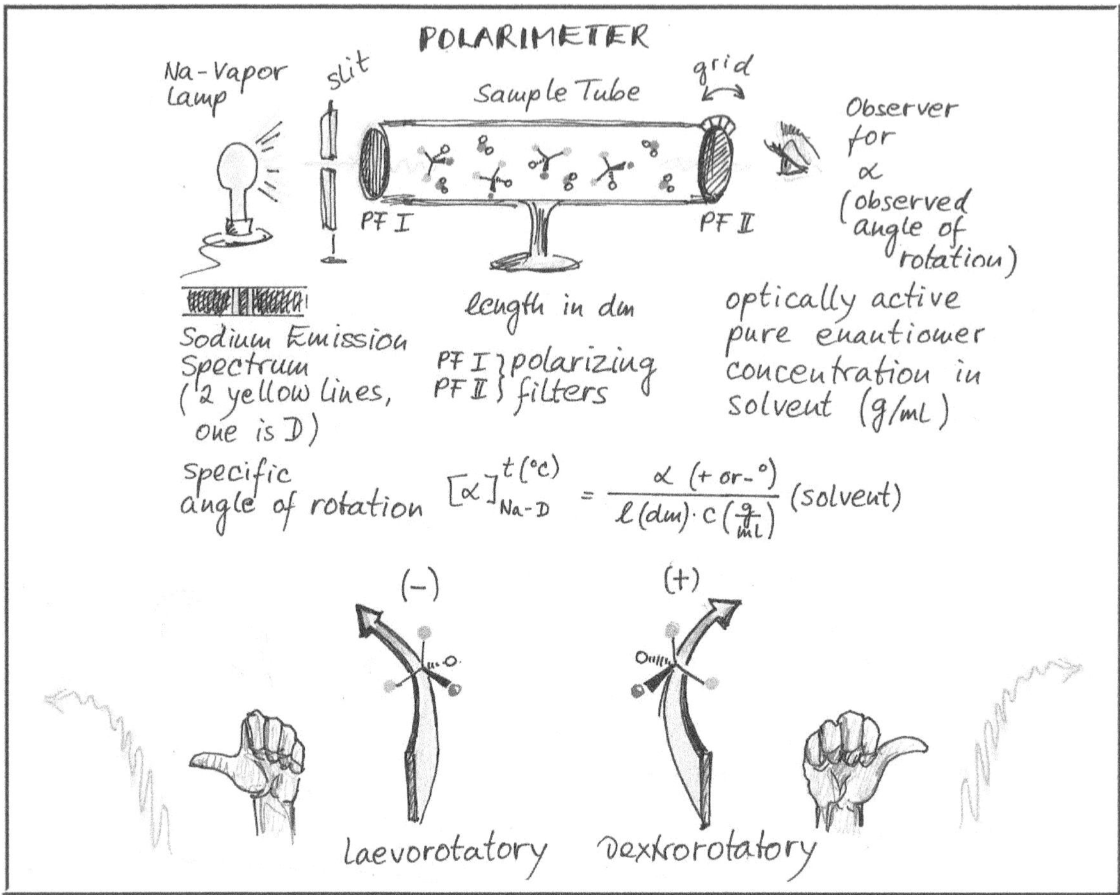


The actual direction of rotation, whether a substance is levorotatory (-) or dextrorotatory (+), cannot be derived from the formula. However, in order to distinguish one molecular form from another by looking at the structure, a classification method, according to a set of established rules for prioritization of the groups, has been developed.

Instead of (+) or (-), which can only be derived by measurement, D / L (dextro / levo) or R/S (sinister / rectus) terminology is applied for a given **configuration**. So, enantiomers can be determined as either a D/R or L/S type by knowing the 3D structure of the molecule. Here the dash-wedge-formalism and the Fischer convention are of extreme importance. D/L classification rules were initiated by Fischer and mainly applied to sugars and amino acids.

**Fig. 9.7**

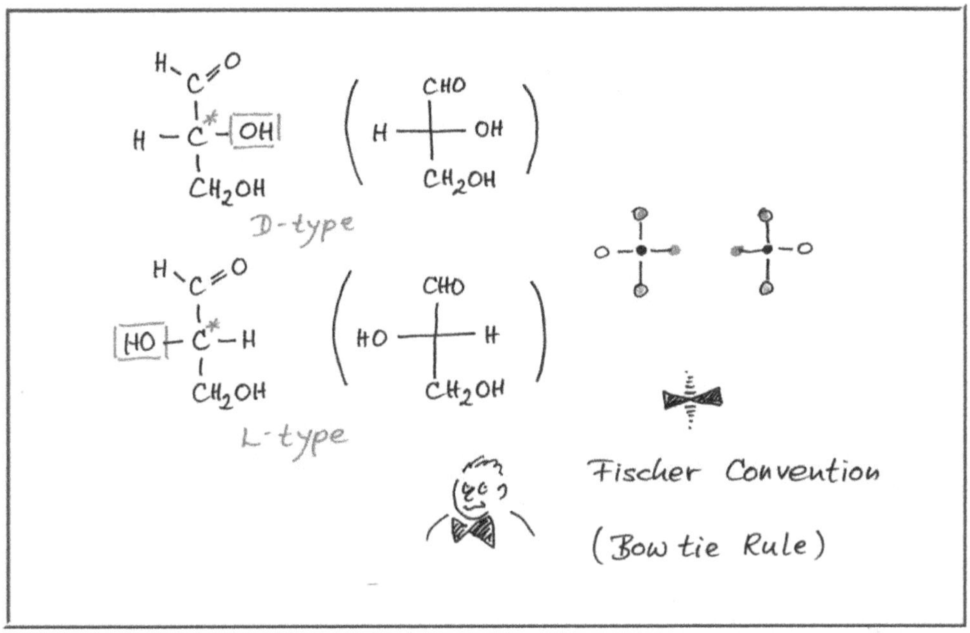


In this molecule, the important group attached to the carbon is literally on the right side and therefore classified as D. If it is on the opposite side, it would become an L-form.
The R/S classification system was introduced by three chemists and is now known by the **Cahn-Ingold-Prelog** set of **prioritization rules**. Here the atomic mass of the atom that is attached directly to the chiral carbon is the basis for prioritization - the higher the mass, the higher the priority. If the immediate atoms happen to be the same, one continues until a difference is found. Atoms attached via double bonds count double and triple bonds count 3 times as much.
Once priority is established, the viewer turns the tetrahedron in such a way that the lowest priority group – usually hydrogen – is turned to the rear, so that only the three remaining groups are faced in the form of the **tetrahedral umbrella**.

**Fig. 9.8**

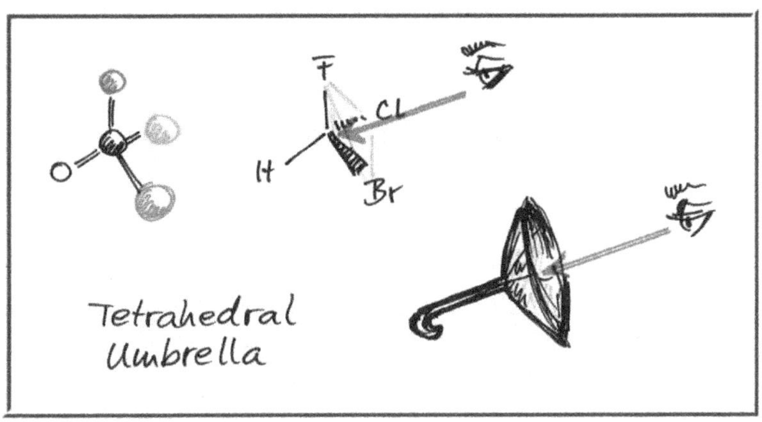


The next step is to connect the "dots" from the highest to lowest priority group. If the umbrella or "wheel" turns to the right (clockwise), it is an R-form and if it turns to the left (counter-clockwise), it is an S-form.

**Fig. 9.9**

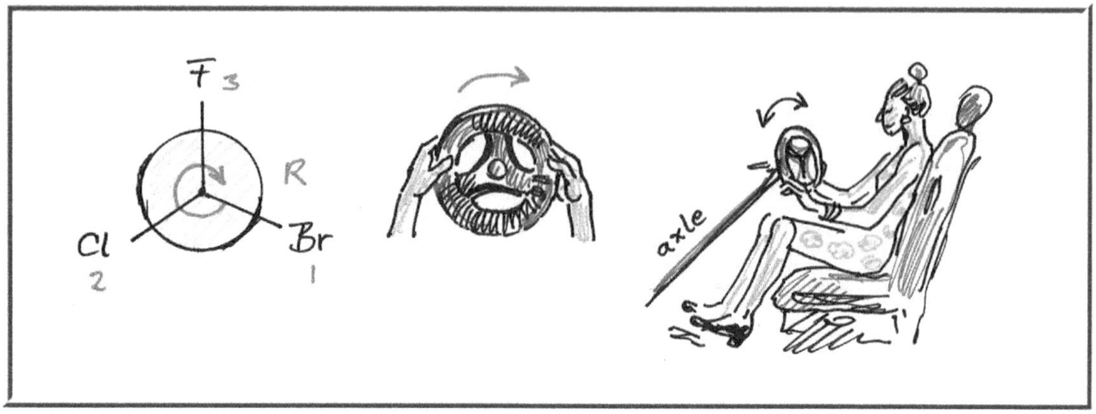


The three remaining groups of a tetrahedral umbrella, with the lowest priority orientated away from the viewer (behind the umbrella) can also be visualized by a **Newman representation or formula**, a circle with three lines going off from a enter point in 120° angles.

When Fischer formulas are used to determine the rotation of either R or S and the lowest priority group such as hydrogen not facing away from the viewer according to the Fischer convention, then the rotation can be determined the normal way, but must then be switched, in order to satisfy the rules. S has to become R and vice versa.

**Fig. 9.10**

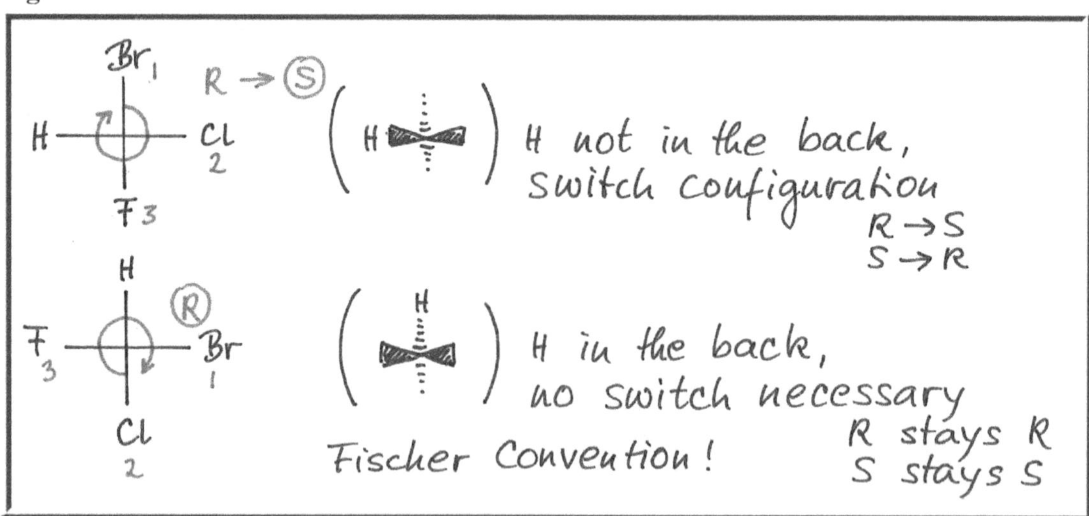


In a regular formula, chiral carbons can be detected, but without actual knowledge of the 3D arrangement of the groups, the configuration of either R or S cannot be determined. Only dash-wedge representations allow an exact classification

**Fig. 9.11**

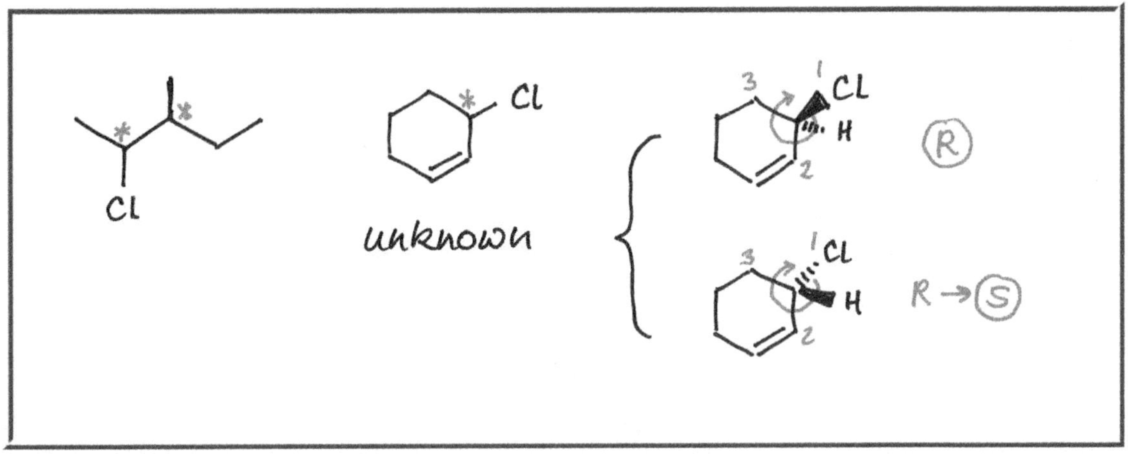


**Example Problem 9.4**
**Convert the following structure into a Newman form:**

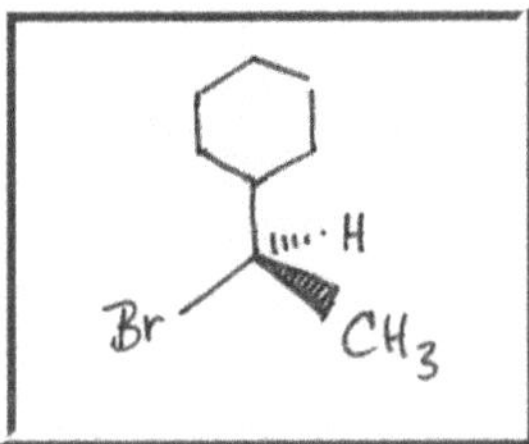


**Example Problem 9.5**
**Classify the following molecules as either R or S configurations**

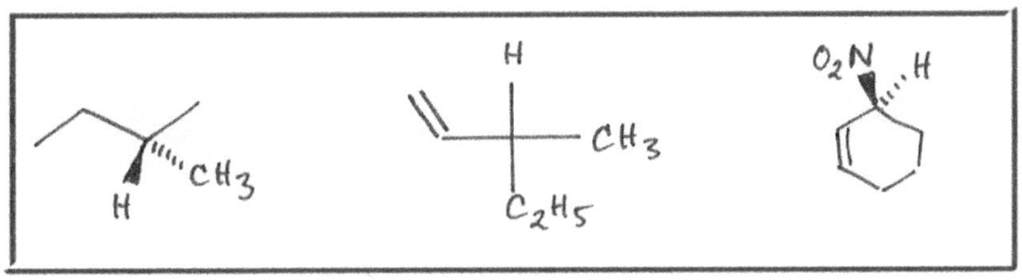


**Example Problem 9.6**
**Classify the following Fischer formulas as either showing a D or L type:**

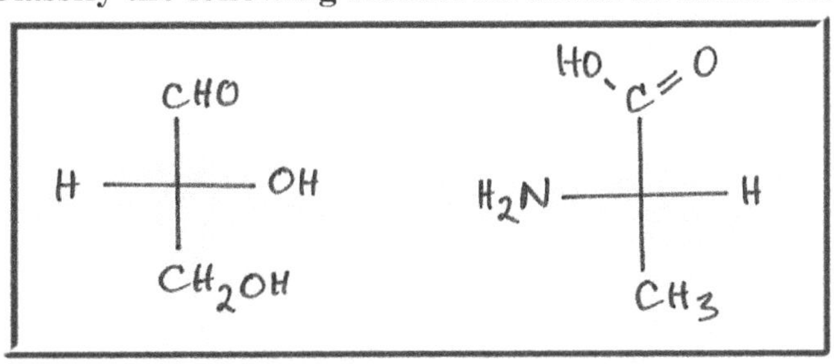


The R/S system allows classification of <u>any</u> organic molecule, whereas the Fischer method is limited. However, once classified by both methods, if possible, the results do not contradict one another. The established results for the configuration can coincide with the actual measurement but is not required to.

R = D, but not necessarily (+)
L = S, but not necessarily (-)

Planar polarized light used for measurements is usually light from a monochromatic source (one wavelength or frequency), such as the light produced by the **sodium vapor lamp**, which can be broken down into two spectral lines, one of which is the so-called D-line (selected to determine optical activity). Natural light is a mix of various wavelengths (polychromatic) oscillating in all angular directions. Even monochromatic light vibrates in all directions and is not polarized yet. It needs to be sent through a polarizing filter, which acts like a parallel grid and is similar to a picket fence, which would only let one plane of oscillating jump rope pass through but reflect off it when coming in from any other angle. Hence, the filter selects only the plane of oscillating waves that are parallel to the grid.

The front view of the wave oscillating in all directions, looking down at the axis of propagation, would look like a star as shown in Fig. 20.5.

This light, when passing through a sample tube that contains the solution of an optically active substance (the solvent should be optically inactive), will either be shifted by a certain angle to the right or to the left. The physical nature of optical isomerism is still not fully understood.

A second, rotational polarizing filter can be manipulated by the observer, who can rotate in such a direction that the light comes straight through, then counter-rotate to block out all light. By reading the grid on a dial, the **observed angle of rotation** ($\alpha$) can be documented. After this is done, a simple mathematical formula is applied to get the specific angle of rotation:
Type equation here.

$$[a]^{T}_{\text{Na-D}} = \frac{\alpha}{\frac{l(dm)g}{ml}} \quad \textbf{(solvent)}$$

**Example Problem 9.7**
**If an observed angle α turned out to +3°, the length of the measured tube 10 cm, the concentration being 0.5 g / 10 ml, and the solvent ethanol, what is the specific angle of rotation, expressed in the standard manner, given that the Na-D line was the light source and the temperature was 20° C?**

$[a]^{20\ \text{degrees C}}_{\text{Na-D}} = +3^{o}/(1\ \text{dm})(0.05\text{g/mL}) = +60^{o}$ **(ethanol)**

When there are more chiral carbons in the molecule, each **stereogenic center** will have a certain configuration and actual optical rotation. When measured, a resultant angle will come about. When the stereochemistry is determined in the formula, each chiral carbon will be assigned either D/L or R/S.

**Fig. 9.12**

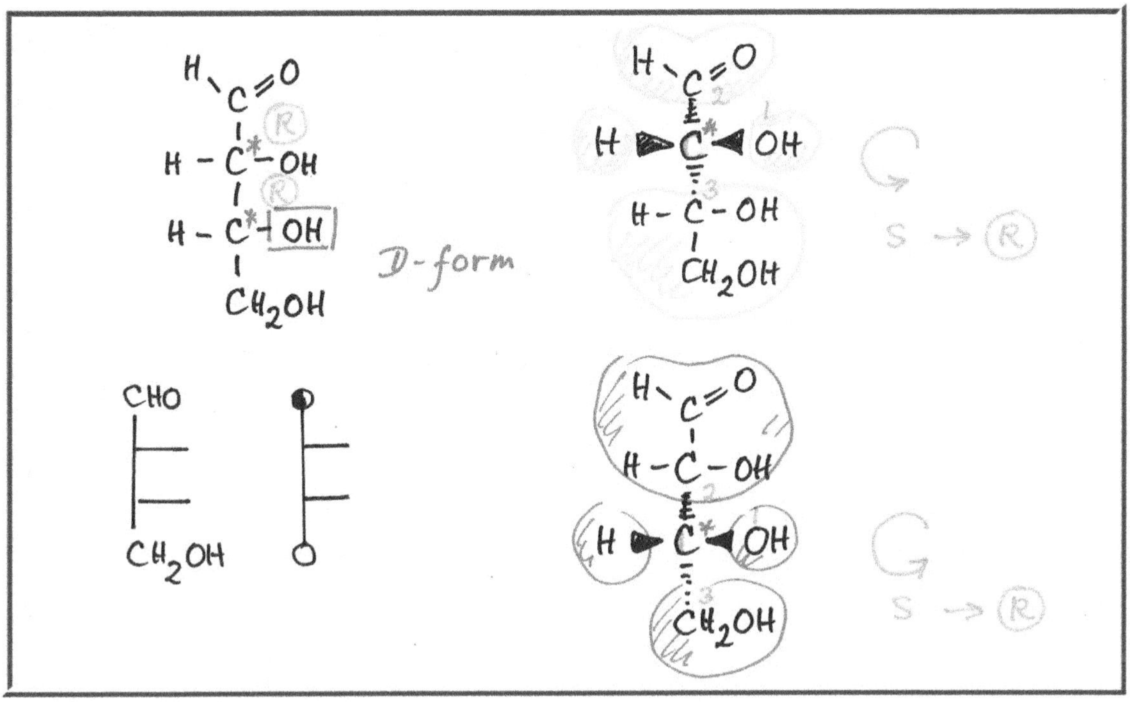


The general formula for **the number of possible stereoisomers is $2^n$**, n = number of chiral carbons, ($2^2 = 2 \times 2 = 4$). Of these 4, two enantiomeric pairs (complete mirror images can be derived.

The number of possible stereoisomers or arrangement patterns grows quickly with each added stereogenic center.

**Fig. 9.13**

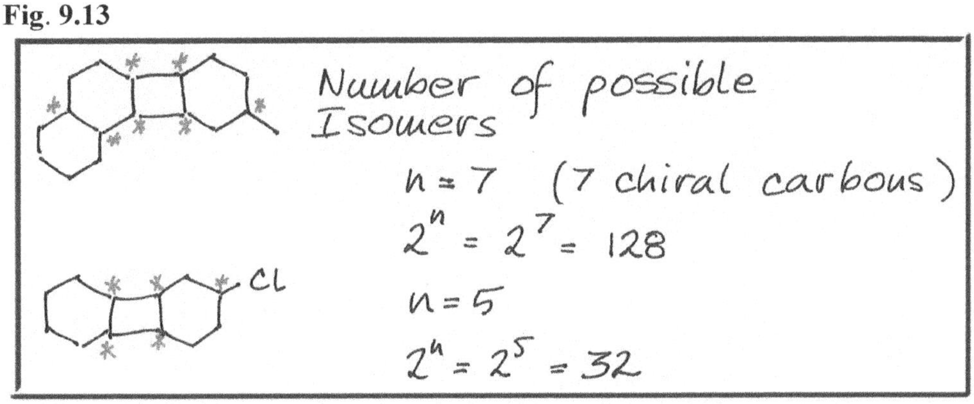


**Example Problem 9.8**

**How many stereoisomers are possible for the following molecules?**

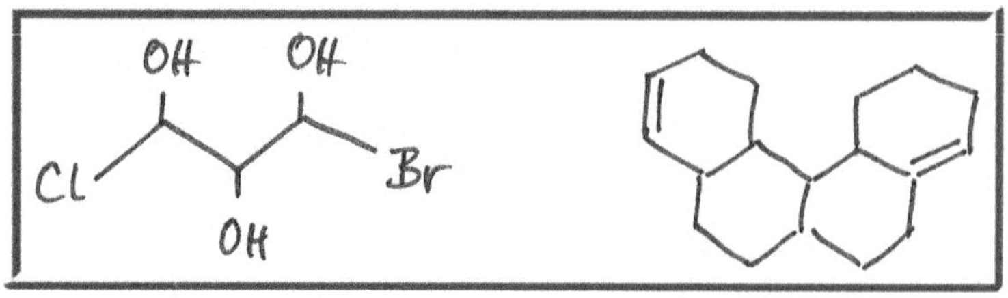


Pairs of molecules, which are **partially mirror images** but otherwise identical (super-imposable or congruent) can also be derived from this set of 4 stereoisomers. These couples are called **diastereomeric pairs**. There are always many more diastereomeric pairs than enantiomeric pairs. A Fischer formula with more chiral carbons sort of resembles a fish (Fischer meaning fisher in German), with the most oxidized group being on top (the head) and the most reduced carbon located on the bottom (the tail). The cross bars resemble the ribs of the fish. Fischer formulas are commonly used in the representation of monomeric, open-chained sugar molecules

**Fig. 9.14**

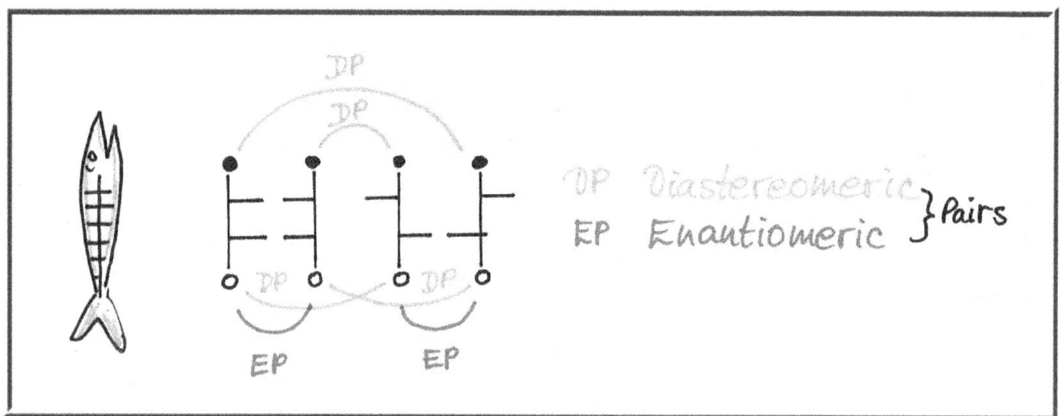


For the molecule $CH_2OH$ CH(OH) CH(OH) $CH_3$ with two chiral carbons, there would be a total of four stereoisomers, of which two are enantiomeric pairs and four are diastereomeric pairs. There are a total of partially mirror-imaged pairs. Fischer representations can sometimes be simplified to just show the left and right orientations of the important groups by lines used as signposts, the most oxidized non-chiral carbon shown as a black dot on top and the most reduced, non-chiral carbon shown as an open circle on the bottom. That way the number of couples can easily be established.

**Fig. 9.15**

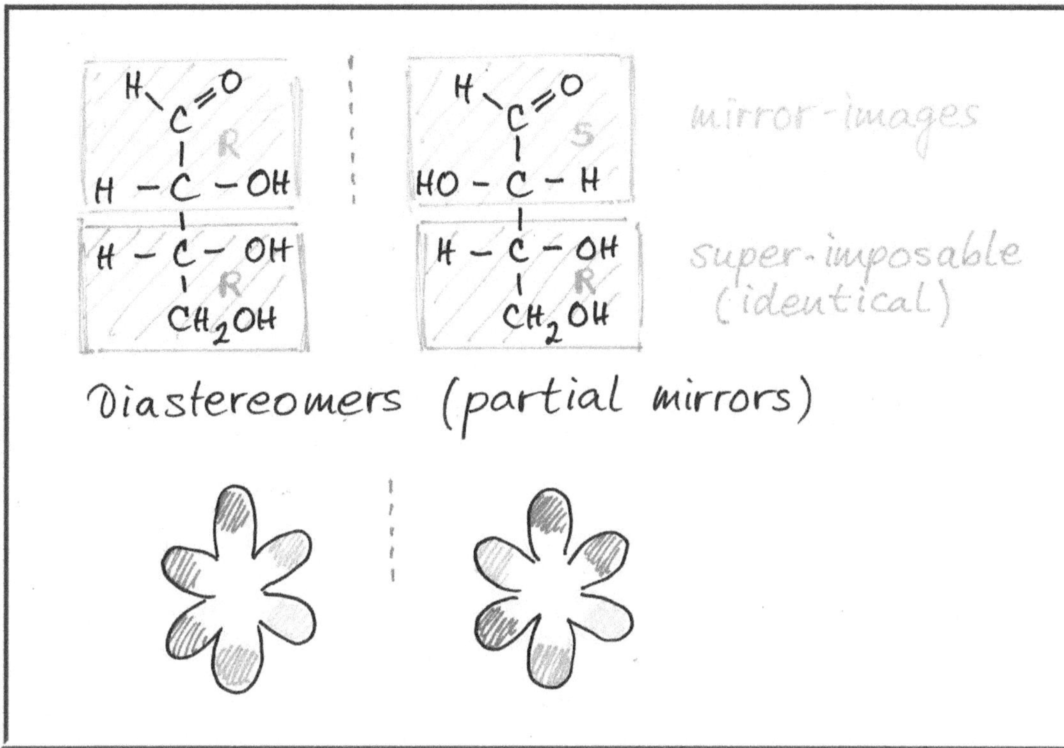


**Example Problem 9.9**
**Draw all the enantiomeric pairs and diastereomeric pairs for the following molecule using the Fischer formulas.**
**$CH_2Br\ CH(OH)CH(Cl)CH_2OH$**

If the formula is not shown in the Fischer presentation, then the total stereoisomers can be grouped as various R/S combinations.

**Fig. 9.16**

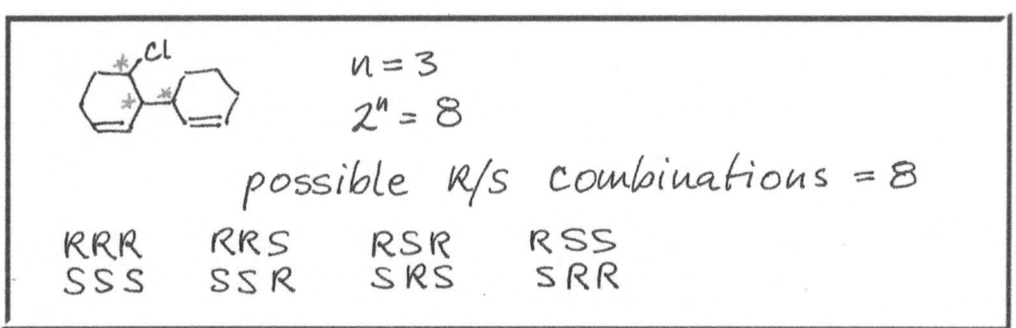


Chirality plays an important role in biochemistry. Many substrate molecules, natural or pharmaceuticals, are made to fit the active site of enzymes for certain reactions to occur. The active site is a pocket that can arrange itself so it complements the surface of the substrate molecule or drug, like a hand in a glove or a key in a key-slot, so the exact 3D-structure has to be involved. One could open a door with a key that is the mirror image of the original.

**Thalidomide** ("Contergan"), an analgesic drug that can exist as the R or S isomer, caused severe birth defects during the time of its use. Drugs were usually administered as **racemates**, since no ways had been found to separate them chemically. Even though optical isomers can now be separated, medication consisting of a pure enantiomer can still be in vain since the body has enzymes (isomerases) that convert some of the pure forms into the other antipode leading to undesired racemization. It was found that the (R) thalidomide can control morning sickness in pregnant women and that the (S) enantiomer causes the birth defect of missing limbs (arms or legs), the latter being a **teratogen**.

**Fig. 9.17**

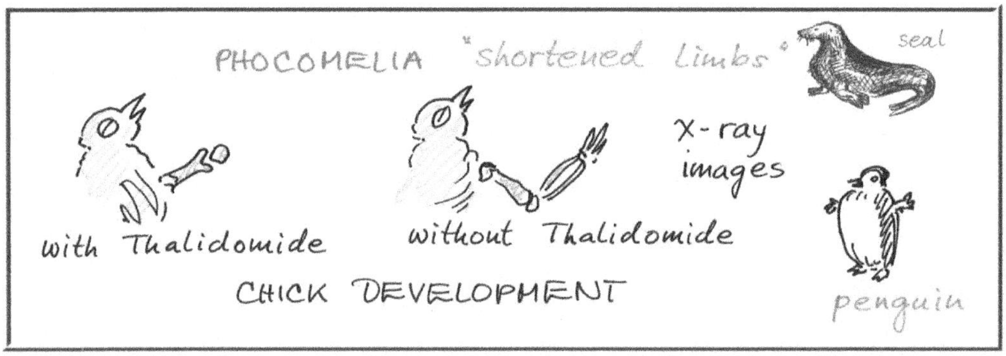


Thalidomide is a two-edged sword. It shows promising results in the treatment and remission of certain cancers, as well as other chronic diseases. It is used effectively in the control of leprosy, but is recommended, to use contraceptives at the same time, since otherwise **embryonal damages** can occur. This time, the formation of the lower extremities was affected, leading to another type of **Phocomelia,** a condition of stunted growth development of arms and legs, resembling the body of a penguin. The term Phocomelia is derived from the Greek and literally means "seal limbs" ("phoke melia").

**Fig. 9.18**

Thalidomide has only one chiral carbon. In the literature a wavy line has come two represent the two possible types simultaneously. In a sense it combines both the dash and wedge possibilities.

**Fig. 9.19**

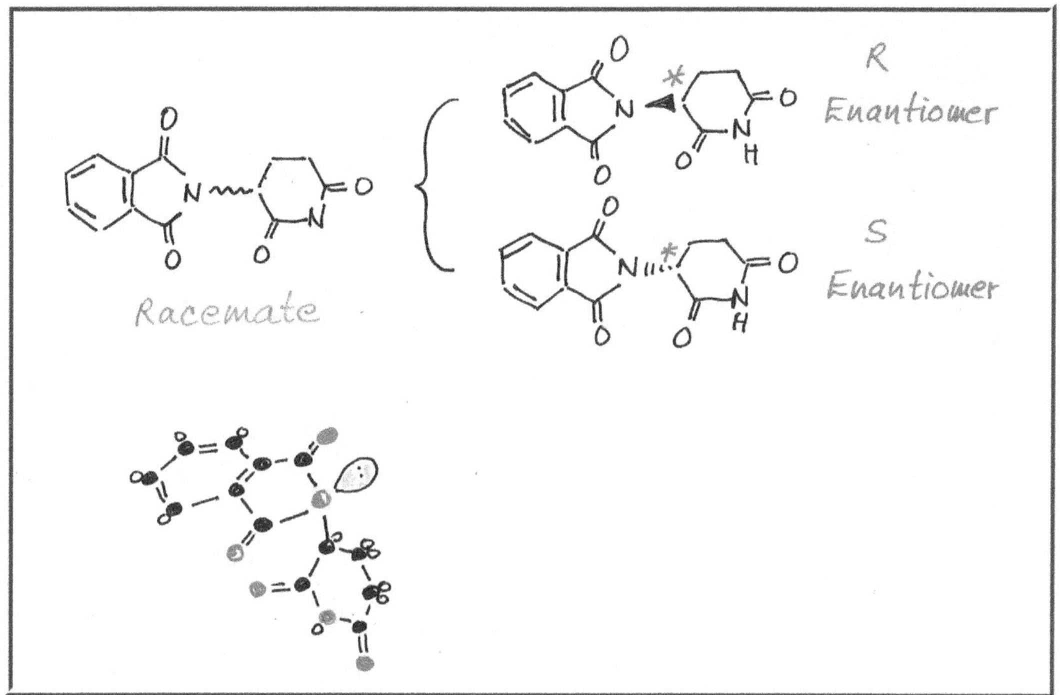


A more exact configurational analysis around the stereogenic center, indicated by the green tetrahedron, of the thalidomide molecule is shown in the following illustration:

**Fig. 9.20**

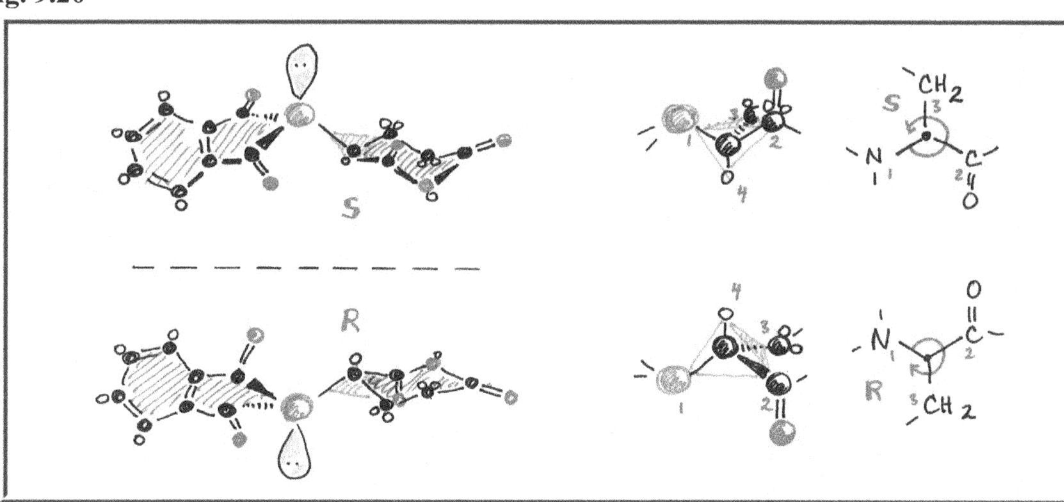


By rotating the molecules 90°, the concept of mirror imaging is more apparent.

**Fig. 9.21**

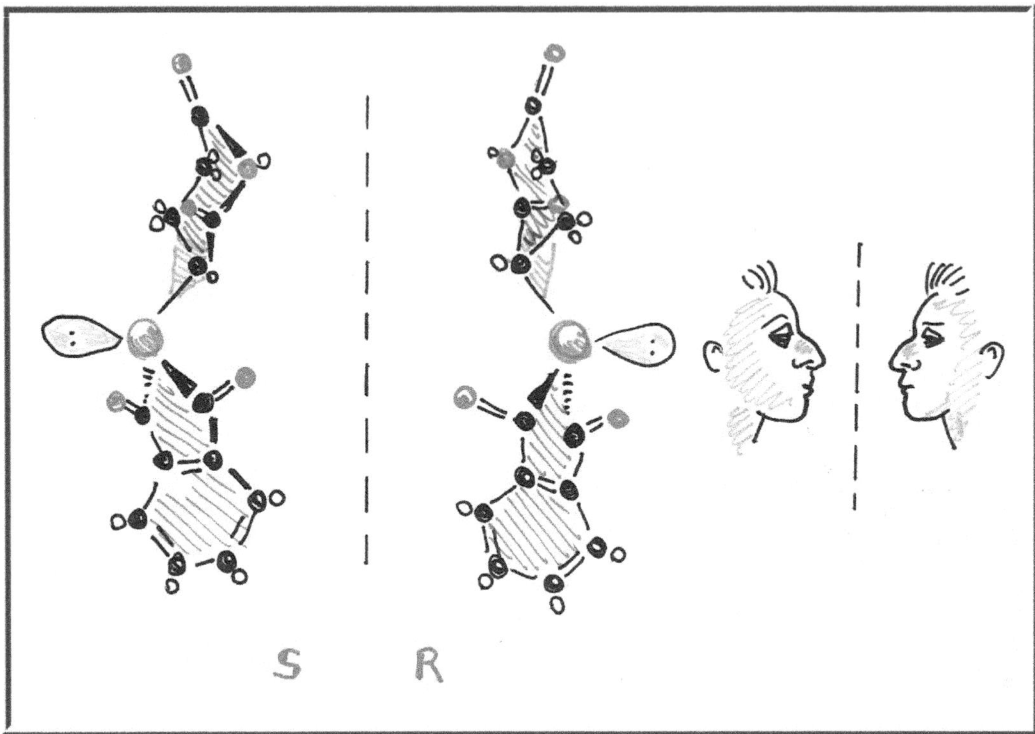


**Optical isomerism** was first observed by Louis **Pasteur** (French, 1822-1895) in 1849, when he noticed that crystals growing in red wine, when isolated with tweezers, showed two types of crystal structures, differing only in the "slant" of their growth direction.

**Fig.9.22**

These crystals turned out to be potassium hydrogen tartrate crystals, a double salt of the organic acid 2,3-dihydroxybutanedioic acid or **tartaric acid**.

Although the phenomenon was first officially observed by Pasteur, the theoretical background information was analyzed and described simultaneously by the French chemist Joseph Achilles **Le Bel** (1847-1930) and the Dutch chemist Jacobus Henricus **van't Hoff Jr.** (1852-1911). Both based their interpretations on the tetrahedral alignment of the four-bonded carbon atom.

**Fig. 9.23**

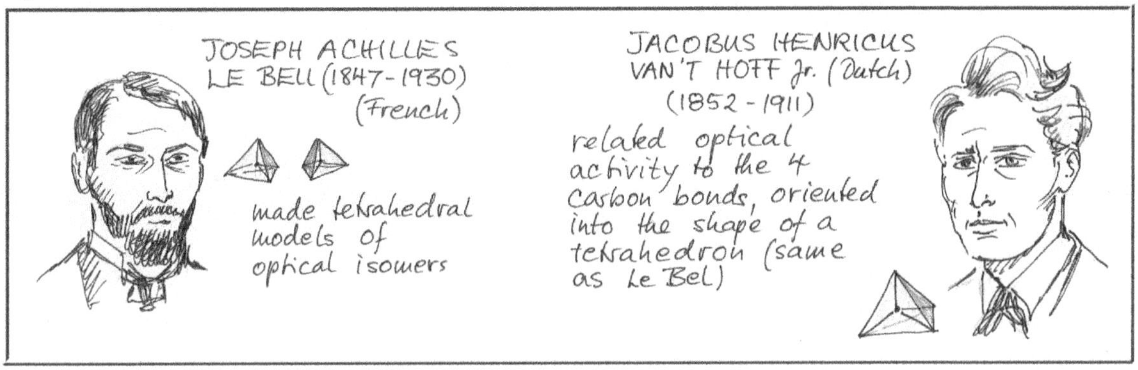


Tartaric acid is a symmetrical molecule, so it can also produce a **meso-form**, which results as internal compensation of an R-part and an S-part of the molecule. It is like a pair of enantiomers conjoined in the same molecule. All molecules that have an even number of chiral carbons with the same set of four different groups attached but with a plane of symmetry, cutting cross the carbon chain, are meso forms – a racemic mixture wrapped in one, like conjoined twins. Meso molecules are as a whole not optically active, although consist of stereogenic centers, which by themselves are optically active.

**Fig. 9.24**

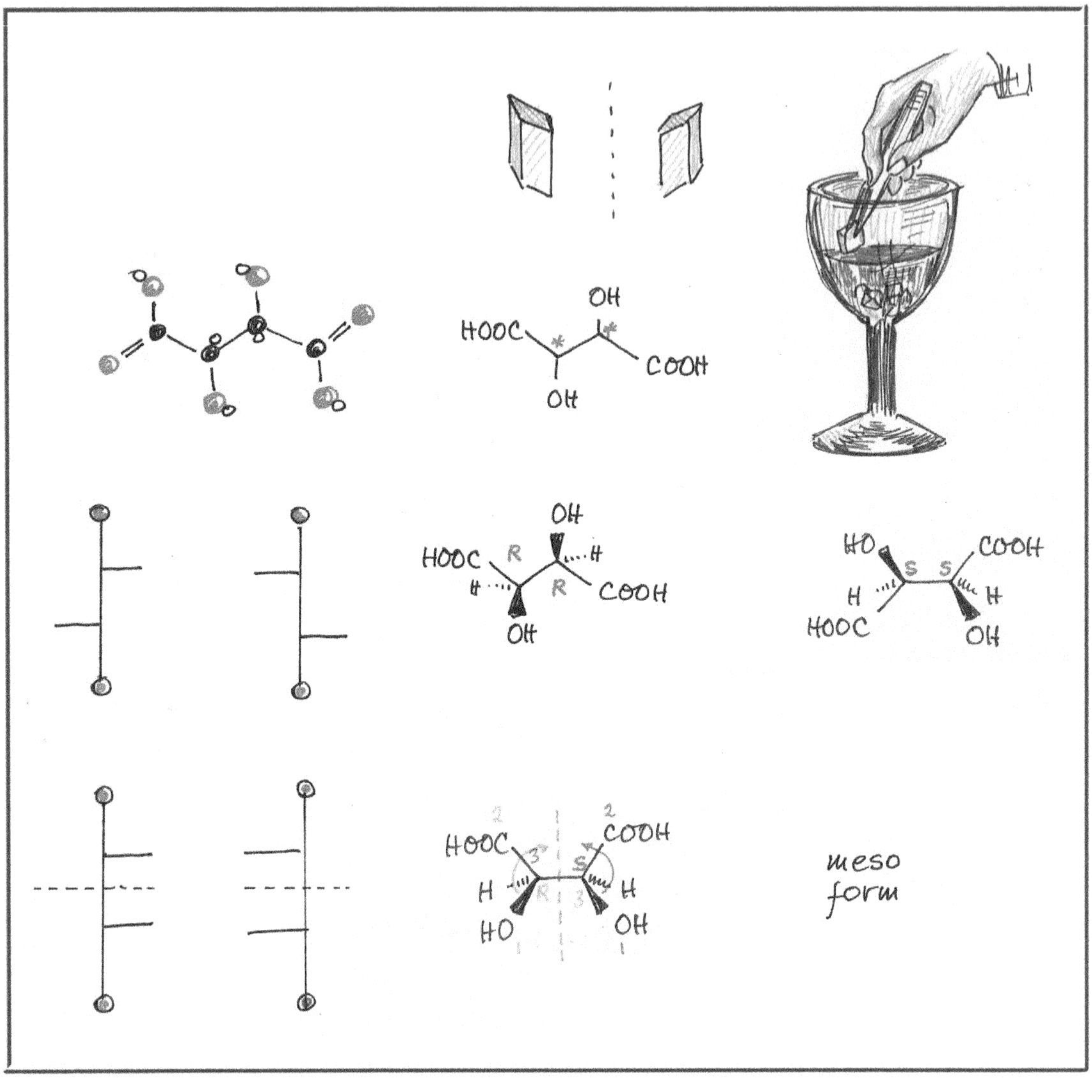


In the molecule shown above in the lower right, where one chiral carbon is an R-configuration and the other in an S-configuration and the non-chiral end groups are identical, a meso form results due to the internal plane of symmetry between the two chiral carbons. The one side of the molecule is the mirror image of the other.

When the Fischer formulas are converted into Newman formulas, the two enantiomeric forms (R,R and S,S) as well as the meso form of tartaric acid can be represented as follows:

**Fig. 9.25**

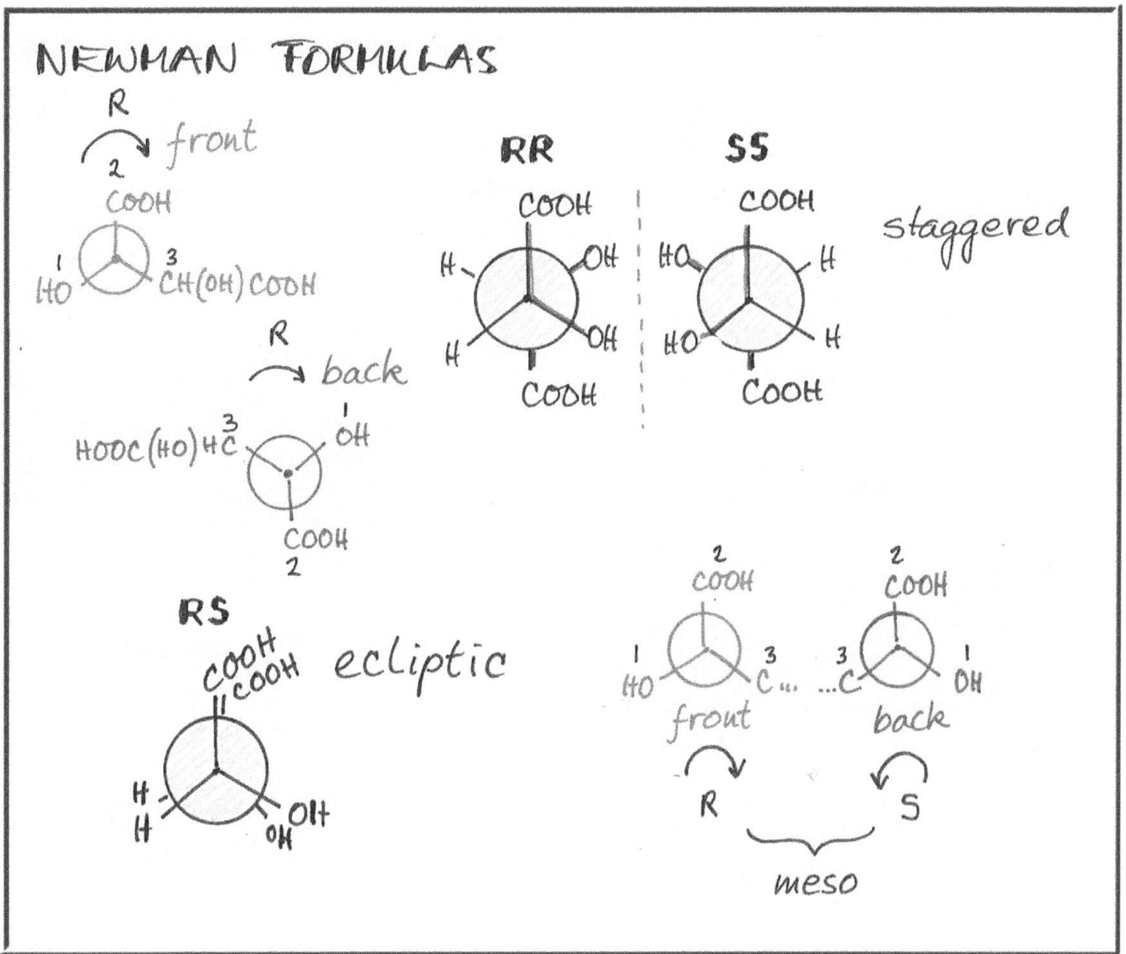


In the following image, the two horse heads facing the same side but reversed with an obvious plane of symmetry between them exemplify the concept of the meso form, whereas the other image resembling a playing card has a point of symmetry between the two heads which does not lead to internal cancellation of the two sides.

**Fig. 9.26**

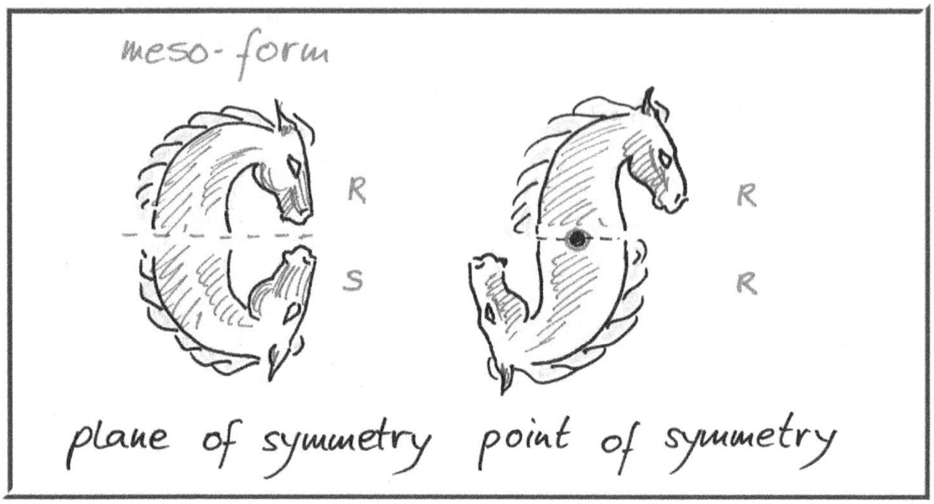


**Example Problem 9.10**
**Which of the following molecules are meso-forms?**

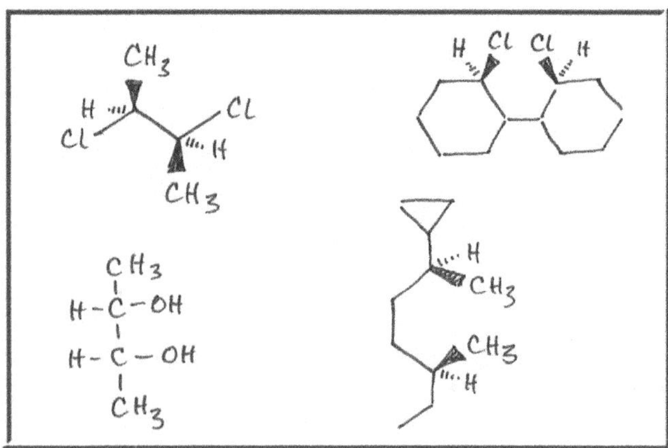


**Example Problem 9.11**
**How many chiral carbon atoms are found in the following statin molecule? What are the configurations (R/S) and how many stereoisomers are possible?**

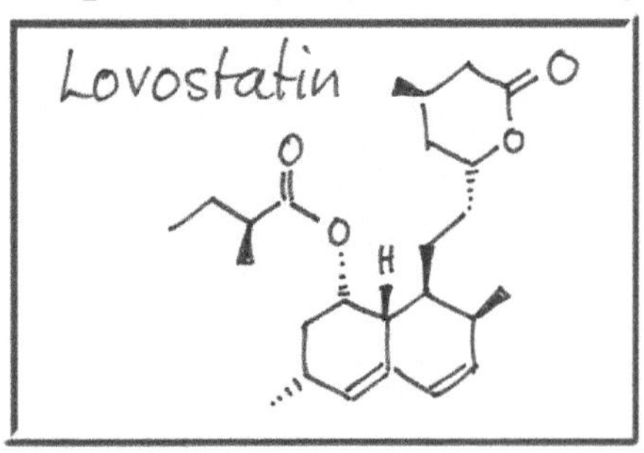


Racemic mixtures cannot be separated by conventional means such as chromatography.

Since all their physical and chemical properties are the same, the exception being the optical property, sufficiently described above, none of the conventional separation techniques can be applied. However, with a trick they can be separated. When both are chemically reacted with the same stereoisomer of another chiral molecule, different products result with different physical-chemical properties, enabling them to be separated. Once isolated, the linked molecule can be removed, leaving the separate stereoisomers in their pure form behind. This method is known as chiral or **enantiomeric resolution**, related to the use of a type of chemical "tweezers"

**Fig. 9.27**

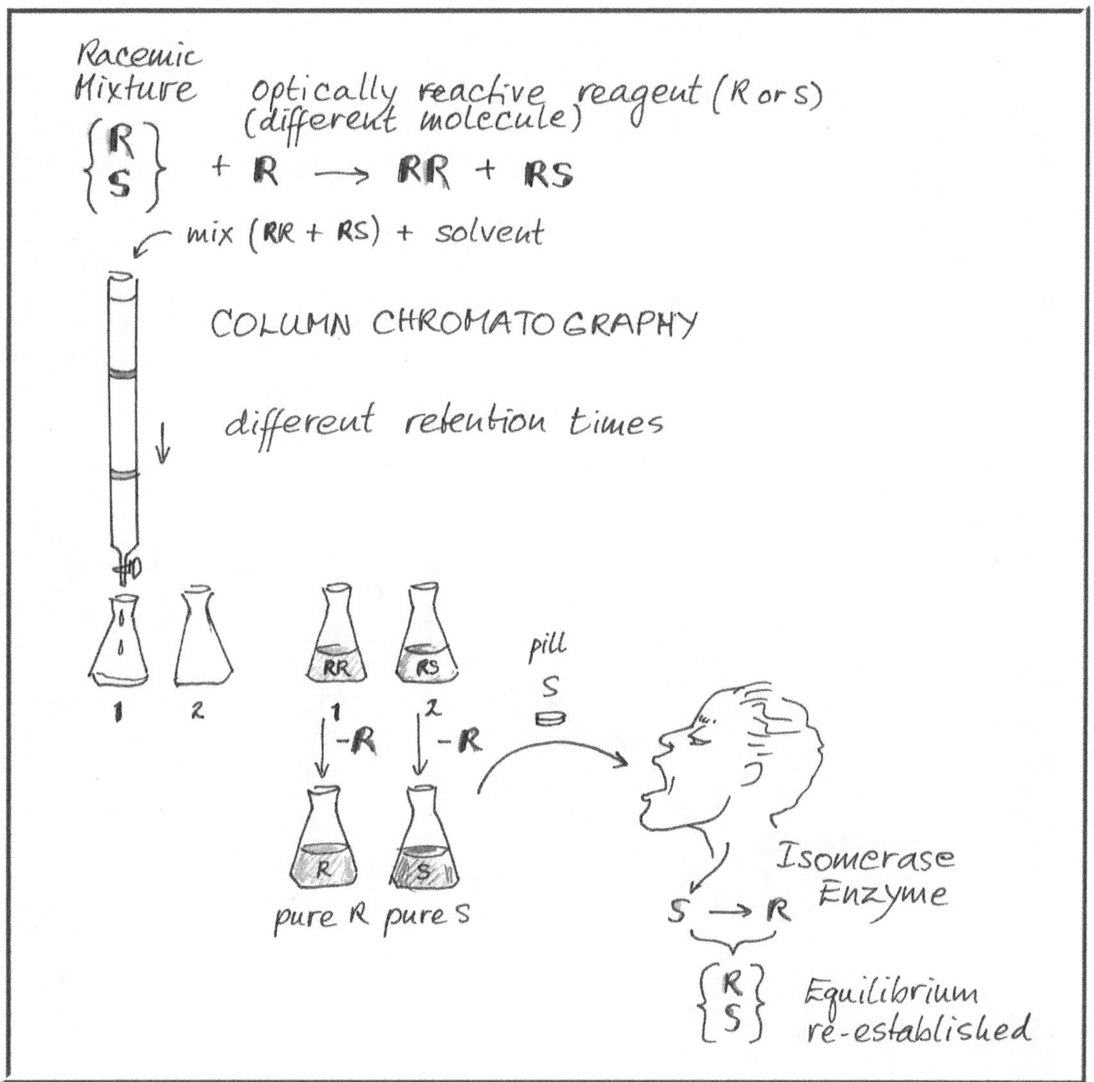


This purification process is elaborate and expensive and can lead to higher prices of valuable

drugs.
On the other hand, the body is not often fooled and a set of **isomerase-enzymes** can convert one optical isomer back into the other, thus leading back to an equilibrium between the two antipodes and possibly making the whole optical resolution process in vain.

# EXPERIMENTS – Ch. 9

## *Stereochemistry*

**Key Terms:** Chiral compounds, stereogenic center, $2^n$ – rule for number of possible stereoisomers, enantiomers (complete mirror images), diastereomers (partial mirror images), racemic mixtures, meso-forms, optical activity, angle of observed rotation α, specific angle of rotation [α], mutarotation, invert sugar, epimeric and anomeric sugars, cyclic sugars (furanoses and pyranoses)

## I Molecular Modeling of Chiral Compounds

### Chemicals:

*none*

### Equipment and Materials:

*Organic chemistry modeling kits (Molymod)*
*1C (black), 1H (white), 1Cl (green), 1 Br (orange), 1I (lavender), 4 single bond pieces*

### Procedure:

1. To a four bonded carbon piece (black) attach a chlorine (green), a bromine (orange), an iodine (lavender) and a hydrogen (white) atom
2. Look into the tetrahedral "umbrella" with hydrogen facing away from you (lowest priority group) and connect the group from highest to lowest atomic number. If the orientation is clockwise or to the right the configuration is an R form, if anti-clockwise or to the left it is an S form.
3. Construct the enantiomer by re-arranging the groups
4. Draw the perspective formulas, the Fischer formulas and the Newman formulas in the the grid below:

### Observations and Results:

| Perspective Formula | Fischer Formula | Newman Formula |
|---|---|---|
| R | D | R |

### Questions and Problems:

1. Why do chiral molecules have to have four different chemical environments?

2. What if the two forms are found in a 50:50 mixture. Would the mixture still be optically active?

## II Molecular Modeling of a D-Glucose Molecule

### Chemicals:

*none*

### Equipment and Materials:

*Molecular model kit*
*6 C (black), 12H (white), 6O (red)*
*22 single bond pieces, 2 double bond pieces*

### Procedure:

1. Connect 6 carbon atoms, then add two single bonds on either side, with the bonds coming out (Fischer convention) for the middle 4 chiral carbon atoms
2. Attach the carbonyl group and a hydrogen atom to C-1
3. Attach an OH group and two hydrogen atoms to C-6
4. Keeping the carbon backbone vertical, attach OH-groups to the C-2, C-3, C-4, C-5 in right, left, right, right positions, respectively. Attach the remaining hydrogen atoms.
5. Sketch the molecule in the Fischer projection formula below.
6. Indicate the chiral carbons with an asterisk.
7. Draw the L-Glucose molecule as mirror image (enantiomer).
8. Draw an example of an epimeric sugar to D-Glucose.

### Observations and Results:

| **Fischer D-Glucose** | **Fischer L-Glucose** | **Fischer Epimeric Sugar *(name)*** |
|---|---|---|
| | | |

**Questions and Problems:**

1. How many possible stereoisomers are possible for a sugar with 4 chiral carbon atoms?
2. How many of these are D-sugars and how many are L sugars?

## III Polarimetry of D-Glucose, D-Fructose and Sucrose

### Chemicals:

*D-Glucose*
*D-Fructose*
*Sucrose*
*Demineralized water*

### Equipment and Materials:

*Beakers (50mL), glass rod, graduated cylinder*
*Top-loading scale*
*Polarimeter (digital Perkin Elmer or other)*
*Volumetric flasks (25mL)*

### Procedure:

1. Turn on the polarimeter by pressing red power button.
2. Turn on sodium lamp, set yellow dial to Na 589, slit width to large (o) and integration time to 20 sec. (warm up time for lamp about 15 min.)
3. Zero instrument by pressing zero-button for about 5 sec. (reading 0.000)
4. Weigh out 0.5g of the sugar, dissolve in 10mL of demineralized water, transfer to a 25mL volumetric flask, fill up to the lower meniscus margin with demineralized water, stopper and gently swirl contents.
5. Calculate concentration in g/mL
6. Pour sample carefully (to prevent air bubbles) into the removed polarimeter cell. Measure D-glucose last and save for next experiment!
7. Insert polarimeter cell
8. Reading of **observed angle of rotation** will occur in about 30 sec.
9. Remove sample and rinse out polarimeter cell.
10. Calculate **specific angle of rotation.**

### Observations and Results:

Tabulate the observed results and add the other information required according to the following table:

| **Sample** | **Temp. (°C)** | **C (g/mL)** | **Length (dm)** | **Na 589** | **solvent** | a | [a] |
|---|---|---|---|---|---|---|---|
| D-Glucose | | | | | | | |
| D-Fructose | | | | | | | |
| Sucrose | | | | | | | |

### Questions and Problems

1. What is the specific angle of rotation for each sample expressed in the proper manner: $[\alpha]^{t}_{\text{Na-D}}$ = ( + or – ) value° (solvent).
2. Does the (+) or (-) orientation of the shift of planar polarized light agree with the D or L configuration of the measured molecule? Explain!
3. What is invert sugar?

4. What would be the angle of rotation or net value of the two individual rotations, if invert sugar would be directly measured (mixture of D-glucose and D-fructose)?

## IV Mutarotation of D-Glucose

### Chemicals:

*D-Glucose*

### Equipment and Materials:

*Polarimeter*

### Procedure:

1. D-glucose can produce two cyclic forms – the anomeric α and β forms which in solution can interconvert via or oxo-cyclo tautomerism ("mutarotation" of open and closed forms).
2. Measure any changes in the observed angle of rotation in 5 min. intervals and record.

### Observations and Results:

| Change in Angle or rotation (Δα) | Time (min.) |
|---|---|
| | 5 |
| | 10 |
| | 15 |
| | 20 |

### Questions and Problems:

1. Were any changes observed? Explain!
2. What are the specific angles of rotation for the pure anomers (α-D-glucopyranose and β-D-glucopyranose)?
3. How do the literature values compare with the actual measurement of the tested D-glucose sample?
4. What is understood by equilibrium between the two forms? Is one favored over the other, which would be reflected in the percentage ratio of the two forms of the measured equilibrium mixture?

# 10 Halogenated Organic Compounds

10.1 Natural occurrence

10.1.1 allopathic chemicals

10.1.2 antibiotics

10.1.3 non-biogenic organohalides in nature

10.2 Organohalides in medicine and drugs

10.3 Organohalides in commercial applications

10.4 Relative toxicity

10.5 Environmental impact

10.5 Properties

10.6 Reactions

10.6.1 Photohalogenation

10.6.2 Elimination and substitution reactions (E1, E2, $S_N1$, $S_N2$)

10.6.3 Aromatic substitution ($S_E$)

10.6.4 Friedel-Crafts reaction (alkylation and acylation of aromatic compounds)

10.6.5 Polyaddition reactions

10.6.6 Grignard and other metal-organic reactions

## Chapter 10 Halogenated Organic Compounds and Reactions

Halogenated compounds have become so important in modern life that a separate chapter is dedicated to them.

With a few exceptions, most halogenated organics a synthetically produced. Nonetheless, in spite of the scarcity, novel natural compounds are being consistently discovered, so the verdict is still not conclusive.

Some natural organohalides are found in the **thyroid hormones** such as **thyroxin**, in mold products such as **grisefulvin**, as well as in some marine animals.

**Fig. 10.1**

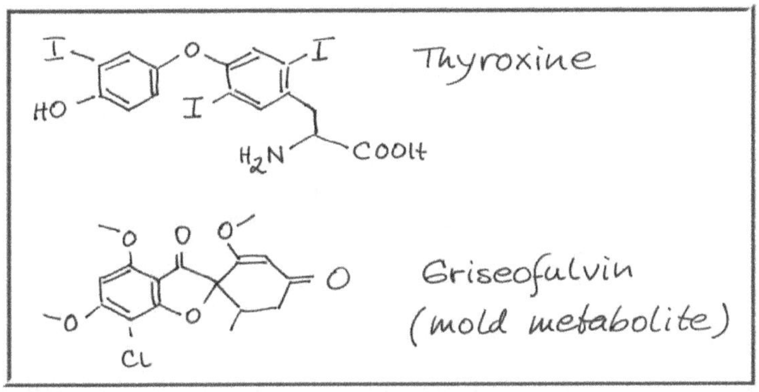


Microorganisms often incorporate a halogen atom into their **cyclic antibiotics** such as the large ring molecules known as the **macrolides**. The *Streptomyces* antibiotic **Chlorothricin,** for example, has a chlorine group in its structure.

**Fig. 10.2**

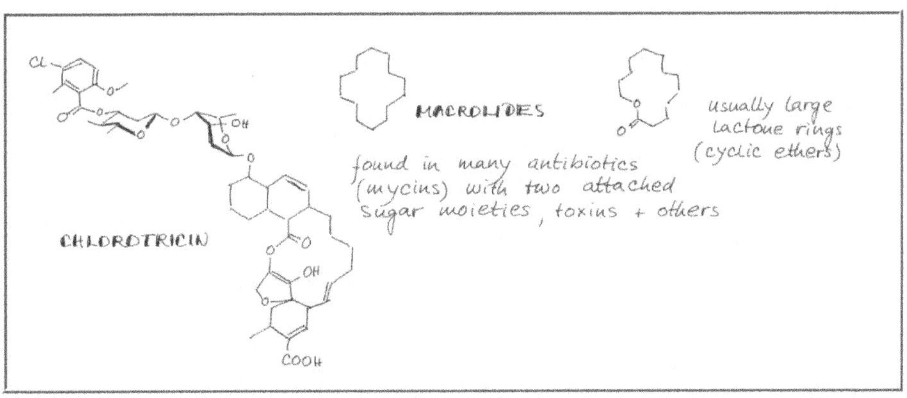

**Vancomycin** is another large ring chlorine containing antiobiotic, first discovered in the soil of Borneo some years ago. It is made by bacteria and used against other bacteria. It is a glycopeptide containing seven amino acids, mostly modified. The image below does not show the intricate stereochemistry.

**Fig. 10.3**

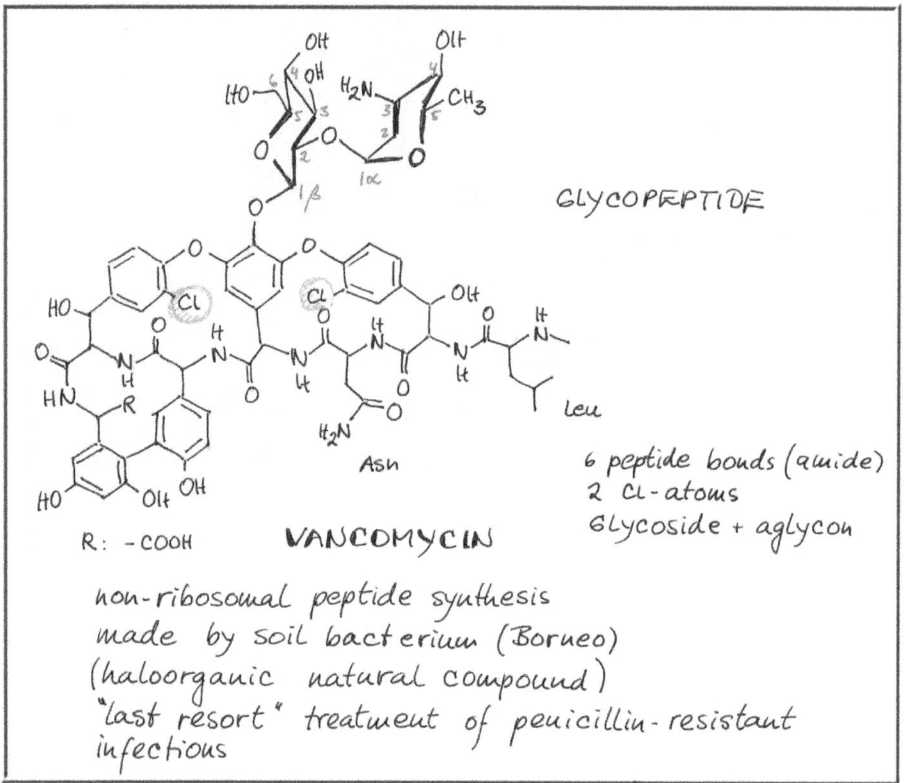


Apparently more natural products contain at least one halogen atom than previously anticipated. **Epibatidine** found in a small Central American frog species after which it is named has a pain-killing effect about 200 times that of morphine. However, it is difficult to monitor as a medicine, sine its toxicity threshold is so close. The molecular structure somewhat resembles nicotine, also a powerful toxin.

Fig. 10.4

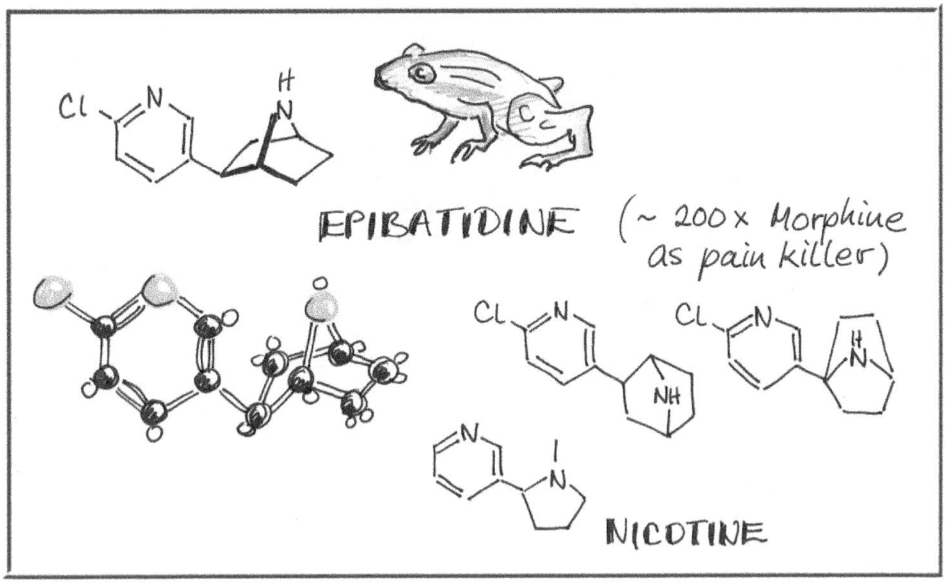


Halogenated organic toxins are found lately in many marine animals such as sea hares and cone snails. **Panacene** is a bromine containing **alkadiene toxin** which helps to ward off its predators such as sharks.

Fig. 10.5

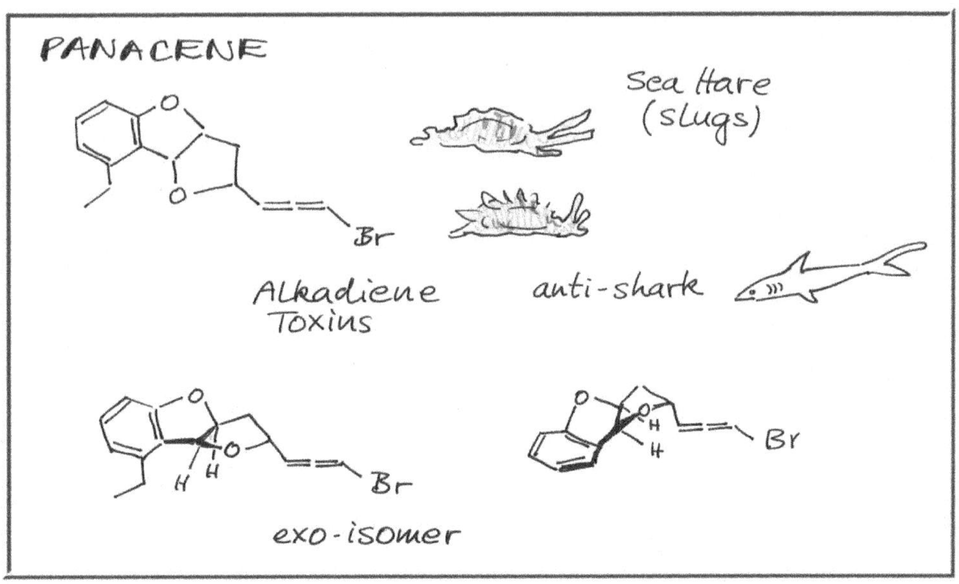

The cone snail has deadly brominated peptides in its arsenal, for which no antidote exists at the time.

**Fig. 10.6**

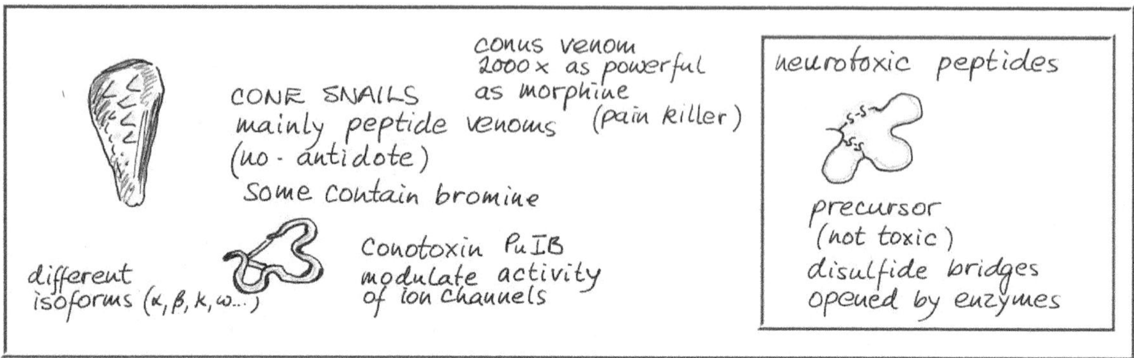


Even plants use certain halo-organic molecules to deter grazing animals or herbivores. **Sodium fluoroacetate** is a relatively simple compound with a most deleterious effect on the victim. It binds with coenzyme A to form fluoroacetyl CoA, blocking the formation of the acetyl CoA, which is vital for the function of the citric acid cycle (TCA or Krebs cycle), being literally the "wheel of life". However, some soil bacteria can come to the aid by providing a life-saving enzyme which is able to remove the fluorine atom from the acetate ion.

**Fig. 10.7**

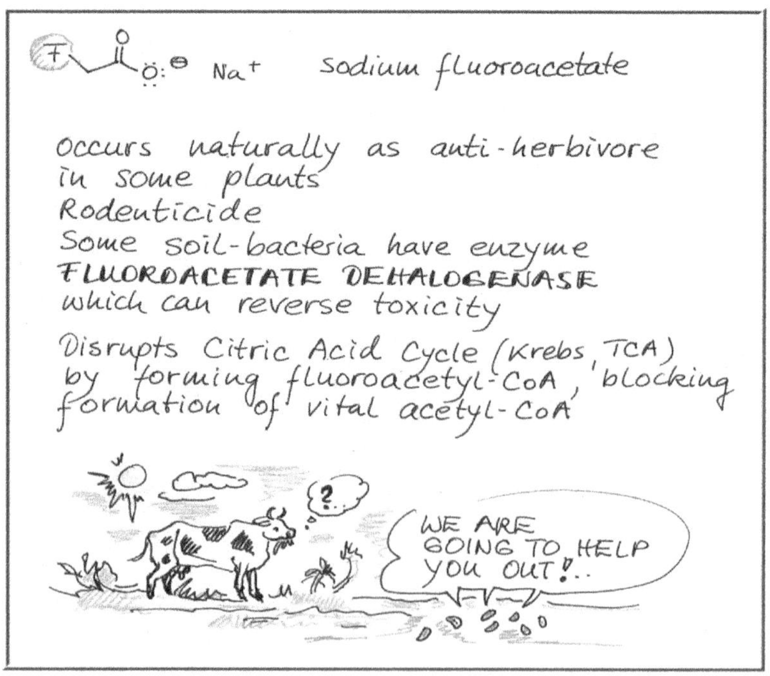

Another chemical that is supposed to keep herbivores at bay is an unusual compound that has both a chlorine and a bromine atom as substituent, produced by red algae.

**Fig. 10.8**

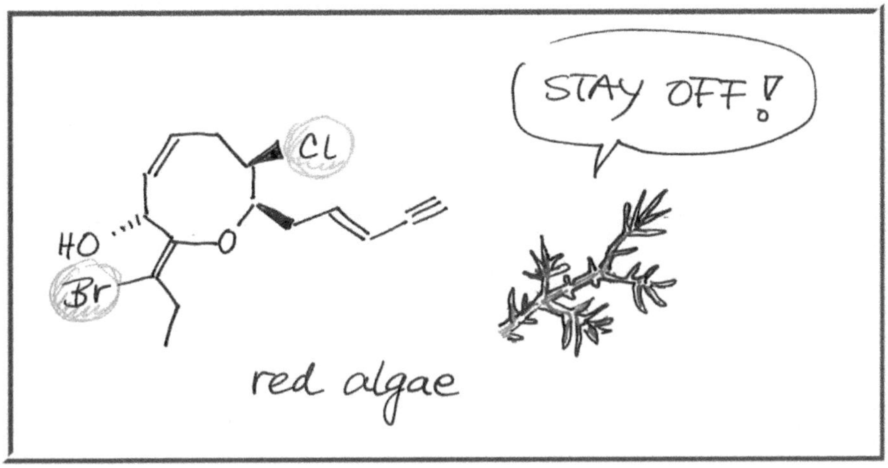


Halogen atoms are also quite often found connected to **pheromone** molecules**.** Cockroaches have up to 150 pheromones of various structures, some containing halogens.

**Fig. 10.9**

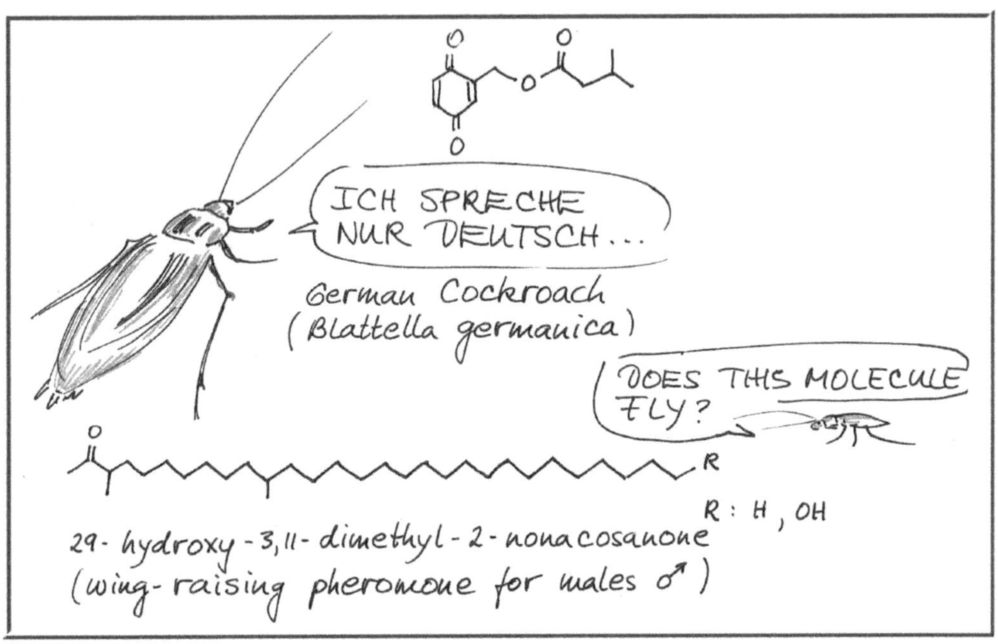

Termites also communicate with a variety of molecular messengers, which not only act as mating calls but also for socializing or assembly meetings (quorum sensing). They release large amounts of chloroform ($CHCl_3$) into the atmosphere, probably as a byproduct from wood consumption. A female tick uses 2,6-dichlorophenol for its signaling to its male counterparts.

**Fig. 10.10**

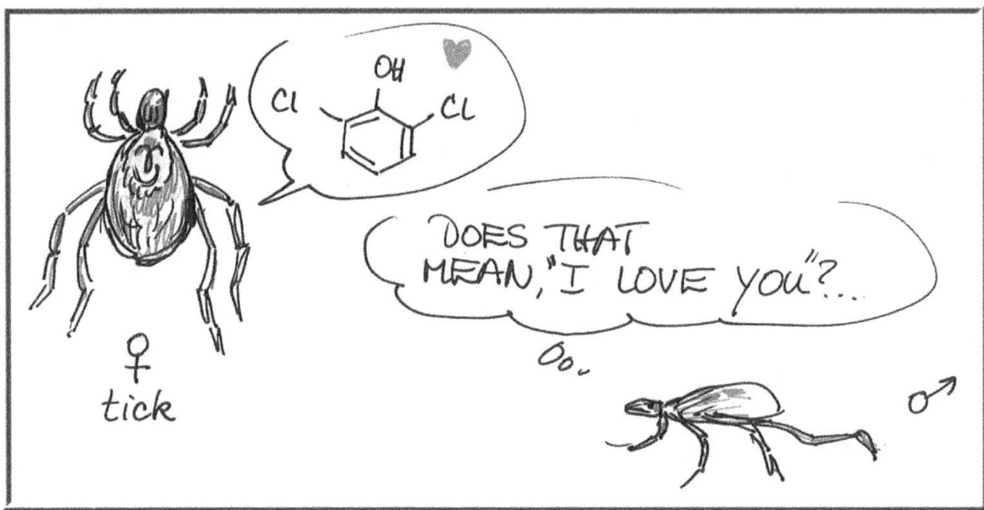


Fluorine is a very small atom with the highest electronegativity of all elements in the periodic table. However the C-F covalent bond is extremely strong, being the strongest of all halogen-carbon bonds. So, organic fluorides are very unreactive, making fluorinated medicinal drugs a method for preventing premature breakdown within the body but also a possible health hazard, still relatively unknown in its consequences.

Many of the most popular drugs contain a fluorine atom. However, halogenated compounds in general are used in all types of medical treatments and environmental pest control imaginable. A fertility drug for horses **Fluprostenol,** a prostaglandin derivative, contains a trifluoromethyl group.

**Fig.10.11**

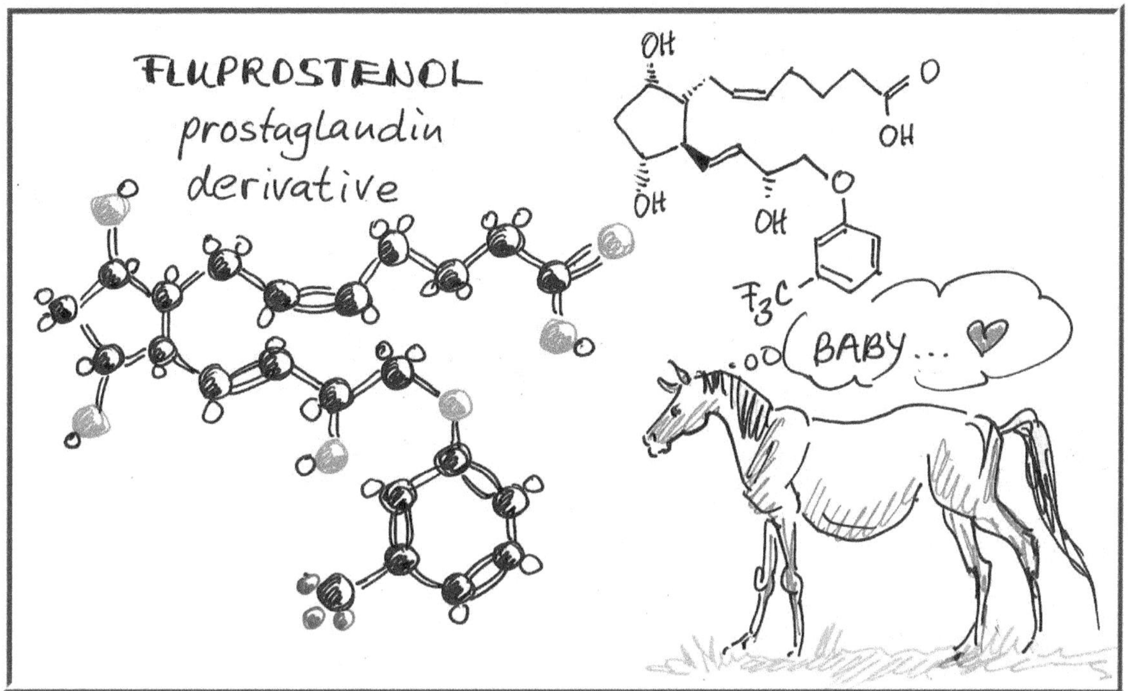


One of the best-known human drugs is the anti-depressant **Prozac** which also contains the trifluoromethyl group.

**Fig. 10.12**

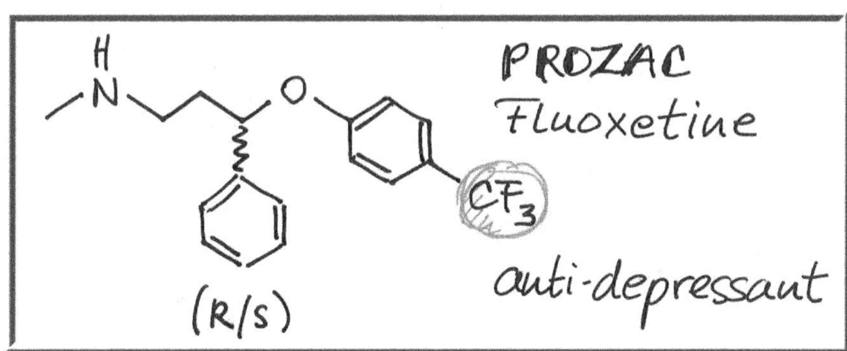


The following table shows the electronegativity of the individual halogens and the bond-strengths of the carbon-halogen covalent bonds. As comparison the carbon-hydrogen bond energy is 413cal/mol. Even though the C-F bond has the highest polarity, it is the most difficult to break apart, partly because the polarity almost creating a type of over-lying ionic bond, strengthening the already existing covalent bond.

**Table 10.1**

| Halogen (X) | Electronegativity (Pauling values) | Carbon-Halogen Bond (C-X) | Bond Strength (cal/mol) |
|---|---|---|---|
| F | 4.0 | C-F | 467 |
| Cl | 3.0 | C-Cl | 346 |
| Br | 2.8 | C-Br | 290 |
| I | 2.5 | C-I | 228 |

Small fluorinated methane derivatives in association with chlorine groups such as $CF_2Cl_2$ molecules were used extensively as aerosol propellants. By drifting into the upper atmosphere they were apparently broken down by cosmic rays into environmentally hazardous molecular species that could possibly interfere with the ozone layer. However, this connection is still controversial, albeit a wise move to prophylactically ban these suspected molecules from further use as propellants. However, these **chlorofluorocarbons (CFC's)** such as $CFCl_3$ and $CF_2Cl_2$ are still very much in use as valuable **refrigerants.**

**Fig. 10.13**

Volatility of simple mono-substituted organohalides increases from iodine to fluorine as substituent. Solubility is mostly confined to non-protic organic solvents.

Most halo-organic molecules are not flammable and therefore valuable **inhalant anesthetics.** They also are fat-soluble and easily cross the blood-brain barrier. Earlier, simple halogenated ethane derivatives such as **halothane** and **teflurane** were used. Nowadays halogenated ethyl methyl ether derivatives are most commonly used, such as

The volatility of liquid fluorinated derivatives of ether make them valuable inhalant anesthetics which also have the benefit of working similar to ether but also being non-combustive. The ideal is to have low blood/gas partitioning coefficients and lipophilic properties so that they can cross the blood brain barrier more efficiently. Halothane was one of the first non-ether halogenated inhalant used for anesthetics but is now more replaced by the less toxic ether halogen derivatives, which mostly contain fluorine atoms.

**Fig. 10.14**

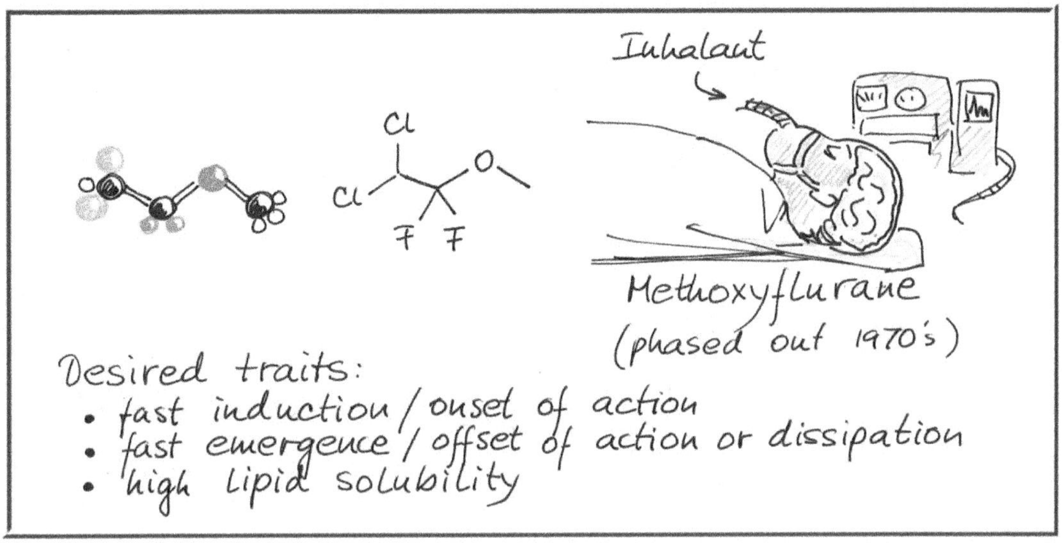


How they work is not fully understood. A list of various inhalant anesthetics is shown below:

**Fig. 10.15**

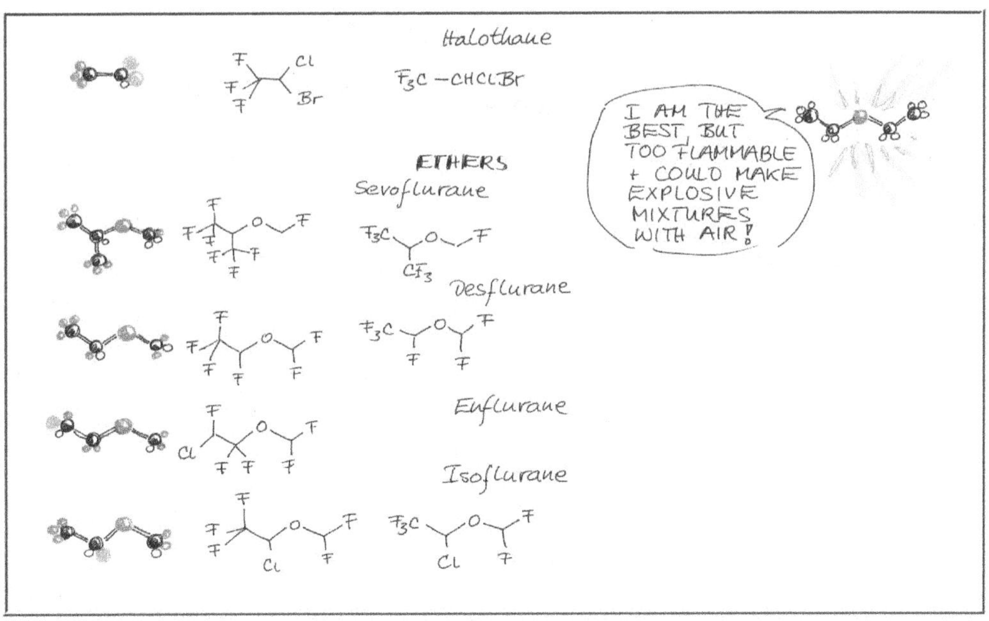


A similar compound to halothane but having a chlorine group by another fluorine atom is **Teflurane.**

**Fig. 10.16**

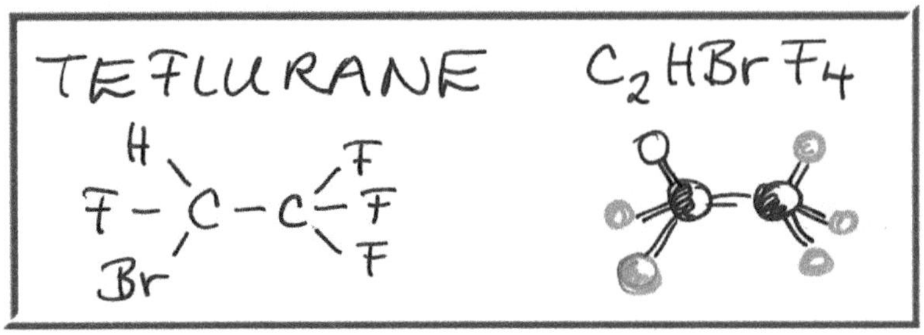


Their possibly damaging side-effects is mainly due to break-down products or metabolites. Chloroform which also is an excellent inhalant anesthetic has the main problem of forming the extremely toxic and noxious phosgene molecule when exposed to the oxygen of the air.

**Fig. 10.17**

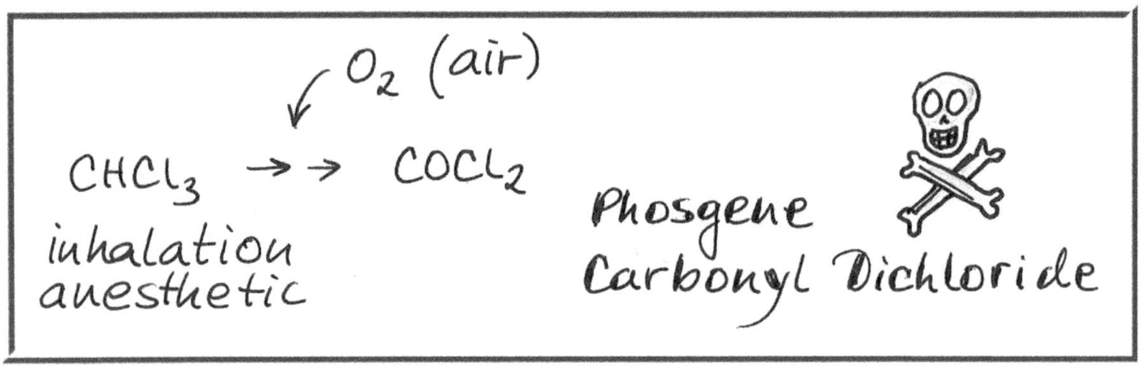


Alkyl halides are themselves excellent solvents of this sort, making them widely used in the dry cleaning business, “dry” meaning not water-based. Here the chlorinated derivatives are of special significance. Tetrachloromethane, however, turned out to be a critical liver toxin and is now replaced by less toxic chlorinated hydrocarbons such as 1,1,1-trimethylethane. Chloroform (trichloromethane, $CHCl_3$) has been replaced by methylenechloride (dichloromethane, $CH_2Cl_2$) as common solvent used in laboratories.

**Fig. 10.18**

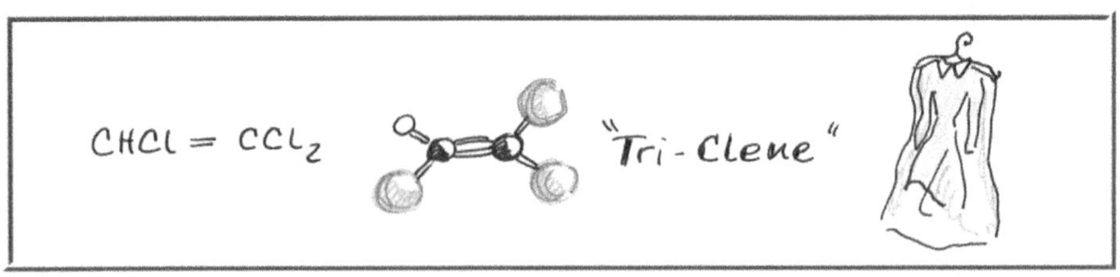

All the halogenated methane derivatives are in some ways narcotic

Methyl iodide is a colorless volatile liquid used for soil fumigation. Chemically it is an excellent tool as a methylating agent sinece iodine is such a good lesving group for nucleophilic substitution reactions, discussed later in this chapter.

**Fig. 10.19**

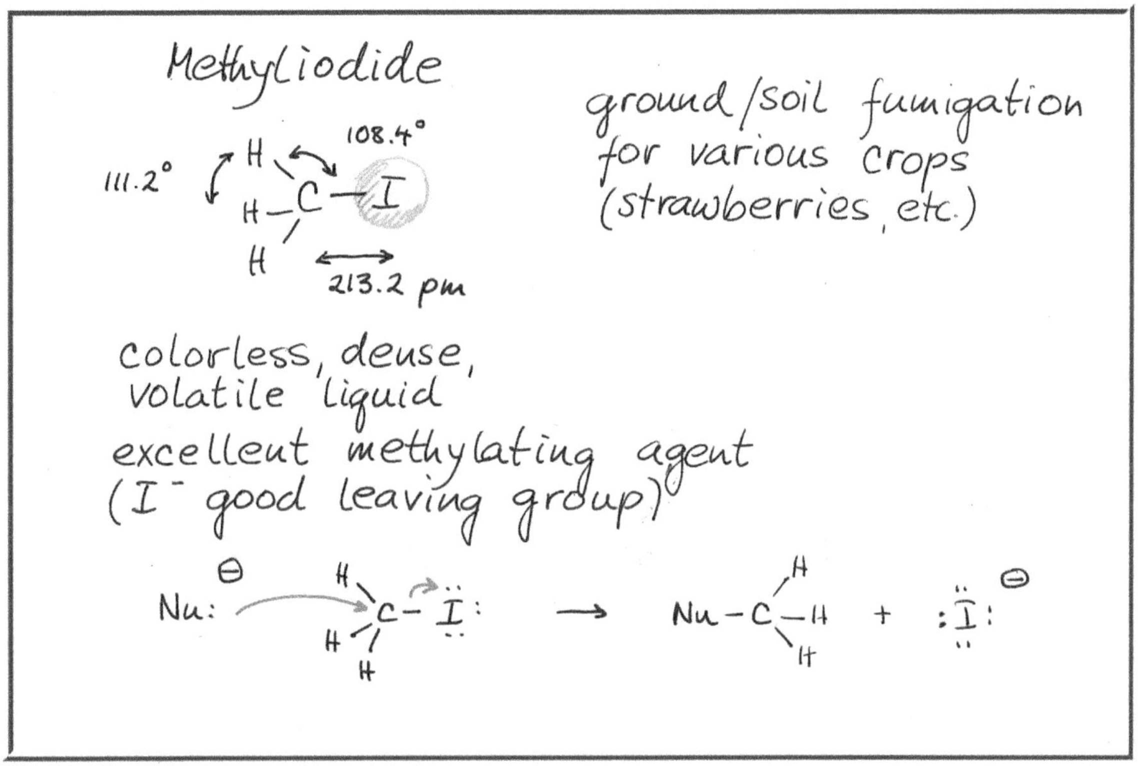


Of special importance is the family of **haloforms ($CHX_3$)**. Fluoroform ($CHF_3$) is the only gaseous representative of the group and is used as a refrigerant for low temperatures. It is a respiratory irritant which can also act as a narcotic in large doses. Chloroform is a well-known and important, but now banned anesthetic, similar in significance to ether, but not as flammable. Due to its health concerns such as listed as a possible carcinogen, anesthesia drugs are now based on a series of new and improved halogenated molecules. Bromoform ($CHBr_3$) acts as a sedative and antitussive. Iodoform is a yellow solid which is used as a topical antiseptic or anti-infective agent. Iodo-organic compounds are often used as radiological contrasting agents, so-called radiopaque media. Bengal Rose is another iodized compound used as a liver diagnostic too

**Fig. 10.20**

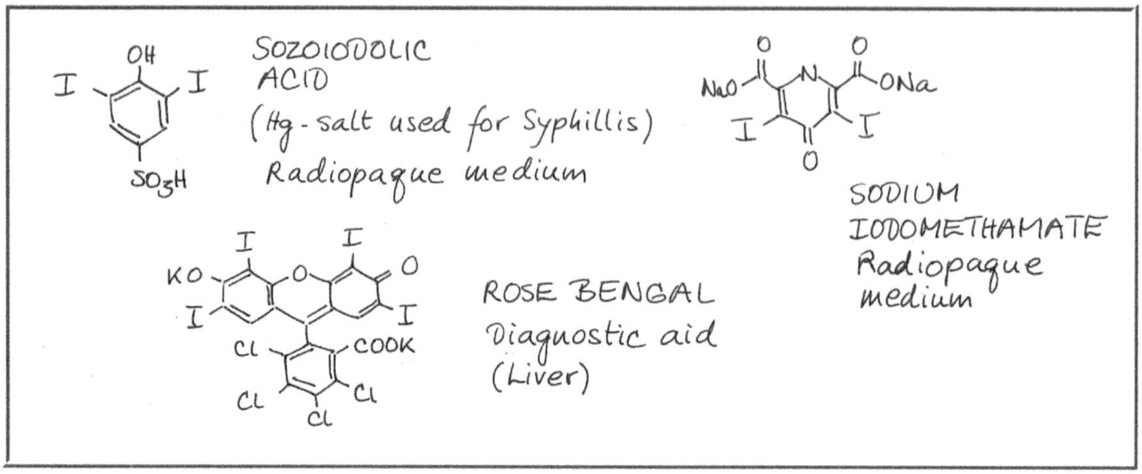


Bromine organic compounds were already known to have distinct sedative properties.

**Fig.10.21**

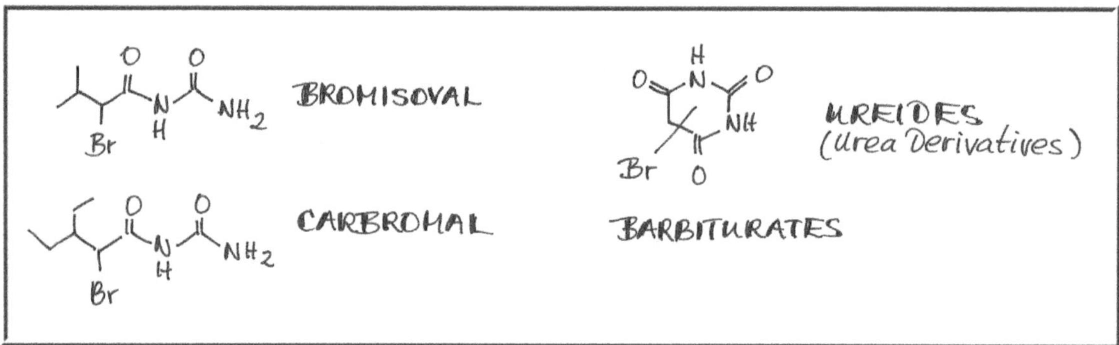


Brominated organics are also especially useful as flame retardants. Simple methane derivatives are often in fire extinguisher such as **Halon-1211** ($CBrClF_2$) or **Halon-1301** ($CBrF_3$). Actually all halogenated methane derivatives serve as fire quenchers such carbontetrachloride or tetra chloromethane, which today can be converted into safer and more efficient bottled extinguishers. It can be systematically fluoridated by hydrogen fluoride with the aid of antimony pentafluoride as catalyst to the important Freon refrigerants.

**Fig. 10.22**

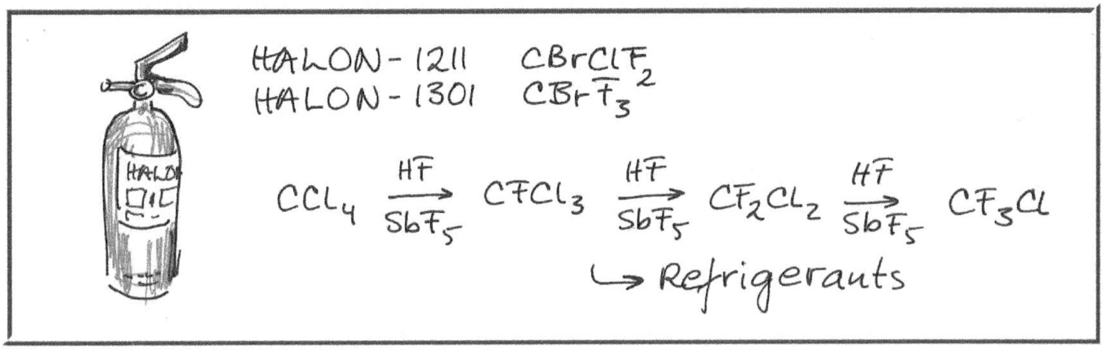


For furniture, carpets, textiles etc. the brominated diphenyl ethers or haloorganic phosphates (tris) are usually used to prevent fires.

**Fig. 10.23**

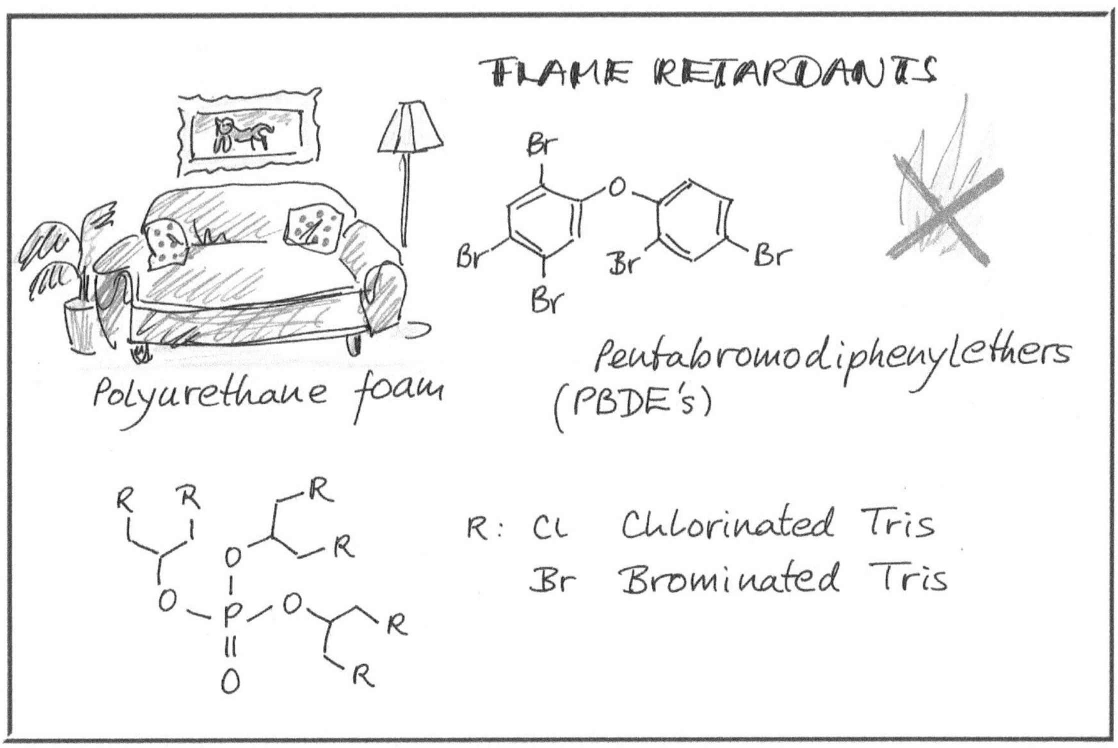


Many of the commercial flame retardants are of the more or less brominated diphenyl ether types.

**Fig. 10.23**

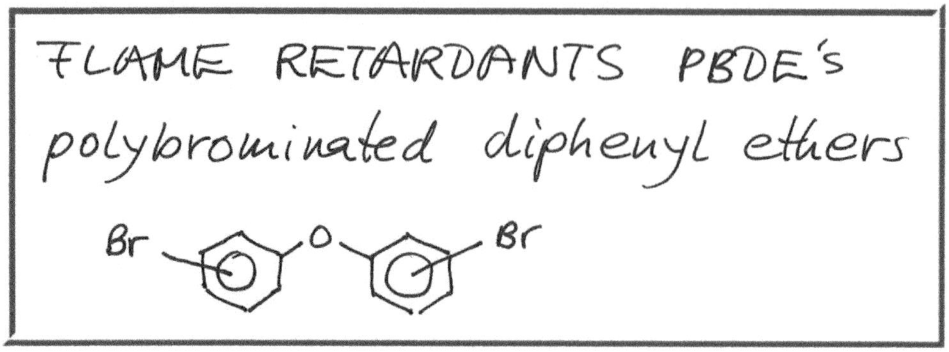


Fluoro-organic molecules have shown special oxygen binding properties and are thus useful chemicals for acting as breathing fluids (1-bromoperfluorooctane, etc )

**Fig. 10.24**

Blood substitutes such as "Fluosol", consisting of a combination of perfluorodekaline and perfluorotripropylamine are already used in medicine and veterinary medicine, including perfluorotributylamine. Toxins, venoms and poisons are harmless when dissolved in artificial blood.

**Fig. 10.25**

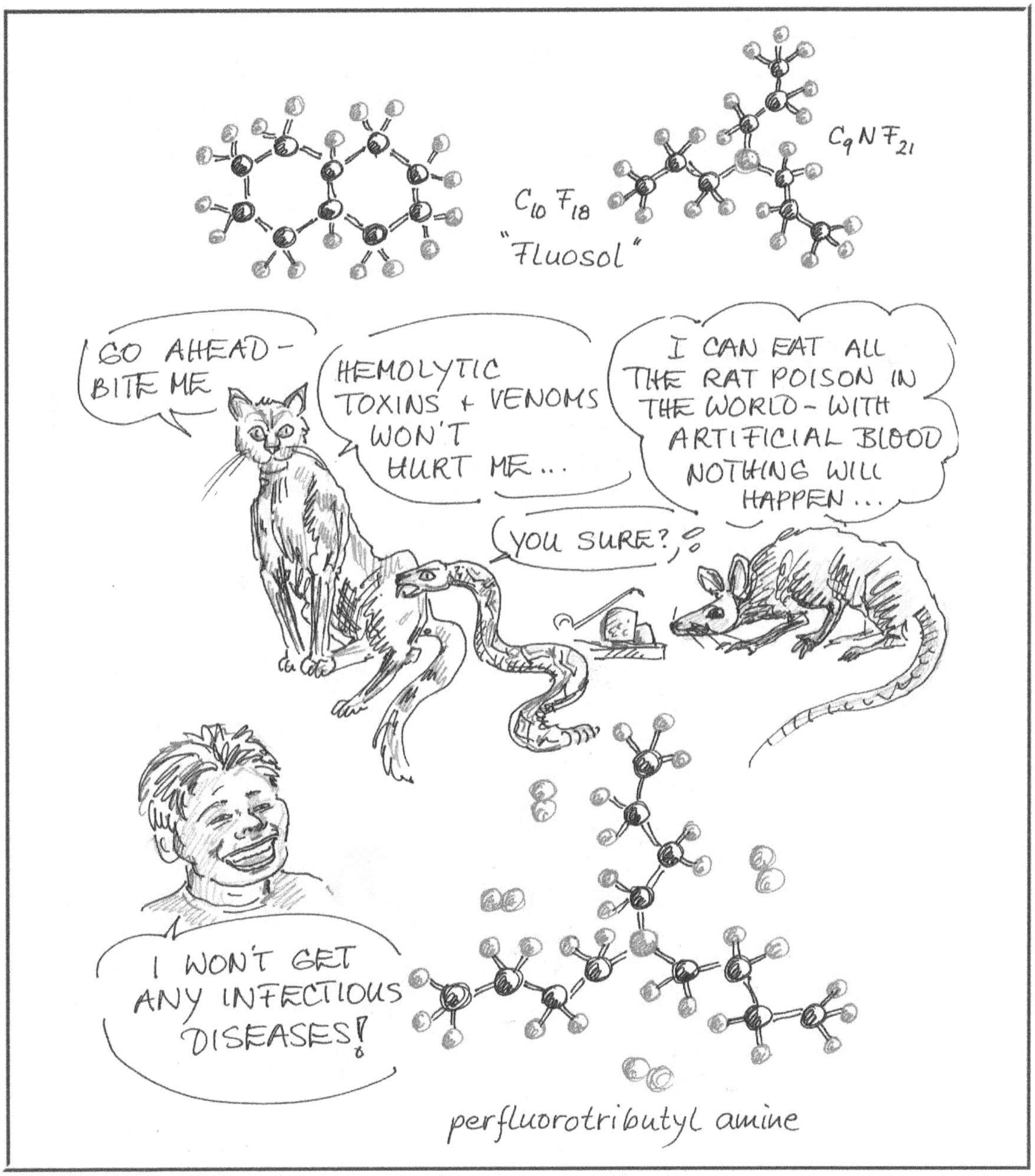


Halogenated unsaturated organics are especially useful in making all kinds of addition polymers.

**Teflon** is formed by polyaddition of tetrafluoroethene (tetrafluoroethylene). Mixed perhalogenated alkenes can lead to other durable plastics such as **Kel-F**, made from 1,1-difluoro-2-chloroethene ($CF_2$=CHCl). Extremely durable elastomers are produced by the polyaddition of different alkenyl fluorides such as the addition of 3-fluoropropene with 1,2-difluoroethene leading to **Viton**.

**Fig. 10.26**

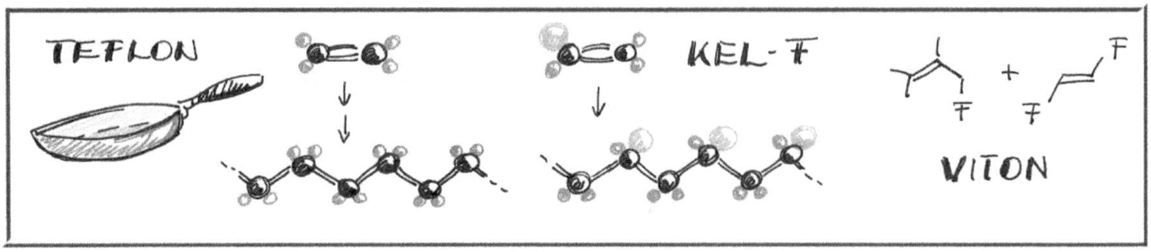


An important halogenated as well as unsaturated addition polymer is **Neoprene,** a very sturdy and durable elastic natural rubber (polyprene) substitute, used for giant parade balloons, engine belts etc..

**Fig. 10.27**

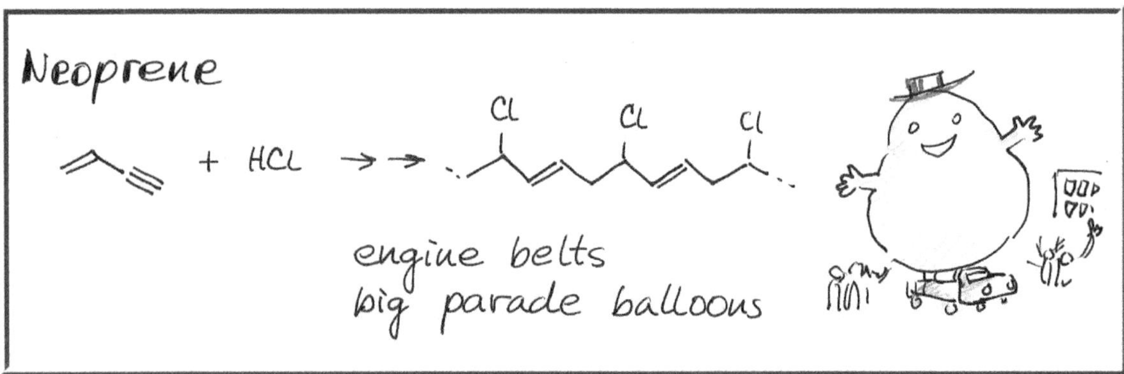


Many halogenated hydrocarbons have received substantial notoriety as environmentally hazardous compounds such as **DDT** and **Agent Orange**. The latter, consisting of the two defoliants 2,4-D and 2,4,5-T, was used extensively during the Viet Nam war. TCDD is a derivative that occurs as a byproduct leading to long-range health effects such as diabetes and others.

**Fig. 19.28**

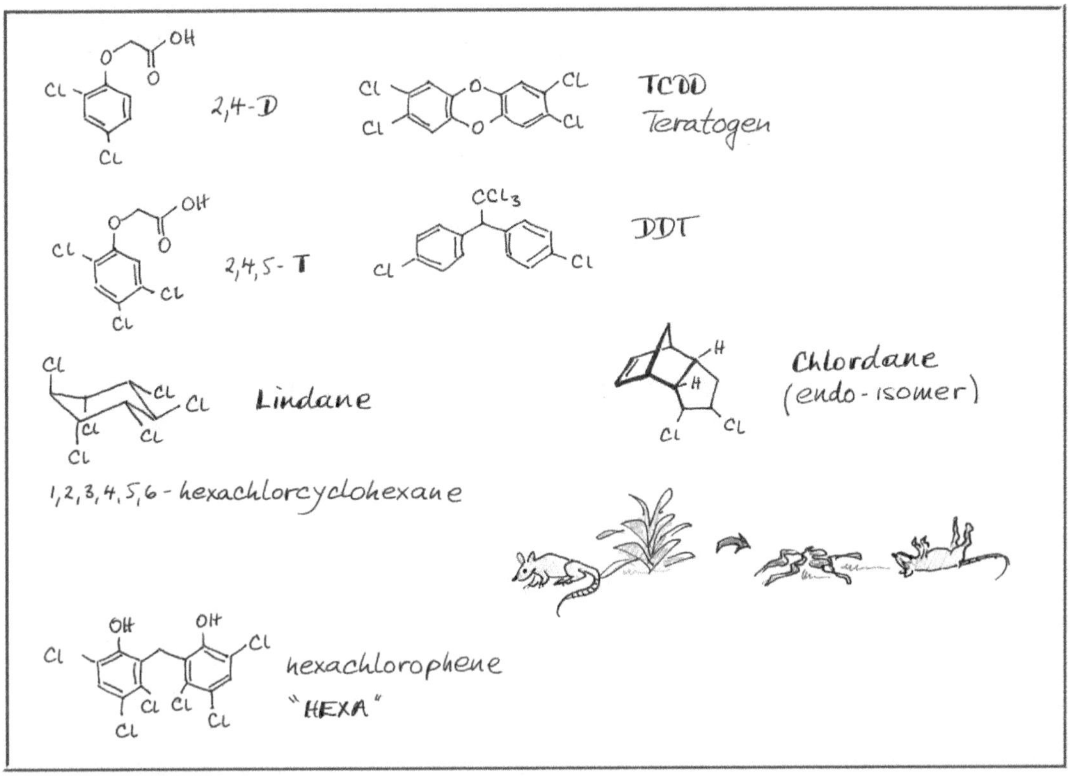


Some pesticides are extremely powerful anticoagulants, acting as lethal Vitamin K antagonists especially for larger mammals such as coyotes, possums and others. **Brodifacoum** represents one of those "super warfarins", which inhibits the enzyme vitamin K epoxide reductase.

**Fig. 10.29**

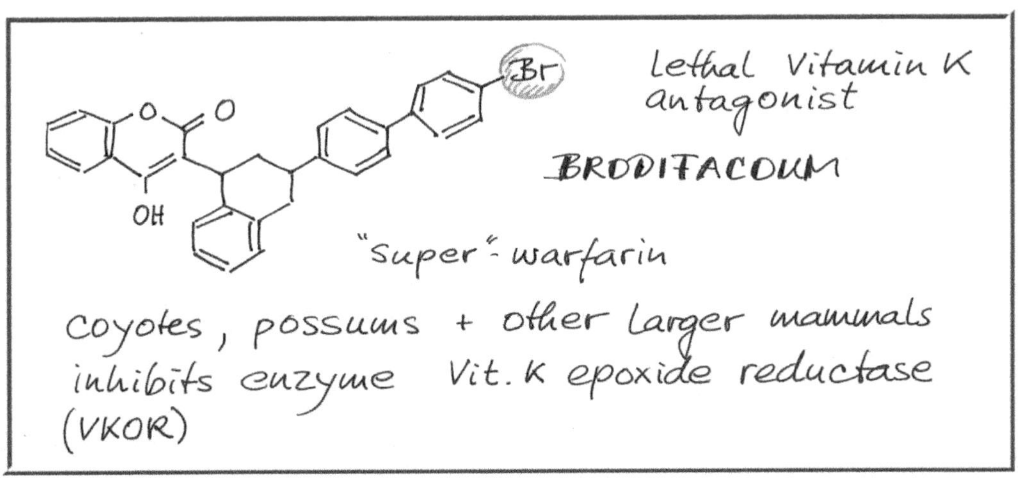

Halogenated phosphor-organic molecules play a special role as warfare gases such as **Sarin** and **Soman.** These are acetylcholine inhibitors that lead to irreversible firing of the nerves.

**Fig. 10.30**

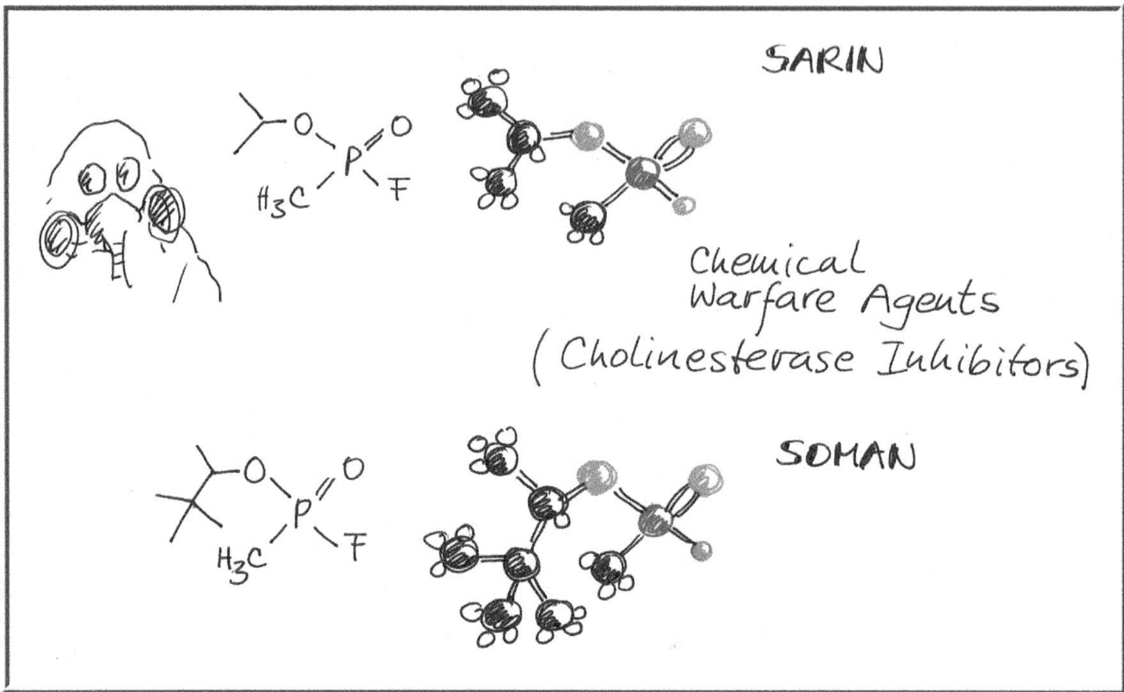


Other phosphor-organic cholinesterase inhibitors containing halogen atoms are used as insecticides such as the mixture of the two **Phosphamidon** configurational isomers E ("entgegen" , opposite in German) and Z ("zusammen", together in German).

**Fig.10.31**

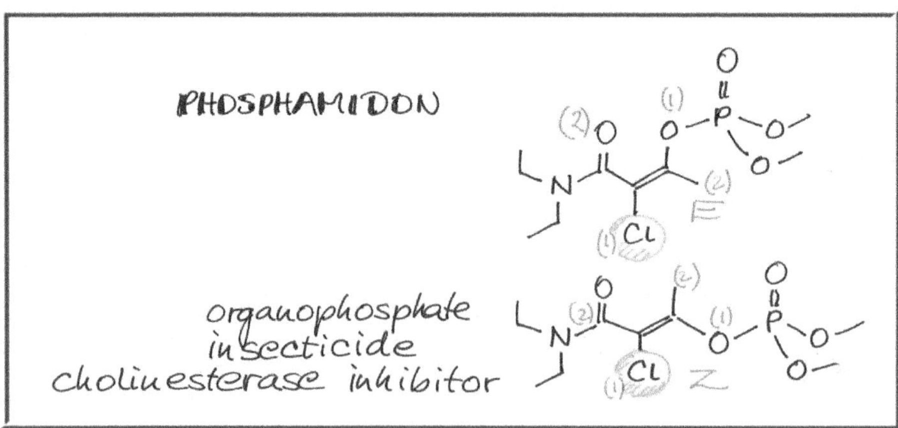

An interesting phosphor-organic compound containing halogen atoms but used for a different purpose is **Defosfamide**, which is a drug used to treat cancer as antineoplastic.

**Fig. 10.32**

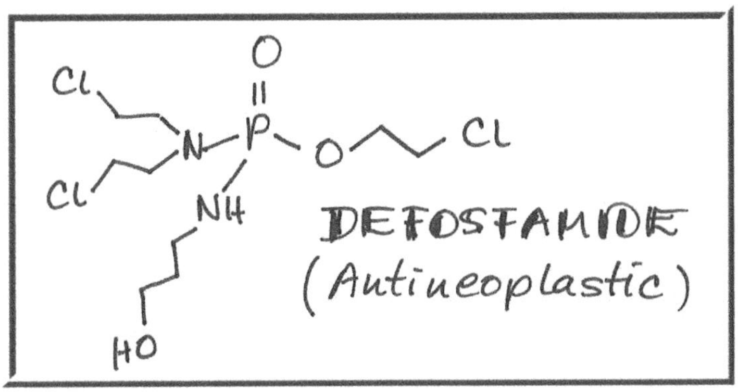


Teargas and other incapacitating agents that work on the lungs and skin are shown below. Others which used to be used in aggressive tactics have found use as a fumigant such as trichloronitromethane.:

**Fig. 10.33**

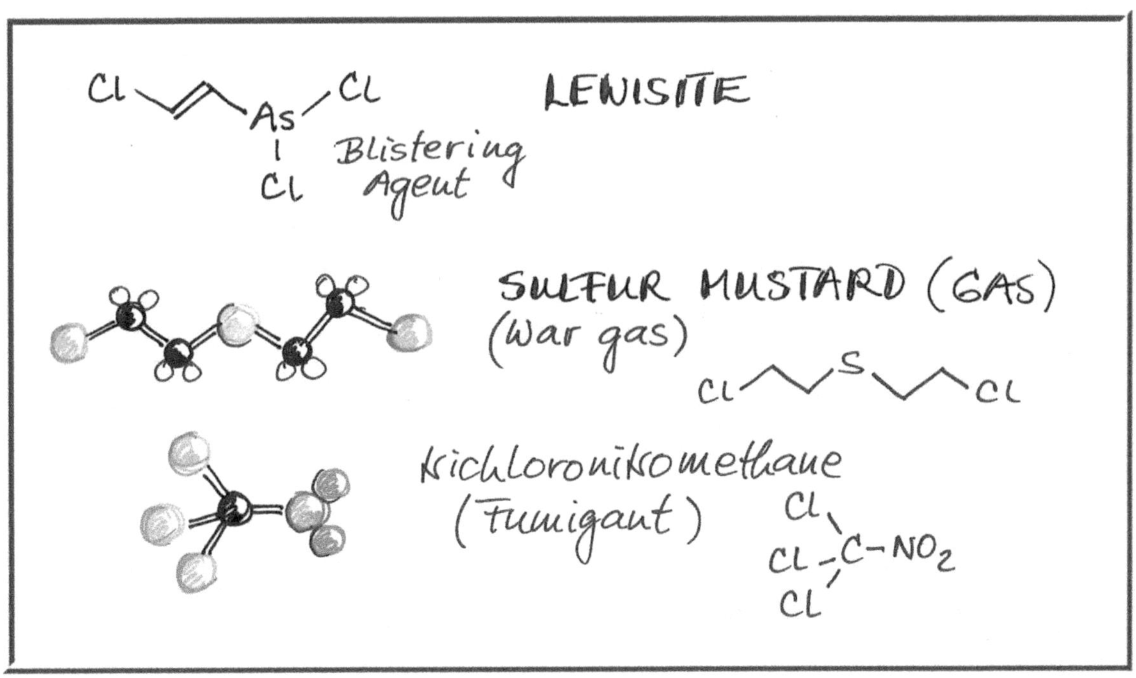


**Sucralose** ("Splenda") is now a commonly used, non-caloric sugar substitute. It is a chlorinated form of the normal table sugar sucrose. As of late, it is now used in many flavored water drinks.

**Fig. 10.34**

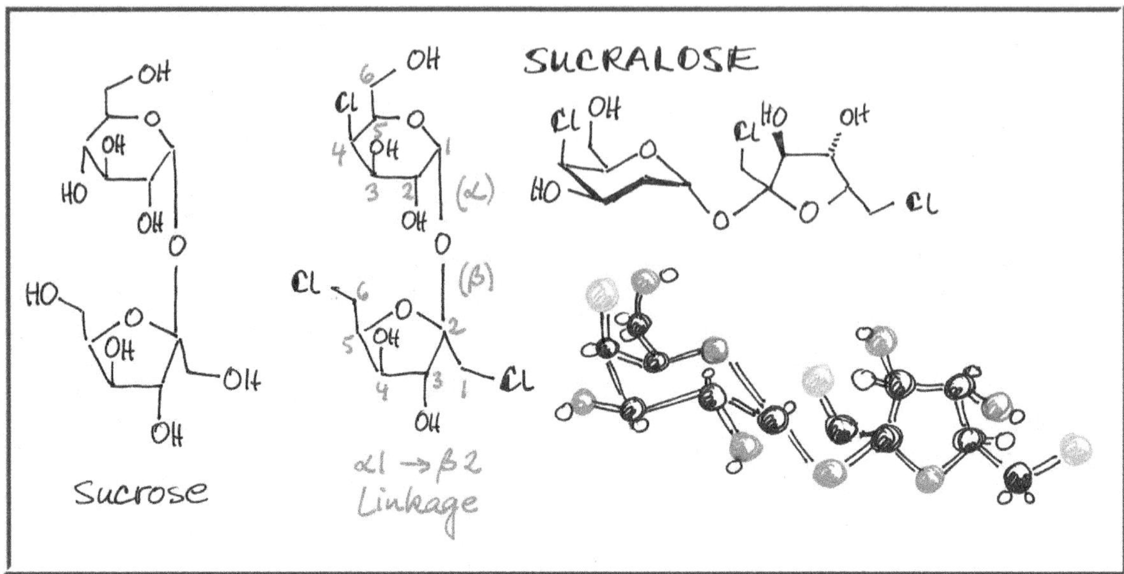


A natural dye found in *Bolinus branderis* shells is a brominated version of indigo that instead of blue adopts the color of purple, a sought after, valuable dye for nobility in the ancient Mediterranean world, known as **Tyrian Purple.**

**Fig. 10.35**

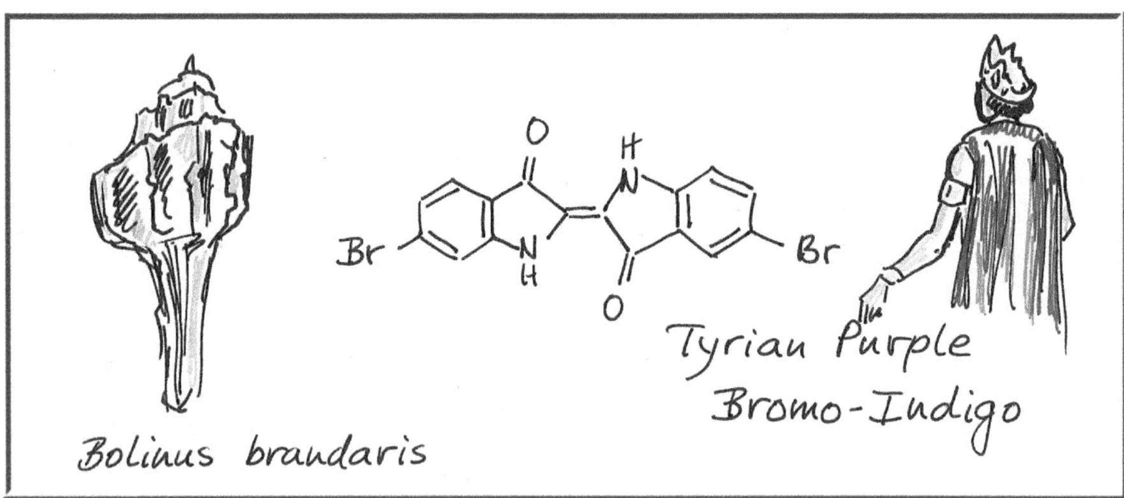


A synthetic halogen compound such **bis(2,4,6-trichlorophenyl) oxalate** or **TCPO** for short is used today as an activator for chemiluminescent dyes in **glow sticks**.

**Fig.10.36**

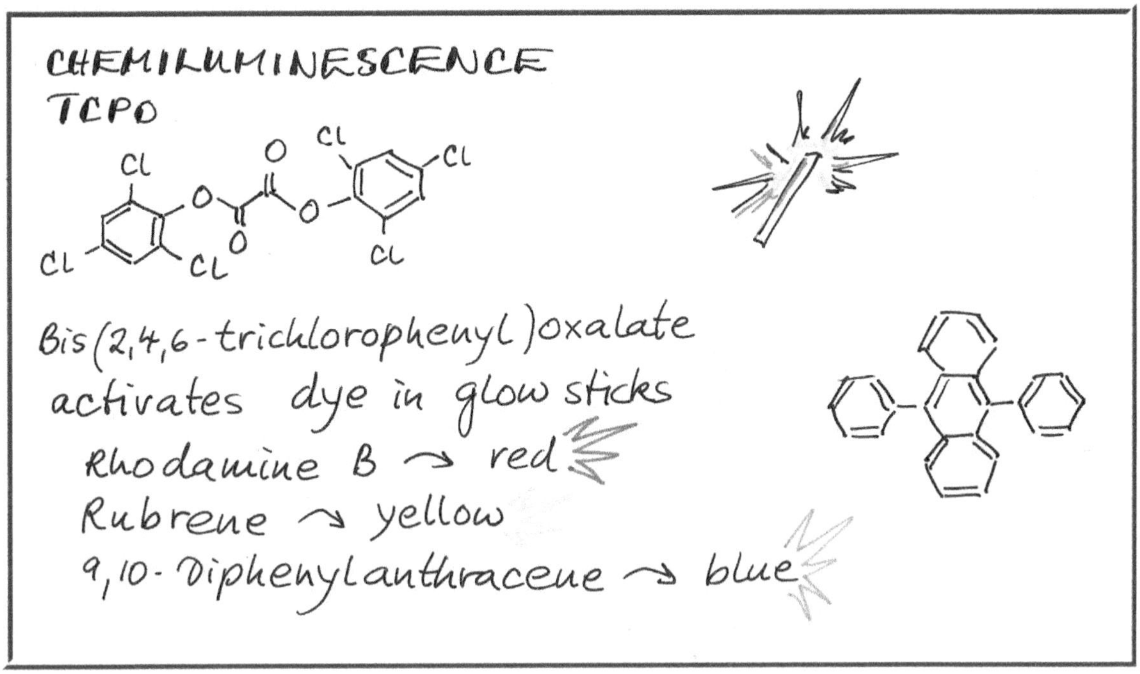


As can be seen, halogen organic compounds are unlimited. Their uses cover the whole spectrum of human endeavor. They seem to be also more represented in nature as previously thought. Each halogen has its specific relevance and character when connected to hydrocarbon frameworks with or without other heteroatoms. The main purpose is to aquaint the reader/student with the multi-dimensional facets of haloorganics. The next part of this chapter focuses on important reactions that utilize the special properties of organic halides and metalorganic halides as synthetic tools for the preparation of many different organic molecules.

Alkyl halides are especially useful in organic synthesis, being used as alkylation and acylation agents for aromatics as is the case of the well-known **Friedel-Crafts reaction,** which with the help of **Lewis acid** catalysts ($AlCl_3$ and $FeCl_3$) allows reactive electrophilic species to form. The alkyl or acyl nucleophile attacks the pi system of the aromatic nucleus such as benzene or its derivatives to form as rate-determining step via transient pi and sigma complexes the intermediate carbocation, which is resonance stabilized.

**Fig.10.37**

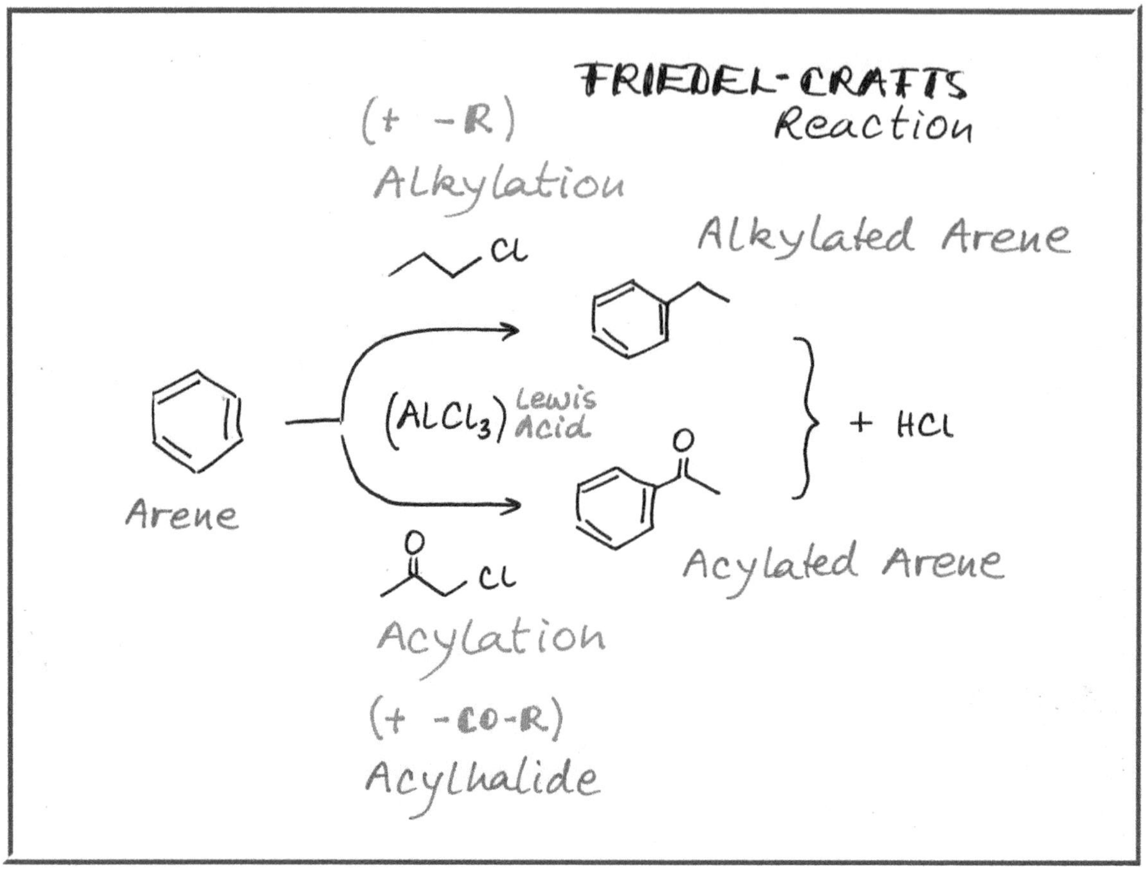


By elimination of a proton the energy-favored aromaticity is restored in a swift follow-up reaction step ($S_E$ mechanism)

**Fig. 10.38**

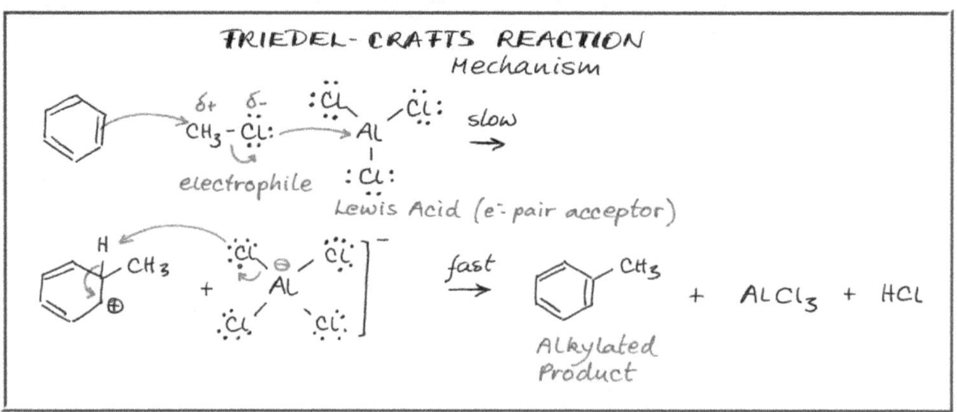

**Example Problem 10.1**
**Complete the following reaction:**

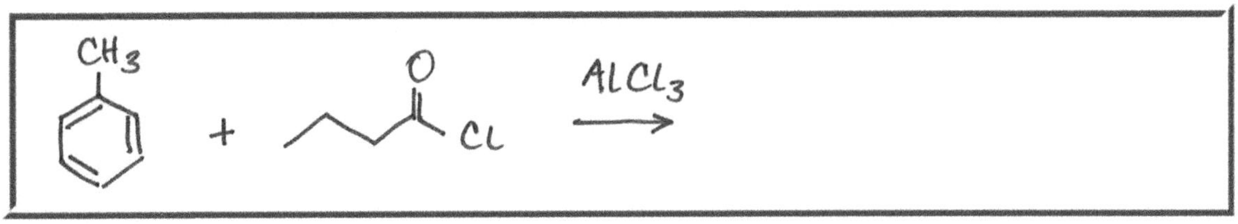


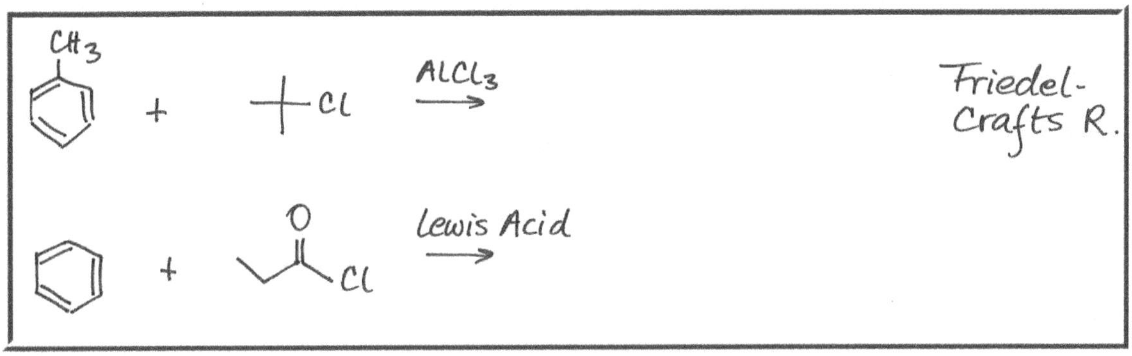


In aliphatic chemistry involving purely aliphatic molecules or aliphatic side-chains of aromatic compounds, halogens play an important role in substitution and elimination reactions ($S_N1$, $S_N2$ and E1, E2, resp.). These reactions rely on halogens acting as good nucleophiles as well as good leaving groups, depending on the reaction conditions. Unsaturated halides do not lend themselves well to substitution reactions however. This is due the resonance structures involving the halide electrons and the pi systems of the multiple bonds.

Of special commercial importance are the unsaturated halogenated hydrocarbons (alkenyl halides), especially those containing the **vinyl ($CH_2$=CH-)** or **allylic ($CH_2$=CH-$CH_2$-)** groups. Vinyl chloride (ethenylchloride) is the monomer leading to polyvinylchloride or PVC. Propenylhalides form via intermediates **epoxide halides**, which serve as building blocks for the manufacture of **epoxy resins**. Alkenyl as well as alkynyl halide bonds are very stable and hard to break. These compounds are inert to nucleophilic substitution reactions such as $S_N1$ and $S_N2$.

Toxicity of halogenated hydrocarbons can vary. Saturated fluorocarbons apparently are non-toxic.

**Perfluoro-2-methylpropene**, however, is exceedingly toxic, much more than the war gases such as sarin or soman.

Carbonylchloride ($COCl_2$) not strictly an organic compound is the toxic phosgene molecule.

2-Fluoroethanol as well as the salts of fluoroacetic acid are very toxic. Sodium fluoroethanoate serves as an effective rodenticide. Strangely enough, sodium trifluoroethanoate is not toxic.

Halogen organic compounds are often involved with metals as **metal-organic compounds.** These types of compounds strictly involve covalent bonds between the metal and carbon atoms and do not include metal coordination compounds. Magnesium and lithium seem to have a special affinity to alkylhalides with reactivity decreasing from iodine to fluorine substituted hydrocarbons, the latter being hardly reactive at all.

Magnesium best reacts with halides containing iodine, bromine and chlorine. Zink prefers bonding with halides containing iodine and bromine as substituents.

An interesting reaction is the **halogen-metal-exchange,** in which a type of double displacement occurs as shown below:

**Fig. 10.39**

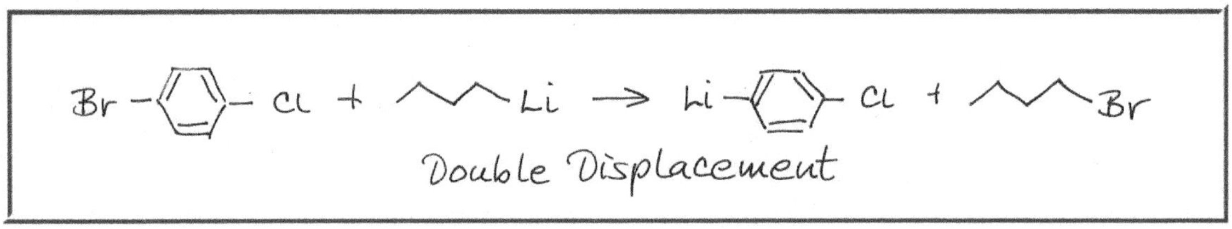


Many iodine organic compounds serve as radiopaque agents (x-ray contrast media) and are thus used as contrast agents in medical diagnostics. A simple iodine derivative of methane for this purpose is methylene iodide ($CH_2I_2$).

The halogen family of elements involved in organic molecules is most useful in every-day life, medicinal and environmental applications, but as is the case for all chemicals, they are a two-edged swords. In other words, "there is no free lunch". For every good use, there is a sacrifice to be made and for each choice a benefit risk analysis needs to be assessed and the "lesser evil" has to be carefully evaluated.

**Grignard reagents** are metal-organic compounds that represent valuable synthetic tools for a variety of chemical compounds. These organomagnesium halides must first be prepared by letting magnesium react with alkylhalides in dry ether solutions. Usually compounds with the general formula **R-Mg-X** form, but also dialkylmagnesium ($MgR_2$ ) compounds by competing reactions can occur as well, which, however, are also reactive. These compounds are very effective in producing new carbon-carbon bonds via nucleophilic attack of the nagativated (δ-) R-group as the result of the electropositive magnesium.

**Fig.10.40**

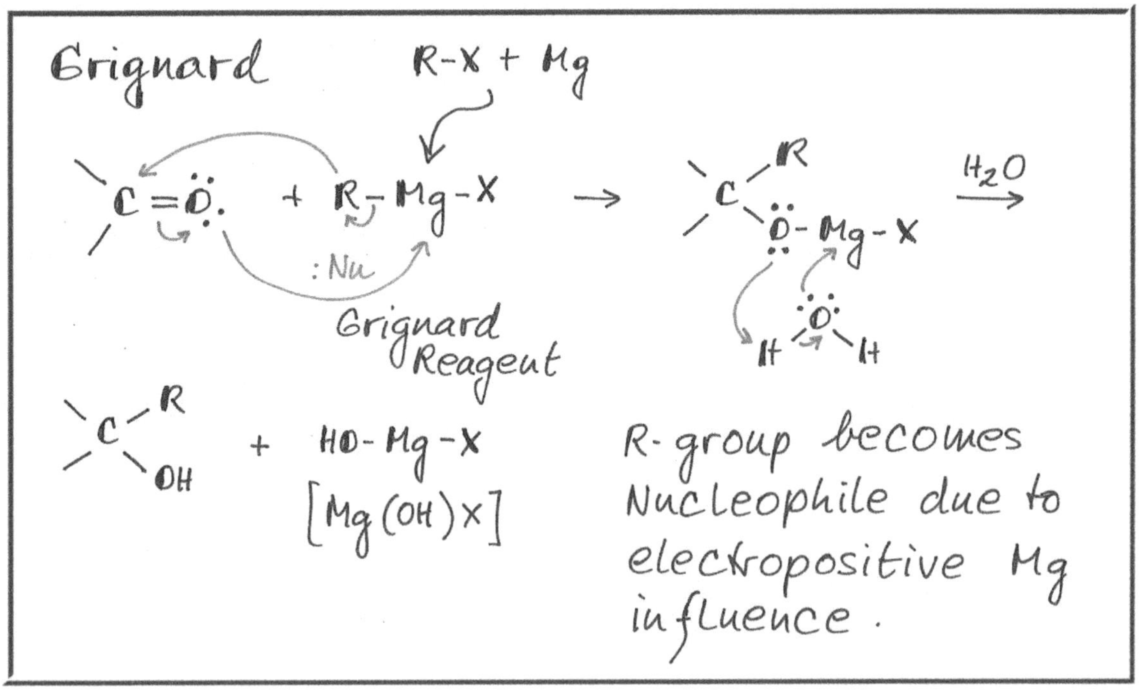


**Example Problem 10.2**
**Complete the following Grignard reactions:**

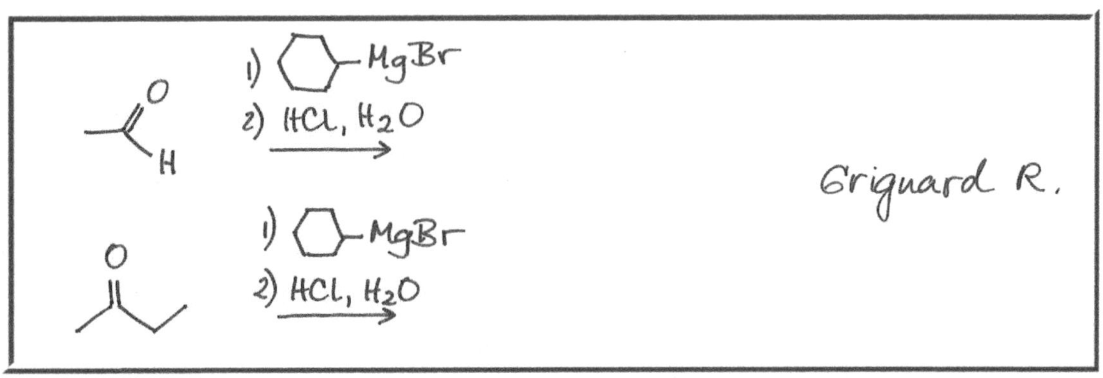


Many chemicals are prepared by nucleophilic aliphatic substitution reactions, which can either proceed in a one step, concerted fashion via an energized **transition state** or in two steps via formation of a **carbocation** as **intermediate** species. The aliphatic substrate molecules usually have a halogen group attached, which will serve as **leaving group.** The leaving group can be "coaxed" to leave with the help of special reaction conditions or structural parameters (heat, special solvent, catalyst, relative strength of nucleophile, steric hindrance etc.) before the nucleophile makes its attack, thereby forming the carbocation as intermediate for this rate-determining step ($S_N1$, two-step mechanism). Other reaction conditions favor the one-step ($S_N2$) mechanism, in which the nucleophile binds to the substrate molecule by the same rate as the leaving group makes its disappearance. This **concerted flow** of reaction leads to the **temporary five bonded carbon**, loosely binding both groups at the same time in the shape of **trigonal bipyramid**.

**Fig. 10.41**

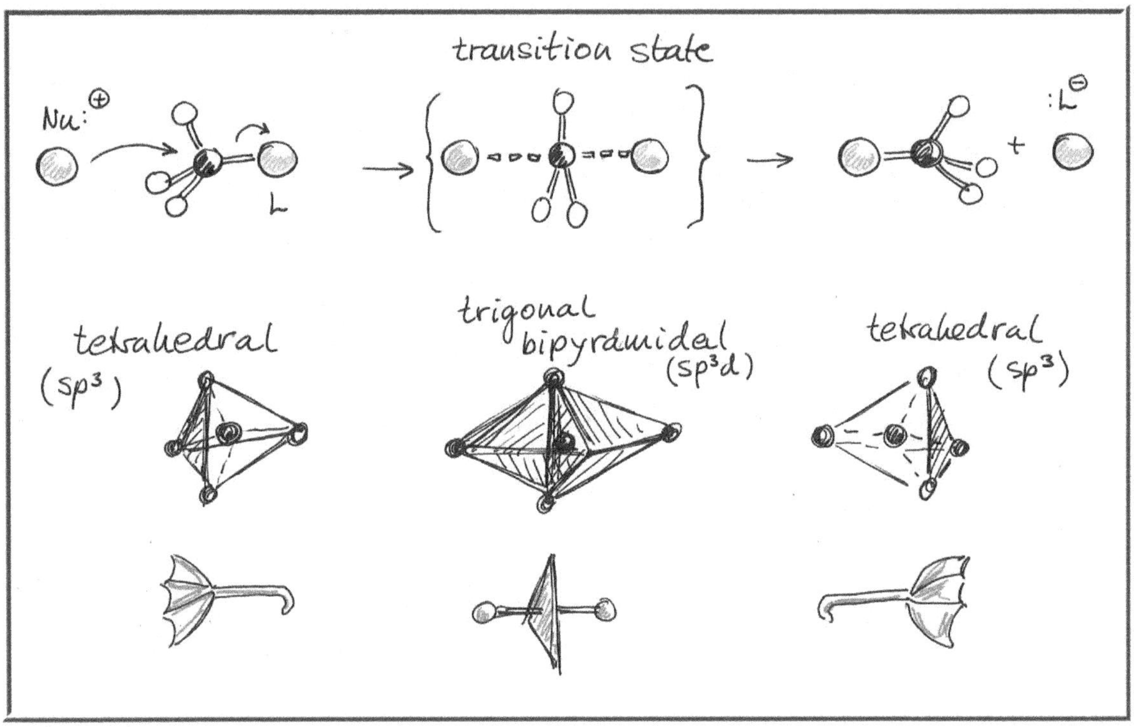


The nucleophile (Nu: or Nu:⁻), either as neutral or ionic species, prefers to enter the substrate from the rear, thereby preventing any steric repulsion with the leaving group. If the tetrahedral umbrella opposite the leaving group is too crowded (3° substitution effects), the nucleophile has to wait for the substrate to drop off the leaving group on its own, thereby creating the **trigonal planar** shaped carbocation plane, which allows the nucleophile to attach itself from either side. In the case of a chiral carbon at the center of the original tetrahedron the $S_N1$ reaction will lead to **racemization.**

**Fig.10.42**

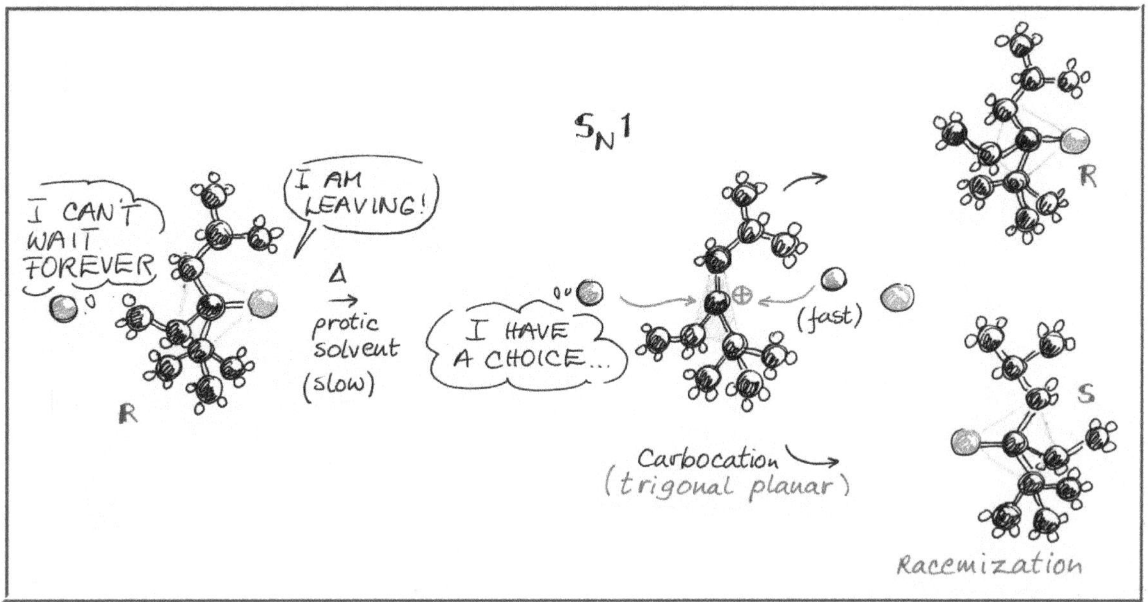


If rear-attack is possible, as is the case for 1° substituted substrate molecules, then the concerted $S_N2$ mechanism can unfold with ease. If chiral carbons are involved in the transaction then **stereochemical inversion** is the outcome – an R substrate will convert into an S substitution product or vice versa. The reason behind this phenomenon is that the tetrahedral umbrella flips to the other side, once the nucleophile has added and the leaving group has left the five-centered transition state (trigonal bipyramid).

**Fig. 10.43**

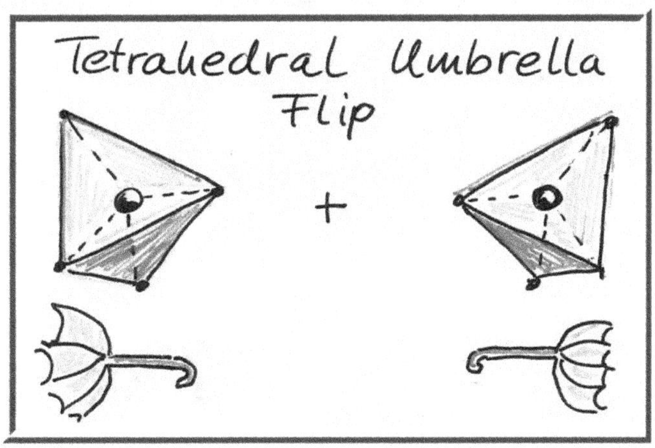

Molecules containing 2° substituted carbon atoms can react by either mechanism. Not only will the degree of substitution affect the reaction mechanism but also, temperature, the relative strength of the nucleophile, the relative efficiency of a group to act as a leaving group, as well as the protic or non-protic character of the solvent. **Protic solvents** can form hydrogen bonds by virtue of their hydroxyl groups, thereby enhancing the stabilization of ionic species such as carbocations or negatively charged leaving groups ($:L^-$) by solvation (hydration) effects. They will help promote the $S_N1$ reaction. A good leaving group is determined by its ability to leave and from an electron-rich :L (:X) or $:L^-$ ($:X^-$) species, meaning a stable, but poor conjugate base to the corresponding acid HL (HX).

**Fig.10.44**

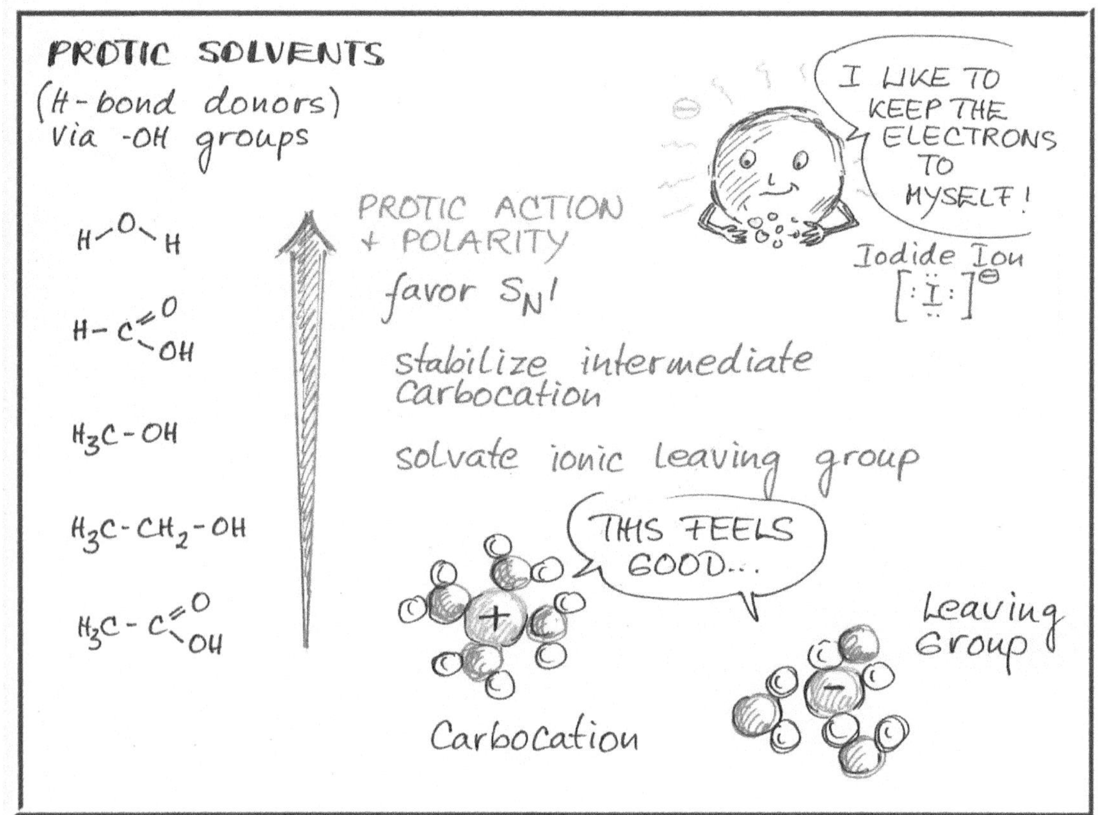


**Non-protic solvents** on the other hand, which happen to be polar, but not able to function as hydrogen-bond donors, favor the $S_N2$ mechanism.

**Fig.10.45**

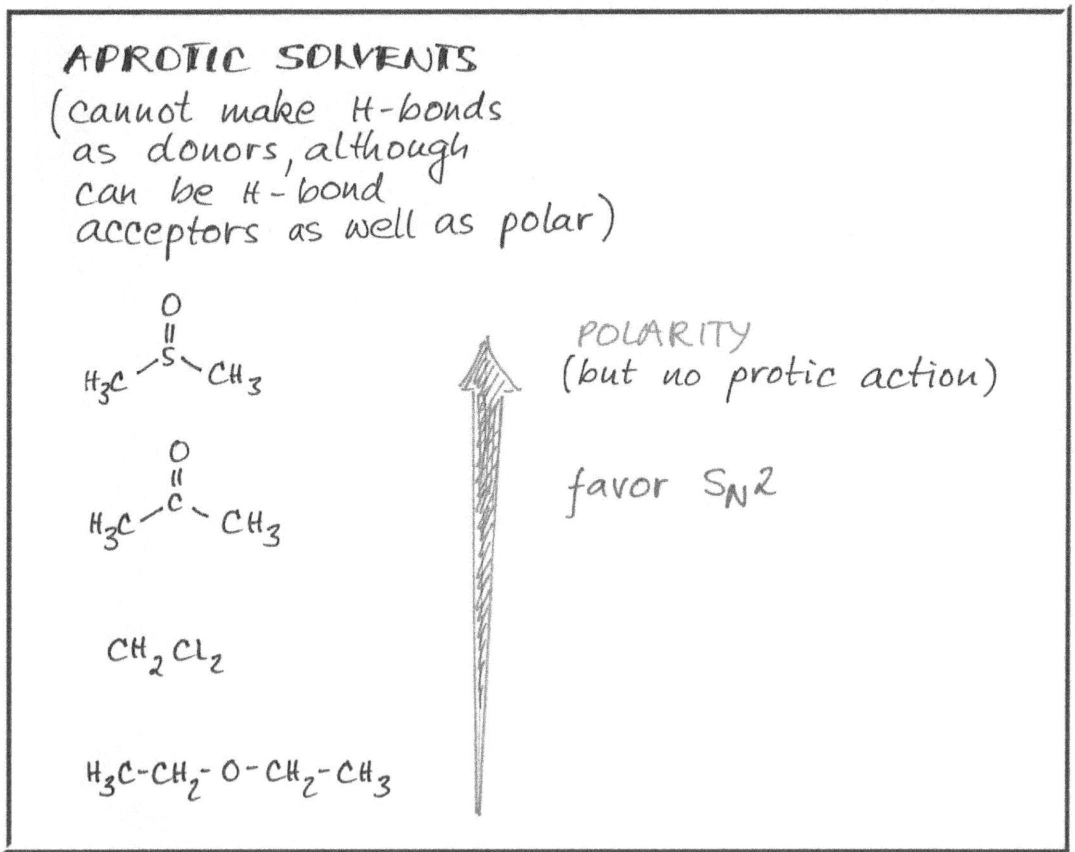


Good nucleophilic tendencies run counter to increased electronegativity values of the atoms involved. Sulfur containing species R-SH or R-S$^-$) are better than oxygen containing nucleophiles. Ions function better than neutral molecules.

**Fig. 10.46**

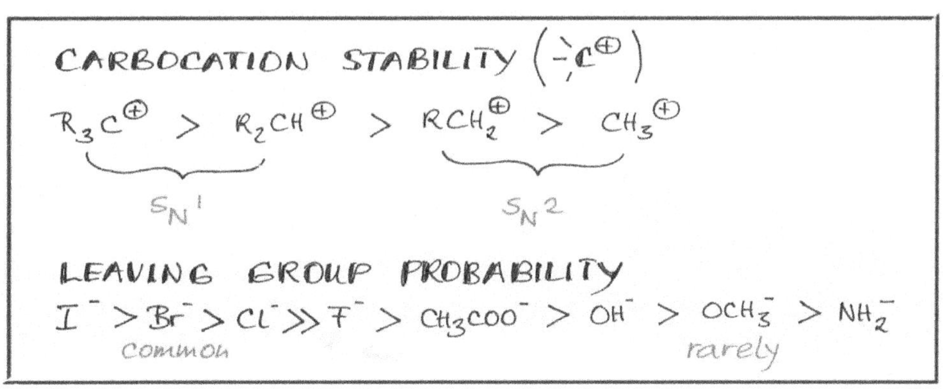

The best leaving groups are sulfonates [$RS(O_2)O^-$], iodide ($I^-$) and bromide ($Br^-$). The best way to determine, whether a leaving group $:X^-$ is of any merit, is to consider how strong the conjugate acid HX would be. Therefore $CF_3COO^-$ is a much better leaving group than $CH_3COO^-$, since the acid strength of the former is much higher than the latter. For alkyl halides the order of reactivity is R-I > R-Br > R-Cl >> R-F. The carbon fluorine bond is so strong and the acidic form HF so weak that fluorine could hardly function as a leaving group at all. The over-all reaction possibilities are shown below:

**Fig.10.47**

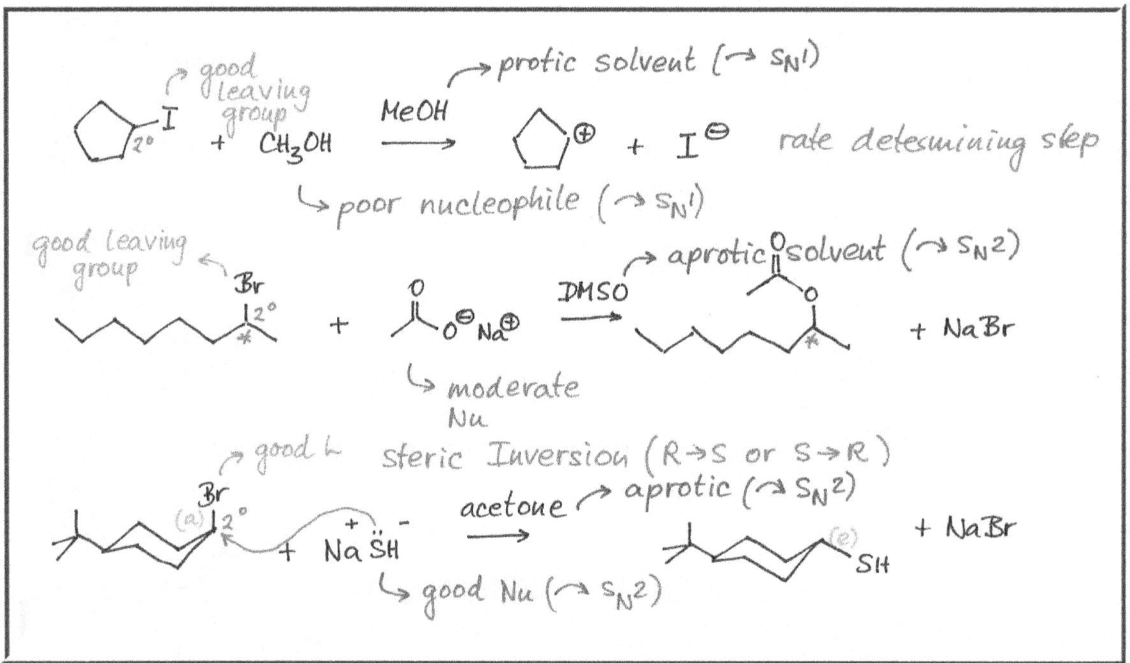


Bromide and iodide ions are some of the best nucleophiles under adequate reaction conditions. However iodide is also considered a good leaving group which seems like a discrepancy.

**Fig. 10.48**

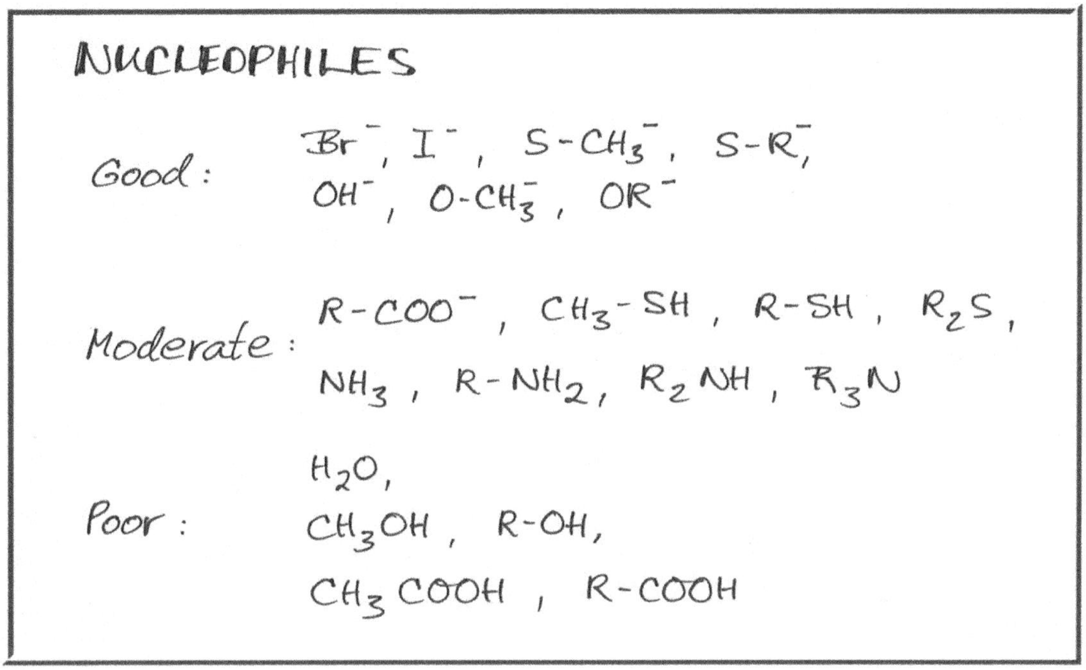


If the substrate molecule contains an **acidic β-hydrogen**, attached to the carbon adjacent to the carbon atom (α) containing the leaving group, then a strong base ($OH^-$, $OR^-$) can lead to a competing reaction, known as **elimination (E).**

**Fig. 10.49**

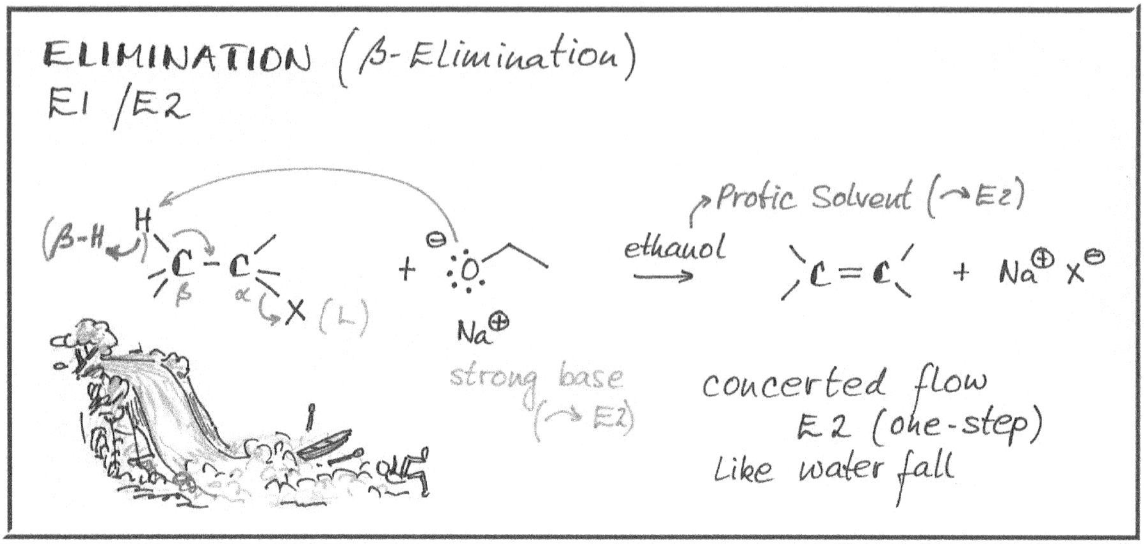


This reaction that can follow **first order (E1)** or **second order** kinetics **(E2)** in a similar manner as the substitution reactions leads to the formation of unsaturated molecules. According to **molecularity,** the unimolecular reaction proceeds via the two-step mechanism (E1) and the bimolecular reaction (E2) according to the one-step concerted flow reaction mechanism.

**Fig. 10.50**

$R_{E1} = k\,[\text{haloalkane}]$ first order, two-step

$R_{E2} = k\,[\text{haloalkane}][\text{base}]$ second order, one-step concerted with transition state

Due to the abstraction of a beta (β) hydrogen atom by the attack of a strong base with moderate to poor nucleophilic activity, the conceptionally reversal of an addition reaction leading to the elimination of small molecules is often also called **beta (β)-elimination.**

**Fig.10.51**

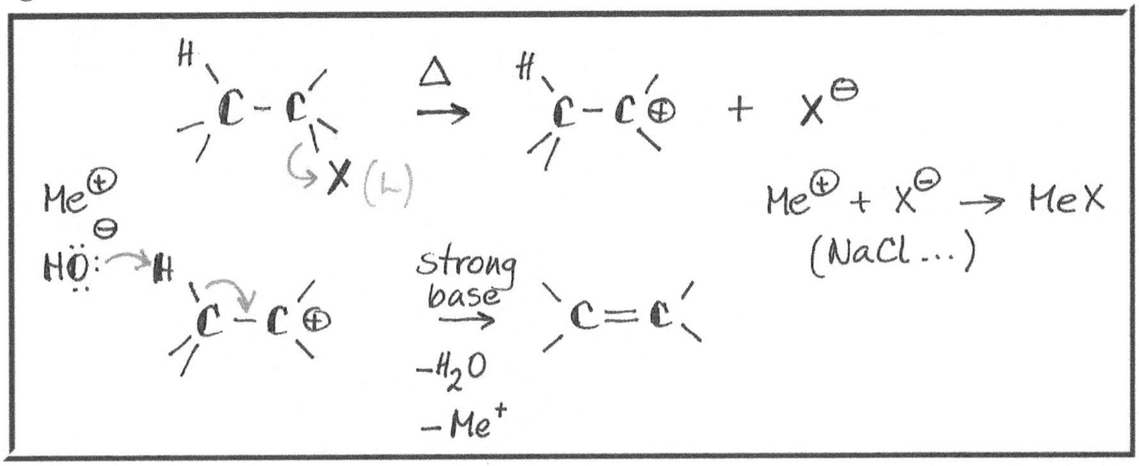

In the end any chemical reaction proceeds via the competing interplay of physical and chemical parameters influenced and provided by various reaction conditions.

**Example Problem 10.3**
**Complete the following reactions and characterize by the mechanism**

**involved:**

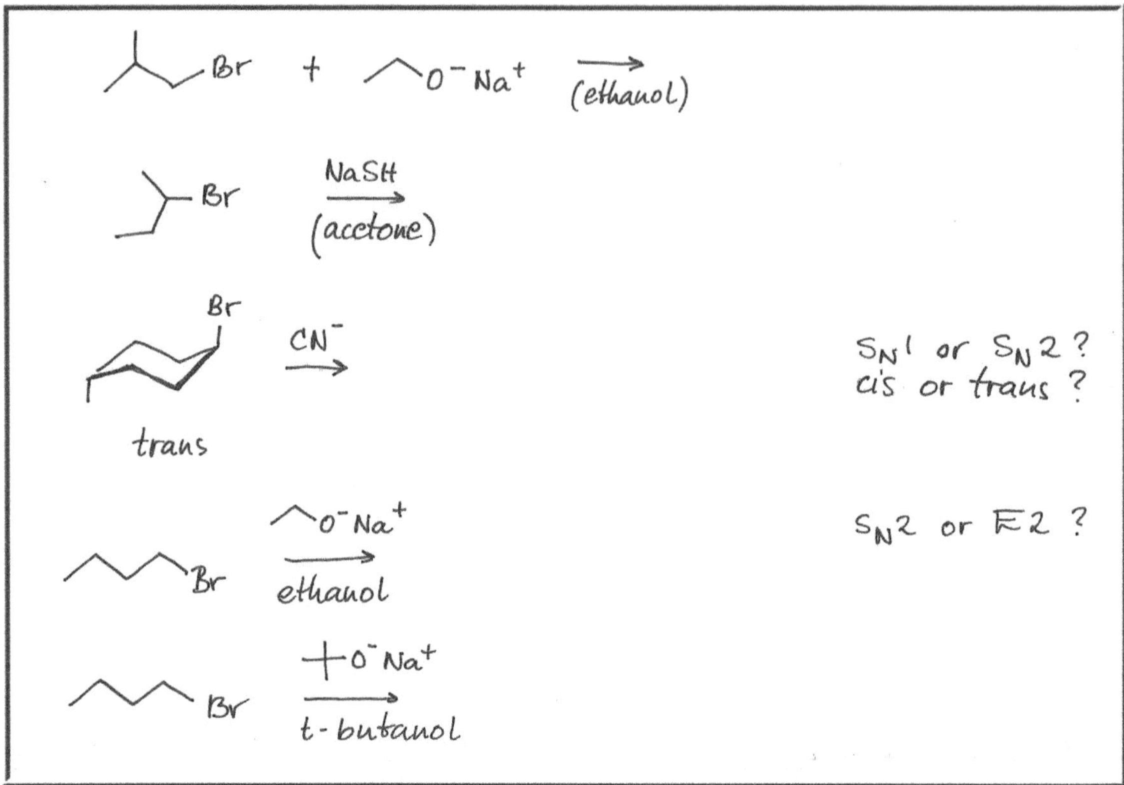

Br
O⁻ Na⁺
(ethanol)
Br
NaSH
(acetone)
Br
CN⁻
trans
SN1 or SN2 ?
cis or trans ?
O⁻ Na⁺
Br
ethanol
SN2 or E2 ?
O⁻ Na⁺
Br
t-butanol

# baEXPERIMENTS – Ch. 10

## *Halogenated Organic Compounds*

**Key Terms:** Halo-organics, organohalides, halogenation via halogen addition to alkenes and alkynes, halogenation of aromatics, halogens as leaving groups in nucleophilic substitution reactions, photohalogenation, Friedel Crafts reaction

### I Nucleophilic Substitution: Halogenation of *n*-Butanol and *tert*-Butanol via $S_N2$ and $S_N1$ Mechanisms, resp.

**Chemicals:**

*n-Butanol*

*tert-Butanol*
*NaBr*
*Conc. HCl (12M)*
*Conc.* $H_2SO_4$ *(16M)*
$CaCl_2$
$NaHCO_3$ *(5%)*
*Water (demineralized)*

**Equipment and Materials:**

*Reflux apparatus, round-bottom flask (250mL), distilling column*
*Separatory funnel*
*Heating mantle or hot plate*
*Stand and clamps*
*Ice-water bath*
*Distillation apparatus*
*Beakers and Erlenmeyer flasks*

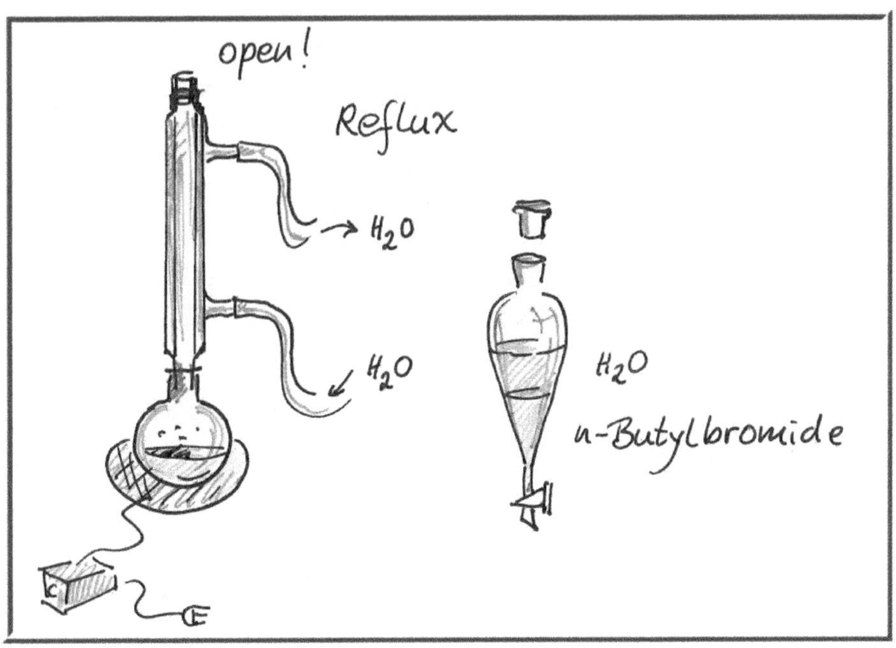

## Procedure:

## $S_N2$-Reaction

1. Place 13.3 g of NaBr, 10mL n-butyl alcohol (1-butanol) and 10 mL of water in a 100 ml round-bottom flask.
2. Cool in an ice-water bath.
3. Slowly add 11.5 mL concentrated sulfuric acid ($H_2SO_4$) to mixture while constantly swirling the flask in the ice-water bath.
4. Transfer the flask to the fume hood and assemble the reflux set-up as shown above. Use a stand and clamps to hold the equipment in place
5. Bring the solution to a steady boil and reflux for about 45 minutes.
6. Once completed, cool the mixture down and replace the distilling column with a distillation set-up ( three-way connecting tube, thermometer, west condenser and receiver)
7. Distill the product that produces an azeotropic mixture with water into the receiving vessel (Erlenmeyer flask or beaker).
8. Pour distillate into a separatory funnel and 10 mL of water.
9. Drain the bottom organic layer into an Erlenmeyer flask and dry with $CaCl_2$.
10. Decant the organic liquid and pour through a fluted filter paper into another clean, dry Erlenmeyer flask.
11. Determine its mass and volume and calculate the density.
12. Save well sealed for experiment 13 to measure the refractive index and NMR spectrum.

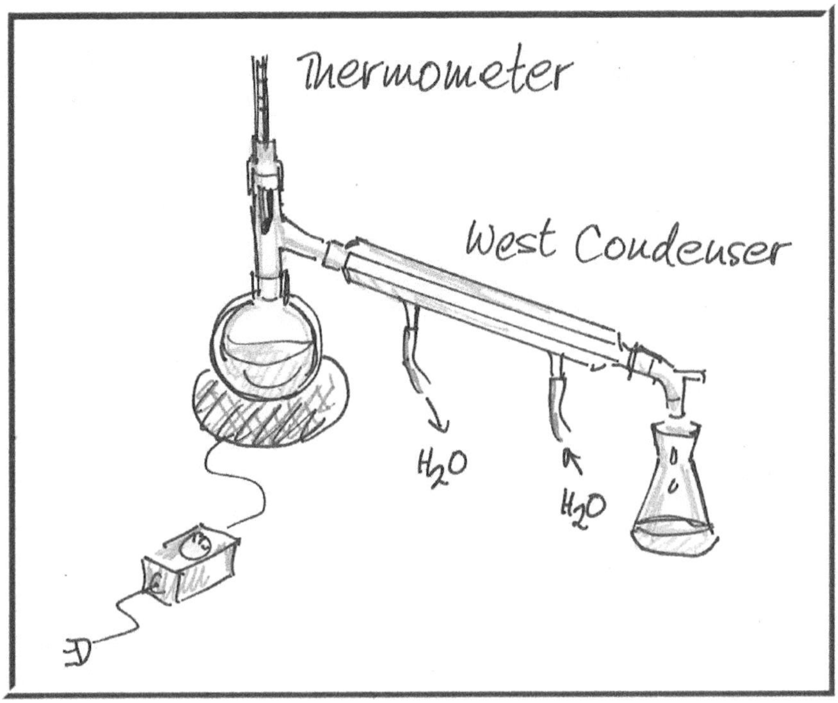


## $S_N1$ Reaction

1. Place 25 mL of *tert* - butyl alcohol (*tert* - butanol) and 60 mL concentrated hydrochloric acid (HCl) in a 250 mL Erlenmeyer flask. Close the flask with a rubber stopper, mix intermittently, each time venting the flask to release excess pressure by removing the stopper for a period of about 20 minutes.
2. Add 6 g of $CaCl_2$ and mix until fully dissolved.
3. Transfer the mixture to a separatory funnel. Shake and "burp" intermittently by opening the stop cock while the funnel is held upside down.
4. Place the separatory funnel in a ring clamp attached to a stand and let the phases settle out.
5. Add 5% $NaHCO_3$ solution to the mixture, shake again and remove the lower aqueous phase.
6. Separate the organic layer, dry it with $CaCl_2$, decant, filter and distill.
7. Weigh and measure the volume of the *tert* - butylchloride and calculate the density.
8. Save part for the next experiment below (II) and part in a well-sealed container for experiment 13 to measure the refractive index and NMR spectrum.

## Observations and Results:

| **Reactants:** | **Molar Mass** | **Moles (mol)** | **Density (g/mL)** | **Bp. (°C)** | **Physical characteristics** |
|---|---|---|---|---|---|

| | (g/mol) | | | | (smell/odor, consistency etc.) |
|---|---|---|---|---|---|
| *Tert-butyl alcohol* | | | | | |
| *Conc. HCl* | | | | | |
| | | | | | |
| **Product:** | **Molar Mass (g/mol)** | **Theoretical yield (g)** | **Density (g/mL) and Bp. (°C)** | **Practical yield (%)** | **Physical characteristics (smell/odor, consistency etc.)** |
| *Tert-butyl chloride* | | | | | |

## Questions and Problems:

1. Write balanced equations and the mechanism for the two reactions.
2. What is the purpose to add $NaHCO_3$ to the solution ($S_N1$ reaction)?
3. Concentrated sulfuric acid has a strong attraction for water, which leads to other possible products. What are these possible side products? Show reactions that lead to these products. Are they contaminants and how could they be removed?
4. Calculate the theoretical and practical yields for both reactions?

## II Friedel-Crafts Alkylation of *meta*-Xylene

### Chemicals:

*Meta-xylene*
*Tert-Butyl chloride (from I)*
*$AlCl_3$*
*Conc. HCl*
*Ice water*
*$MgSO_4$ (drying agent)*

### Equipment and Materials:

*Filter flask, rubber stopper and rubber hose*
*Glass funnel, large beaker for water sink/trap*
*Ice bath*
*Test tube, small beaker, graduated cylinder*
*Separatory funnel*
*Stand and clamps , digital top-loading balance*

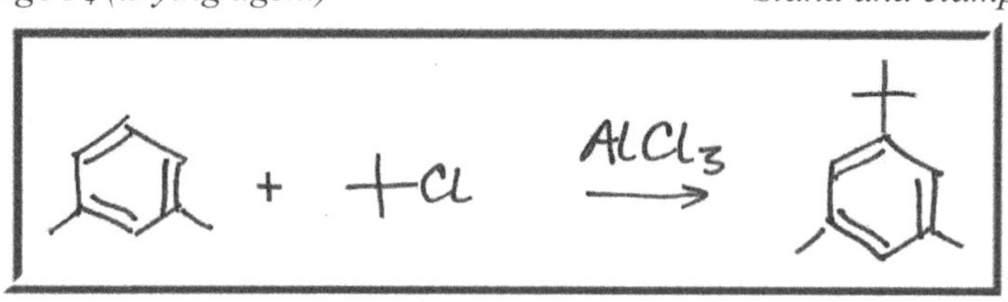


### Procedure:

1. Prepare the set-up according to the following image:

2. Place 12.2 mL of m-Xylene and 12.5 mL of *t*-butylchloride in the filter flask
3. Cool the solution below 5°C in an ice bath.
4. Weigh out dry $AlCl_3$ in a tarred out small beaker (dry), transfer to a dry test tube with rubber stopper and add about 1/3 of the crystals in three 5-7 min intervals, sealing the test tube with the stopper during each interval.
5. Swirl the contents of the reaction flask during each addition of the catalyst.
6. When the release of HCl gas into the gas/water trap has subsided (no more bubbles), the reaction is completed.
7. Warm to room temperature and add 20 mL of ice water and 5 mL conc. HCl.
8. Transfer the solution to a separatory funnel and isolate the organic layer containingg the product.
9. Dry the organic phase over $MgSO_4$, decant and weigh product and determine its volume
10. Calculate the theoretical yield and the density of the product

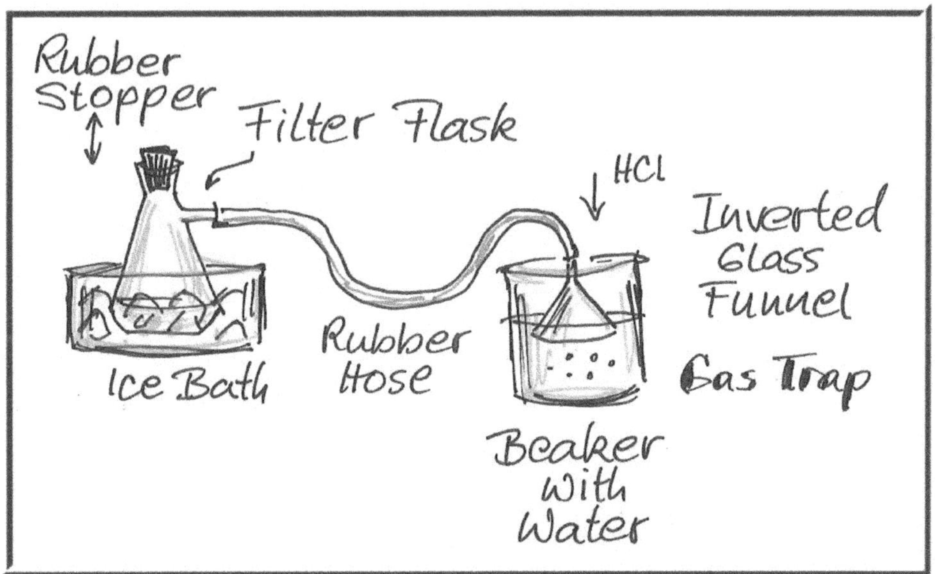


## Observations and Results:

| **Reactants:** | **Molar Mass (g/mol)** | **Moles (mol)** | **Density (g/mL)** | **Bp. (°C)** | **Physical characteristics (smell/odor, consistency etc.)** |
|---|---|---|---|---|---|
| ***Meta-xylene*** | | | | | |
| ***Tert-butylchloride*** | | | | | |
| | | | | | |
| **Product:** | **Molar** | **Theoretical** | **Density** | **Practical** | **Physical** |

| | **Mass (g/mol)** | **yield (g)** | **(g/mL) and Bp. (°C)** | **yield (%)** | **characteristics (smell/odor, consistency etc.)** |
|---|---|---|---|---|---|
| ***1-tert-Butyl-3,5-dimethylbenzene*** | | | | | |

## Questions and Problems:

1. What is the reaction mechanism for this alkylation reaction?
2. Why does the tertiary butyl group go to the *meta* location on the benzene ring even though the two methyl groups are *ortho/para* directors?
3. Why does the $AlCl_3$ catalyst have to remain free from moisture or completely dry?

# III Synthesis of Fluorescein and its Brominated Derivative Eosin

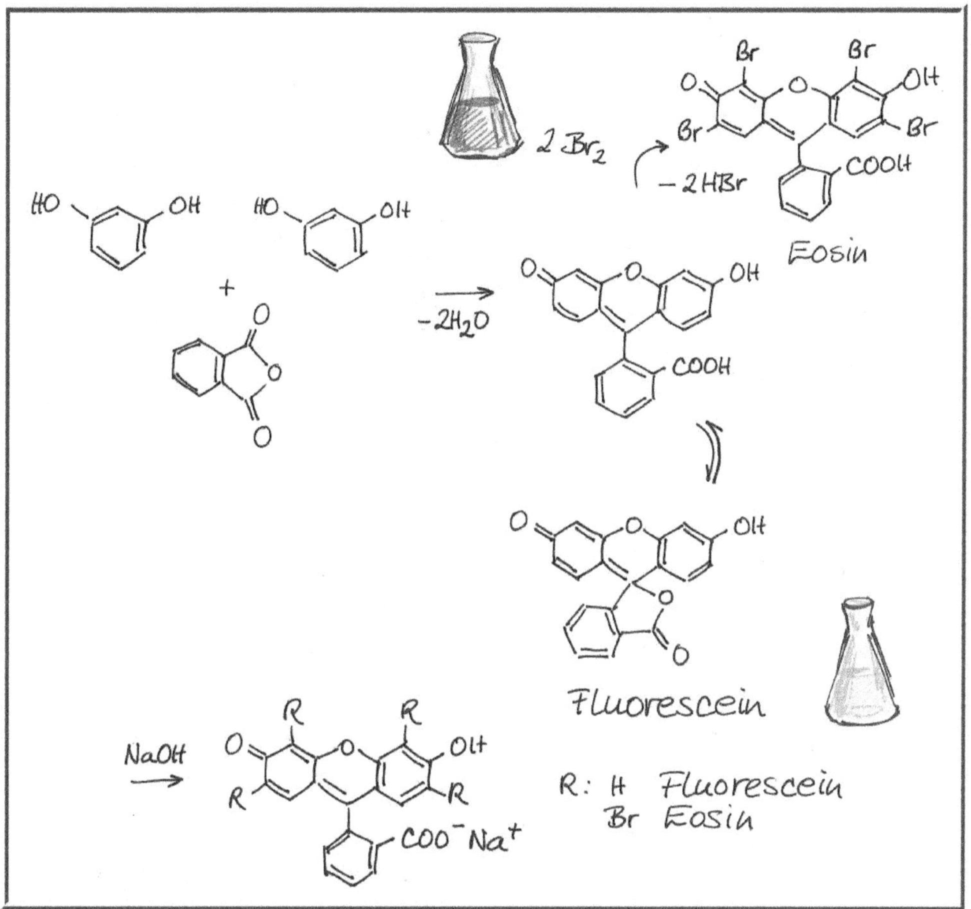


## Chemicals:

*Resorcinol*
*Phthalic anhydride*
*Conc. $H_2SO_4$*
*Bromine ($Br_2$/ $CH_2Cl_2$)*
*Methanol*
*Ethanol*
*2M NaOH*
*Red ink (pens Pilot G-2 07 etc.)*
*Developing solvent*

## Equipment and Materials:

*Small beakers (250mL),*
*Hot plate*
*Spatula, Pasteur pipet*
*Vacuum filtration apparatus with Hirsch funnel*
*Ice bath*
*UV lamp*
*Top loading digital balance*
*TLC set-up with developing chamber*

## Procedure:

### *Fluorescein*

1. Place a spatula tip each of resorcinol and phthalic anhydride in a small beaker

2. Heat the dry mix on a hot plate
3. Save half of the residue for preparing eosin.
4. Dissolve the other half

***Eosin***

1. Place 100 g of fluorescein or the saved solid residue from above in a small beaker and dissolve in 0.4 mL of ethanol.
2. Add drop-wise the bromine solution, made up of 215 mg $Br_2$ dissolved in 0.5 mL $CCl_4$ or $CH_2Cl_2$. At first the di-bromo derivative is formed, which is still soluble in the solution, but later the tetra-bromo derivative forms as a precipitate.
3. After 15 min. cool the mixture in an ice bath and remove the supernatant with a Pasteur pipet.
4. Suction off the precipitate in a Hirsch funnel and rinse the residue with ice-cooled methanol.
5. Dissolve the precipitate in a random amount of dilute NaOH and observe its color in daylight and its fluorescence under a UV light.
6. Prepare a TLC plate (silicagel) and spot it with fluorescein, eosin and red ink and record results after development.
7. Save fluorescein and Eosin for UV-spectroscopy for Experiment 13 (Chapter 13).

## Observations and Results:

| **Chemicals:** | **Color/Fluorescence** | $R_F$ |
|---|---|---|
| ***Fluorescein*** | | |
| ***Eosin*** | | |
| ***Red ink*** | | |

## Questions and Problems:

1. How do the structures of fluorescein and eosin relate to phenolphthalein? To what class of compounds do all three belong to? What other dyes belong to this class?
2. What structural characteristics cause the colors to form?
3. What are some of the most important uses for these two compounds prepared in this experiment?

# 11 Reactions of Organic Compounds

11.1 General types and patterns

11.1.1 radical and polar mechanisms

11.1.2 arrow formalism

11.1.3 Kinetic molecular theory

11.1.4 Thermodynamics of organic reactions

11.1.5 Rates of reaction and rate laws

11.1.6 Intermediates and transition states

11.1.7 Radicals, nucleophiles and electrophiles

11.2 Chain reactions /Polymerization reactions

11.2.1 Three step radical mechanism

11.2.2 Stability of radicals

11.3 Polyaddition and polycondensation

11.4 Role of catalysts and other reaction conditions

11.5 Reactions involving acidic alpha carbons

11.5.1 Michael addition

11.5.2 Aldol addition

11.5.3 Cannizzaro disproportionation and absence of alpha hydrogens

11.6 Addition reactions involving pi-systems

11.6.1 Simple additions

11.6.2 Asymmetric additions and Markovnikov rule

11.6.3 Hydroboration

11.6.4 Cycloaddition reactions, [4+2] (Diels-Alder) and others

11.6.5 Addition involving conjugated dienes (1,2 1nd 1,4 addition)

11.7 Oxidation reactions

11.7.1 Simple redox reactions and oxidizing agents

11.7.1 Combustion

11.7.2 Epoxidation

11.7.3 Ozonation

11.8 Substitution and Elimination reactions

11.8.1 Aromatic substitution ($S_E$)

11.8.2 Aliphatic substitution ($S_N1$, $S_N2$)

11.8.3 Aliphatic elimination (E1, E2)

11.9 Overview of Organic Reactions (Selection)

## Chapter 11 Reactions of Organic Compounds

Reactions can be divided into two major classes, depending on the type of substrate molecule:

(1) Reactions involving the carbon framework of the organic molecules directly, meaning dealing with the hydrocarbon itself as substrate without involvement of hetero-atoms, to produce molecular transformations
(2) Reactions involving carbon frameworks interlaced with hetero atoms or connected to specific functional groups containing hetero-atoms.

Three **major reaction types** are repeated in many forms:

**(1) Addition (A)**
**(2) Elimination (E)**
**(3) Substitution (S)**

These can be characterized by the type of **mechanism** involved, which can be initiated by a:

**(1) Radical reagent**
**(2) Electrophilic reagent**
**(3) Nucleophilic reagent**

They can also be defined by the order of reaction, if kinetically governed be different **rate laws**, such as:

**(1) First order**
**(2) Second order**

All these factors are combined to distinguish different reaction types, such as:

**$A_R$, $A_E$, $A_N$, E1, E2, $S_E$, $S_N1$, and $S_N2$**

The main molecule being acted upon is referred to as the **substrate** molecule. The usually smaller molecule, compound or ion is considered the **reagent**. The latter species is the one that actively attacks the substrate, provided the reaction conditions are opportune and/or the proper catalyst is present. Chemical reactions involving covalently bound organic molecules usually occur via a variety of steps, two-step mechanisms being the most common. Since according to the kinetic molecular theory, effective collisions between the substrate and reagent are essential to propagate a chemical change, the initial step is often the slowest and therefore called the rate determining step. This step leads temporarily to an intermediate species of high energy, but with a certain stability, usually a carbocation that initiates a swift follow-up reaction by combining with a residual ion produced during the first part of the reaction, concluding the second step.

The general pattern of the two-step polar mechanism is illustrated in the following scheme:

**Fig. 11.1**

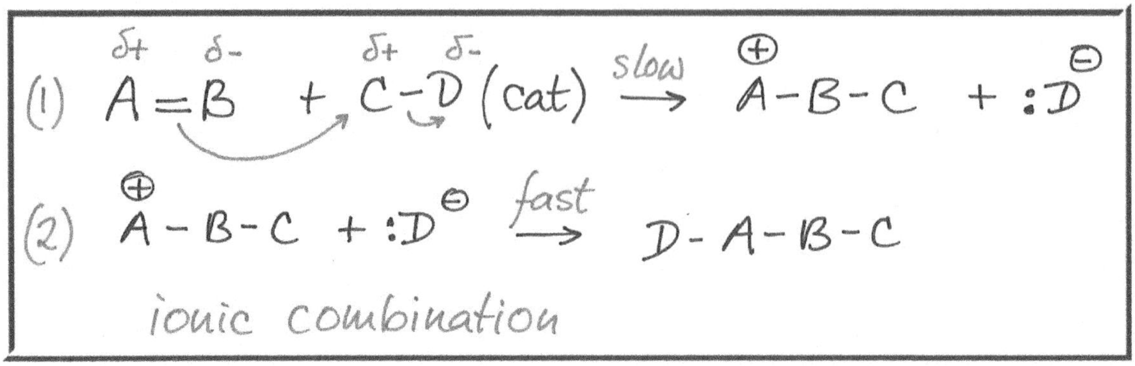


Organic reactions involve shifts of electrons, usually in pairs, so the **curved arrows** are very handy in following and demonstrating the dynamics of a reaction sequence. The arrow always points away from an electron pair, whether shared or unshared.

**Fig. 11.2**

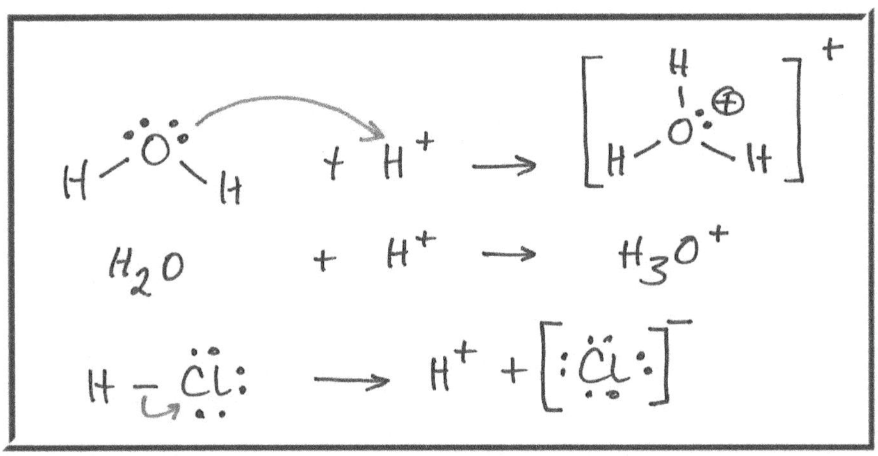


The polar mechanism can either rely on an electrophile or a nucleophile as initiator for the attack on the substrate molecule. In the first case the substrate is electron-rich and the reagent electron-poor, so in need of acquiring electrons, thereby "loving" electrons as an **electrophile**. If the substrate molecule is electron-poor, then the attacking reagent is willing to spread its own sufficient electron density around, thereby being a nucleophile, searching out positivated (δ+) de-shielded areas ("nuclei") of the substrate molecule. These reactions are usually of second order, which means that the concentration of both reaction partners, substrate and reagent, affect the rate of the reaction, each one to the first degree. Reactions that involve electrophiles are found in the $A_E$ and $S_E$ mechanisms. Nucleophilic reactions are usually found in the substitution types $S_N$. Polarized double bounds such as the carbonyl group can lead to a nucleophilic addition ($A_N$) reaction, the target of the attack being the δ+ carbon atom.

**Fig. 11.3**

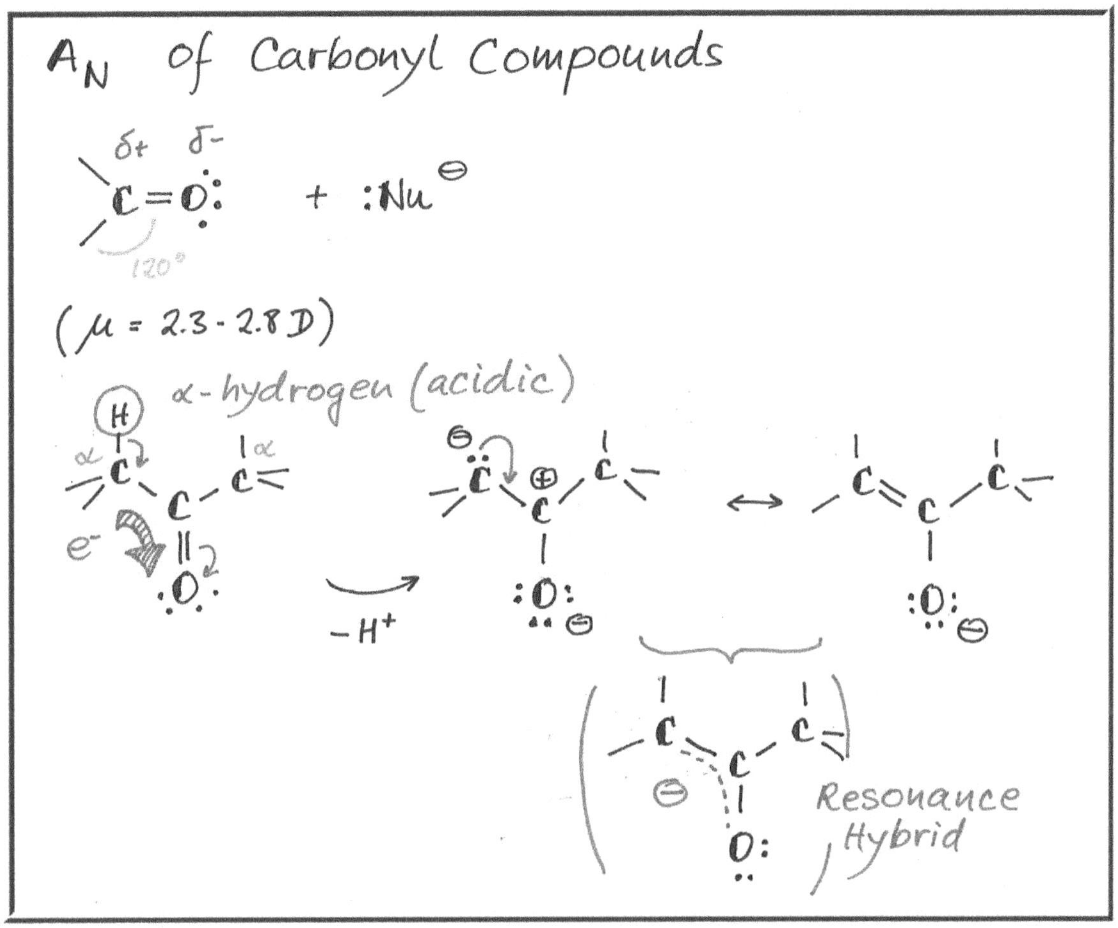


A similar concept to the **order of reaction,** which is strictly based on the concentration of the participants, is **molecularity**, which relies on the number of colliding molecules to initiate the first step or rate-determining step of a reaction.

**Fig. 11.4**

$r = k[A]^1$ or $r = k[B]^1$ 1st Order

$r = k[A]^1[B]^1$ 2nd Order

$r = k[A]^2$ or $r = k[B]^2$

$r = k[A]^2[B]$ or $[B]^2[A]$ or $[A]^3$ or $[B]^3$ or $[A][B][C]$ rare! 3rd Order

First order reactions are not as common but can happen, when the substrate itself undergoes a chemical change prior to a follow-up reaction. It can release a part or segment of its structural framework and then become vulnerable to an attack by the reagent. So only the concentration of the substrate determines the rate of the reaction and the reagent has to wait around, until the substrate shows a sign of weakness.

**Fig.11.5**

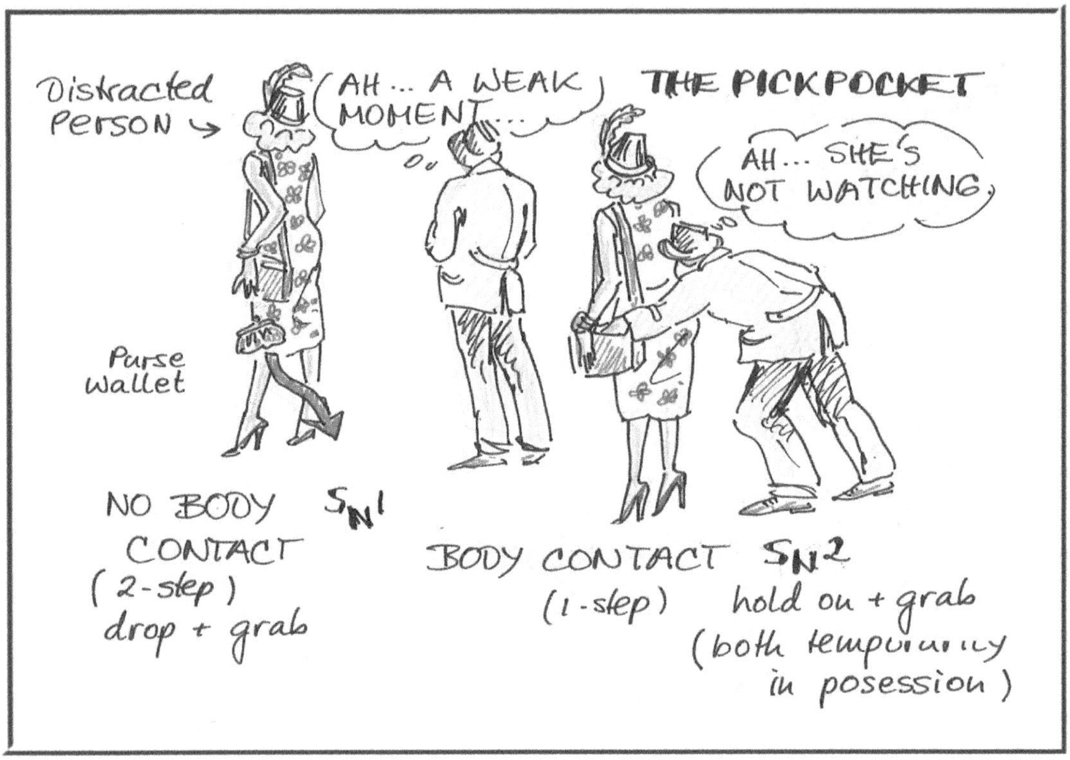


Higher orders of reactions are extremely rare.

The number of steps does not have to be consistent with order of reactions. Sometimes a reaction can be second order but involving one continuous step, known as a **concerted reaction** proceeding via a **transistion state**. On the other hand a first order reaction can proceed via two steps without the occurrence of an intermediate or transition state. The substrate molecule in a sense forms its own intermediate without first reacting with the reagent. In this case the reagent is passive, waiting until it can get involved with the second step. Examples are the $S_N1$ and $S_N2$ as well as the E1 and E2 reactions, which are mainly discussed in the previous chapter (Ch. 10).

All organic reactions are accompanied by energy changes, expressed as the Enthalpy of reaction or $\Delta H$. This means any heat change at constant pressure ($\Delta Q$ at p=const.) If the degree of disorder is also affected (entropy change, $\Delta S$) then the Gibbs-Helmholtz equation plays a role, with which the free energy $\Delta G$ (also referred to as **Gibbs free energy**) can be determined.

$\Delta G = \Delta H - T\Delta S$

The Gibbs free energy is the actual qualifier as to whether a reaction is spontaneous or not. If the value is smaller than zero, then the reaction proceeds spontaneously. If entropy does not change then the term $T\Delta S$ becomes zero and $\Delta G = \Delta H$.

Although combustion can occur with many organic compounds, the most well-known reaction is the combustion of hydrocarbons. Combustion is a highly exothermic reaction producing heat and light and in case of an explosion even sound or sonic booms. Since the degree of disorder or entropy changes as well, the term exergonic should rather be used ($\Delta G \leq 0$).

Complete combustion mostly leads to the final oxidation products $CO_2$ and $H_2O$. Incomplete combustion where oxygen is not sufficiently available or supplied, can lead to all other kinds of oxygen containing products. For instance, if ethane is completely combusted the ration of alkane to diatomic oxygen is 1:3.5 as the following balanced equation shows:

$$2C_2H_6 + 7O_2 \rightarrow 4CO_2 + 6H_2O$$

For the controlled preparation of ethanol via oxidation of ethane, on the other hand, the ratio should be held at 2:1.

$$2C_2H_6 + O_2 \rightarrow 2C_2H_5OH$$

Ethanol is an octane booster for gasoline delivering an extra amount of energy per gallon consumed.

**Fig.11.6**

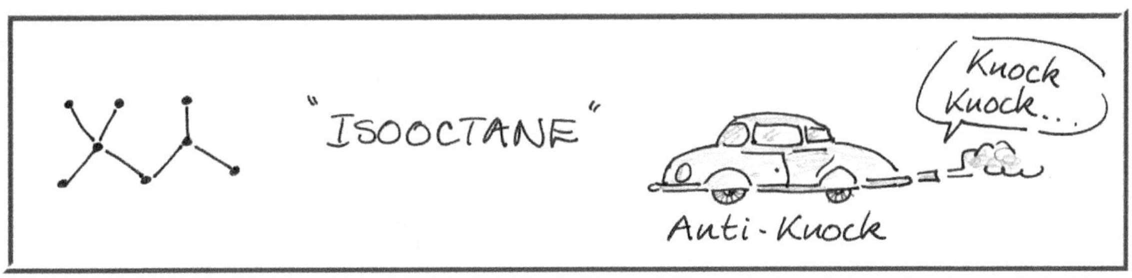


Gasoline's performance is rated by the so-called **octane rating**, which actually involves "isooctane" or the specific structural isomer of many possibilities, **2,2,4-trimethylpentane**, its octane number arbitrarily set as 100. A good octane rating means that smooth combustion is ensured and no "knocking" occurs.

Saturated hydrocarbons can also be oxidized by specific oxidizing agents other than free diatomic oxygen, but these often require specific catalysts, often metal-organic compounds.

Another important reaction for aliphatic hydrocarbons is **photohalogenation**. Aromatic molecules with hydrocarbon side-chains can also react. Single bonds between diatomic halogens can easily be broken by intense light produced in the laboratory by a high wattage light bulb. The cleavage proceeds homolytically and produces two reactive radical species, which will instantaneously attack other covalent bonds, especially those involved in hydrocarbon frameworks, leading to the formation of new radicals. This is basically a random process leading to a variety of free radicals. Once the radical producing photo energy is removed, the radicals pair off and result in many new molecules many of which are halogenated derivatives but also more complex hydrocarbons including many isomeric forms. Intense heat can lead to similar reactions. The overall reaction can be subdivided into an **initiation** stage, a **propagation** step and a **termination** stage. The reaction is "quenched" by free **radical coupling** reactions.

The radicals that form most readily are also the ones with increased stability. The higher the **degree of substitution** of the carbon radical, the more the lone electron can be spread around, literally making the radical more comfortable, thereby increasing its longevity. The longer the lifetime of a radical, the more it can get involved in various reactions

**Fig. 11.7**

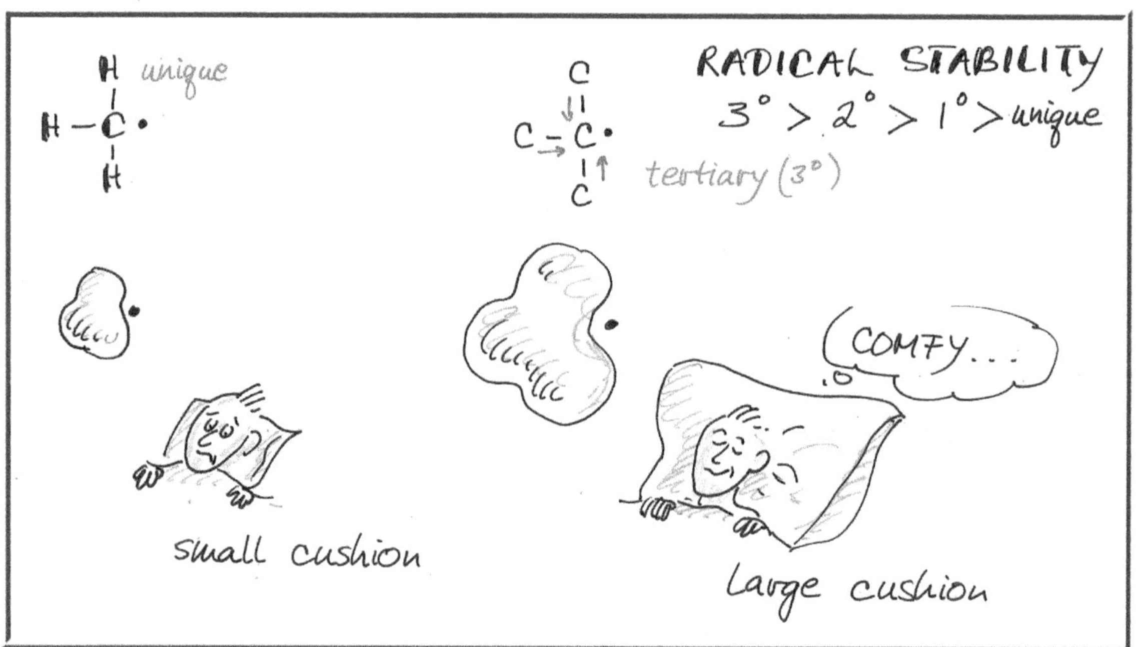

The same tendency is found with carbocations where the positive charge will be better delocalized.

Chlorine is more reactive than bromine producing a larger variety of halogenated products. Bromine is more selective in that it prefers to form more derivatives with higher substituted carbon atoms. Bromine mainly reacts with higher substituted carbon radicals (3°)

**Fig. 11.8**

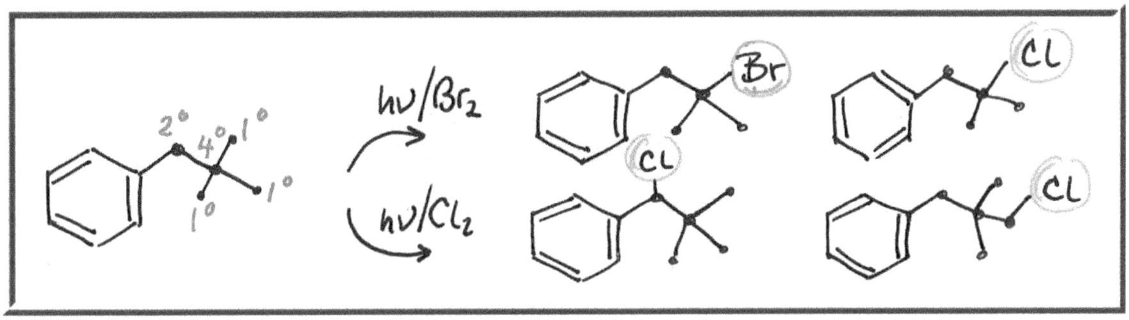


The mechanism for **photohalogenation** of hydrocarbons can be viewed as a type of addition via the action of free radicals or **A$_R$ reaction.**

A similar reaction mechanism involving radicals is **polymerization** in the true sense, where monomers line up to form polymers via radical coupling of identical radical subunits. Here the initiation step is provided by the action of reactive molecular species such as peroxides or superoxides. These initiators begin forming radicals of the substrate molecules, which in turn produce more radicals so that a type of chain-reaction ensues (propagation). This process can continue indefinitely unless the reaction is brought to a standstill by the addition of radical quenching molecules or terminators. In the polymer industry, the chain length can be controlled by modification of reaction conditions.

**Fig. 11.9**

In most cases linear, un-branched macromolecules are formed. However, occasionally, by the action of "back-biting", some branching can be observed. Technically, this is referred to as a **chain-transfer reaction**.

**Fig.11.10**

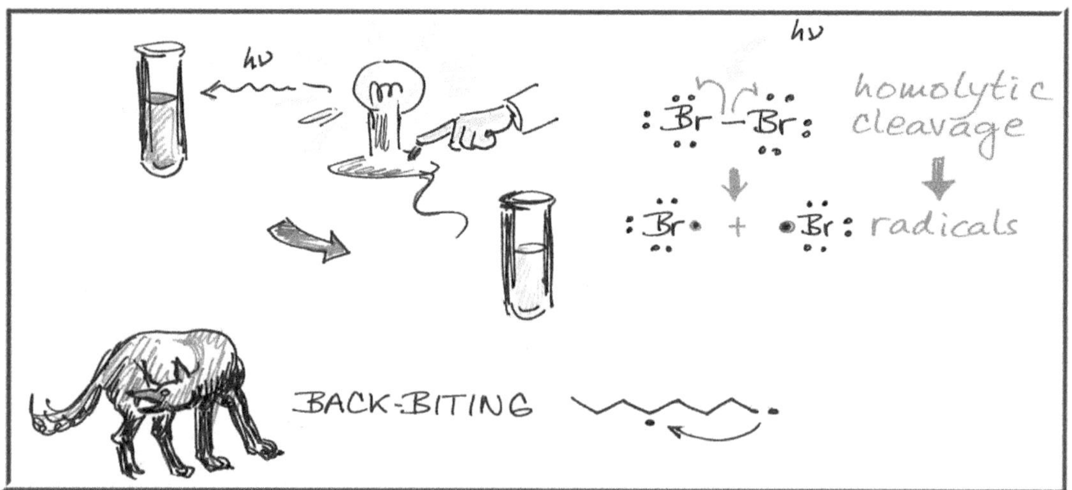

Unsaturated hydrocarbons are converted to macromolecules by the **polyaddition** reaction. The mechanism can involve free radicals ($A_R$) or proceed via the electrophilic route ($A_E$). Head to tail linkages produce linear polymers.

**Fig. 11.11**

Monomers with specific functional groups can lead to higher structures by elimination processes such as **polycondensation**. Here a small molecule such as water can form between two reacting monomers leading via oligomers to eventual polymers. Heat and/or condensation agents such as phosphoric acid can help the process along. Bifunctional molecules lead to linear structures and tri or multifunctional monomers to branching and the establishment of networks

**Fig.11.12**

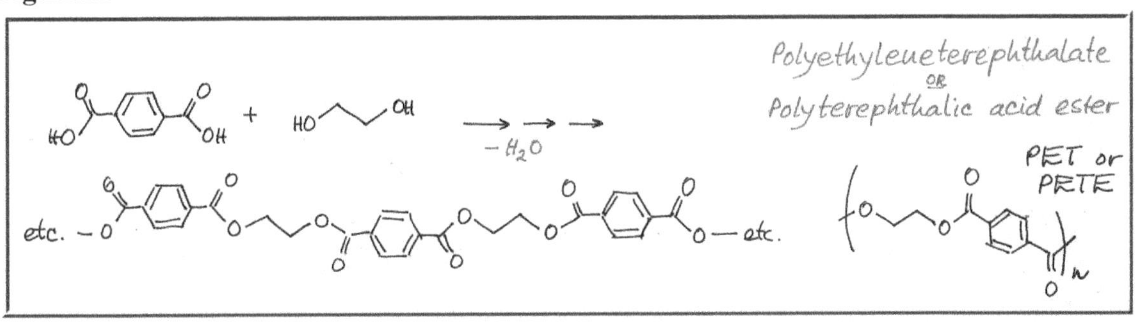


Polymers are highly important materials for modern existence, so much research is invested in finding new and better catalysts. For instance, the so-called Ziegler-Natta catalysts helped produce the much stronger HDPE (high-density-polyethylene) plastics than the low density varieties (LDPE). The catalysts are formed in situ on a solid support which helps regionalize the reactants and thus a more focused and effective polymerization process.

**Fig. 11.13**

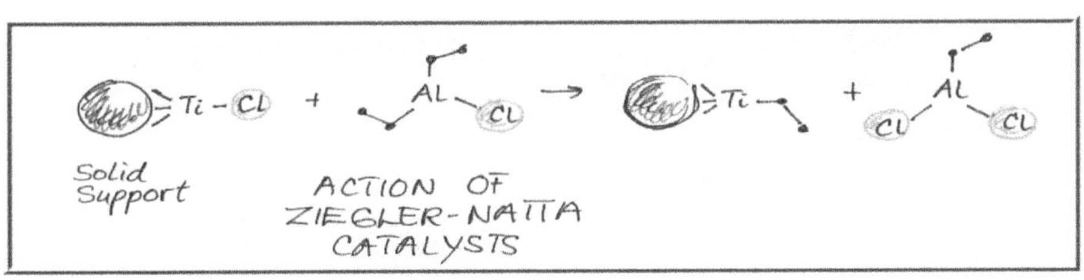

The most widely used plastic has become PETE (polyethylene terephthalate or terephthalic acid ester), which is used as a very sturdy container material for a variety of purposes. In biotechnology it has literally replaced glass as laboratory equipment.

**Biopolymers** are the foundation of life, although some scientists prefer the term macromolecule rather than "polymer", the latter implying the connection of repeated identical elements or subunits. Only polymeric sugars such as starch and cellulose as well as derivatives such as chitin can be considered true polymers. Proteins and nuclei acids are made of varying sequences of 20 (amino acids) or 4 (nucleotides) building blocks respectively. Peptides and proteins are therefore called polyanhydro-amino acids and nucleic acids also referred to as polynucleotides.

Unsaturated hydrocarbons are understandably more reactive than their saturated counterparts. The pi system is very attractive for many reagents, especially electron deficient species such as electrophiles. The most important reactions for aliphatic unsaturated compounds are:

1) Air-oxidation with oxygen (combustion), leading to similar products as produced by saturated hydrocarbons
2) Oxidation with specific oxidizing agents to epoxides, ozonides, and glycols
3) Electrophilic addition reactions ($A_E$)
4) Nucleophilic addition reactions ($A_N$)
5) Cycloaddition reactions [4+2] $\pi$ (diene + dienophile) and 1,2 /1,4 addition to conjugated dienes
6) Hydroboration

When dealing with aromatic unsaturated carbocyclic compounds the reactions are governed by the special conditions of aromaticity. The aromatic pi systems will be contained at all costs, even if temporarily disrupted. Here the important reactions are:

1) Aromatic substitution via the action of electrophiles ($S_E$)
2) Alkylation and acylation via **Friedel-Crafts reaction** using Lewis acids as catalysts (a type of $S_E$ reaction), discussed in chapter 10
3) Hydrogenation (reduction)

Carbonyl compounds due to the highly polarized condition of the carbon oxygen double bond are instrumental for a variety of special reactions concerning an **acidic proton on the α-carbon atom**. The addition of a strong base such as $OH^-$ will lead to its abstraction and the formation of a lone electron pair on that particular carbon atom. This can lead to certain addition reactions based on the newly formed **carbanion** as nucleophile. The reaction mechanisms of the **aldol addition/condensation** and **Claisen addition/condensation** will fall into this category. The former will produce β-hydroxyaldehydes (aldols) or β-hydroxyketones (ketols), the latter will lead to the formation of β-ketoesters.

**Fig. 11.14**

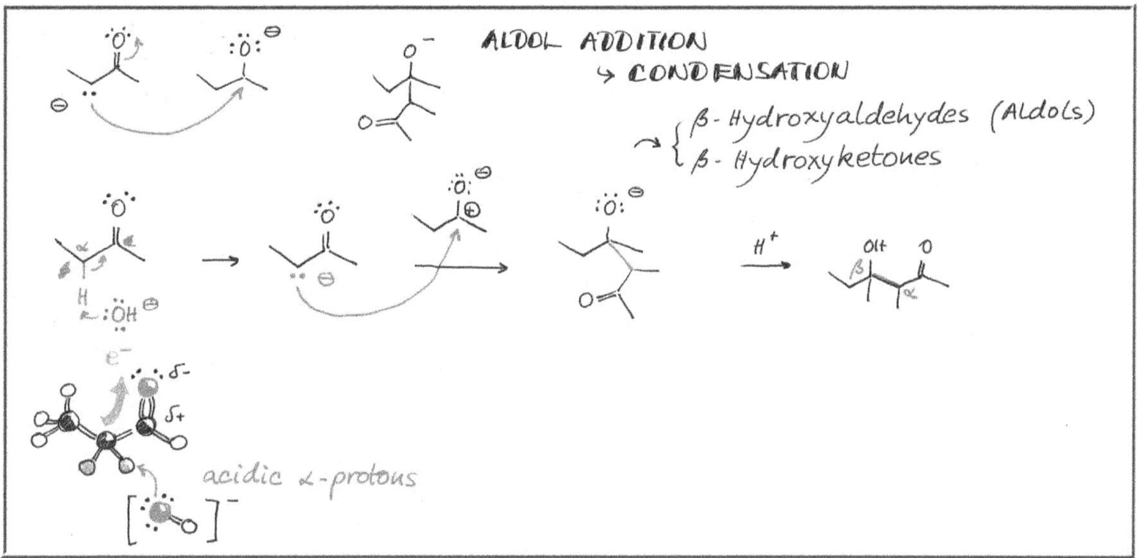


**Fig. 11.15**

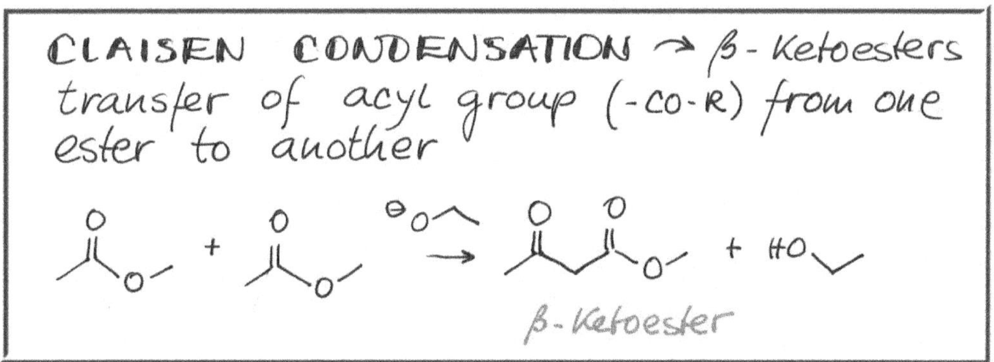


The **Michael addition** reaction also relies on the formation of a carbanion that will act as a nucleophile, which will enable the formation of a new carbon-carbon bond via connection of two carbon units, each of which contain one or more carbonyl groups. A strong base will abstract an acidic methylene proton of molecule that will react as a "**Michael donor**"**.** This will usually involve the carbonyl compound with more "pulling power" or electron withdrawing tendencies like a bi-functional carbonyl compound or a β- diketone that readily converts into the carbanion as nucleophile. This will then attack the "**Michael acceptor**" molecule leading to an adduct with a newly established C-C bond. The acceptor should be an α,β unsaturated carbonyl compound, where the carbonyl group will pull electrons out of the neighboring C=C double bond, making the β-carbon susceptible to the attack.

**Fig.11.16**

MICHAEL ACCEPTORS

α, β- unsaturated carbonyl compounds

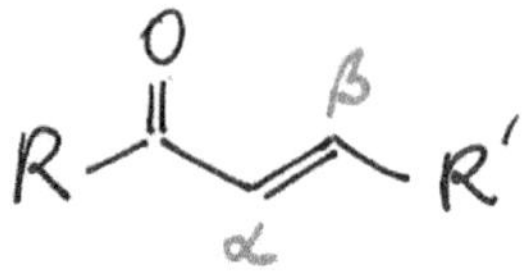


MICHAEL DONORS

stabilized anions

Carbanions, best: β-diketones

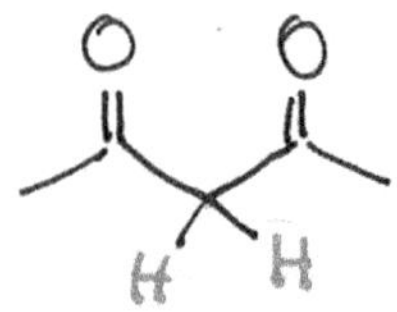

**Fig. 11.17**

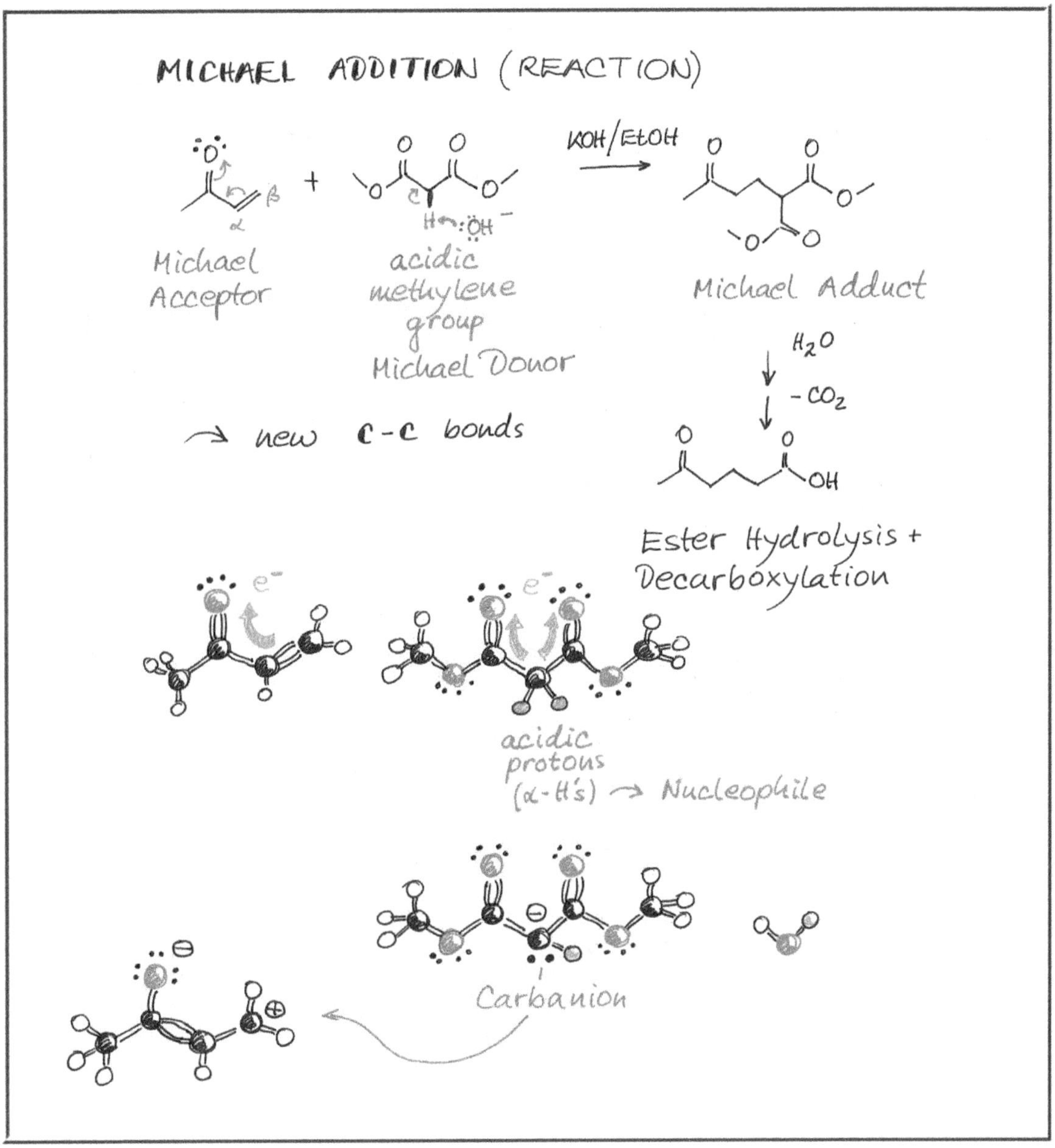


All these reactions are considered nucleophilic addition reactions involving the action of carbanions as nucleophiles.

An interesting variation is the **Cannizzaro reaction** that involves aldehydes as carbonyl compounds that do not have a methylene acidic proton. Aldehydes **missing the α-hydrogen**

**atom** will in the presence of strong base lead to a compound of higher oxidation state and one with lower oxidation state. This **auto-redox reaction** is referred to as **disproportionation**. One aldehyde molecule will form an alkoxide ion when treated with a base. This will then lead via hydride transfer (:H⁻) to the second aldehyde molecule to an alcohol, the remaining aldehyde having made a carboxylate salt with the cation of the base. Hydrolysis can produce the free carboxylic acid.

**Fig.11.18**

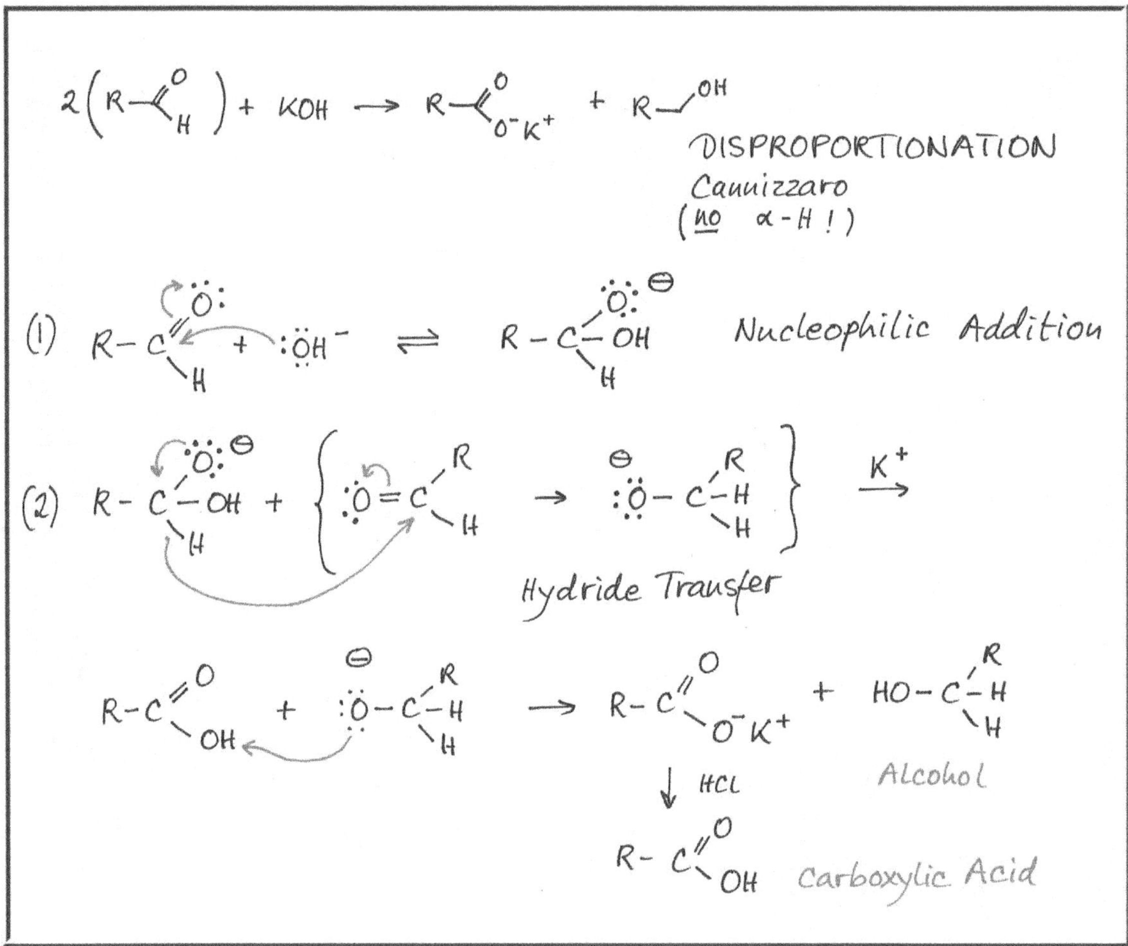


The multiple bond of unsaturated aliphatic hydrocarbons is an electron-rich area, perfect for the attack of electron-seeking electrophilic reagents. Often the electrophile has to first be produced in order to cause sufficient electron deficiency within the molecule. For instance, bromine is a nonpolar diatomic molecule with no desire to search for electron-rich areas. However, once water is added the strong dipole molecules will induce polarization within the bromine, causing a shift in electron density. The δ+ polarized end of the molecule will search for compensation of its electron density loss and feel attracted to the unsaturated substrate molecule. At first a loose attraction, like a bee hovering over a flower, will establish a so-called pi(π)-complex. Via rearrangement the pi-complex converts into sigma(σ)-complex via a **bromonium ion** (-$Br^{+}$), in which a small three-membered ring is formed, in which bromine atom carries the positive charge. The bromonium ion opens up to form a regular three-bonded carbocation, adjacent to the $sp^{3}$ –hybridized carbon, carrying the bromine as a substituent. The carbocation formation represents the first uphill reaction and therefore the rate-determining step of this two-step $A_E$ mechanism. After separation of the δ+ bromine, a bromide ion ($Br^{-}$) is left behind. This now in a swift move will attach to the carbocation as an instantaneous ionic combination, establishing the second bromine to be bonded to the original multiple bond of the substrate molecule. The overall mechanism is shown in the scheme below:

**Fig.11.19**

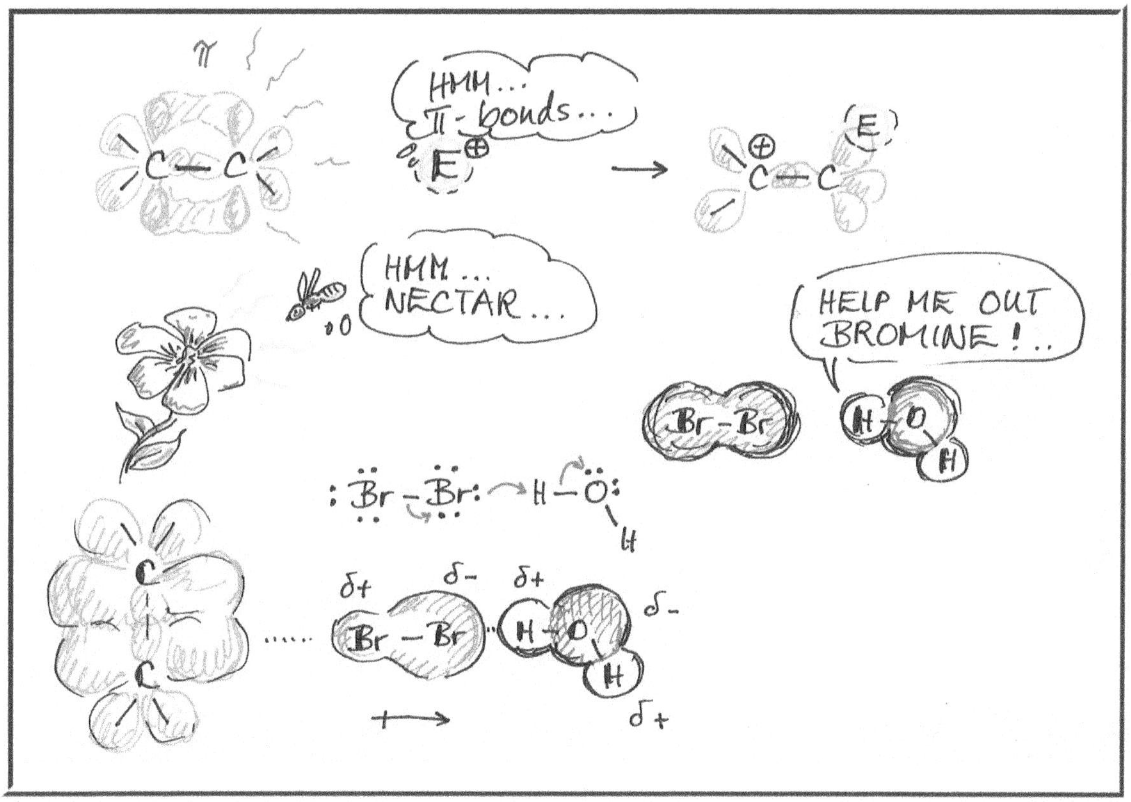


If the substrate molecule, as well as the electrophilic reagent is asymmetric in structure, certain stereo-selective or stereospecific selections will apply. Theoretically two possible products would form via different modes of alignment.

**Fig. 11.20**

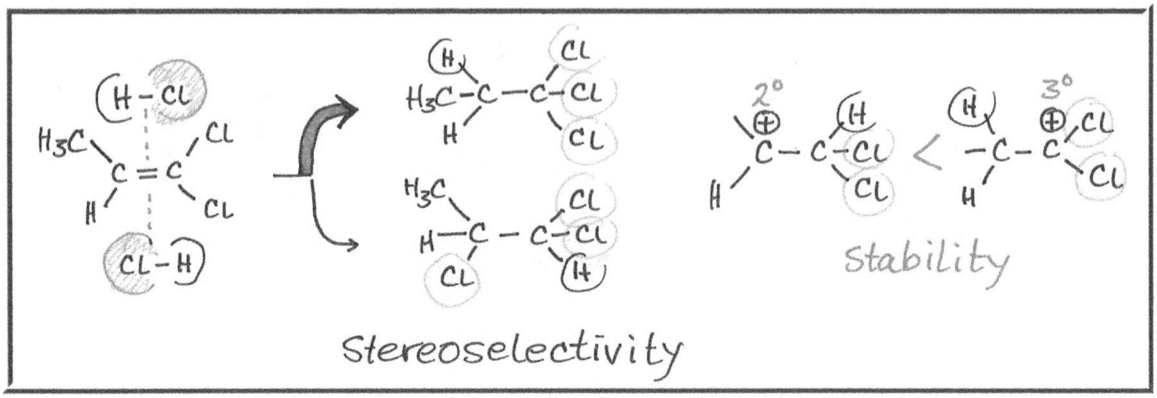


However, one adduct (addition product) seems to be favored over the other. This phenomenon was first observed by the Russian chemist Markovnikov and is now explained by the **Markovnikov Rule,** which generally implies that **hydrogen goes to hydrogen**, meaning that the hydrogen group of the electrophile will add to that side of the double bond that already has more hydrogen atoms attached to it. The basis for this selective behavior is the different **stability of the intermediate carbocations**. The higher the degree of substitution, the more favored it will be as intermediate of the rate determining step. The alkyl groups of higher substituted carbons provide a better electron-cushion to the positive charge, thus increasing its stability and longevity.

**Example Problem 11.1**
**What molecule is mainly formed when the following reactions occur?**

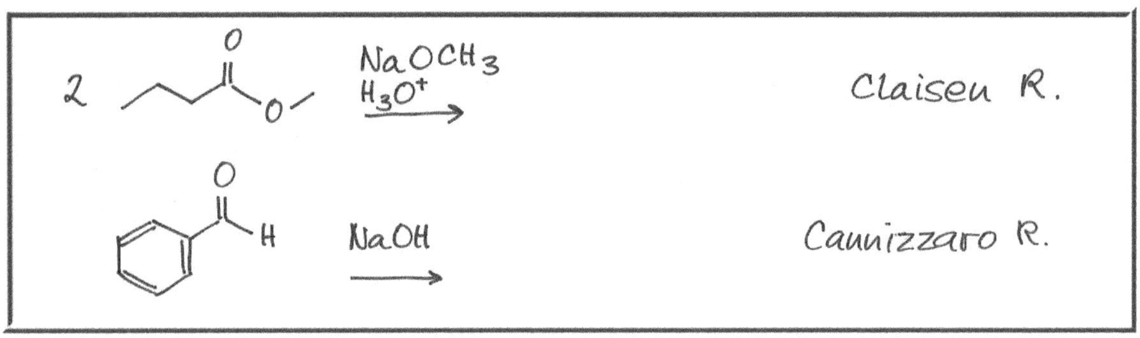


A way to bypass Markovnikov-based stereo-selective tendencies is by way of initial **hydroboration** of the multiple bond. As already discussed above, any addition to a multiple bond at first leads to a three-centered π-complex, which quickly stabilizes itself to a σ-complex of quadratic arrangement via **hydride migration**, both carbons now becoming $sp^3$ hybridized. The fairly large boron molecule will prefer to attach itself to the less crowded carbon of the multiple bond, thus ultimately favoring **anti-Markovnikov** addition products. Boron is also less electronegative than hydrogen and therefore more electrophilic. Since no intermediate carbocations could be detected during the hydroboration process, it appears that the 4-centered aggregate represents a transition state and that the reaction is the result of a concerted one-step mechanism. This would explain why alkynes react more rapidly than alkenes, contrary to the normal two step reaction mechanism.

**Fig.11.21**

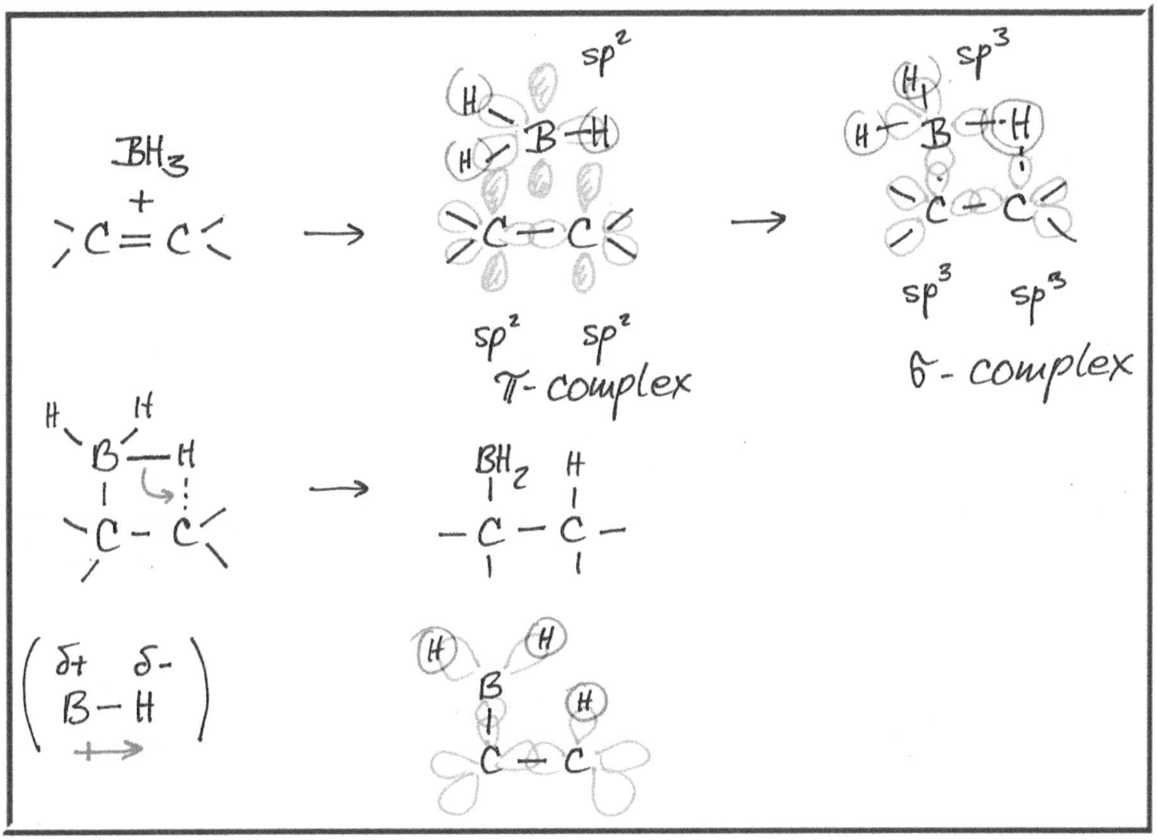


After the **organoborane** ($BH_2R$) molecule is formed, repeated hydroboration reactions of the same substrate molecule can follow leading via second substitution ($BHR_2$) to a fully substituted borane ($BR_3$).

**Fig.11.22**

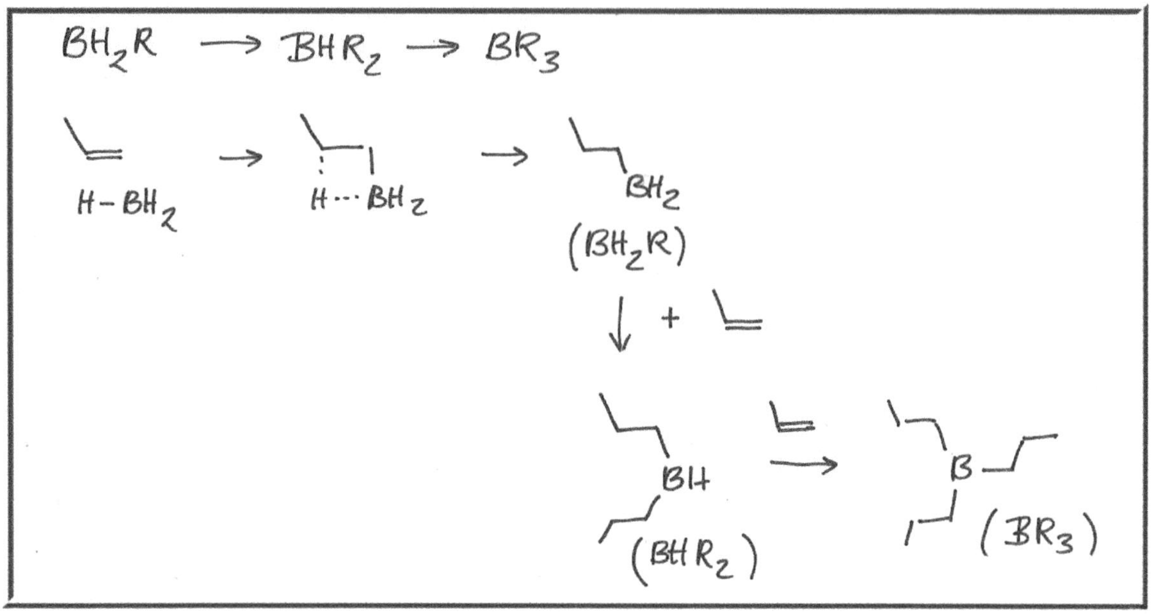


The substituted boranes can convert it to a variety of new derivatives via hydrolysis and other reactions as shown below:

**Fig.11.23**

Acidic Hydrolysis ($H^+/H_2O$) → Alkanes

$BR_3 \rightarrow 3RH + B(OH)_3$

Basic Oxidation ($H_2O_2/OH^-$) → 1°-Alcohols

$BR_3 \rightarrow 3R(OH) + B(OH)_3$

Treatment with $H_2N\text{-}OSO_3H$ (hydroxyamino-Sulfonic acid) → 1° Amines

$BR_3 \rightarrow 3R\text{-}NH_2 + B(OH)_3 + H_2SO_4$

**Example problem 11.2**

**Complete the following hydroboration reactions:**

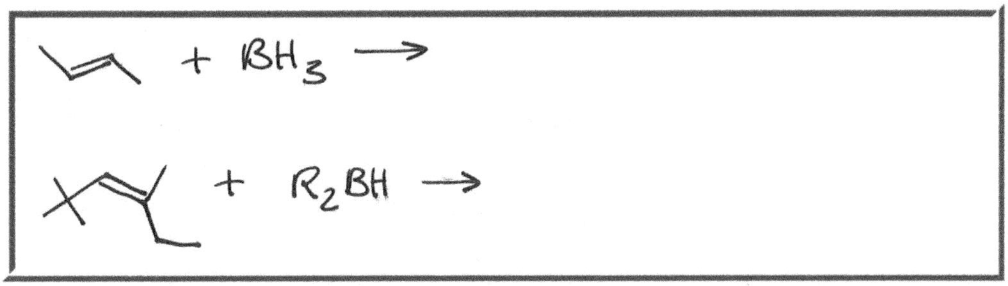


The multiple bond of unsaturated substrate molecules can easily be oxidized when the following reaction conditions are applied

1) Oxidation to **epoxides (oxiranes)** with $H_2O_2$ or peroxidic reagents (peroxy acids)
2) Oxidation to **ozonides** with ozone ($O_3$)
3) Oxidation to 1,2 substituted **glycols** with $KMnO_4$

**Epoxidation** can take place when using peroxidic products such as peroxiacids or hydrogen peroxide according to the following reactions:

**Fig.11.24**

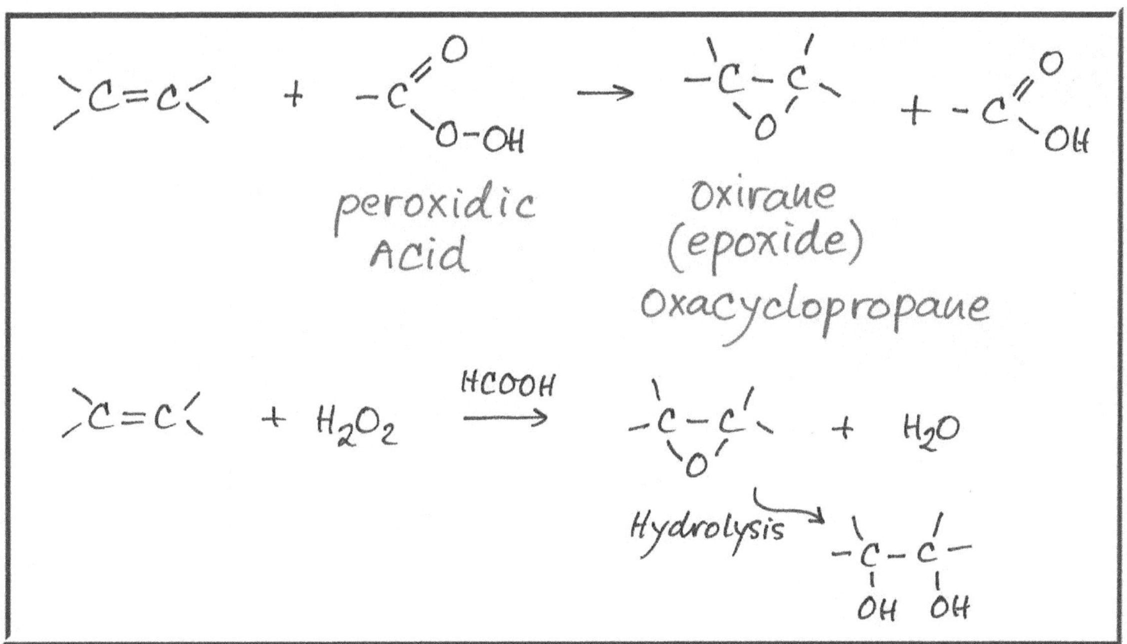


**Example Problem 11.2**
**Complete the following oxidation reactions:**

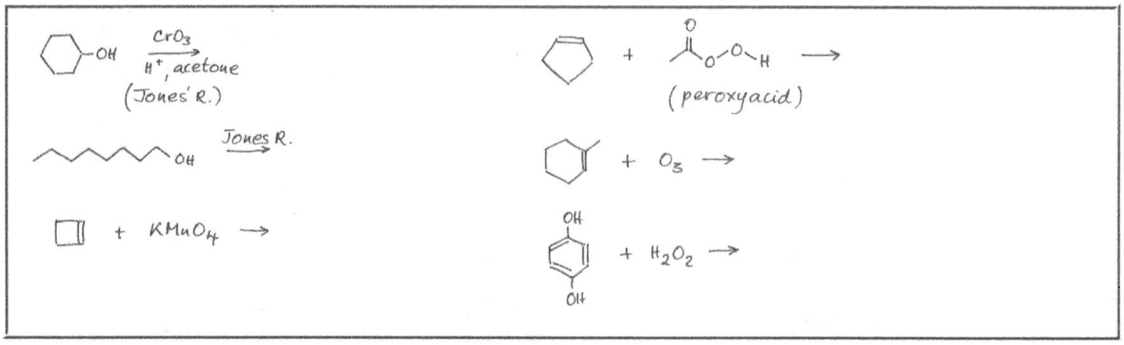


**Ozonation** proceeds faster with alkenes than with alkynes. Explosive conditions warrant caution and care when performing this reaction, first leading to a **molozonide**, which via intra-molecular rearrangement forms the more stable **ozonide.**

**Fig.11.25**

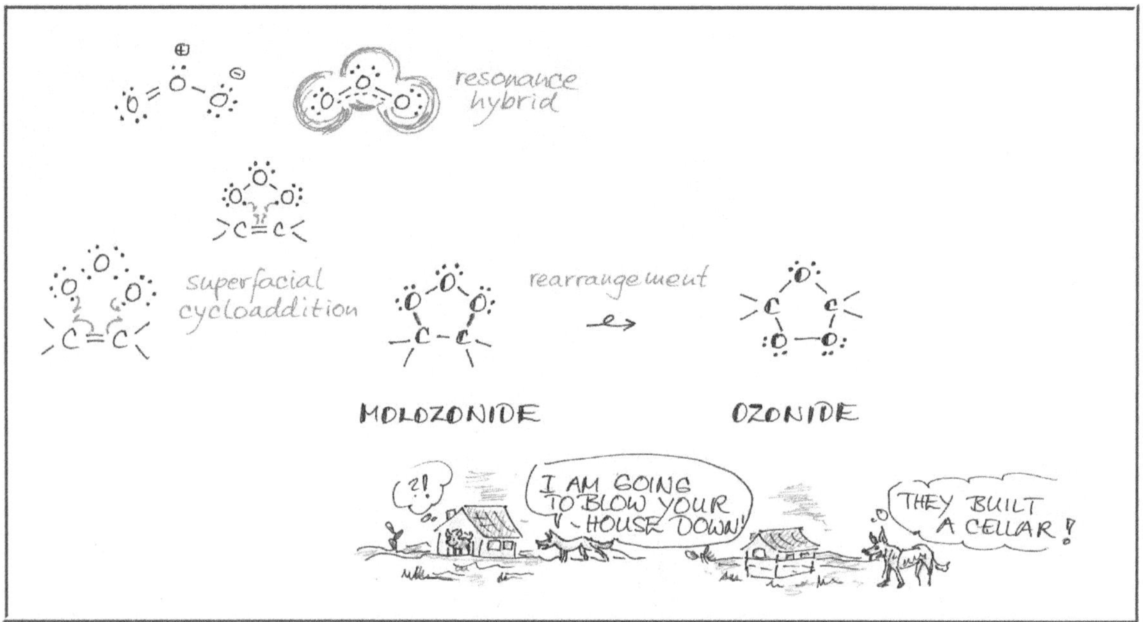


In situ hydrogenation (reduction) as a result of zinc reacting with HCl destroys the ozonide ring by producing two carbonyl groups (aldehyde or ketone depending on the degree of substitution of the two involved carbon atoms). By analyzing the formed break-down products, one can piece together the carbonyl groups or ends and thus have an idea at which site of the molecule the original double bond was located.

**Fig. 11.26**

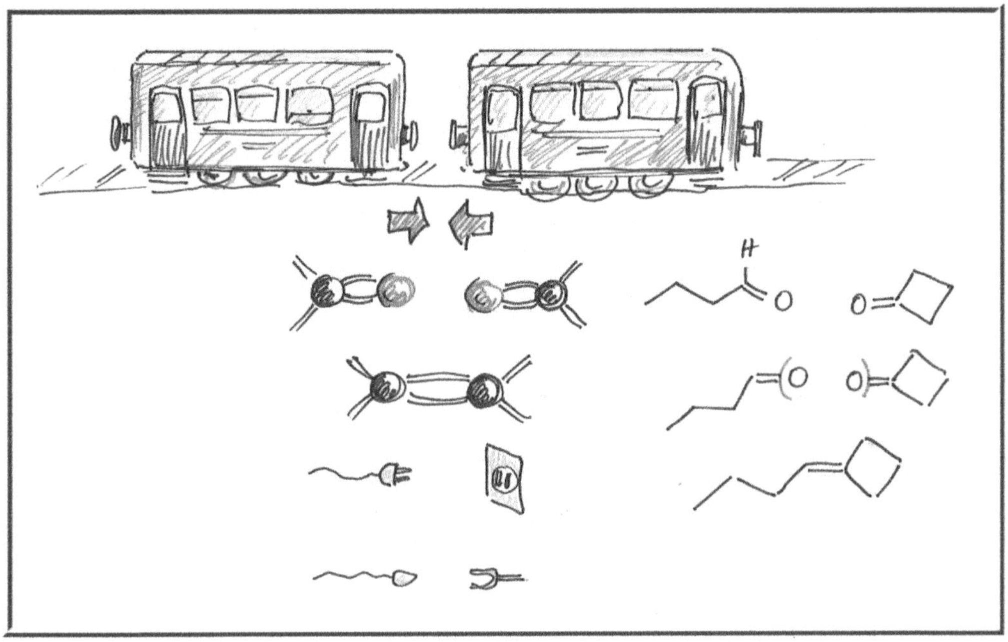

So **ozonlysis** (bond separation via ozone reaction) represents an analytical tool for establishing the exact position of multiple bonds within a molecule.

**Fig. 11.27**

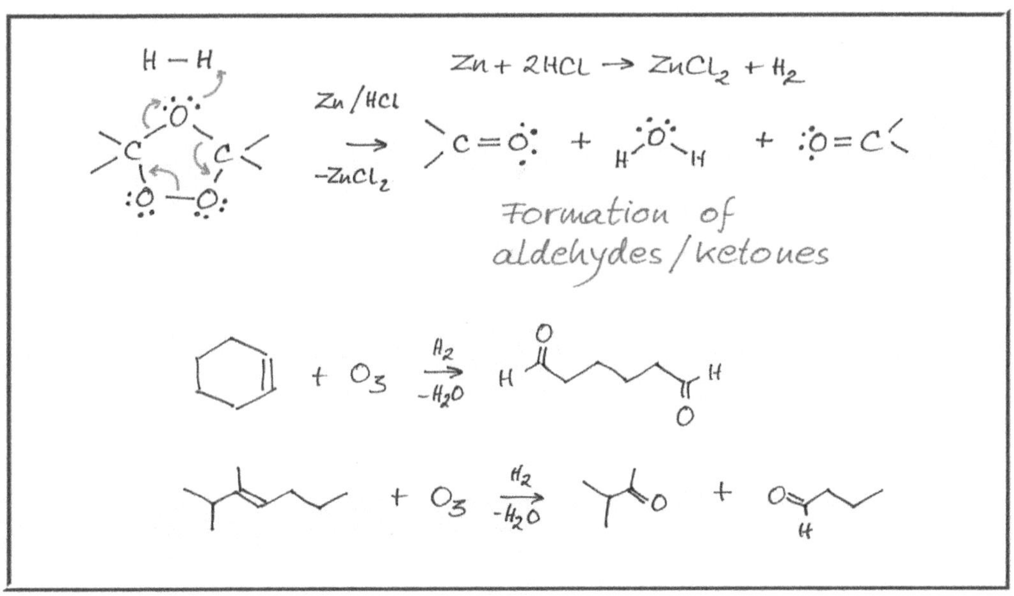

**Example Problem 11.3**

**Where was the original double bond located in the parent molecule before ozonolysis if the following products were produced.**

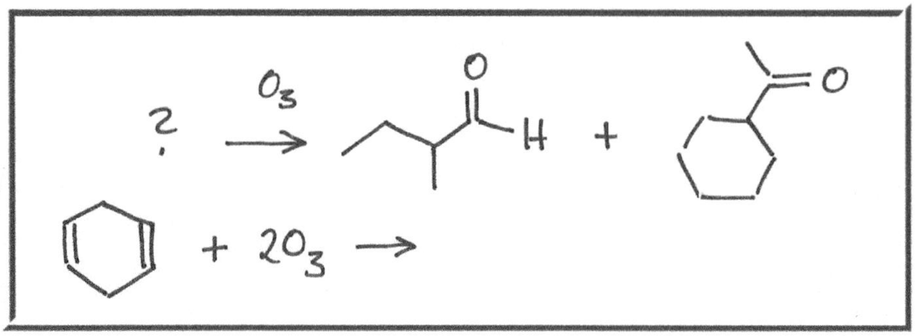


Purple colored potassium permanganate ($KMnO_4$) adds in a similar fashion to a multiple bond as ozone, at first forming a five-membered ring with the bond via **suprafacial** cycloaddition, which means approaching the pi-system from the same side. The opposite approach is known as **antarofacial** addition.

**Fig. 11.28**

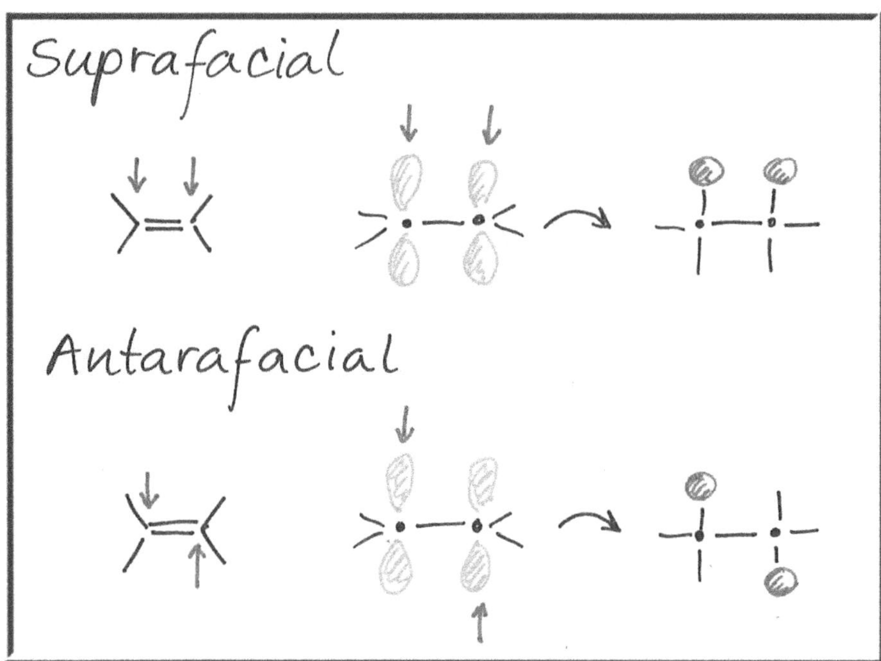

Final reaction with water leads to the reduction of manganese from oxidation state of +7 to that of +5, releasing the element in the form of $MnO_3^-$ ions and leaving a glycol behind. The reaction is completed with the precipitation of brown $MnO_2$ (oxidation state +4) and of course discoloration of the purple reagent.

**Fig.11.29**

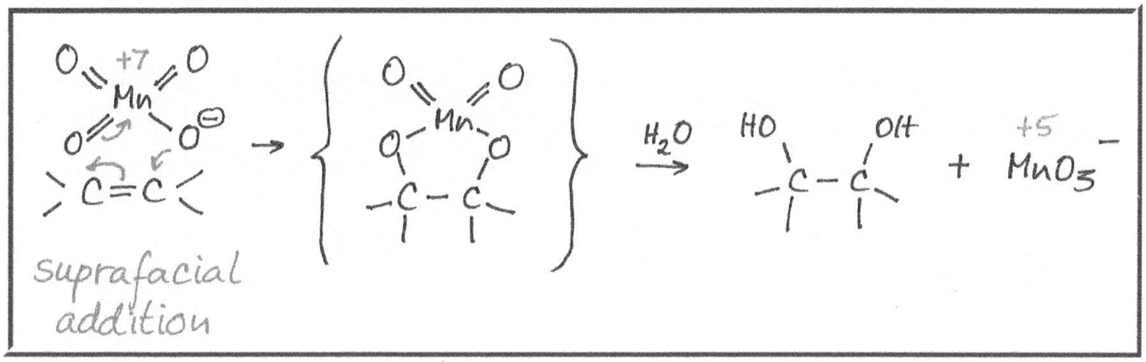


Molecules containing **two conjugated double bonds** can undergo two important reactions:

1) **Cycloaddition** or **Diels-Alder** reactions with other aliphatic unsaturated molecules leading to the formation of six-membered rings
2) **1,2 and 1,4 addition**

Any molecule that has at least two conjugated double bonds in its structure serving as a **diene** can undergo cycloaddition reaction with any molecule with at least one multiple bond (double or triple), which will act as **dienophile** or the reagent that will attack the diene. The two participants of this **pericyclic reaction** should be arranged such a manner that a type of **"CI" (C + I)** is formed.

**Fig.11.30**

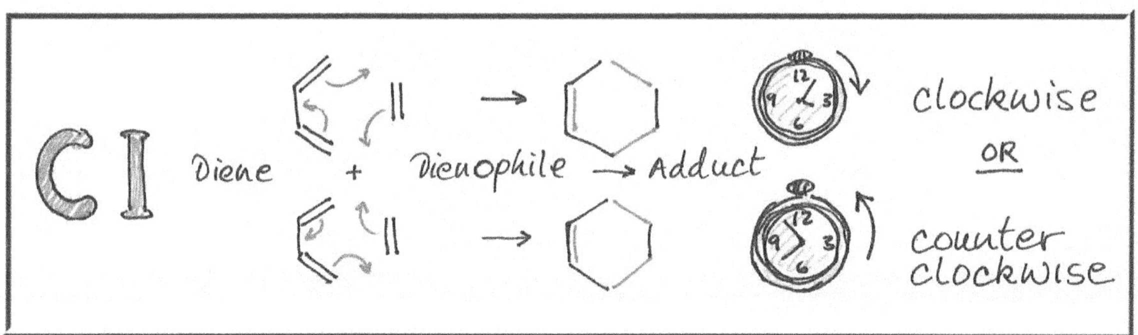

Pericyclic ("peri" – around) addition reactions are all reaction types that proceed via a one-step, synchronized or concerted fashion involving a cyclic transition state.

**Fig. 11.31**

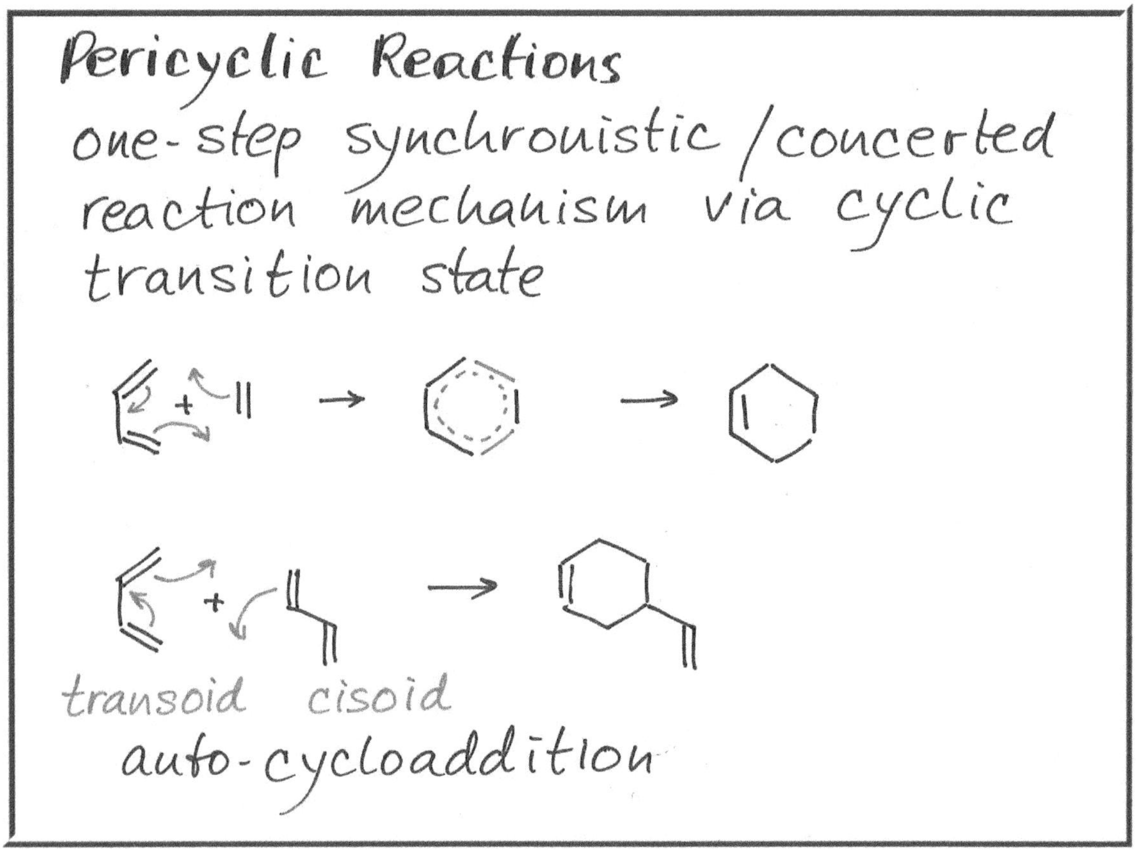


Actually the diene is always arranged in a **cisoid conformation**. A diene can also function as a dieneophile. Here it might be better to select the corresponding **transoid** conformation.

**Fig.11.32**

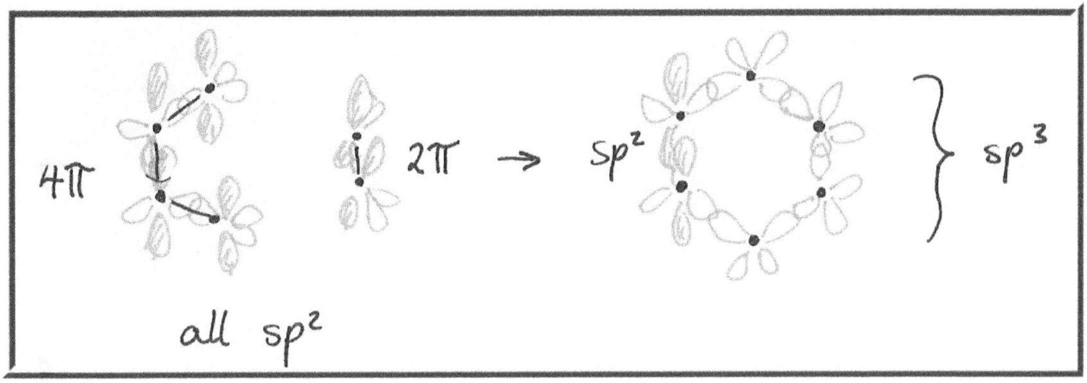


That way it is easy to position the three curved arrows starting from one of the multiple bonds, clockwise or anti-clockwise but always in sequence leading to ring closure of the cycloaddition-product. The adduct always becomes a six-membered ring with one double bond (cyclohexene) or with two double bonds in parallel position (1,4-cyclohexadiene), if an alkyne derivative was used as the dienophile.

Moving the bonds around can be compared with the moves of a windshield wiper, where the blades pivot around an anchor point.

The Diels-Alder reaction represents the special case of **[4+2]** $\pi$ cycloaddition, since there also other types such **[2+2]**$\pi$ and **[4+1]**$\pi$ ring-forming reactions.

**Fig. 11.33**

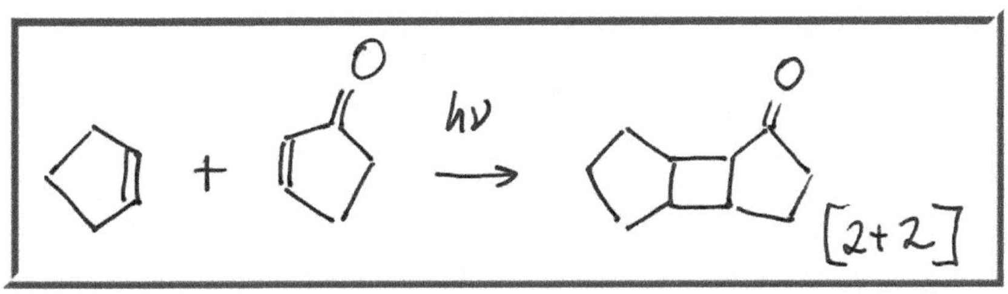


**Fig. 11.34**

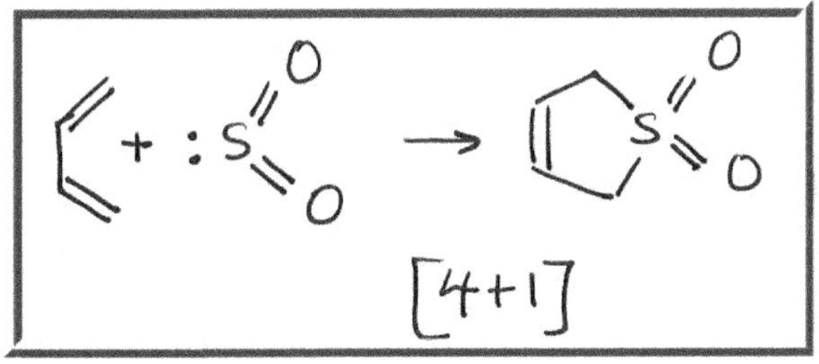

The diene as substrate molecule utilizes the total of 4π-electrons and the dieneophile only uses 2π-electrons for the cyclization reaction, even if coming from a triple bond. The reaction can occur spontaneously since the ring structure seems to be form of stabilization for both molecules. The reactions can be also successfully performed with hetero-atoms being involved in the diene as well as dienophile. Sometimes even oxygen itself can be a dienophile, as well as nitrogen or its derivatives.

**Fig.11.35**

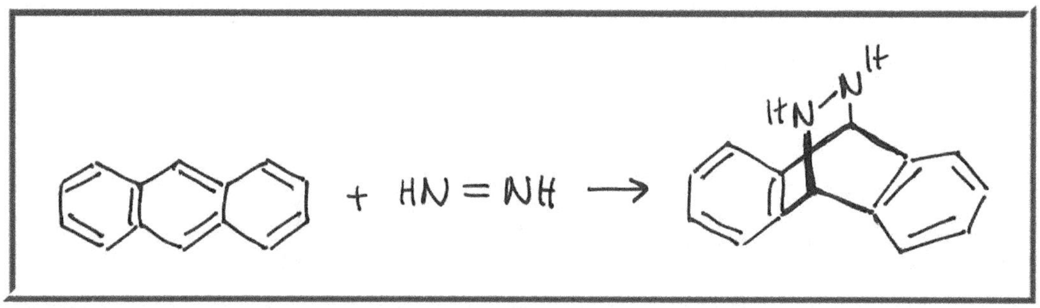


An example of how the procedure of cycloaddition can easily performed using the CI pattern is shown below:

**Fig. 11.36**

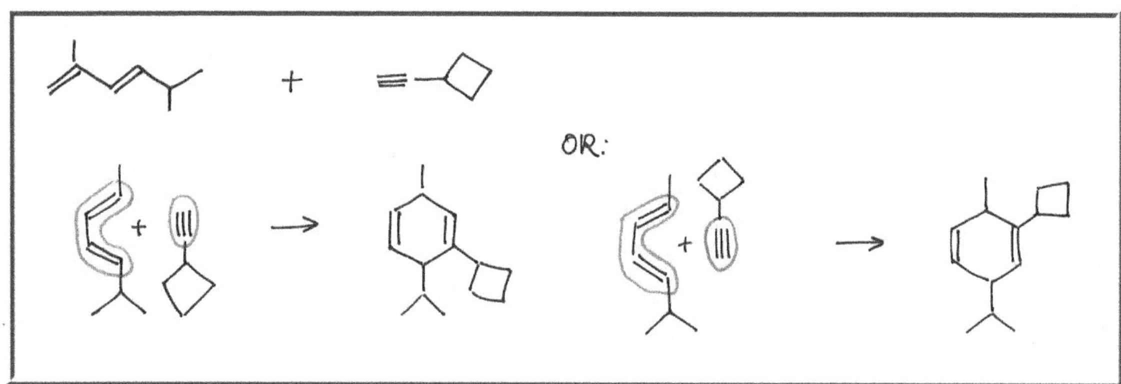

**Example Problem 11.5**
**Complete the following cycloaddition reactions:**

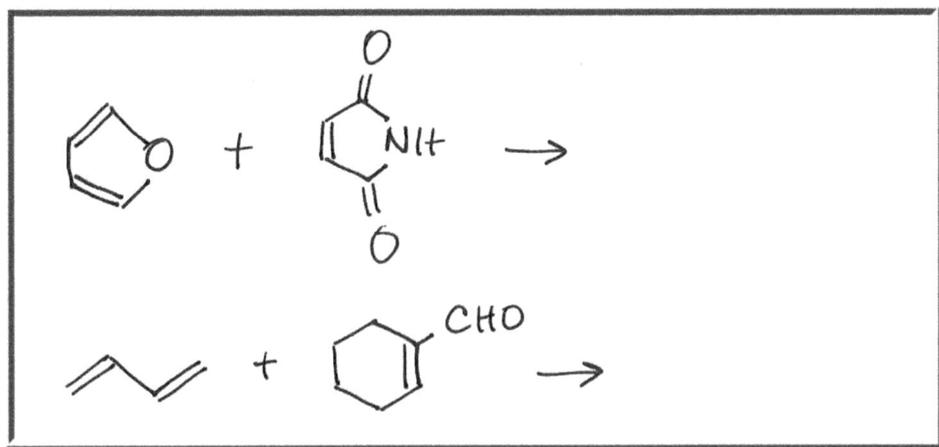


As any forward addition reaction can be reversed leading back to the products, so it is the case for cycloaddition. By reversing the arrows that led to the cyclic adduct, the "merry-go-around" can be reversed as **retro-cycloaddition** reaction (retro-Diels-Alder).

**Fig.11.37**

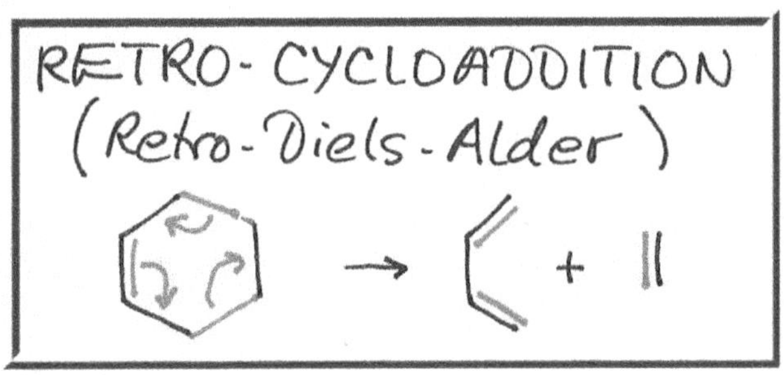


**Example Problem 11.6**
**Complete the following retro-cycloaddition reaction:**

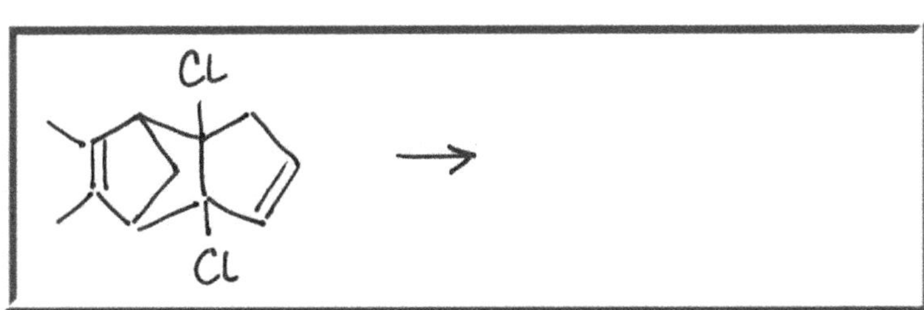

Many natural and commercial products are formed via cycloaddition reactions and naturally occurring products that seem to have been bio-synthesized in such a fashion can be better understood and possibly lead to synthetic pathways for the chemist, if analyzed by "reverse technology, in this case meaning retro Diels-Alder reactions. The highly complex and deadly "Tetradotoxin" found in the puffer fish and other species was partially synthesized in the lab by applying Diels Alder reaction mechanisms.

Cycloaddition reactions that lead to fused ring systems can result in two types of configurational isomers, the endo- and exo- types, a type of cis and trans isomerism such as cis/trans or E/Z isomerism. Depending on whether the cyclic dienophile attacks from above or below the plane of a cyclic diene, an endo or exo isomer can form. Viewed from the side the two isomers resemble a chair or boat conformation, even though they are strictly configurational variations, "frozen" into their specific shape.

**Fig. 11.38**

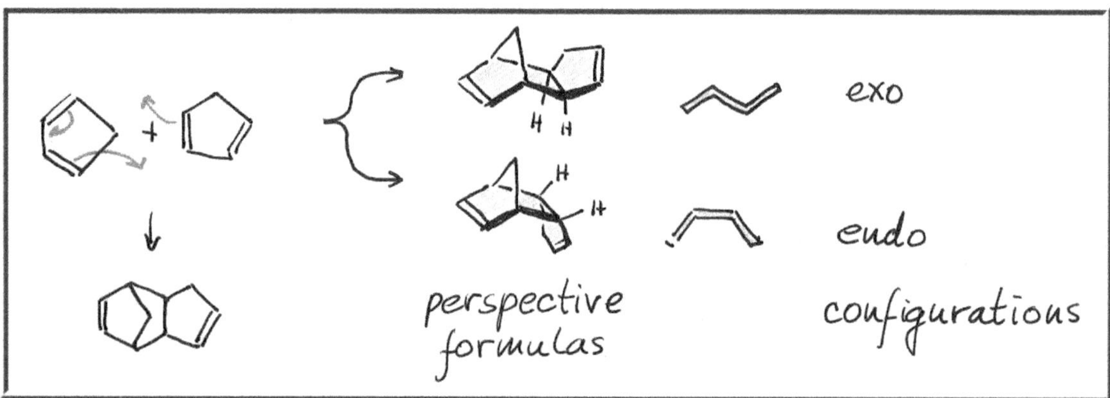


Another type of pericyclic reaction is the Claisen rearrangement, in which one sigma bond is broken but turns into another new one three atoms away, known as a **[3.3] sigmatropic** conversion. Intramolecular rearrangements are quite common in organic chemistry and are symbolized by a looped arrow as is seen below.

**Fig. 11.39**

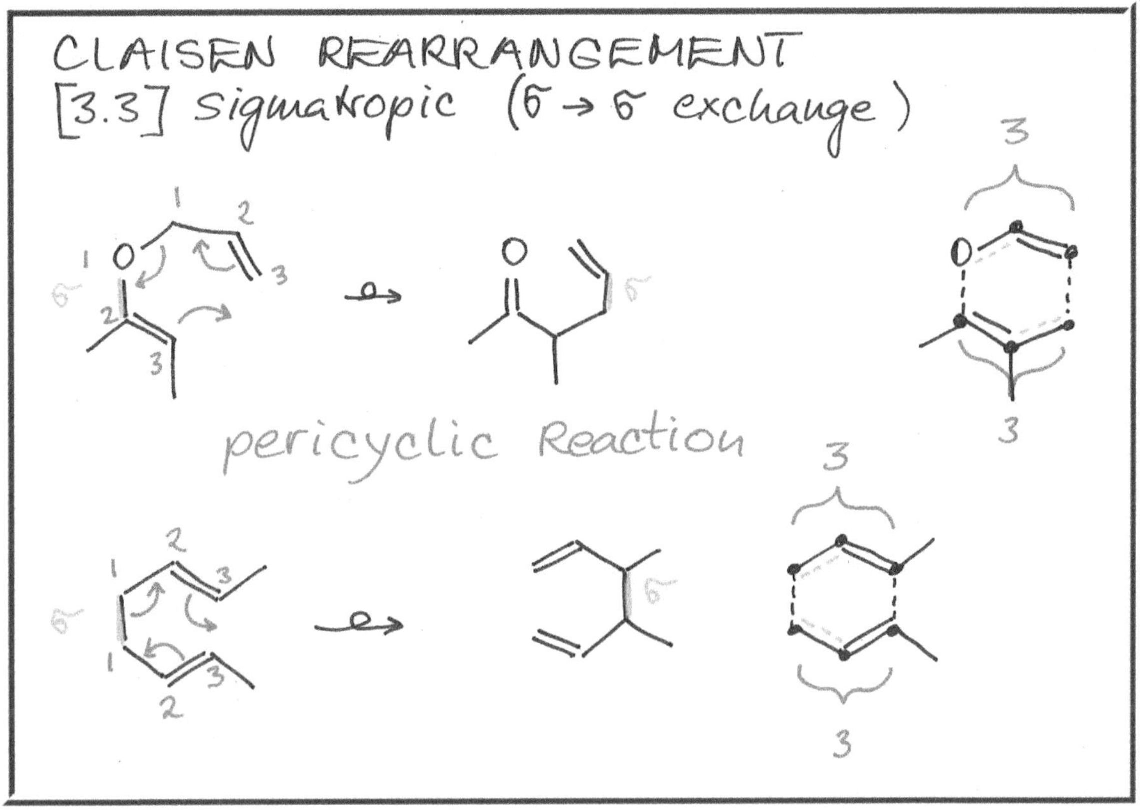


The other type of diene addition is similar to simple addition to a multiple bond, only that two conjugated multiple bonds are involved. They can be either double or triple bonds, since no ring will be formed. The reagent is not a dienophile but a simple symmetric or asymmetric addend such as $Br_2$, HCl, $H_2O$ or others that can be patterned into an A-B section. The A-B type molecule can add to either one of the conjugated multiple bonds or both if the molar ratio is 1:2 respectively.

However conjugated systems have resonance structures, as can be seen for 1,3-butadiene shown below, consisting of two polar types and one bi-radical form, in which only one multiple bond appears in the middle, connecting C2 with C3, as well as the normal structure with two multiple bonds separated by one single bond.

**Fig.11.40**

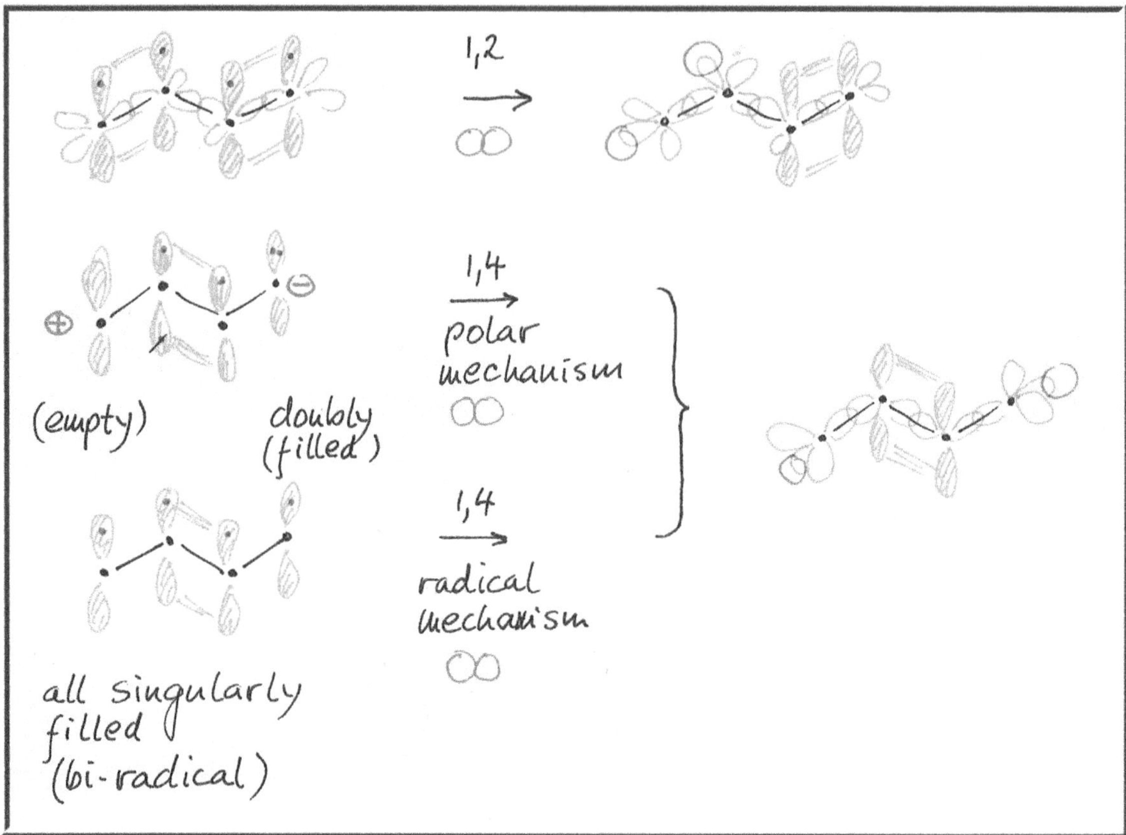


Therefore, the middle bond can remain positioned in the reaction product with the A and B part of the addend connected to ether end of the substrate, namely at C1 and C4.

The resulting possibilities for 1:1 molar ratio addition products are thus increased since not only 1,2 additions but also 1,4 additions can take place. As in regular addition reactions Markovnikov rules apply if both substrate and reagent are not symmetrically structured.

**Example Problem 11.7**

**What are the possible addition products for the following set of reactants:**

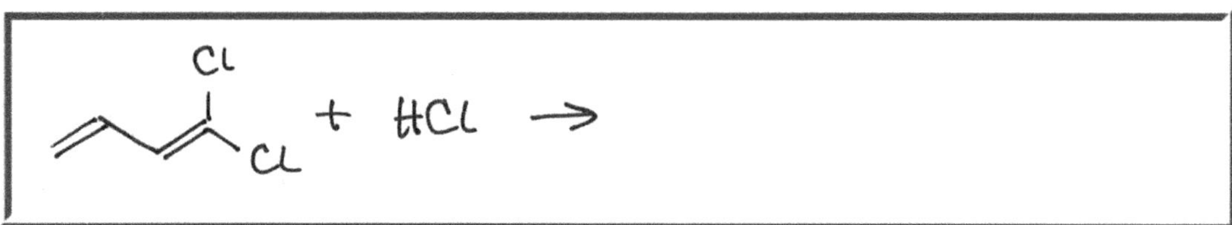

Aromatic compounds do not undergo addition reactions although they are unsaturated systems. The driving force of trying to retain their aromaticity status leads them into **substitution ($S_E$)** reactions instead. The process is initiated by a slow, rate-determining step that leads via π-complex and consecutive σ-complex with the added electrophile to the ultimate **carbocation** as **intermediate**.

**Fig. 11.41**

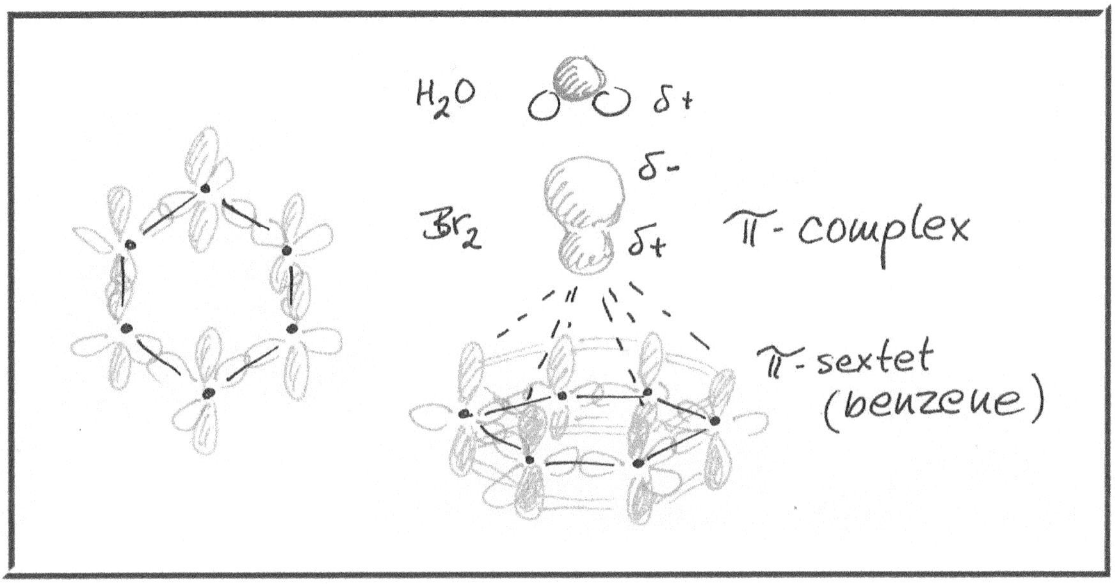


The addition of the electrophile to the ring temporarily causes the carbon atom to become $sp^3$ hybridized, but by elimination of the attached hydrogen atom, $sp^2$ hybridization and re-organization of a pi-bond will re-establish the π-sextet and thus aromaticity.

**Fig.11.42**

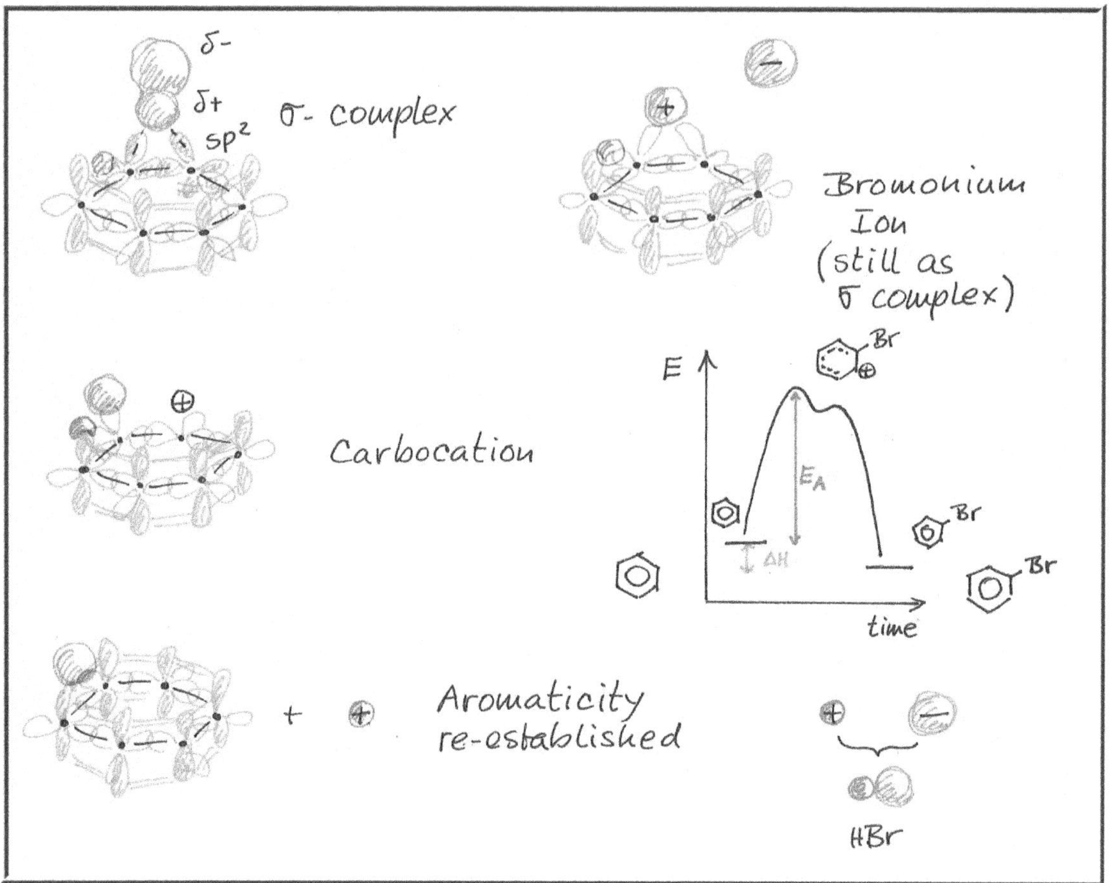


Alkyl, hydroxyl, amino, as well as halogen groups belong to this class. However, halogens can be ring-deactivating and still be *ortho/para*-directors. So, essentially a hydrogen atom is substituted by the electrophile as new group.

Once a first substituent is attached to the benzene "nucleus", second substitution can proceed. However, depending on its chemical nature, the first substituent can either be ring activating or ring de-activating, therefore leading the second substituent into selected positions in reference to its own. If the substituent is made up of single bonds with or without lone pairs they predominately lead into the *ortho* and *para* positions. These o-/p-directors can be considered as electron pushers, increasing the electron density of the aromatic system (+I effect or positive induction). On the other hand, groups with their own multiple bonds or pi-systems, along with their resonance capabilities, can compete with the pi system of the aromatic rings. They seem to be electron pullers, depleting the ring of its electron density (-I effect or negative induction).

**Fig. 11.43**

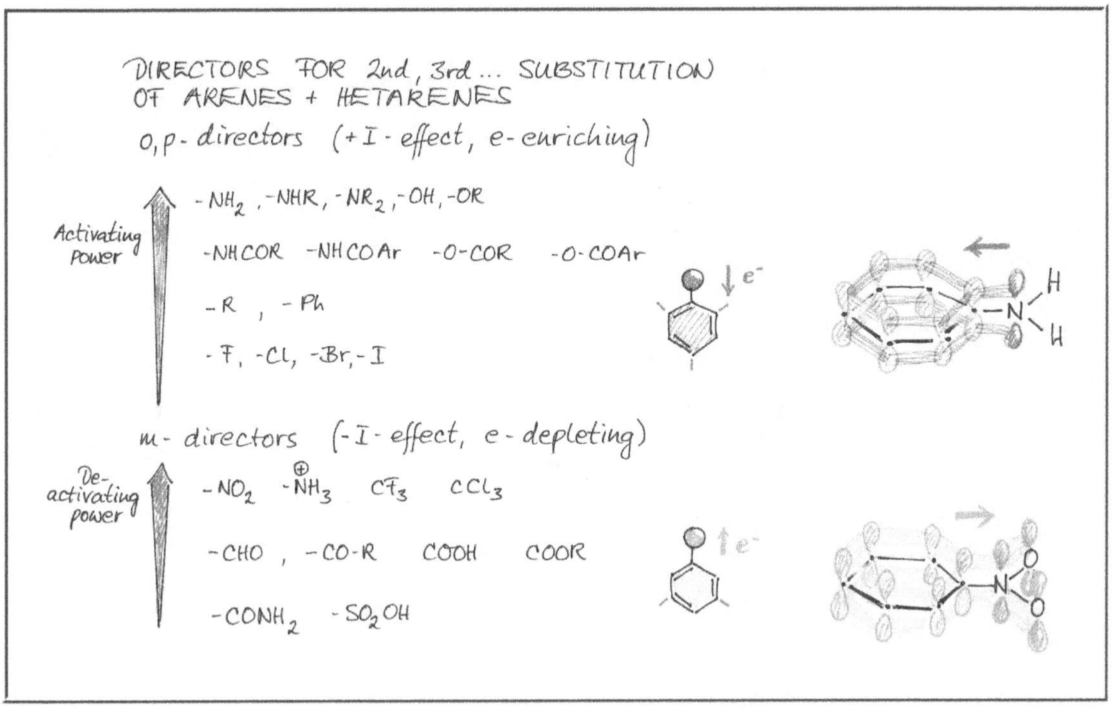


An analogy would be either joining a party or competing with a party going on by starting one's own party, enticing people to leave and join the merriment "next door".

**Fig. 11.44**

As usual in two step mechanisms, the relative stability of the intermediate carbocation often determines in which direction a reaction will go. *Ortho/para* directors have the intermediate positive charges at the bridge point carbon and the two equivalent *meta* positions, resulting in two secondary and one tertiary carbocation **(2°, 2°, 3°)** which is the more favored combination than the **2°, 2°, 2°** - substitution pattern if meta direction would have been preferred. *Meta* directors have the same choice, however, the bridge-point 3° is in direct proximity to the positive charge of the connecting atom of the unsaturated substituent, that it is in one of its resonance states. This creates a problem of **like-charge repulsion**. So, its intermediate carbocation is preferably resonating in the *ortho/para* positions as a 2°, 2°, 2° – combination. Carbocations within the ring are resonance-stabilized.

**Fig.11.45**

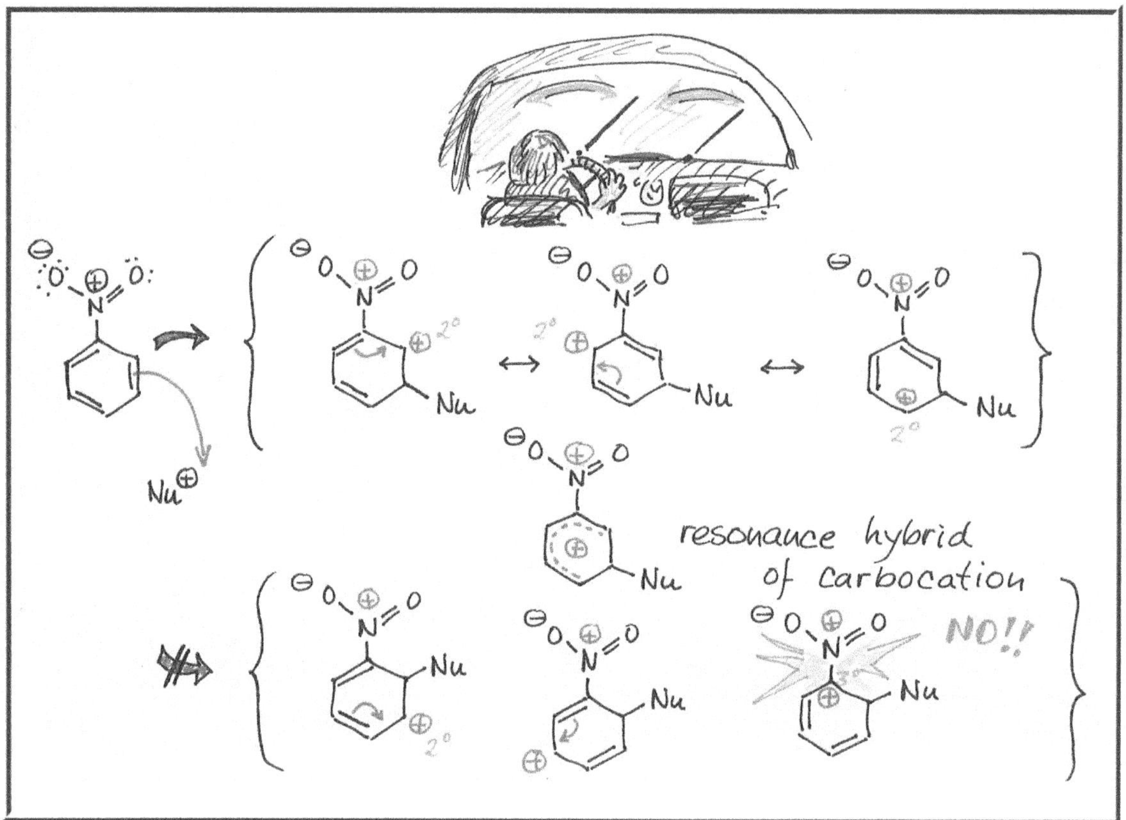


For a third substitution on the benzene ring, the first two groups will both influence the substitution site of the incoming group. Knowledge of the directing behavior will allow the set-up of a most effective sequence of preparatory steps for the synthesis of a target molecule.

**Fig. 11.46**

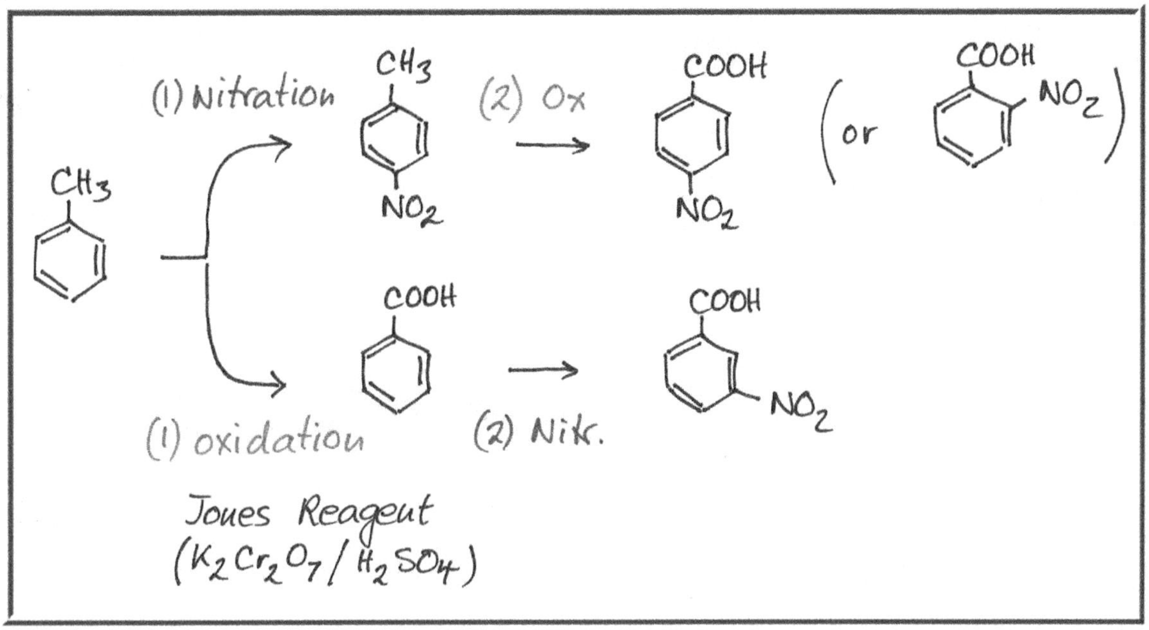


**Example Problem 11.8**
**What sequence of reactions allows the synthesis of the following molecule?**

Nucleophilic aromatic substitutions also exist, but will not be discussed here.

The following table shows an overview of some important reactions discussed in this chapter:

**Table 11.1**

## Overview of Reaction Types

| Reaction Type: | Subcategories: | Substrate: | Reagent: | Reaction conditions: | Mechanism: | Outcome: |
|---|---|---|---|---|---|---|
| **Addition** | $A_E$ | Aliphatic alkene, alkyne | Electrophile | Polarizing agent (catalyst) | Markovnikov rule applies for asymmetric addends | Various adducts, depending on addend |
| | Hydroboration | Aliphatic alkene, alkyne | Boranes ($BH_3$, $BH_2R$, $BHR_2$, $BR_3$) | | Anti-Markovnikov additions | Various adducts, depending on addend |
| | Poly-addition | Aliphatic unsaturated molecule or aromatic molecule with unsaturated side chain | Unsaturated monomers | Initiating catalyst | Initiation, propagation, termination sequence | Macromolecules Poly-adducts ("polymers") |
| | $A_R$ | Hydrocarbon | Radical | Light (photo), Radical initiator | Initiation, propagation, termination sequence | Macromolecules ("polymers") |
| | $A_N$ | Carbonyl compound with de-shielded carbon (δ+) | Nucleophile | | | Various adducts, depending on addend |
| | Conjugated cyclo-addition $[4+2]\pi$, **Diels-Alder** and other pericyclic reactions | Conjugated **diene** | **Dienophile (unsaturated molecule)** | | | Cyclohexene or 1,4-cyclohexadiene derivatives |

| | | | | | | |
|---|---|---|---|---|---|---|
| | [2+2} and [4+1] | | | | | |
| | Conjugated 1,2 or 1,4 addition | Conjugated diene | Electrophile | | | 1,2 and 1,4 adducts |
| | **Aldol-Addition** ($A_N$) Addition to carbonyl group (δ+ carbon) | Aldehyde, Ketone | Nucleophile (carbanion) enolate anions | Strong Base NaOH, KOH, ROH (alkoxide) α – hydrogen abstraction | | β-hydroxy aldehydes (aldols) or ketones (ketols), R-CH(OH)-CHO or R-CH(OH)-CO-R', new C-C-bond, follow-up elimination to alkenals or alkenones, resp. |
| | **Michael Addition** ($A_N$) Addition to polarized double bond (δ+ carbon), special type of conjugated addition | "Michael acceptor", Conjugated enone α,β-unsaturated carbonyl compound (-CN, $-NO_2$ also work as electron-withdrawing groups) | Nucleophile (carbanion) such as enolate anions "Michael donor", molecules with acidic α-hydrogens | Strong Base NaOH, KOH, ROH, (alkoxide) α – hydrogen abstraction | | New C-C bond, 1,5-diketones and others |
| | **Claisen** | Ester | Ester carbanion as nucleophile | Strong Base NaOH, KOH, ROH (alkoxide) α – hydrogen abstraction | Addition, followed by condensation (elimination) | β-keto esters $R-CO-CH_2-COOR'$ New C-C bond |

| | | | | | | |
|---|---|---|---|---|---|---|
| | **Grignard** Reaction ($A_N$) Carbonyl compounds | Grignard reagent: organo magnesium halide (R-Mg-X) $R:^-$ carbanion as nucleophile | Grignard formed in situ: Mg + R-X | | Follow-up by acid hydrolysis of adduct ($H_2O/H^+$) | Alcohols: $1^0$ (formaldehyde), $2^0$ (aldehydes), $3^0$ (ketones) New C-C bond |
| Disproportionation (self -oxidation-reduction) | **Cannizzaro** | Aldehyde | aldehyde | Strong base NaOH, KOH, **absence** of α-hydrogen | Auto-redox reaction | Alcohol and carboxylic acid |
| Substitution | $S_E$ | Aromatic compound (arene or hetarene) | Electrophile | | $2^{nd}$ and $3^{rd}$ substitution is governed by directing properties of first substiuent(s); o,m,p directors | Nitration, sulfonation, hydroxylation, amination etc. of aromatic nucleus |
| | $S_N$ | Aliphatic compound with leaving group (L) or aromatic compound with side-chain containing leaving group (L) | Nucleophile | Steric hindrance of substrate, reagent and solvent determines mechanism; Protic solvents favor two step ( $S_N1$), aprotic favor one step ($S_N2$) mechanism | $S_N1$ - two step with carbocation intermediate If chiral substrate - racemization $S_N2$ –one concerted step with transition state If chiral substrate – chiral inversion | Substitution products |

| | | | | | | |
|---|---|---|---|---|---|---|
| | **Friedel-Crafts** R. ($S_E$)-alkylation and acylation | Aromatic compound (arene or hetarene) | Alkyl halide or acyl halide | Lewis acid as catalyst ($AlCl_3$, $FeCl_3$, etc.) | | Alkylated or acylated arenes/ hetarenes |
| | $S_R$ photohalogenation | Alkane | Halogen radicals | Light (hv) | | Halogenated hydrocarbons |
| Elimination | E (β-elimination) | Aliphatic compound with leaving group (L) | Strong base $OH^-$ or $OR^-$ (hydroxide or alkoxide) | α – hydrogen abstraction Steric hindrance of substrate, reagent and solvent determines mechanism; Protic solvents favor two step ( E1), aprotic favor one step (E2) mechanism | E1 - two step with carbocation intermediate<br>E2 –one concerted step with transition state | Unsaturated molecules, Multiple bonds formed |

# EXPERIMENTS – Ch. 11

## *Reactions of Organic Compounds*

**Key Terms and Concepts:** Addition ( $A_R$, $A_E$, $A_N$), polyaddition, stereospecificity and selectivity, Markovnikov rule, hydroboration, substitution ($S_E$, $S_N1$, $S_N2$), elimination (E1, E2), pericyclic addition (cycloaddition, Diels-Alder, diene and dienophile), conjugated addition (1,2 and 1,4 addition), acidic hydrogens in α-positions, aldol addition, Michael addition (Michael donors and acceptors), Claisen condensation, Cannizzaro reaction, polycondensation, polymerization, kinetics and thermodynamics, energy profile diagrams, endo/exothermic reactions (enthalpy change, $\Delta H$), ender/exergonic reactions (Gibbs free energy change, $\Delta G$; influence of entropy changes, $\Delta S$), Gibbs-Helmholtz Law $\Delta G = \Delta H - T\Delta S$, role of catalysts, spontaneous reactions, equilibrium, theoretical and practical yield determination

## I Cycloaddition of Anthracene with Maleic Acid Anhydride

### Chemicals:

*Anthracene*
*Maleic (acid) anhydride*
*Xylene (dry)*
*Decolorizing carbon (1-2 grains)*

### Equipment and Materials:

*Round-bottom flask (50mL) with reflux condenser*
*Hotplate and magnetic stir bar*
*Short-stemmed with fluted filter paper*
*Vacuum desiccator*
*Büchner vacuum filtration set-up*
*Ice bath*
*Top-loading digital balance*
*Graduated cylinder*
*Stand and clamps, spatula*
*Cork ring*
*Fingerlings (rubber pieces for finger protection)*

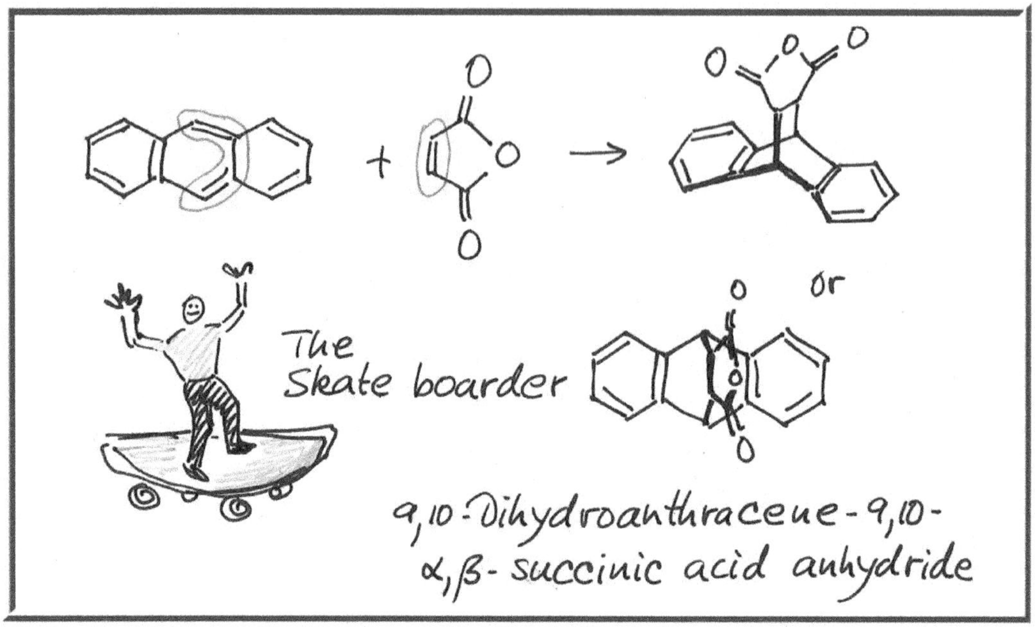


**Procedure:**

1. Weigh out 2.0g of anthracene and 1.1g of maleic anhydride and dissolve both in a round-bottom flask in 25mL xylene (dry, no water as contaminant).
2. Add a stir bar to the solution, attach the flask with reflux column to a stand and position it over the surface of a hotplate.
3. Bring the solution to a boil and continue heating while constantly stirring (magnetic stir bar) for about 20 minutes.
4. Remove and detach the flask with protected fingers (fingerlings) and place it on a cork ring.
5. Add a few grains of carbon powder, swirl and quickly pour the contents while still hot over a fluted filter paper, situated in the short-stemmed funnel.
6. Transfer the clear filtrate into another clean, dry round bottom flask, add the cleaned stir bar and reassemble the reflux apparatus over the hot plate.
7. Heat to boiling for another 5 minutes.
8. Dismantle the reflux set-up and place the flask into an ice bath.
9. Vacuum filter the crystals with a Büchner apparatus.
10. Rinse with ice-cooled water and let air dry a few minutes.
11. Remove the filter cake residue including the round filter paper gently with a spatula and place on a watch glass. Weigh out the crude product.
12. Place the residue containing watch glass in the desiccator and save for the next glass period (determination of yield, Mp. and spectroscopic characteristics)

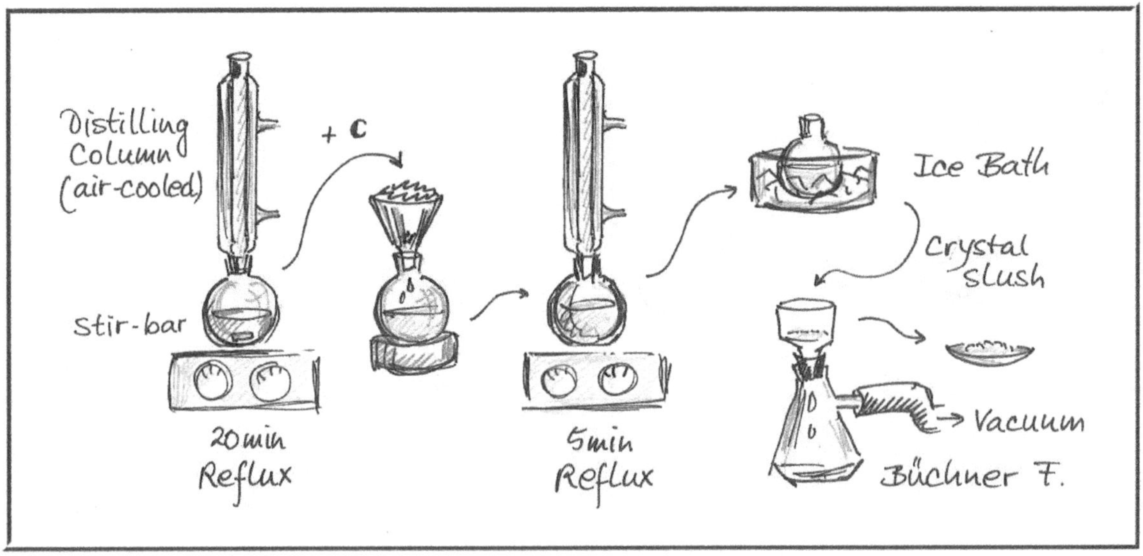


## Observations and Results:

| Color and texture of crystals | Mp. (°C) | Theoretical yield (g) of dried crystals | Practical/experimental yield (%) |
|---|---|---|---|
| | | | |

## Questions and Problems:

1. Describe the course of the reaction. What observations were made?
2. What was the purpose of adding decolorizing carbon? Was it effective?
3. What mechanism is involved in this reaction?
4. What other dienophiles could be used with anthracene as diene?
5. Anthracene can act as a diene in two ways. Which parts of the molecule can be involved leading to what adducts? Why is one adduct preferred over the other? Explain.
6. Cycloaddition products can occur as endo or exo configurational isomers. Were different isomers possible for anthracene? Why or why not?

## II Simulation Experiments for Cycloaddition (Diels Alder) and Retro-cycloaddition Reactions

**Chemicals:**

*None*

**Equipment:**

*Modeling Kit ©Heinz*

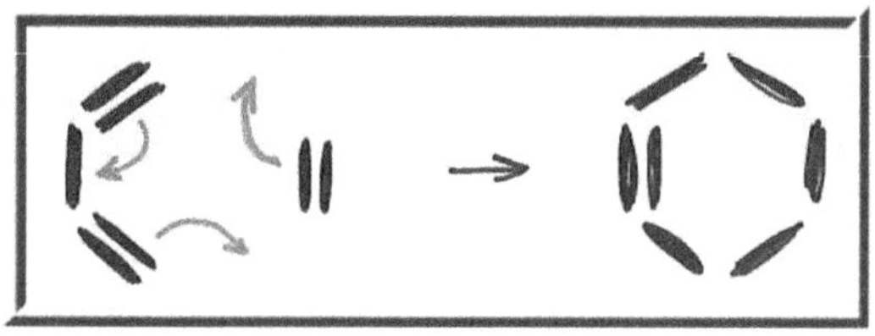

**Procedure:**

1. With black bond pieces arrange various dienes and dienophiles on a platform (table, foam sheet etc.).
2. Move the bonds around according to a pericyclic fashion as illustrated below to obtain the cycloaddition product (adduct).
3. Reverse the movement of the bonds to open up a ring leading back to a diene and dienophile.
4. Repeat the process for more or less complicated systems, including cyclic molecules functioning as diene and/dienophile as well as molecules that can add to themselves.

## Questions and Problems:

1. Give the structures and the associated reaction schemes for the chosen cycloaddition reactions.
2. What retro-cycloaddition reactions were studied with this model system?

# III Simulation Experiments for the Influence of *ortho/para* and *meta* – Directors on the Higher Substitution of Aromatic Compounds

## Chemicals:

*None*

## Equipment and Materials:

*Aromatic Substitution kit/game ©Heinz*
*(10 benzene rings, 30 functional group pieces, front side for right positioning, backside for left, reflected, positioning), box out of which the functional group cards can be selected*
*Wooden bond pieces for attaching functional group pieces*

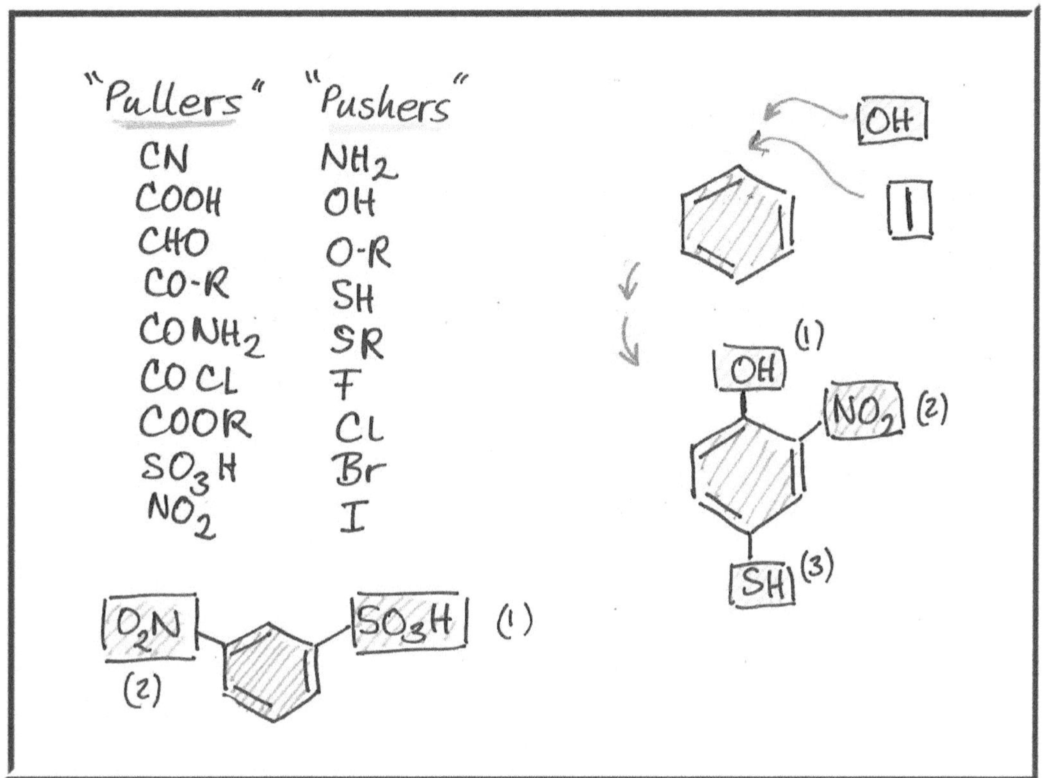


## Procedure:

1. The kit consists of 10 benzene ring pieces (cards) that can be extended to higher fused arenes or hetarenes to extend the concept of o,m,p directive properties, if possible.
2. In addition, 30 pieces (cards) of various functional groups are included. If needed, some show lone pairs and formal charges, in order to illustrate the directing mechanism of the two types of groups: the "pushers", electron enriching, o/p-directing groups and the "pullers", electron-withdrawing, m-directing groups.
3. A card depicting the aromatic nucleus of a benzene molecule is placed on the table. A functional group is randomly selected out of a box and attached to the benzene nucleus as first substituent.
4. A second functional group is randomly selected as well and positioned onto the spot directed by the first substituent. If more products are possible a second or third ring can be laid out.
5. The same procedure can be repeated for the third substituent coming in and leading to positions directed by the first two groups, so more derivatives can be possible. This, of course, can be continued until all the substitution sites are filled, if possible.
6. When played as a game, the players will take turns in selecting a functional group card out of a box and positioning it onto a correct site on the aromatic ring. Each correct product earns a point for the player. Instead of choosing a functional group card the player also has a choice of laying out another ring and starting a "new" molecule. All

rings can be played on simultaneously. The game is over when the last functional group card is laid out.

## Questions and Problems:

1. What reaction sequences leading to which products were laid out using the kit as a model or as a game?
2. Sketch a "family tree" of all the molecules leading from the first substituent to a multiple substituted derivatives.
3. Can benzene substitution rules be adapted to higher, fused aromatic systems such a s naphthalene and anthracene? Explain!

# 12 Separation Techniques

## Chapter 12 Separation Techniques

Whether investigating reaction products obtained synthetically or molecular components derived from natural sources, mixtures must be separated by various means and purified, in order to apply further techniques for structure elucidation such as spectroscopic analysis, fully discussed in the next chapter. Liquid mixtures can be separated by **distillation** with more or less resolving power. **Refraction columns** placed between the boiling components and the condenser can increase separation of liquids with similar boiling points considerably, especially when multi-component mixtures are involved such as the case for the separation of crude oil in refineries that represent an upscale version of a laboratory set-up. Many types of refraction columns exist for laboratory use, as shown in the following image:

**Fig. 12.1**

For a binary liquid-liquid separation, it is important to study the phase **diagram** that involves not only the relative liquid compositions but also the relative vapor compositions at various temperatures. Both curves enclose an area called the **boiling lens**.

**Fig. 12.2**

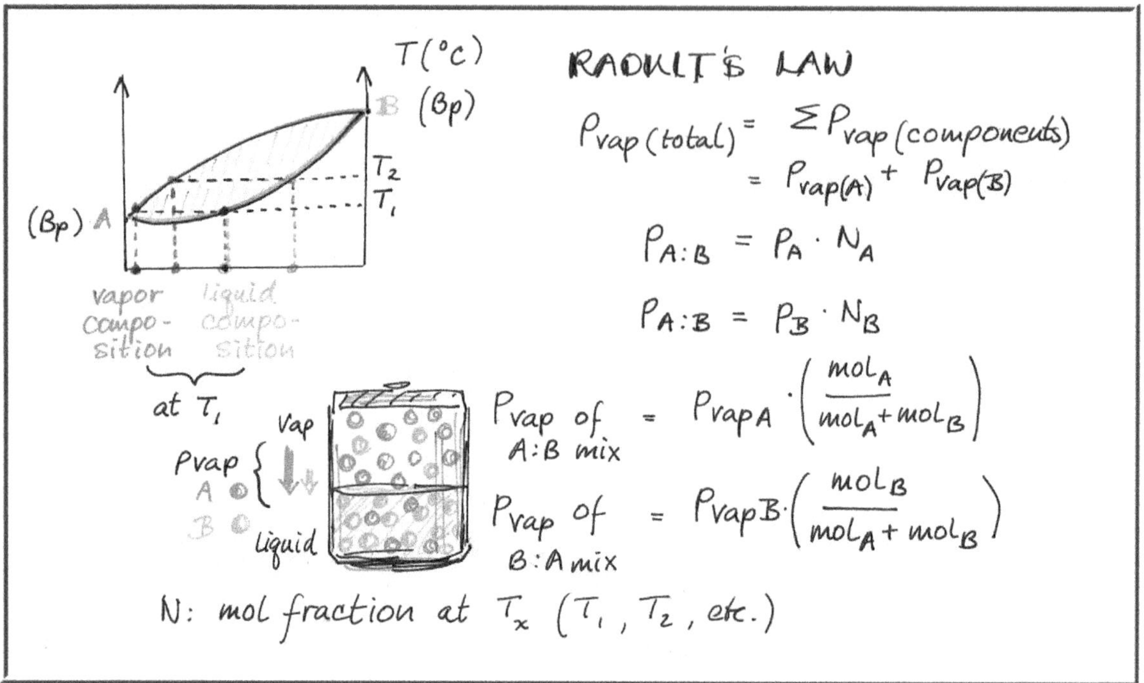


Each temperature will allow the relative liquid/vapor ratios of each of the two components.

**Example Problem 12.1**

**What percentage A vapors are involved in the binary mixture of A/B at 65°C? What % B vapors are involved? What is the corresponding liquid ratio in % at that temperature?**

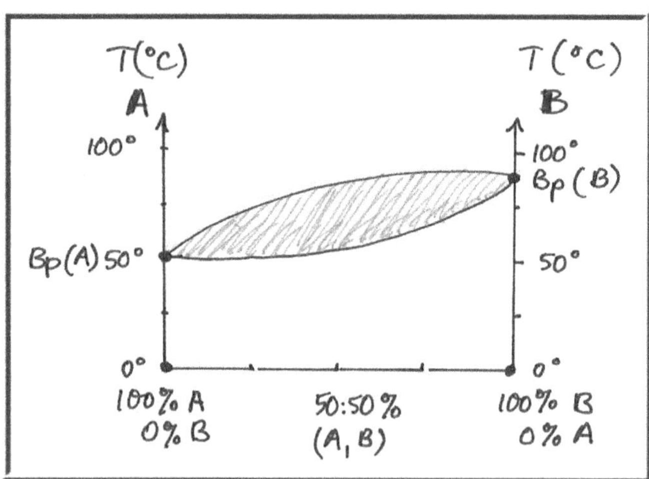

A simple distillation set-up is shown as follows;

**Fig.12.3**

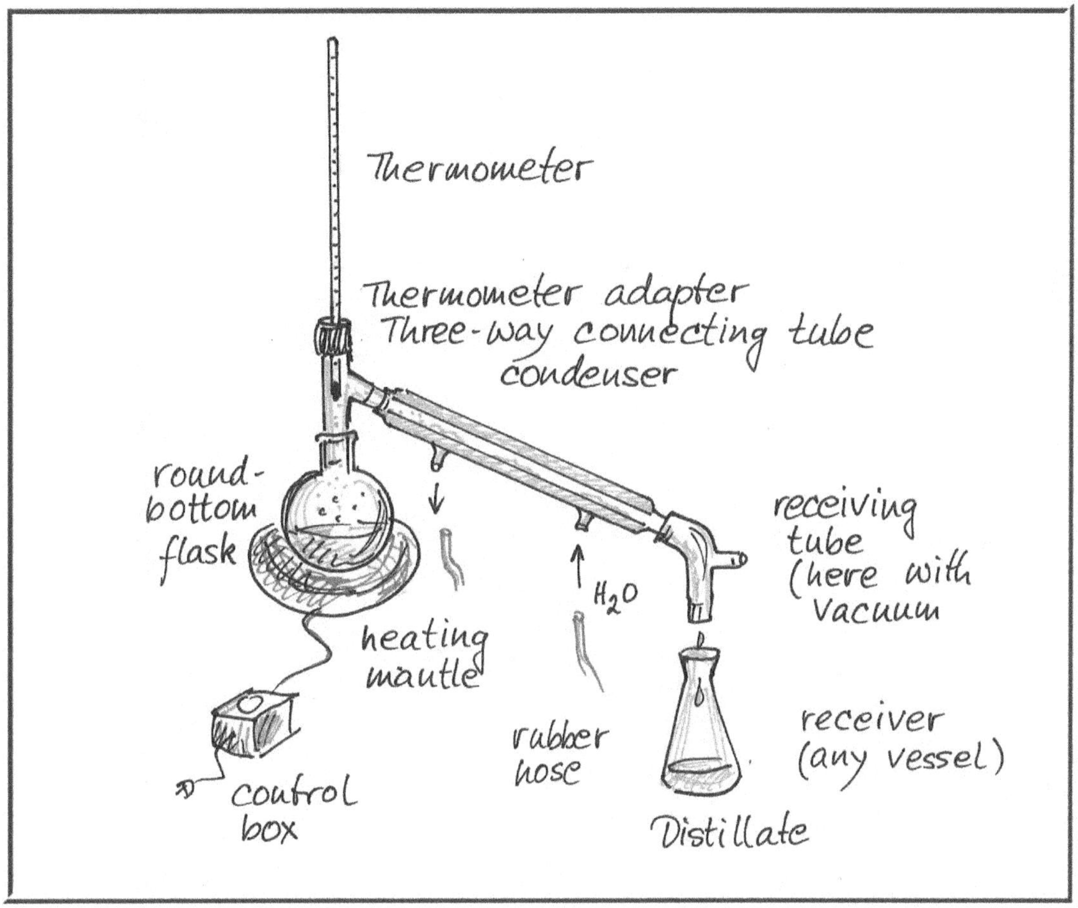


The molecules of the lower-boiling fraction reach their escape velocity at lower temperatures, so they can make it over the "hurdle" earlier than those of the higher boiling fraction. However according to the kinetic energy distribution curve of a population of molecules (Maxwell-Boltzmann), some will always have an excess amount of energy ("speeders") and some will have a below average velocity ("slow pokes") than the majority of the molecules.

**Fig. 12.4**

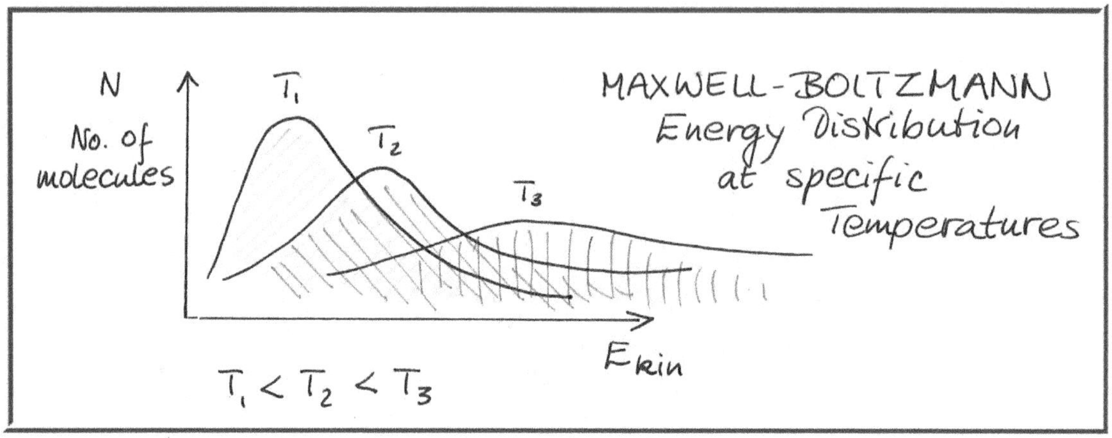


To prevent the faster molecules of the higher boiling fraction to cross over too early and thus contaminating the low-boiling fraction excess hurdles have to be built in to "push" them back down in to the "pot" and let them wait their turn. This is done by inserting a fractionating column into the distillation set-up. These allow successive re-distillations expressed as the **number of theoretical plates**. The more "hurdles are built in the vertical distilling or fractionating column, the better the resolution power of the distillation.

**Fig. 12.5**

The impediments can be of various designs. A simple distilling column can also be filled with different granular material, glass beads or cut glass tubing called "Raschig rings". A comparison can be seen in the steeple chase races for horses, so that the best horse rider group may out-perform the others in the race.

**Fig. 12.6**

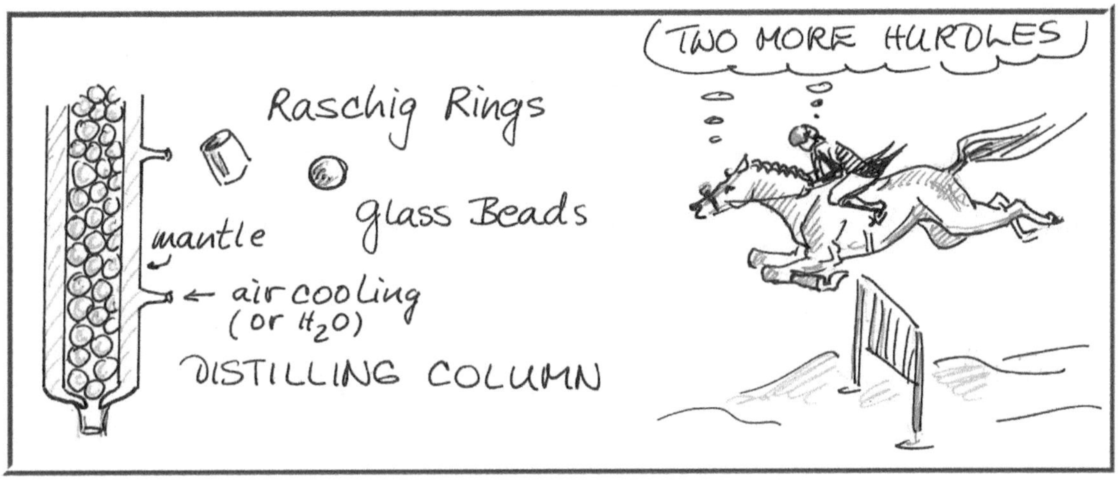


The fractions as they appear in the order of their separation in the **receiver** or collecting vessel are called first cut, second cut etc.

Fractionating columns, which are inserted between the flask and the three-way connecting tube, allow for better resolution in the separation of liquid mixtures. The portion of the mixtures having both the lower and higher boiling components together becomes narrower as can be seen in the following illustration:

**Fig. 12.7**

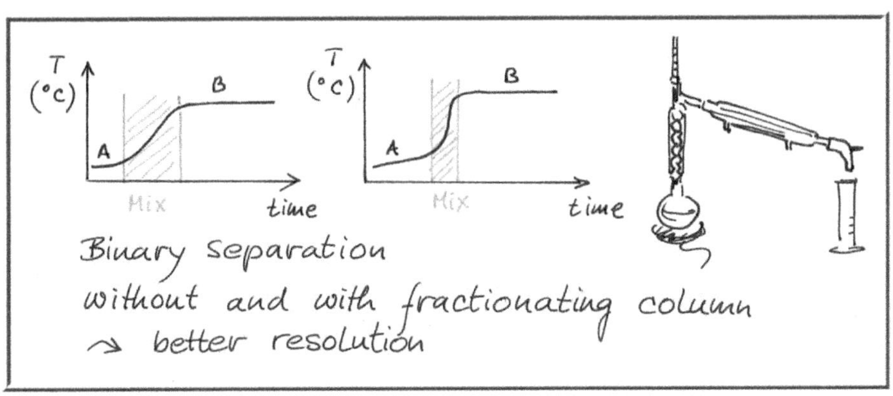

In the petroleum industry, refineries depend on large, tower-like versions of refractionating columns, since many components of natural oil need to be separated. Many theoretical plates within the columns allow many evaporation and re-condensation cycles to resolve the many molecular fractions and their isolation.

Sometimes miscible two miscible liquids cannot be separated by simple distillation. At a certain composition they reach a condition at which they act as if they would be one single chemical, correlating to a corresponding temperature at which they reach a "dead end". In the boiling lens it represents a point at which the two curves seem to be tied together. These liquid pairs are referred to as **azeotopes** or **azeotropic mixtures.** The point in the graph at which the **constant boiling mixture** is reached is called the **azeotropic point**.

**Fig. 12.8**

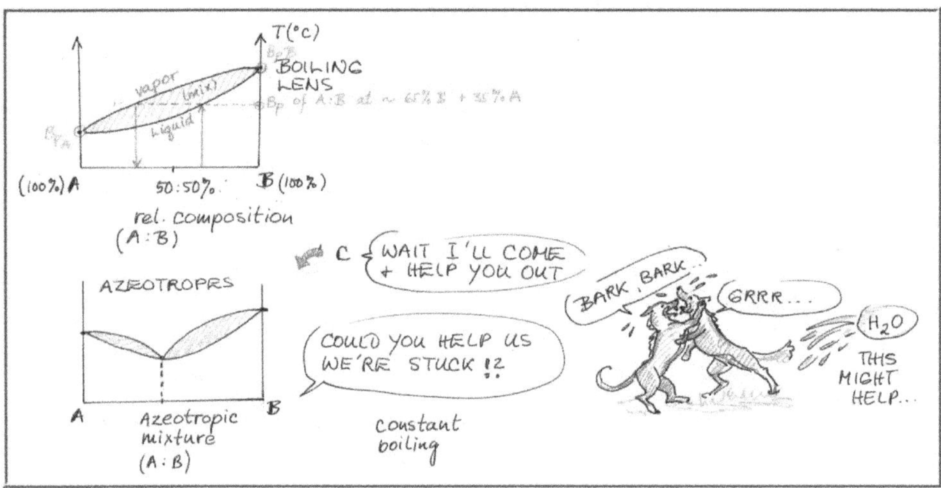


The best remedy is to add another liquid that changes the whole dynamics of the mix leading via changes in intermolecular forces to different phase diagrams. Even solids like sulfur or salts can break up an azeotropic mixture. It is like breaking up two fighting dogs by way of distraction.

Many **essential oils**, the volatile, non-polar components of plants can be separated by a modified form of distillation, namely **steam distillation**. The crude plant extract is suspended in water and the water then distilled. Although hydrophobic the essential plant oil is literally dragged by mechanical forces of the distilling water molecules over the "hump" in the receiver flask, where it usually appears as a turbid, milky emulsion (highly dispersed immiscible liquids). Extraction with a proper organic solvent such as dichloromethane (methylene chloride, $CH_2Cl_2$) can help remove and after solvent evaporation isolate the pure essential oil, such as eucalyptus oil or clove oil (eugenol) etc.

**Fig. 12.9**

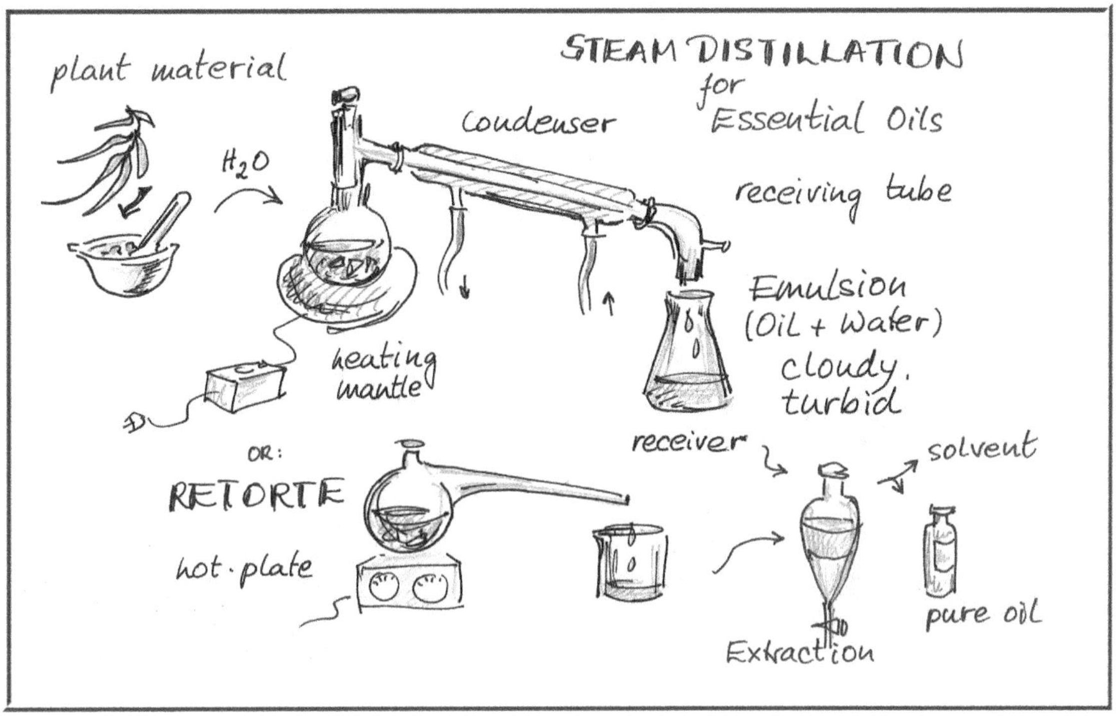


Another important method for separating or isolating a chemical from a mixture is **extraction**. Here, the different solubility of solid solutes in immiscible (non-miscible) solvents is utilized. The glass device used for this purpose is the **separatory funnel**, available in various sizes.

Some chemicals convert directly from the solid phase into the gaseous phase, a transition known as **sublimation.** Such compounds, as is the case for caffeine, can be separated by a type of "cold finger", consisting of two concentric tubes. The inner tube is filled with crushed ice and the outer mantel is connected to vacuum, so that the sublimation point can be conveniently lowered. Pure sublimed crystals can later be collected from the outer surface of the inner, cooled tube.

**Fig. 12.10**

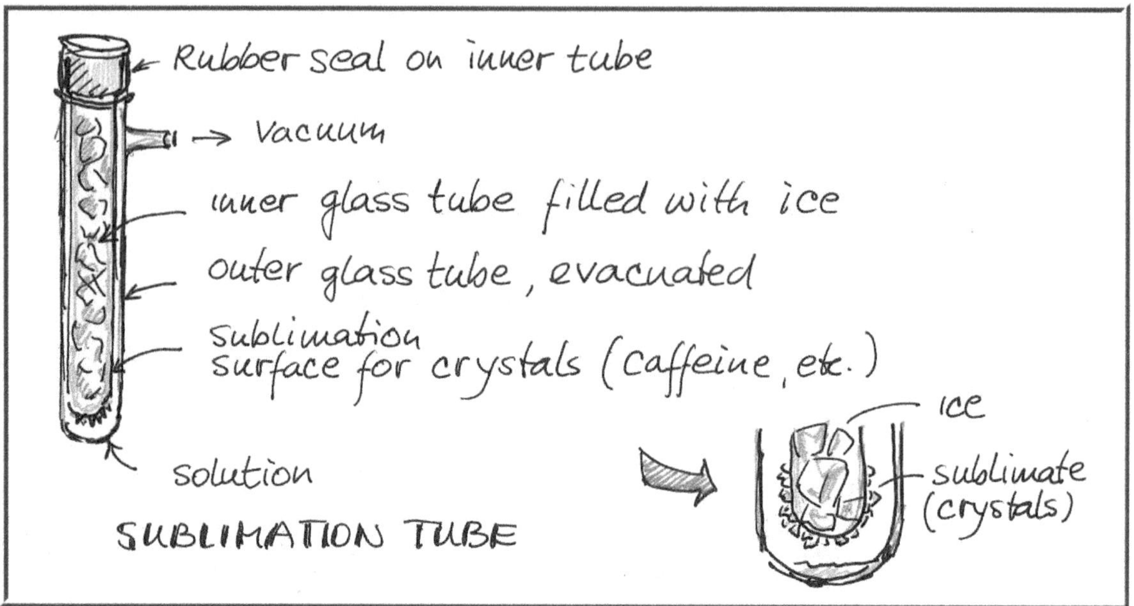


Most commonly, a chemical is extracted out of an aqueous solution by a non-protic organic solvent, in which it is more soluble. Depending on the density, the "heavier" liquid will always be the lower phase, when partitioning in a separatory funnel. The portioning ratio between the two phases when equilibrium is reached is defined by the **distribution coefficient $K_D$.** This can easily be calculated by dividing the solubility of the molecule in the organic solvent by the solubility of the same compound in water, making sure that similar environmental conditions and the same units are used. **Solubility** is usually defined as g solute in 100g solvent, but can also be expressed as a ratio of parts such pro mille or ppm etc. in the literature or other reference sources

$$\mathbf{K_D = S_{organic\ solvent} / S_{aqueous\ solvent}}$$

Once the $K_D$ is known, calculations can be carried out to solve for various unknowns such as the amount of solute that can be extracted using a specific volume of extraction solvent, according to the equation:

$$K_D = \frac{\frac{x}{V2}}{\frac{m-x}{V1}}$$

Multiple extractions lead to a higher yield than a one-time extraction using an equivalent amount of volume, due to three equilibria being established with each portioning event rather than one. For instance three separate extractions of the same aqueous solution with 50mL of fresh extraction solvent each time will be more effective than using 150mL at once.

**Fig.12.11**

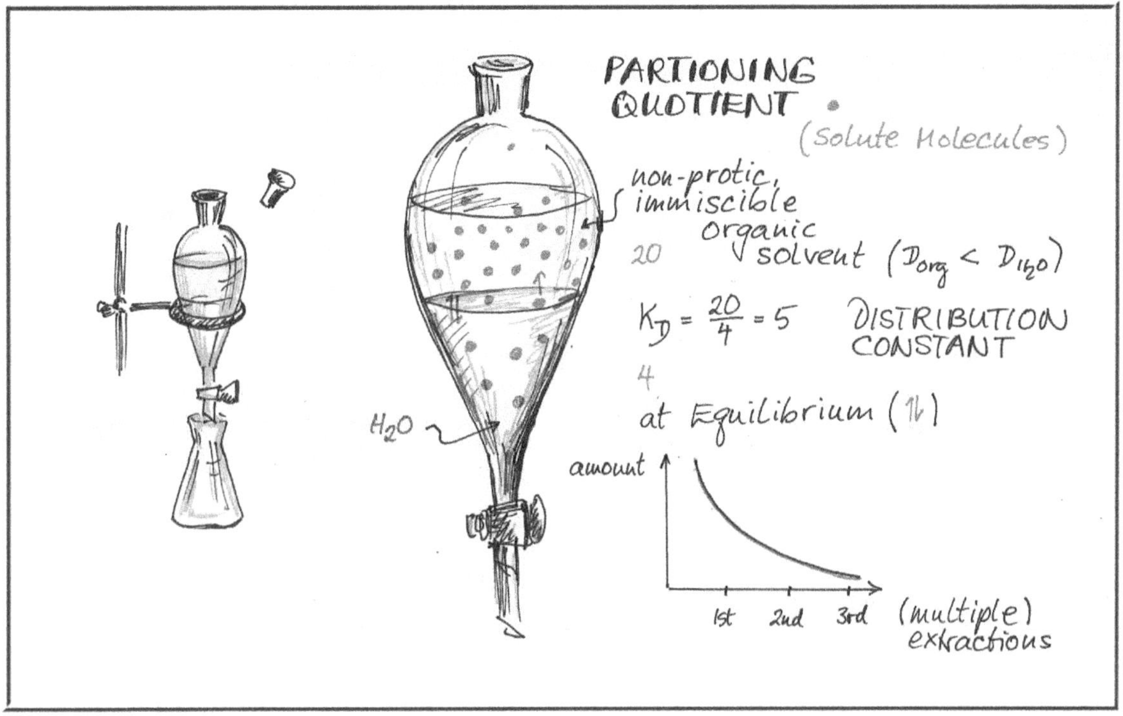


**Example Problem 12.2**
**With $K_D$ being 4, how much more solute can be extracted out of an aqueous solution containing 8.0g in 100 mL with 3 x 50 mL than 1 x 150 mL extraction solvent?**

Many organic molecules, such as those from natural sources, have basic (amino groups) or acidic functional groups (carboxylic, phenolic etc.), that, however, not necessarily guarantee good solubility in water. Their solubility can be improved by converting them into salts by adding strong inorganic acids (HCl, $H_2SO_4$, etc.) to the former and strong inorganic bases (NaOH, KOH, etc.) to the latter. Once isolated the soluble counter ion making the salt can be neutralized by adding a proper amount inorganic base to the former and acid to the latter forming the inorganic salt such as NaCl and "freeing" the base or acid respectively. This procedure then allows the free base or acid as organic molecules to be extracted with an adequate non-protic organic solvent such as ether. After removing traces of water with a drying agent such as $CaCl_2$, evaporation of the "dry", usually volatile solvent, the extracted chemicals can be isolated in their pure form and recrystallized if necessary. The inorganic salts will be left behind in the aqueous phase and can be discarded. A flow chart can thus be set-up and followed according to the following general pattern for two component and three-component separations, respectively.

**Fig. 12.12**

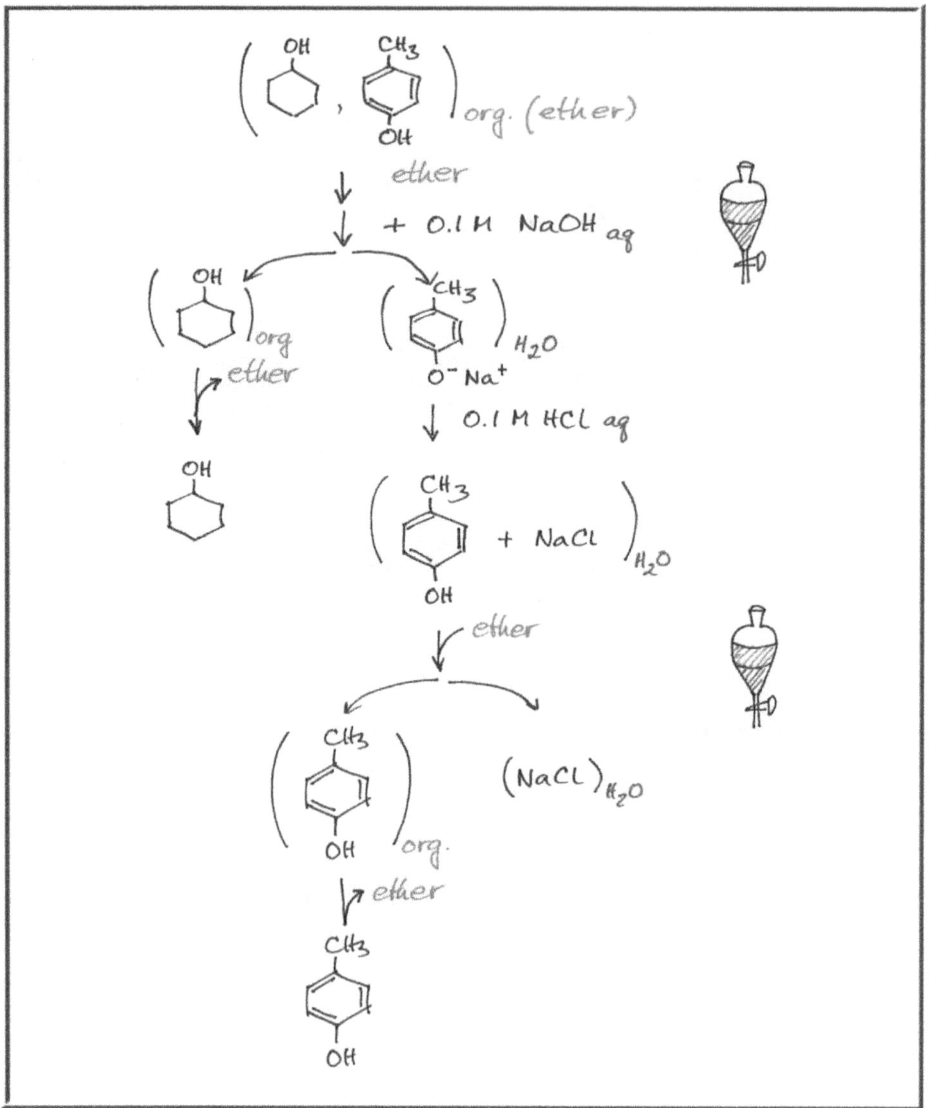


**Example Problem 12.3**
**Design a flowchart showing the conditions and steps for the three-component separation of benzoic acid, aniline and naphthalene.**

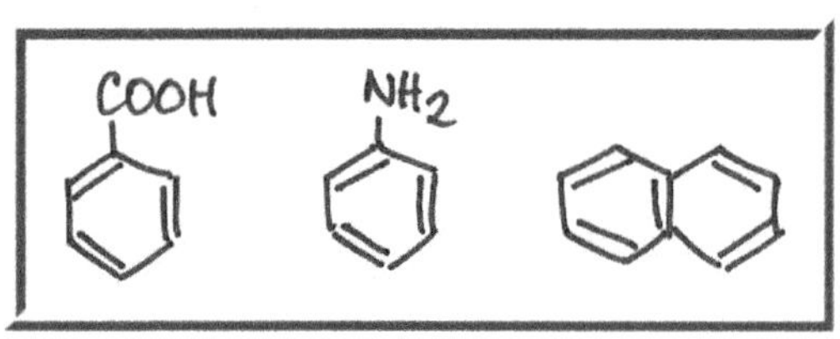

**Recrystallization techniques** exploit varying solubility of a solute in two different solvents at constant temperature or the same solvent at different temperatures. The first method is based on two different, but miscible, protic solvents, in one of which the chemical is well soluble such as methanol and the other one providing poor solubility such as water. The chemical is first completely dissolved in the "good" solvent and then practically titrated with the "bad" solvent until the solution turns slightly cloudy (turbid, opaque). This is the solvent ratio at which the specific molecule will selectively precipitate out. The idea is always to remove the selected molecule from contaminants and/or impurities. Water solubility is often much improved when the solution is heated (hot/cold-one solvent method). The solution is heated with just enough water to make it clear, insuring that all molecules are dissolved (supersaturated solution). Slow cooling at room temperature or placing the solution into an ice-bath will lead to the formation of crystals. Slow recrystallization will produce bigger crystals.

**Fig. 12.13**

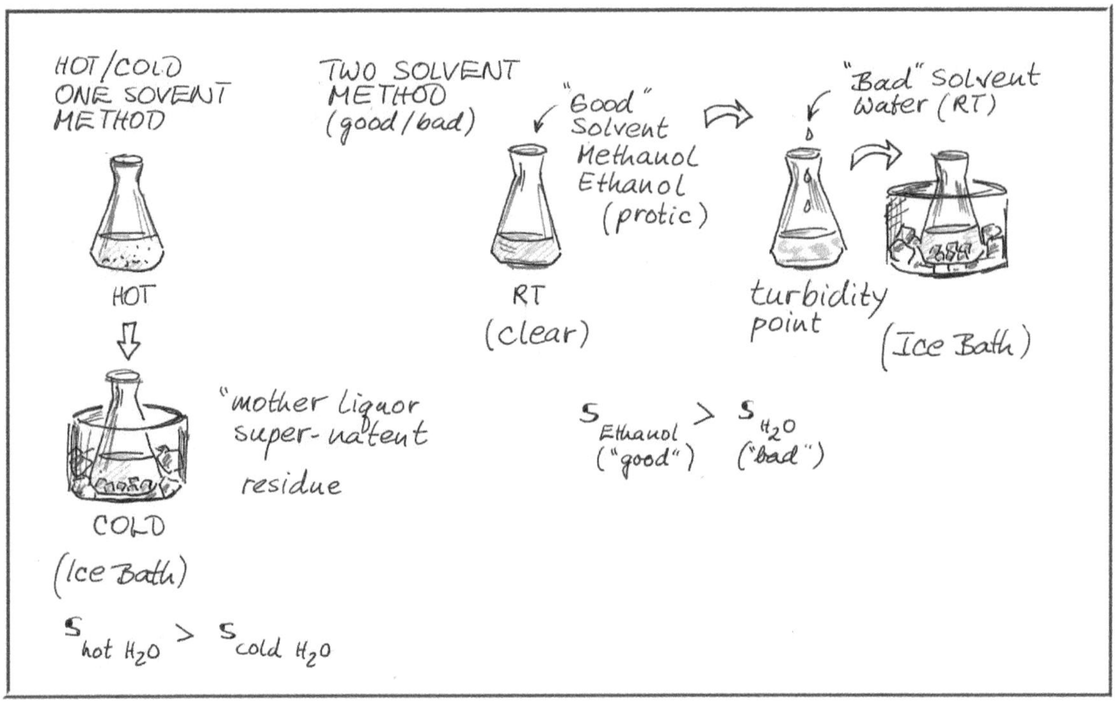


Recrystallization can be a finical process and sometimes can be initiated by "seeding" the solution with a tiny crystal of the same compound already at hand (if available) or rubbing the sides of the glass container with a glass-rod, where supersonic vibrations might help the recrystallization process along.

**Fig. 12.14**

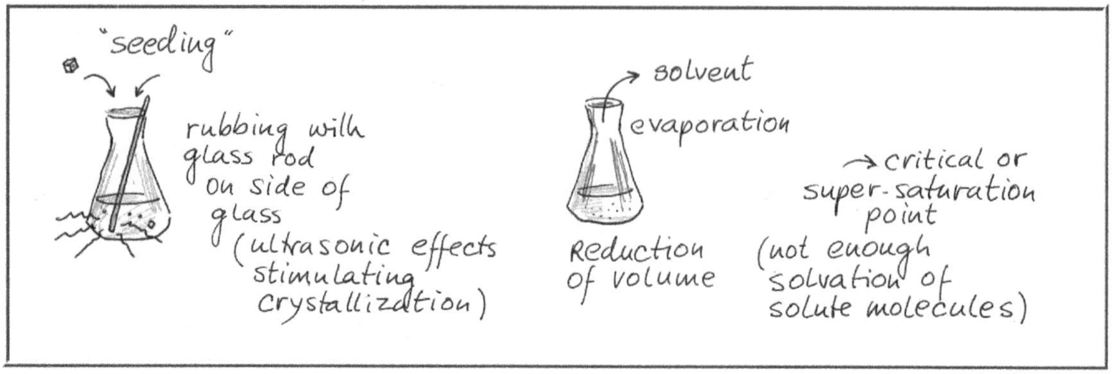


Besides normal gravity filtration depending of a regular glass funnel fitted with a conically folded filter paper which could become a long and tedious process, the much more efficient **vacuum filtration** has become a most common practice in the separation of solids from liquids. Depending on the type of funnel used, one distinguishes between the **Büchner** funnel for larger quantities and the **Hirsch** funnel for smaller amount to be separated by applied vacuum. Both funnels can be made up of porcelain or Teflon and come in different sizes. The Büchner funnel has a perforated platform that must be covered by a snuggly fitting round filter paper in order to hold back the residue. The Hirsch funnel is conically shaped and fitted with a permanent fritte, a type of mesh with a small pore-size of usually 20 microns made up of polyethylene, which can be used with or without filter paper. At times the fritte can removed for cleaning purposes or fitted with another disk of different pore-size.

**Fig. 12.15**

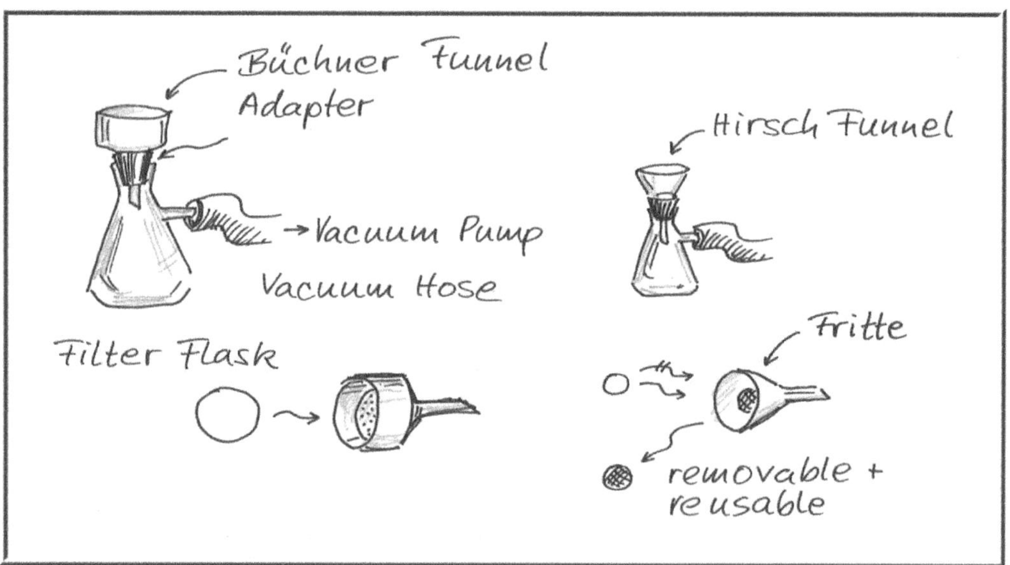

Chromatography is another type of separation based on partitioning within different phases, so it is also governed by chemical equilibrium reactions. For **liquid-solid chromatography**, the phases involved are liquids driving components through a solid. The phase on or within which portioning occurs is in this case usually a solid and referred to as the **stationary phase.** The phase that drives the partitioning forward is either a pure liquid solvent or a mixture of solvents (effluent/s) when dealing with liquid chromatography. However the so-called **mobile phase** can also be a gas as is the case for gas-chromatography.

Anybody who has seen a stained coffee filter or the ugly muddy stains of a stucco house after the rain will have an idea of what chromatography is all about. No material is solid throughout. Tiny pores, crevices and channels exist that can act like capillary tube able to absorb fluids and spread them against gravity throughout the material. Observations in natural color separations such as the **adsorbates** (soluble components) "bleeding" into an **adsorbent** material led to the term "chromatography", literally meaning "color writing".

**Fig. 12.16**

**Ascending chromatography** relies on **capillary forces**, in which adhesive attraction to the capillary walls exceeds the downward pull of gravity. Thin **layer chromatography (TLC)** belongs to this category. This technique is quick and efficient, allowing the chemist to monitor reactions in the laboratory, especially when multiple steps in a synthesis scheme are involved. The thin coating of the adsorbing material (adsorbents: silica gel $SiO_2$, cellulose powder, or alumina $Al_2O_3$), adhered to a thin, inert support material such as aluminum or plastic film with the help of an adhesive, only provides qualitative results.

**Fig. 12.17**

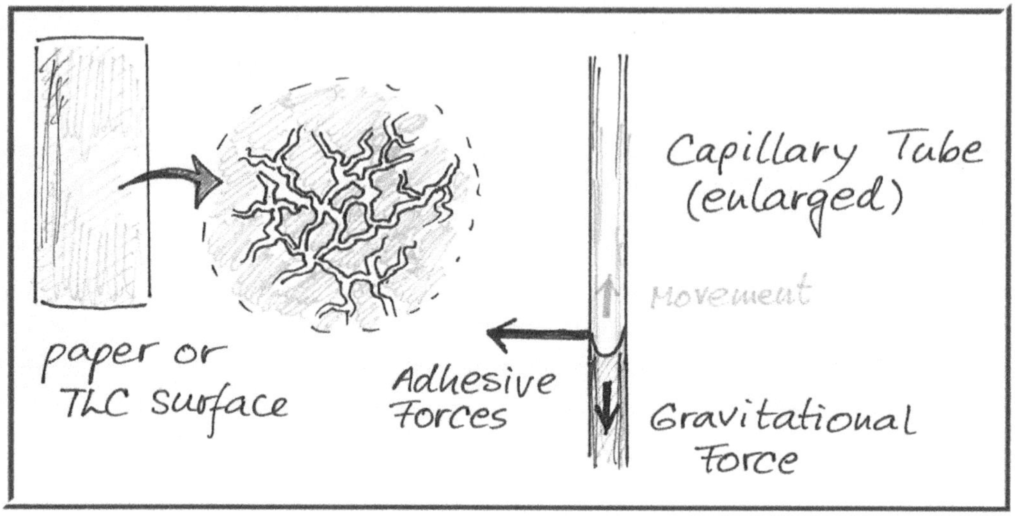


For quantitative or "preparatory" purposes a thick layer has to be used, from which the separated components can be selectively removed and eluted with a proper solvent, allowing further analytical investigation (spectroscopy, etc.). The chemical mixture is applied with the help of a micro-pipette (often a drawn out, open-ended capillary tube) as solution to the plate about 1cm from the bottom edge (start-line). After drying the spot(s), which ideally have a diameter of 3-5mm, the TLC-plate is placed into a developing chamber or tank, filled with developing solvent system (solvent mixture as effluent and mobile phase), well below 1cm height. The solvent mixture will drive the components of the solute mixture upward and thus separating them out on various positions below the solvent front (the "wetness" line), as function of their molecular structure and relative polar properties. The solvent front should never reach the upper edge of the TLC plate. Once the separation is complete, the plate is removed from the tank, dried and viewed under a UV lamp. Since most organic compounds are colorless and only a few fluoresce, TLC plates treated with a fluorescent indicator are available for better visibility. The plate then, when viewed beneath a UV lamp, shows up with dark spots on yellow-green fluorescent background. The chemical components will usually "quench" the fluorescence of the indicator background. With a pencil the dark spots can be encircled and later the center point of the encircled spots measured for $R_F$ value determination. Any marking should always be with soft pencil, since graphite does not "bleed" into the solvent such as inks.

**Fig. 12.18**

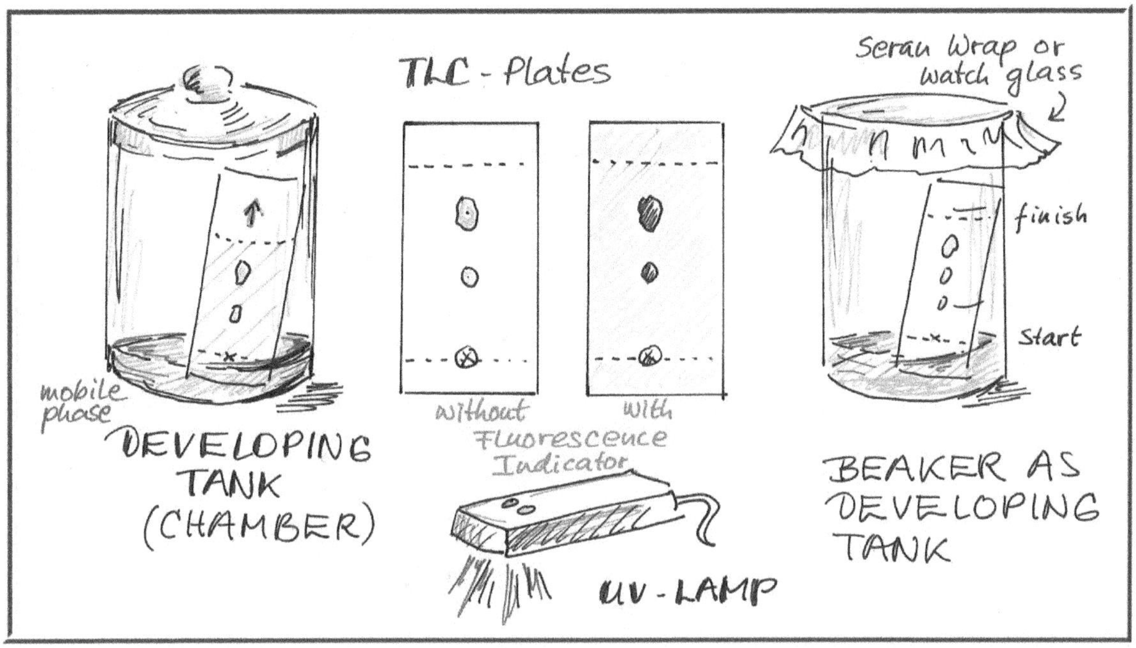


Common to all types of chromatography is the term **retention time ($R_T$)** and **retention factor ($R_F$).** The latter is mostly used for any type of layer chromatography and defined as the distance travelled by a component (seen as a spot on the chromatogram), divided by the distance of the travelled solvent front. The result is a numerical value between 0 and 1 without any units. If the conditions are given along with the value, the $R_F$ can be a valuable tool for the identification of an unknown, especially for forensic drug analysis.

$$R_F = \frac{d\,(spot)}{d\,(solvent\;front)}$$

Retention actually means how well the component or particular solvent is retained by the solvent system (effluent or eluent). Two forces compete with one another: the attraction to the solvent system (mobile phase) in a way of "drifting along" and the attraction towards material of the stationary phase as a type of staying power. The following image shows a rough vector analysis of the separation phenomenon, which can be applied to all other types of partioning chromatography.

**Fig.12.19**

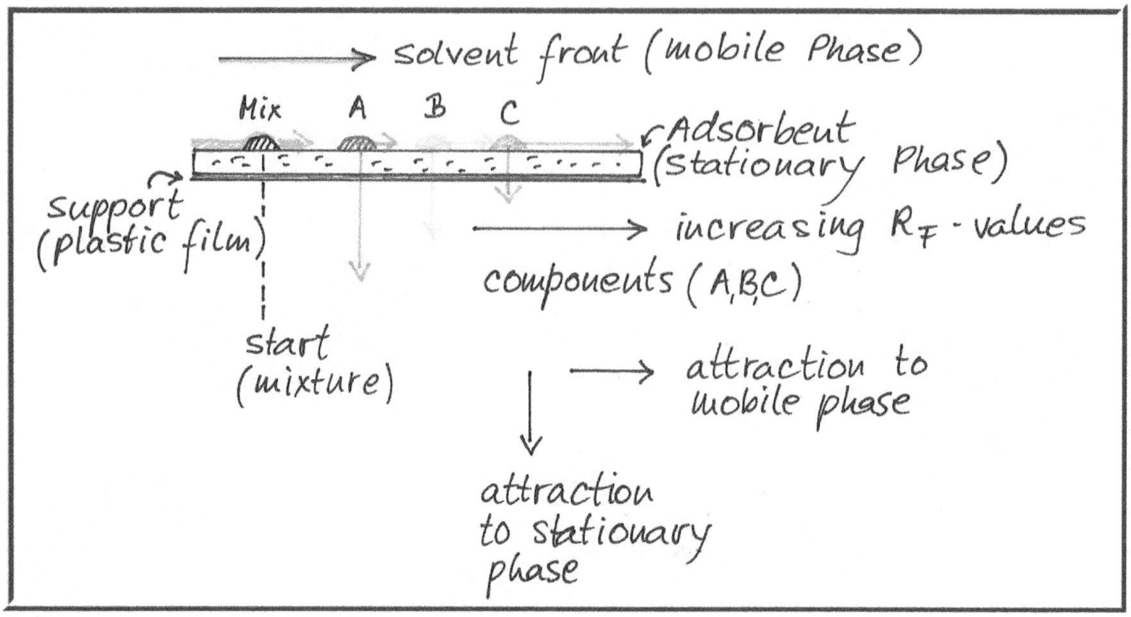


**Example Problem 12.4**
**Consider the following TLC chromatogram and calculate the $R_F$ values for the components A, B and C. Which is the most polar, which the least? Explain!**

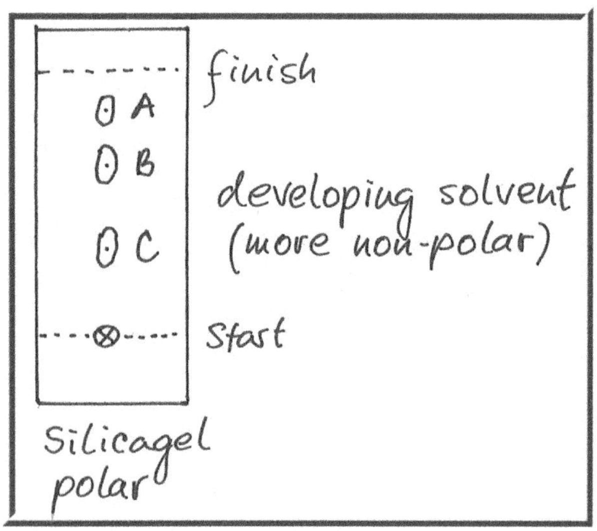


The other type of chromatography is **descending** in nature. Here gravity forces drive the chemical constituents downward through the adsorbing material, which can be made up of any porous material, usable in dry form or in a pre-wetted more or less thick paste (starches, resins, silica gel etc.). In the oldest form of chromatography, namely **paper chromatography (PC),** the porous paper can be used in the dry or pre-wetted state. Here the chemical component also has the choice of either being more attracted to the stationary phase or the mobile phase, ultimately involving the interplay of relative polarity between the three parties. Physical properties, however, can also determine the migration tendencies of a component such as drag. All interacting physical and chemical dimensions make a chemical unique in its chromatographic behavior. Sometimes a n adsorbent surface can be coated with a special chemical to change its polar properties to make it more non-polar. This technique is called **reverse phase chromatography.**

Chromatography in various forms involving partitions of a chemical component in different media such as column (**CC and HPLC – high performance or high pressure liquid chromatography**), **layer chromatography (PC, based on paper** and various **adsorbent** materials applied as thin or thick coats on different supports - glass, plastics - as **TLC**), **gas chromatography (GC)**, in which a gas partitions chemical components, can be carried out by specifically engineered instruments, automated and computerized. These techniques are helpful in the separation of mixtures and identification of specific components and/or impurities/contaminants. Sometimes different techniques are combined within the same instrument such as **GC-MS or HPLC-MS**, in which the separated components are directly processed and possibly identified by the fragmentation trail they leave behind.

Column chromatography represents a valuable technique for the physical separation of components, since solutions coming out of the column can be separately collected as fractions and isolated for further studies. Automated fraction collectors working like a rotating carousel are time-saving devices. Here the retention time plays an important role, since with the help of an automated fraction collector, equipped with an UV-sensor, the sequence of the eluted products as well as their relative amounts can be recorded graphically and computerized. The isolated fractions migrating the fastest and thus arriving earlier have the lowest retention times.

Fig.12.20

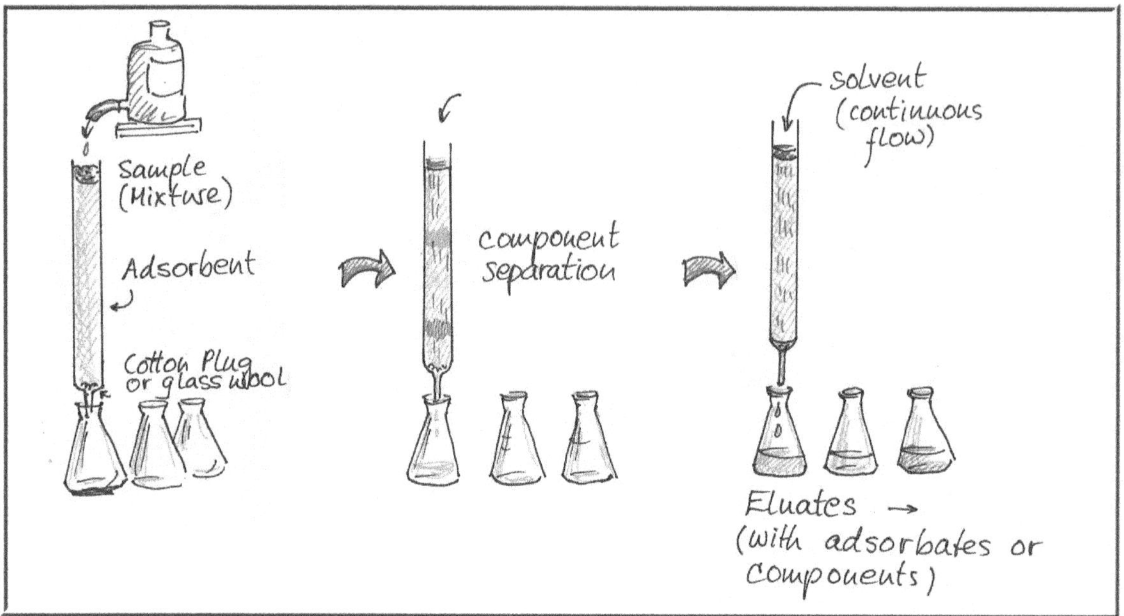


High performance or **high pressure liquid chromatography (HPLC)** allows separations within short time intervals. It usually consists of long thin columns, through which the mobile phase is pushed by outside pressure. Attached fraction collectors and detectors can monitor the separation process. The fractions can be diverted into a mass spectrometer and directly analyzed. The technique is more commonly used in biochemical laboratories, where it plays an essential role in protein chemistry.

Another more analytically used separation technique relies on gas as the mobile force in driving molecules apart. **Gas chromatography (GC)** as a form of **gas-liquid chromatography**, since the actual stationary phase is a viscous, high boiling liquid and the mobile phase is a gas. The stationary phase consists of a long thin, heated metal tube filled with solid, porous support material such as crushed firebrick, impregnated with a non-volatile liquid such as silicone oil, through which an inert carrier gas, usually helium or nitrogen, is pumped through by pressure. The carrier gas drives the heat vaporized molecules through the column. A detector at the end of the column will produce signals for a chart recorder or computer screen visible as a spectrum-like chromatogram in the order of the various components passing through the system with more or less retention times. Subtle changes of relative thermal conductivity of the carrier gas due to the effect of the vaporized components will be the ones registered.

**Fig.12.21**

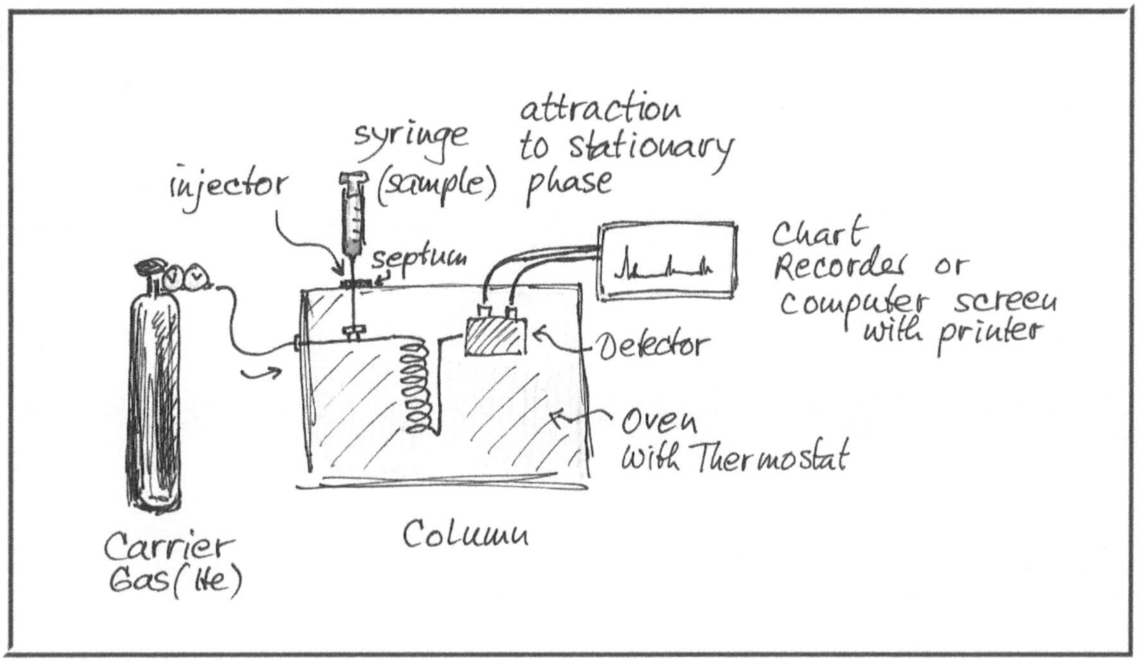


The technique is more reserved for volatile chemicals, which are, as solutions or as neat liquids, filled into a syringe and directly injected through a septum into the instrument. The technique is fully computerized and so sensitive that its resolution power allows the separation of chemicals with minute boiling-point differences. The layout of a GC-system and the chromatogram of the four separated isomers of butanol molecules are shown below:

**Fig. 12.22**

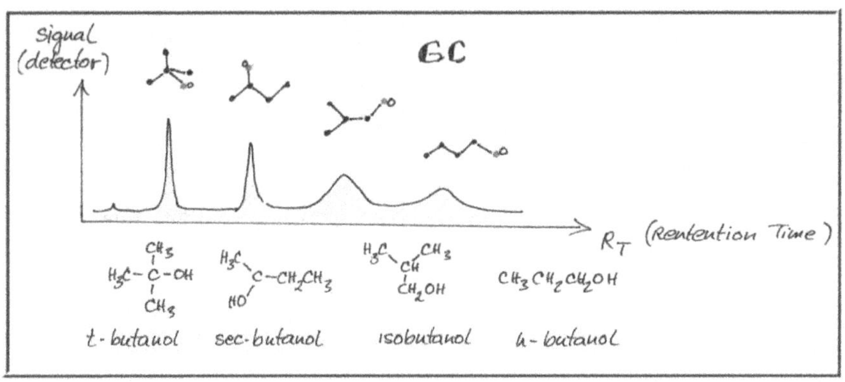


**Example Problem 12.4**

**Considering the sequence of retention times for the gas-chromatographic separation of the four butanol isomers, why would *tertiary*- butanol come out the fastest?**

Analogous to HPLC-MS, gas chromatography can also be coupled with MS analysis, allowing the swift separation and characterization of volatile materials.

The following image shows in overview of the various techniques used for the separation and isolation of a specific target molecule:

**Fig. 12.23**

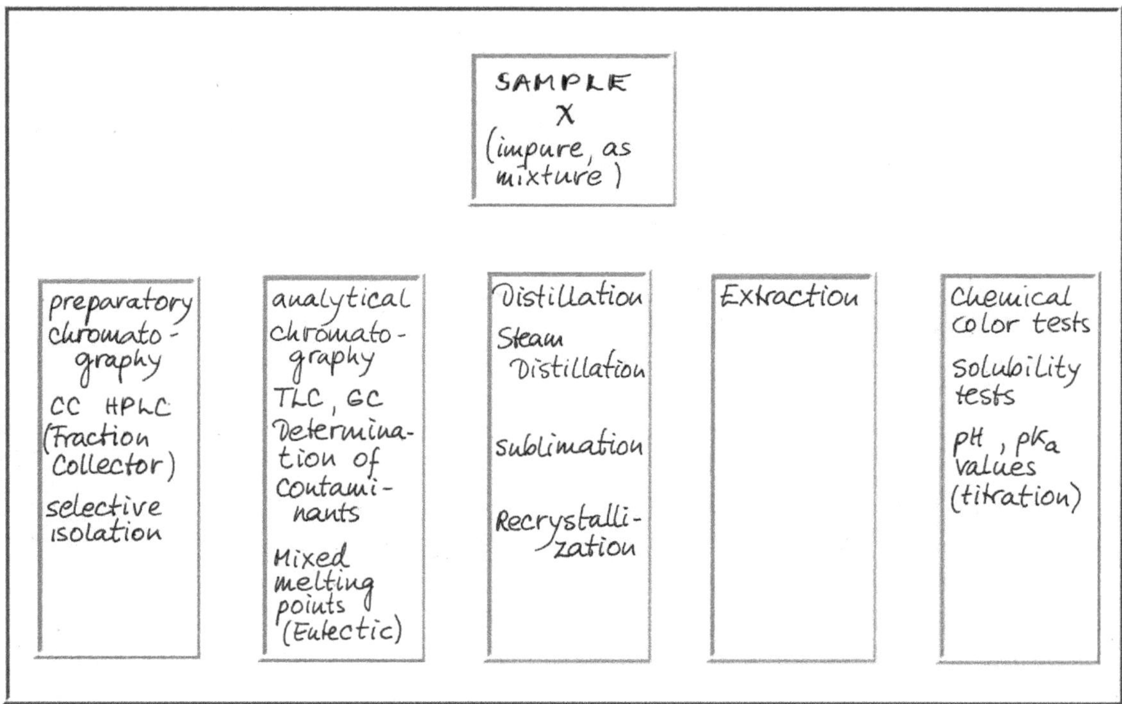


Chemicals can also be separated according to charge. But this technique is more commonly applied in biochemistry, especially for the separation of amino acids, peptides and proteins. **Electrophoresis** is thoroughly discussed in chapter 10 of the corresponding **Biochemistry text ("Topics in Biochemistry, B. Heinz,** published by **Kendall-Hunt, 2013).**

Once a chemical is separated and purified it can undergo further structure analysis applying the methods underlined and explained in the next chapter.

# EXPERIMENTS – Ch. 12

## *Separation Techniques*

**Key Terms and Concepts:** Partitioning of molecules, physical phases, adsorption, adsorbate and adsorbent, elution, eluents and effluents, retention times or factors, stationary and mobile phases, distillation, azeotropic mixtures, boiling point and vapor pressure, number of theoretical plates, fractionating columns and fractional distillation, boiling lens, extraction, distribution coefficients ($K_D$), drying agents, multiple extraction, chromatography (CC, HPLC, TLC, GC etc.), filtration (gravity, vacuum assisted – Büchner ), developing tanks/chambers and solvents, elutropic series

### I Thin-layer Chromatography of Analgesic Drugs

#### Chemicals:

*Ethanol*
*Aspirin, Advil, Tylenol, Ibuprofen, Excedrin*
*Naproxen, Acetaminophen, Caffeine*
*Developing solvent (95% ethylacetate/5% acetate)*

#### Equipment and Materials:

*Test tubes and vials; mortar and pestle*
*TLC plates with fluorescent indicator*
*Developing chamber and micropipettes*
*UV-lamp*

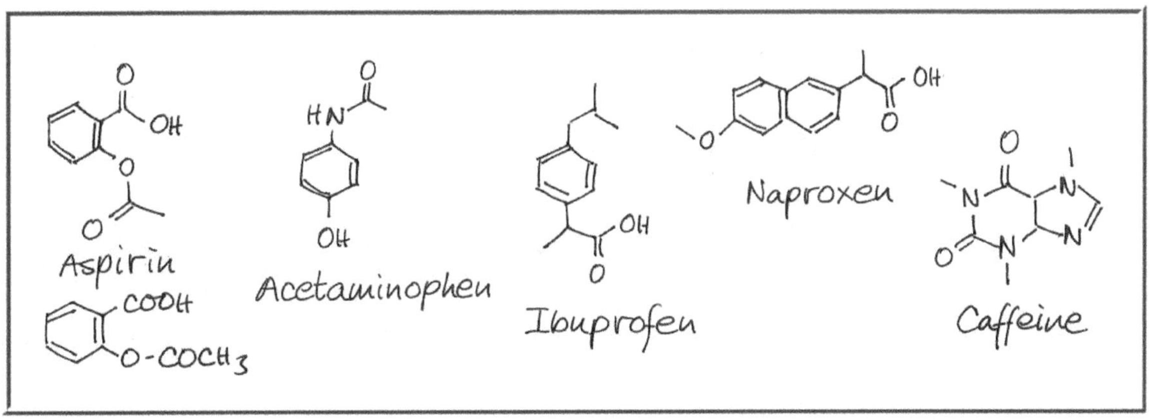


#### Procedure:

1. Prepare the test solutions by dissolving a crushed tablet or the contents of a capsule with sufficient amount of ethanol. Store the solutions in labeled vials. This can be done for the whole class. Caffeine (either pure or from "Nodoz" tablets) is used as a reference.
2. Prepare two TLC plates, each spotted with 4 different drugs.
3. When dry, place the two plates back to back, only touching at the top of their rims into the developing chamber.

4. After developing and drying the plates, view under the UV lamp in a darkened room and circle the observed spots lightly with a regular No. 2 pencil.
5. Calculate the retention factors ($R_F$) of all the components observed

## Observations and Results:

| Name of Drug | Constituent chemicals | $R_F$ – values of components |
|---|---|---|
| | | |
| | | |
| | | |
| | | |
| | | |
| | | |
| | | |

### Questions and Problems:

1. Evaluate the constituents of each drug tested. Where the drugs homogeneous or heterogeneous?
2. Can the intensity of the spots give information about the relative quantities of different constituents used for the drug's formulation?

## II Column Chromatography of Printer Inks and Dyes (Isolated from Cartridges)

### Chemicals:

*Printer inks and dyes*
*Ethanol*
*Silica gel*
*Elution solvents: petroleum ether, ethyl acetate etc.*

### Equipment:

*Printer cartridges*
*Chromatography column or glass tube with narrow opening,*
*Stand and clamps*
*Rubber hose, pinch or screw clamp*

*Glass wool or cotton*
*Separatory funnel or glass funnel*

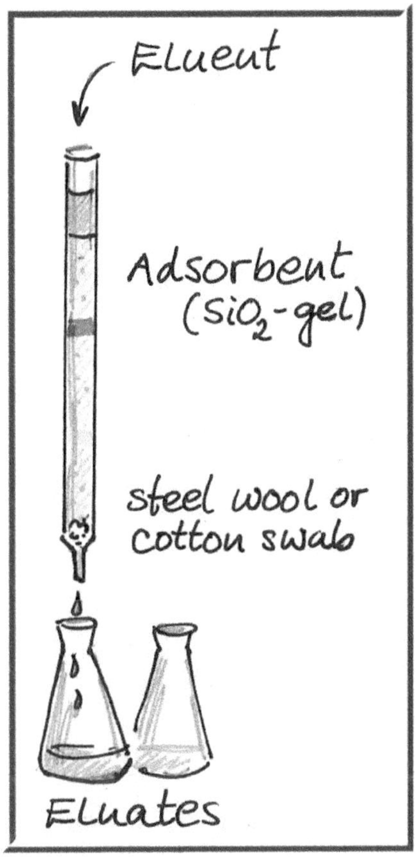


**Procedure:**

1. Prepare the chromatography set-up according to the images shown.
2. Prepare the silica gel slurry with petroleum ether (mixture of isomeric pentanes) and fill into the column (bottom opening closed). The column can also be filled with dry silica gel and then the first eluting liquid poured through.
3. Drain out the liquid but do not run dry. Let the liquid layer sink in a bit from the top, then add the sample solution carefully on top of the adsorbent layer with a Pasteur pipette or dropper, let sink in just barely below the surface and finally pour more eluting solvent over the column.
4. Observe the separation tendencies. If the components are not separating well enough use more polar eluents such as ethyl acetate.
5. Drain the individual components into receiving vessels such as Erlenmeyer flasks.
6. Identify the different bands by color and relative retention times.

**Observations and Results:**

| Type and make of cartridge | Color: | Order of separation | Relative retention times ($R_T$) | Type of compound, Name, if available |
|---|---|---|---|---|
| | | | | |
| | | | | |
| | | | | |
| | | | | |

**Questions and Problems:**

1. Evaluate the chemical make-up of the inks used in the various printer cartridges. Where the inks homogeneous or heterogeneous? Compare ink compositions of different brands, tested by other student groups.
2. What is the difference of dyes and pigments when referred to in printer terminology? There are printers based on dyes and inks.
3. Evaluate the relative polarity of the different ink components.
4. What is the chemical nature of the inks used in the printer industry. Are they more the same or different? (Research)
5. Some prints coming freshly out of the printer smell like acetic acid or vinegar-like. What solvent is usually used in ink-jet printers? (Research)

## III Radial Paper Chromatography of Ink and Marker Pens

### Chemicals:

*Inks from pens and markers*
*Developing solvent (95% ethyl acetate, 5% sodium acetate*

### Equipment:

*Round filter paper*
*Petri dishes with lids*
*Open-ended capillary tubes*

**Procedure:**

1. Prepare a round filter (10 -12 cm diameter) paper for each color of ink or marker pen used.
2. Mark the center of the round filter with a dot of the ink to be tested (0.5-1.0 cm) and cut a tiny hole in the center (0.2-0.3cm).
3. Prepare the petri dish by filling it 1-2mm high with developing solvent.
4. Prepare a small cone out of adsorbent paper (filter or chromatography paper), place it in a center position into the developing solvent and attach the round filter paper onto the tip of the cone (like a wheel on an axle).
5. Cover the petri dish with its lid and allow the solvent front to migrate close to the rim.
6. Remove the round chromatogram and let dry
7. Evaluate the chromatograms.

**Observations and Results:**

| Type and make of Ink Pen or Marker | Color of Ink | Number and Color of components of separation | Order of separation and Relative retention times ($R_T$) | Type and Name of compound, if available |
|---|---|---|---|---|
| | | | | |
| | | | | |
| | | | | |
| | | | | |

**Questions and Problems:**

1. Characterize the chemical make-up of the inks used. Where they homogeneous or heterogeneous.
2. How do the inks of pens and markers compare with inks of printers?
3. What solvents are usually used for pen and marker pens? Characterize any odors coming from the pens.
4. Compare the relative polarity of the tested inks.

The solvents for different types of chromatography can be selected from the elutropic series, which lists the solvents with increasing polarity. The concept "like dissolves like" helps in the selection of adequate solvent systems to reflect the nature of the solutes.

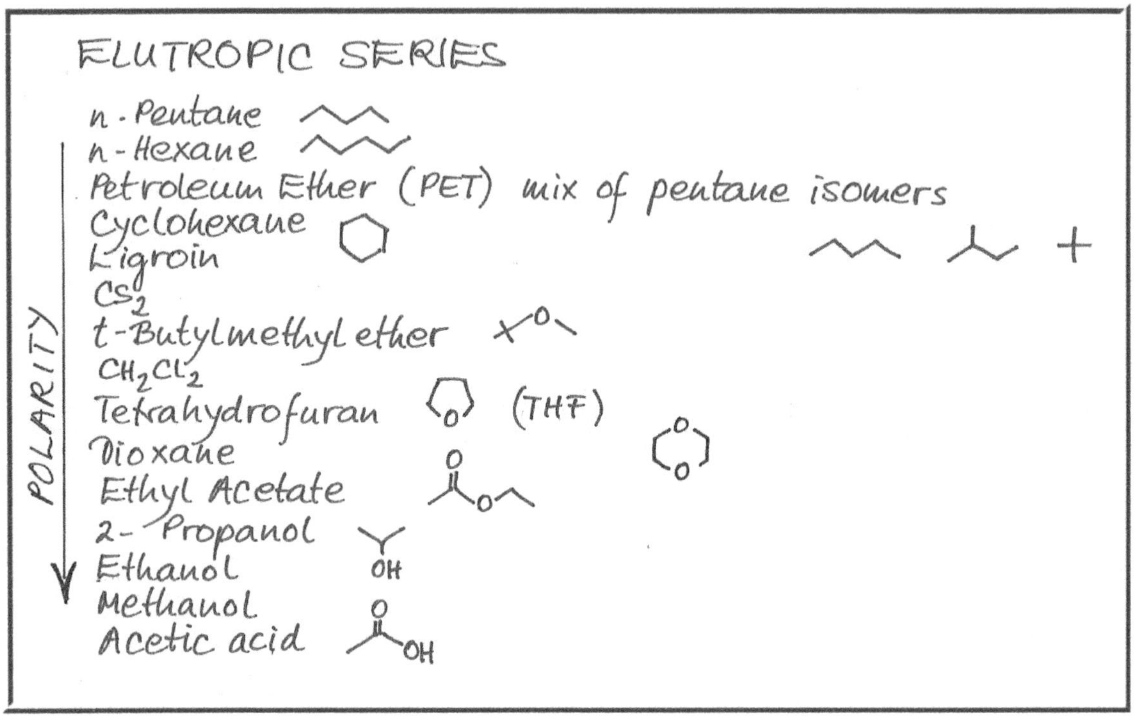


## IV Steam distillation of Essential Oils

### Chemicals:

*Plants, spices or herbs*
*Water (demineralized)*
*Dichloromethane*
*$CaCl_2$ (drying agent)*

### Equipment and Materials:

*Mortar/pestle or blender, scissors or knife*
*Distillation apparatus without thermometer,*
*Separatory funnel, rotavap*
*Brown bottles with label*
*Graduated cylinder*

### Procedure:

1. Cut up the plant material, grind down with mortar and pestle or homogenize in an electric blender.
2. Place the plant material in a large round bottom flask, fill half full with water, and set up distillation apparatus according to the following image:

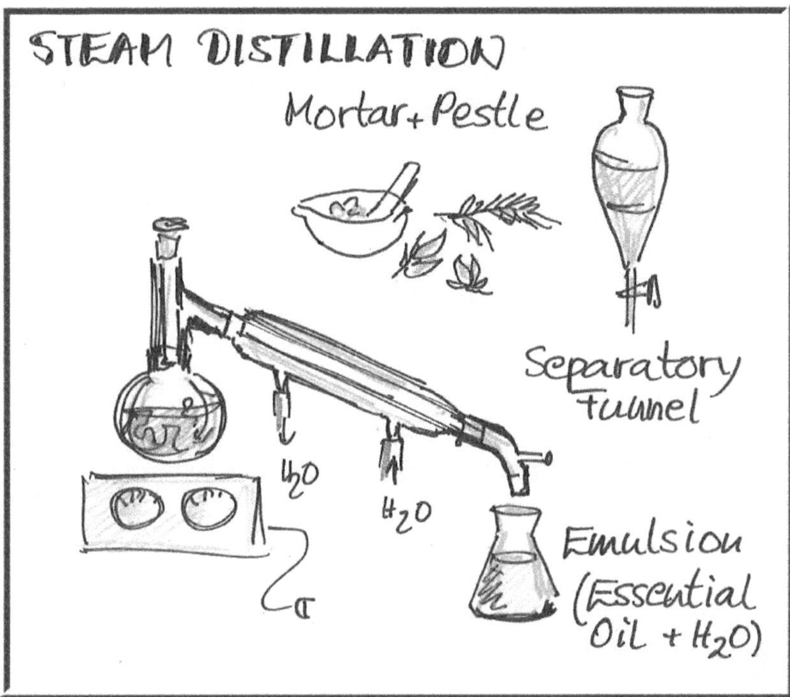


3. Heat the solution over a hot plate or heating mantle to a steady boil.
4. Collect the emulsion which could be more or less opaque, in the receiver and transfer to a separatory funnel.
5. Extract two to three times with dichloromethane (methylene chloride) and combine the organic fractions.
6. Dry the extracts with $CaCl_2$ for 5-10 minutes, decant and transfer through a fluted filter into the conical flask of a rotavap.
7. Reduce volume by distilling off the solvent.
8. Transfer the oily residue into a brown bottle and allow the rest of the solvent to air-evaporate.
9. Determine the volume of the essential oil obtained from the source material and save for measuring the refractive index for exp. 13.

## Observations and Results:

| Plant material: | Color and consistency of distillate: | Aroma of plant material: | Aroma of isolated essential oil and amount obtained (mL): | Possible chemical ingredients: |
|---|---|---|---|---|
| | | | | |

| | | | | |
|---|---|---|---|---|
| | | | | |
| | | | | |

## Questions and Problems:

1. What is an essential oil? Do all plants produce essential oils? Explain!
2. How do essential oils compare with perfumes?
3. Can perfumes be collected via steam distillation? What are some better methods for sensitive aroma molecules?
4. Compare some prices for some essential oils (research).
5. How are essential oils used in everyday life, in medicine and other areas?
6. What role does water play in the process of steam distillation?
7. What is an emulsion? How can it be cleared?
8. Why are essential oils soluble in organic solvents and not in water?

## V Extraction of Caffeine from Coffee via Sublimation on a "Cold Finger"

**Chemicals:**

*Instant Coffee*

**Equipment and Materials:**

*Sublimation Apparatus*
*Hot plate, crushed ice, spatula*
*Vial*

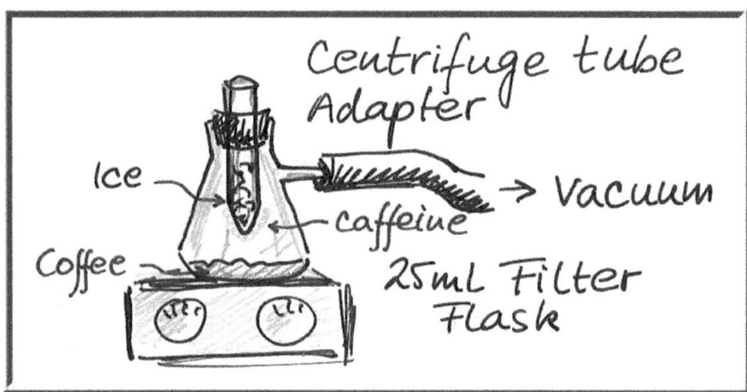


## Procedure:

1. Weigh out 1-2 scoops of instant coffee.
2. Place in the filter flask, insert the cold finger filled with crushed ice and position it above the coffee powder or grains.
3. Attach the vacuum and start heating the coffee.
4. Wait till a white powder settles onto the outer wall of the cold finger.
5. Scrape off the sublimed caffeine and weigh the amount recovered from the coffee.

6. Save the crystals and place in a labeled vial.

## Observations and Results:

| Sample: | Amount of coffee used (g) | Recovered caffeine (g) | Structural formula of caffeine: |
|---|---|---|---|
| *Instant coffee* | | | |

## Questions and Problems:

1. What is sublimation?
2. Does caffeine have a melting point if it sublimes? 3.
3. How can identity of caffeine as a molecule be verified?
4. How is decaffeinated coffee prepared technically?

# VI Separation of α- and β-Pinene found in Turpentine by Gas Chromatography (GC)

## Chemicals:

*Turpentine*
*Alpha and beta pinene*

## Equipment:

*Gas chromatograph*
*Hypodermic syringes*

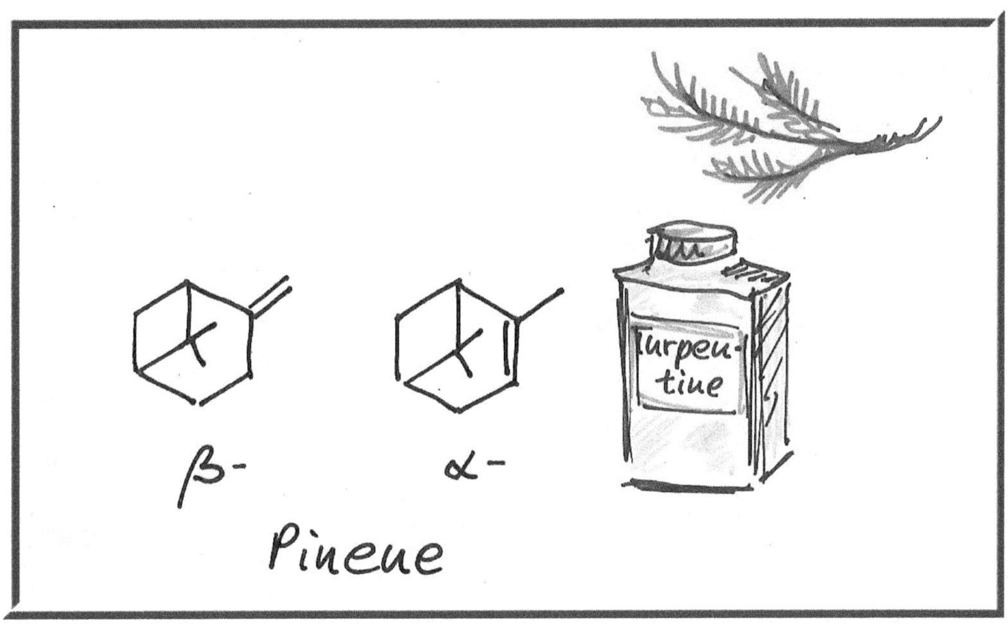

## Procedure:

1. Turn on the gas chromatograph and wait until the oven as reached its operating temperature.
2. Fill the hypodermic syringe with 1-25 microliters (μL) with the liquid sample and inject the contents carefully through the rubber septum of the intake valve.
3. Test turpentine by itself and then an equimolar mixture of α and β pinene.
4. The samples pass quickly through the GC column and by the detector, so a printed diagram of the spectrum is almost delivered simultaneously.
5. Evaluate and differentiate the gas chromatograms of the two tested sample.

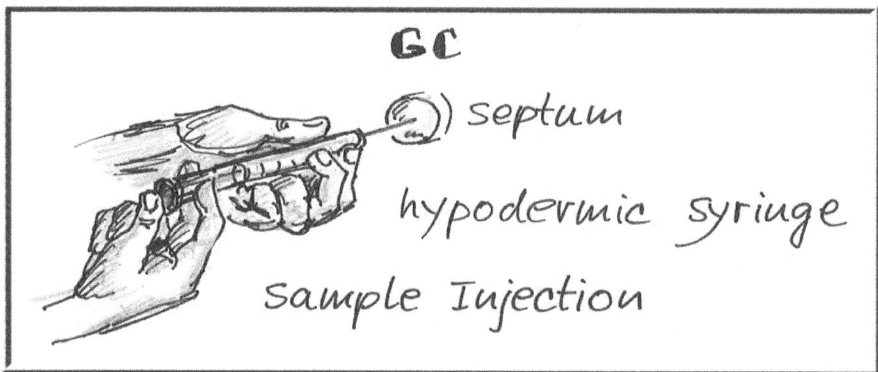


## Observations and Results:

| Sample: | Components: | Retention time ($R_T$) |
|---|---|---|
| *Turpentine* | | |
| *α,β - pinene mixture* | | |

## Questions and Problems:

5. How many peaks were observed for turpentine?
6. Turpentine is derived from pine trees. Could the two pinene isomers be detected in the sample? If so were the retention times similar to the pure sample?
7. Why do the two configurational isomers of pinene have different retention times? Use structural considerations for explain the difference.

# 13 Analytical Methods

## Chapter 13 Analytical Methods

Besides elementary analysis of an organic compound in question, which gives the percentages of the elements involved from which an empirical formula (lowest molar ratio formula) can be derived and the determination of conversion points, melting and boiling point (Mp and/or Bp), physical analytical methods are most commonly applied. Usually they are of the spectroscopic sort such as ultraviolet/visible (UV/vis), fluorescence and phosphorescence, infra-red (IR, FTIR), Raman, nuclear magnetic resonance (NMR) instrumental devices. Chromatography in various forms involving partitions of a chemical component in different media such as column (CC and HPLC – high performance or high pressure liquid chromatography), layer chromatography (PC, based on paper and various adsorbent materials applied as thin or thick coats on different supports - glass, plastics - as TLC), gas chromatography (GC), in which a gas partitions chemical components. These techniques are helpful in the separation of mixtures and identification of specific components and/or impurities or contaminants. Sometimes combinations of different techniques are determined by the same instrument such as MS-GC or MS-HPLC. Mass spectroscopy (MS) analyses a single, purified compound by ionic fragmentation and passing the fragmental ions through a curved passage affected by magnetic fields. With the help of a detector at the end of the "race track" the incoming ions are separated according to their m/e ratios and graphed. When combined with a type of chromatography, the eluted components will be directly passed through a mass spectrometer and analyzed. This process saves much time and is of great value where a fast turn-around is of the essence such as in forensic science and medicine.

Optically active molecules are analyzed by polarimeters, optical instruments that measure the angle of observed rotation $\alpha$, which can be converted mathematically into the specific angle of rotation $[\alpha]$. This method is discussed in chapter 9, dealing with stereochemistry.

Properties in circular dichroism and the measurement of the refractive index lead to valuable molecular characterizations as well. Of special significance is refractometry, which allows closer identification of liquids, which will be discussed at the end of this chapter.

The most commonly applied spectroscopic methods for the elucidation of molecular structure are based on the effects of:

(1) Electromagnetic energy (hν, photons) on bond movements (IR, FTIR, Raman)
(2) Electromagnetic energy (hν, photons) on electron transitions (UV, UV-vis, fluorescence, phosphorescence)
(3) Magnetic fields on nuclear spin alignment ($^{1}$H-NMR, $^{13}$C-NMR)
(4) Electronically induced fragmentation of molecules into ions of different mass, variously deflected within a magnetic field (MS)

**Infrared spectroscopy** has become one of the most important spectroscopic methods for the organic chemist. It relies on the mobility of the covalent bonds connecting two atoms. Molecules are not only in constant motion as a whole but also each and every bond that makes up its molecular framework is subject to some kind of kinetic activity. Bonds can vibrate, oscillate, bend, rotate etc., each activity leaving a tell-tale signal in the microwave and infrared region of the electromagnetic spectrum behind. A detector can convert specific signals, unique for every part of the molecule and transfer the information to a recorder or graphic program in the computer so a spectrum of a combination of peaks with varying sharpness and intensities typical for the molecule investigated can be obtained and studied. A linear three-atom molecule, for instance, can mainly undergo stretching (linear) or bending (angular) vibrations. The oscillations can be in or out of plane, symmetrical or asymmetrical.

**Fig. 13.1**

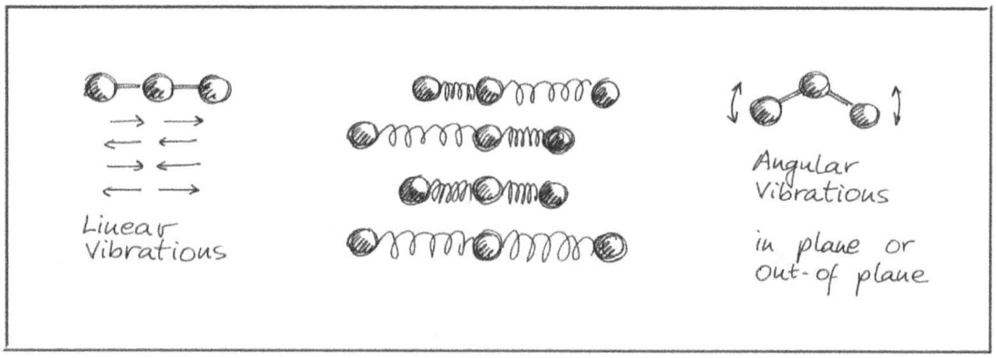


The main types of bond movements are shown in the figures below:

**Fig. 13.2**

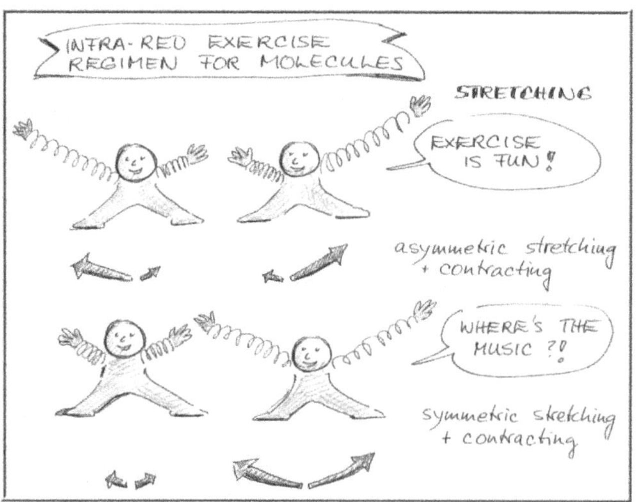

**Fig. 13.3**

Homo-diatomic molecules like oxygen are transparent to IR analysis and show no signals. The more symmetrical a molecule is the weaker the signal becomes for the region that represents the point or plane of symmetry within a molecule. Their vibrations are electrically symmetrical and cancel each other out. Other substances such as glass are opaque towards IR radiation.

**Fig. 13.4**

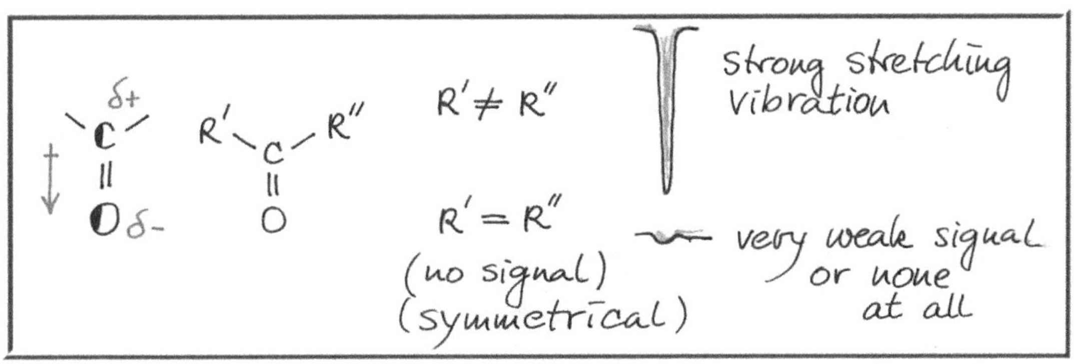


The more symmetry is found around a bond in question the weaker the signal becomes. Asymmetric surroundings of a bond on the other hand give quite intense signals. Purely symmetrical molecules as already stated above give no signals at all. Also, strong pi-bond polarization such is the case for the carbonyl group (-CO-) give stronger signals than weakly polarized bonds.

**Fig. 13.5**

Samples are usually ground up with NaCl or KBr and pressed into a pellet or tablet that is then inserted between the IR source and detector for measurement. This is still the only viable method for chemicals with poor solubility. These salts transmit the IR photons but do not interfere with the signal. So as a glass like substance they act as an inert medium or solvent for the IR-active analyte or solute. Chemicals with good solubility in volatile organic solvents can be directly placed on a specific film card, which contains a circular window of a transparent polymeric film. After evaporation of the solvent the card can be placed into a card slot between the IR beam and detector of the instrument. However, the background spectrum of the polymeric film material must be cancelled out to receive a usable spectrum for the molecular sample.

**Fig. 13.6**

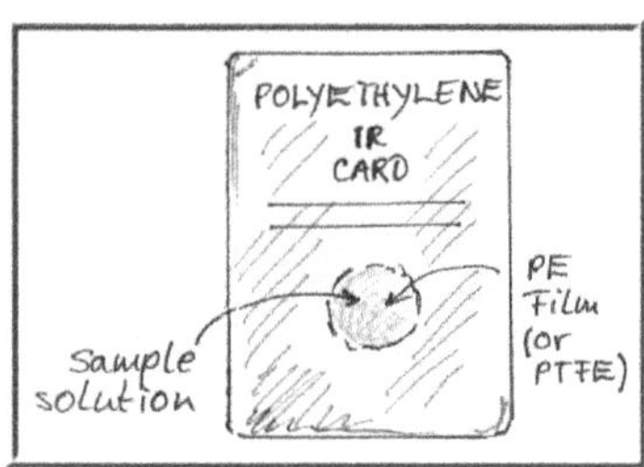


Most stretch-vibrations occur in the high frequency region with **wavenumber ($\bar{\nu}$)** between 3600-1200 cm$^{-1}$. A reason for the near IR absorption is that more energy is needed for stretching than bending of covalent bonds. The mass of the connected atoms can facilitate or impede bond mobility. Single bonds connected to hydrogen atoms (O-H, N-H, C-H) are easy to activate, thus causing higher vibrational energies. Carbon–bromine (C-Br) or carbon–iodine (C-I) bonds vibrate so sluggishly that they don't even show up in the spectrum. Bond lengths and strength also influence vibrational frequencies. Multiple bonds make stronger springs and therefore their vibrational energies are at higher frequencies than single, non-hydrogen bonds such as C-O. The triple bond frequencies (C≡N, C≡C) are higher than the double bond frequencies (C=O, C=C).

The most important molecular environments can be separated into individual regions or segments of the molecule such as purely saturated hydrocarbon fractions, H-bonding groups with heteroatoms, multiple-bond groups with or without heteroatoms. Even signals based on the molecule as a whole vibrating entity with absorption bands in the far IR region (downstream), often referred to as the **"fingerprint" region** can be obtained.

The specific group vibrations are of course influenced and modified by their immediate chemical or electronic environment, but usually they are to be found in a consistent part of the spectrum. The carbonyl group as one of the groups producing one of the strongest absorption bands is usually found in the vicinity of wavenumbers 1815-1650 $cm^{-1}$. Depending on the number of atoms and whether a simple molecular is angular or linearly shaped the number of possible vibrational modes can be determined by a simple mathematical calculation as shown below:

**Fig. 13.7**

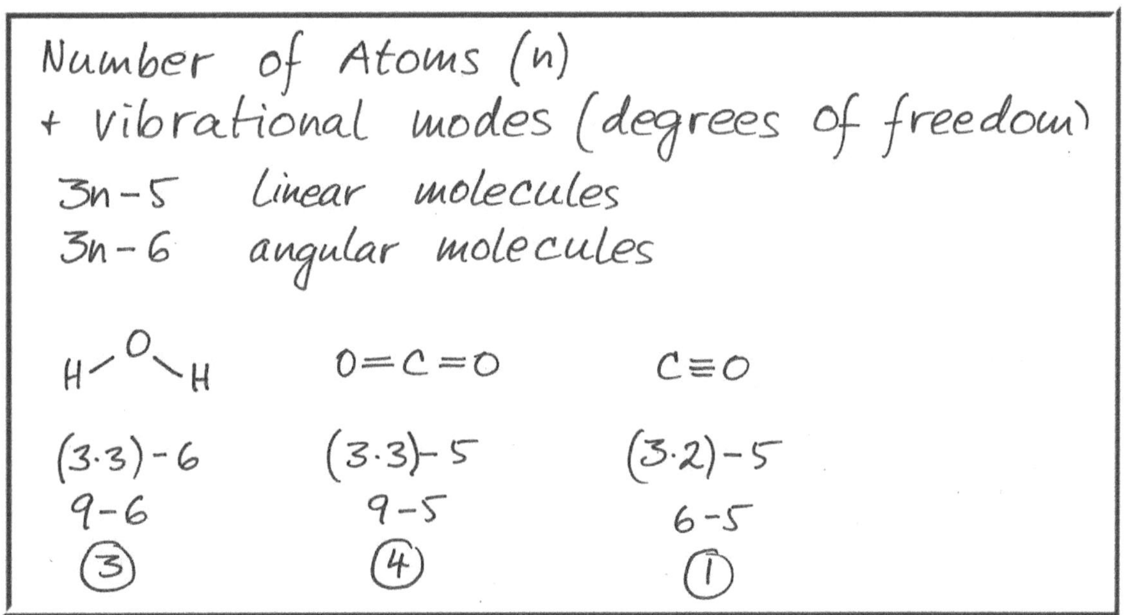


The main signal regions for specifically selected groups are represented in the scheme below.

**Fig.13.8**

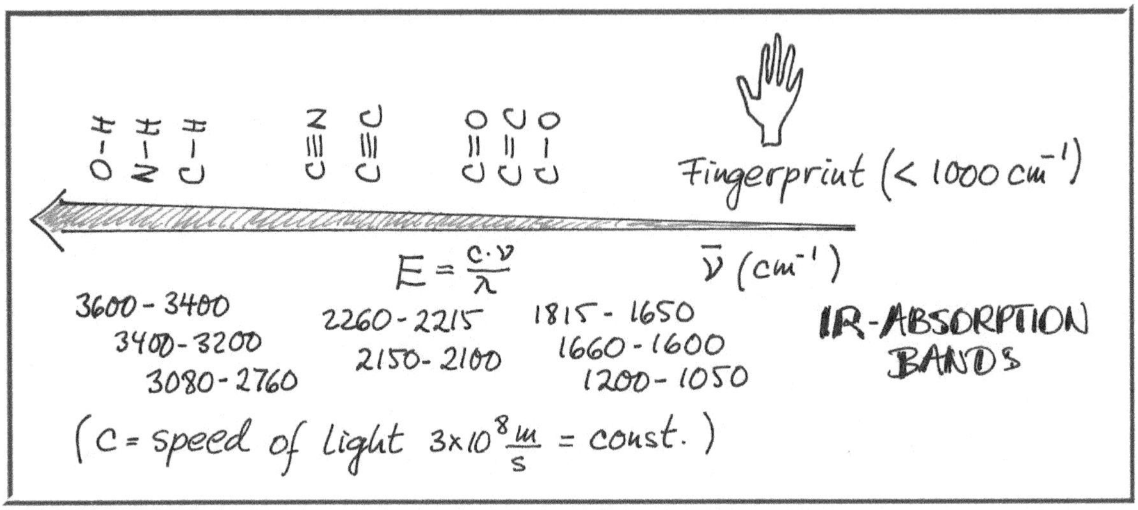


A typical spectrum, representing a specific, unidentified organic molecule is shown below:

**Fig. 13.9**

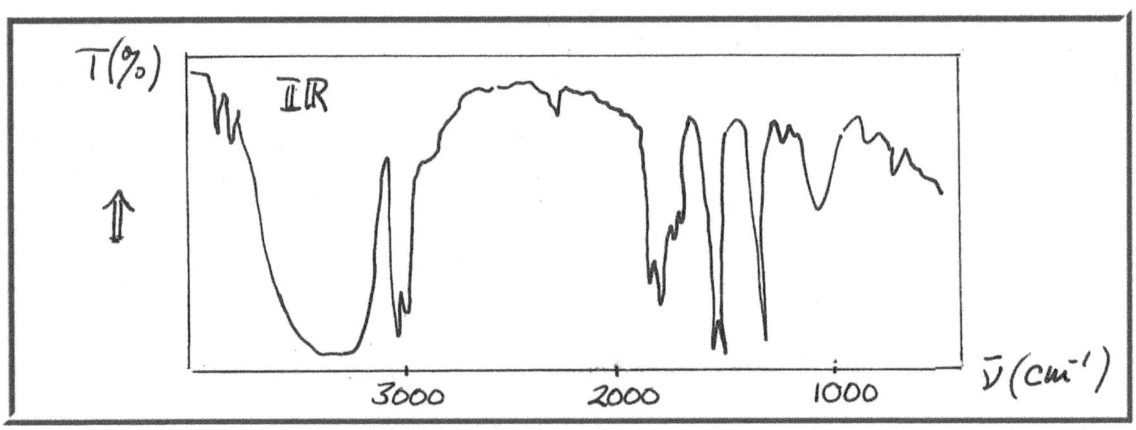


Analytically the whole molecular structure cannot be completely ascertained, but structural elements within the molecule can be verified with a considerate amount of certainty. **Fourier Transfer Infrared spectroscopy (FTIR)** is now the most commonly used computerized method in the laboratory. Fourier transfer analysis is a mathematical method that allows many digitized scans at once, converting the digitized signals via computer into a usable spectrum based on frequencies, which are given in Hz (cycles per second). However, the actual spectra are shown in the wavelength (λ) unit **micron** (μm, $10^{-6}$m) shown on top of the spectrum and a type of modified frequency (ν) unit, the **wavenumber** $\bar{\nu}$ ($cm^{-1}$) shown in the scale below the spectrum. As infrared radiation passes through the sample, the bonds characteristic to the molecule will selectively absorb their specific vibrational and rotational energies, which will be reflected as absorption bands in the spectrum. IR-spectra usually range from 4000 $cm^{-1}$ (2.5μm) as the high end energy range to 500 $cm^{-1}$ (20μm), the low end energy range.

**Example Problem 13.1**
**Which of the following IR-spectra represents a carboxylic acid, which a primary alcohol? Explain!**

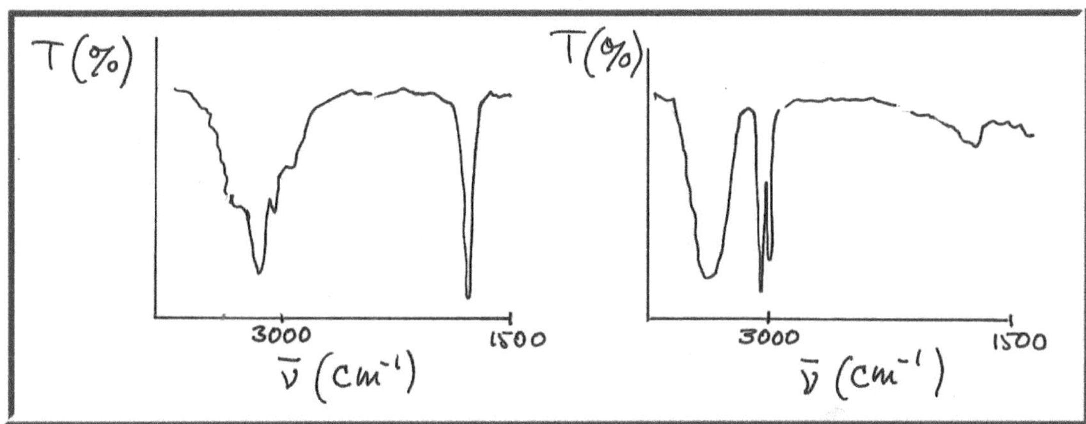


Another important type of spectroscopic analysis, namely **nuclear magnetic resonance spectroscopy (NMR)** can shed more light on molecular investigations. Here the energy comes from the alignment of the normally randomly orientated nuclear magnetic spin vectors of the hydrogen atoms with an outside magnetic field in either parallel or antiparallel fashion. These alignments create changes in energy ($\Delta E$) that can be measured.

**Fig. 13.10**

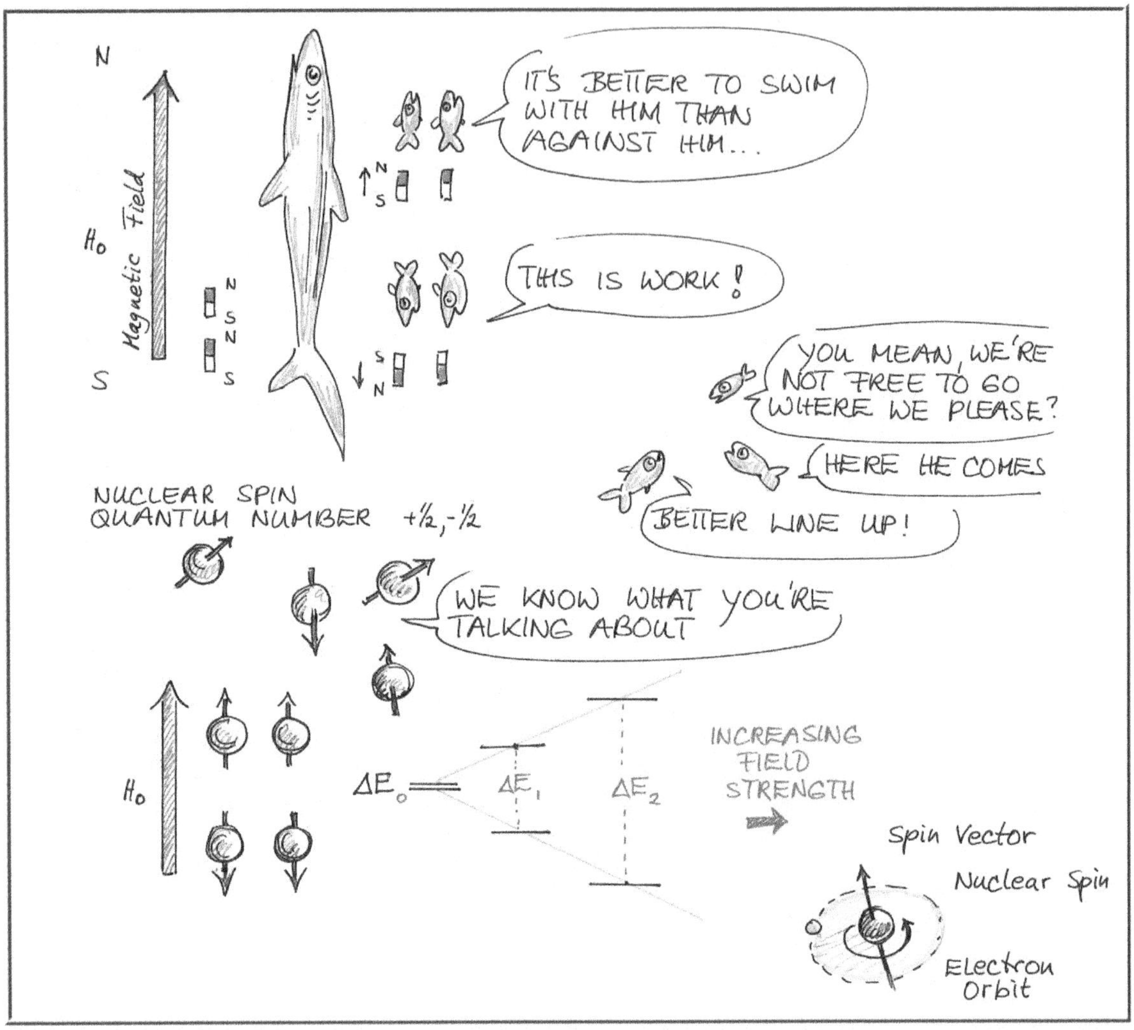


Atomic nuclei act as tiny magnets that create their own magnetic field. However, magnetic properties are only found in nuclei with odd-numbered masses such as the following:

**$^{1}H$, $^{13}C$, $^{15}N$, $^{17}O$, $^{19}F$, $^{31}P$, etc.**

But even nuclei that have even mass numbers but odd atomic numbers give NMR signals such as

**$^{2}{}_{1}H$, $^{10}{}_{5}B$, $^{14}{}_{5}N$, etc.**

All other with both even mass as well as atomic numbers do not respond, such as

**$^{12}{}_{6}C$, $^{16}{}_{8}O$, $^{32}{}_{16}S$, etc.**

The nuclear spin of magnetic nuclei can either be $I = + ½$ or $I = - ½$. For non-magnetic nuclei the spin is $I = 0$. Once an outside magnetic field is applied the electrons surrounding the magnetic nuclei offer a shield to the influences of the magnetic field lines. According to the principles of electromagnetism, moving electrical charges produce their own magnetic field. The three-finger rule or Lenz' rule would point the magnetic field vector into a direction opposite the super-imposed magnetic field produced by the instrument. This weakens the effect of the outside field. Of course if electron-withdrawing atoms are in the vicinity, the shielding effect of the hydrogen electrons is compromised.

**Fig.13.11**

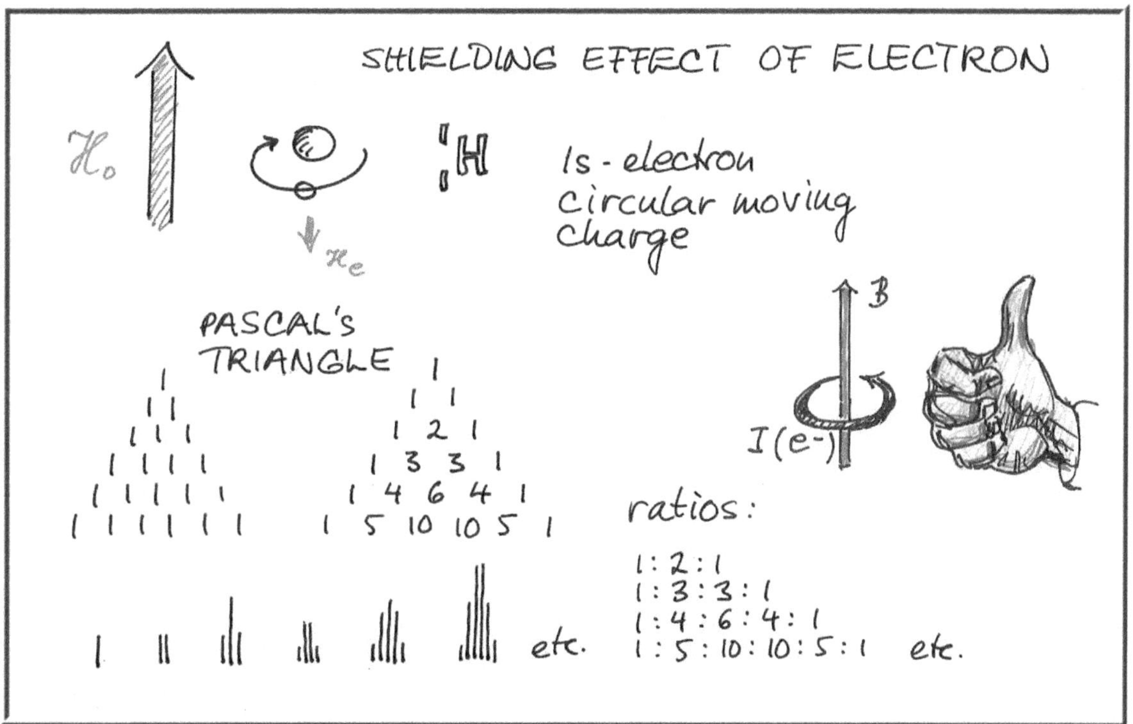


Disregarding the shielding effect of the orbiting electrons, which is an independent process, the nuclear magnetic spin vectors can align with the outside magnetic field lines either in parallel or antiparallel fashion. The actual activity of the alternating alignment attempts constitutes the basis of nuclear magnetic **resonance**. The larger the gap between the two energy levels of these transitions the more resonance energy is required.

The field strength of various instruments can vary between 30 and 300 MHz. The more powerful the field the more expanded the bandwidth for the chemical shift values become offering better resolution and separation of otherwise super-imposed signal groups.

Since organic molecules are mainly made up of carbon and hydrogen atoms, $^{1}H$ NMR and $^{13}C$ NMR based spectroscopic analysis is the most valuable.

The principle of the NMR measuring device is quite simple and outlined in the following diagram:

**Fig.13.12**

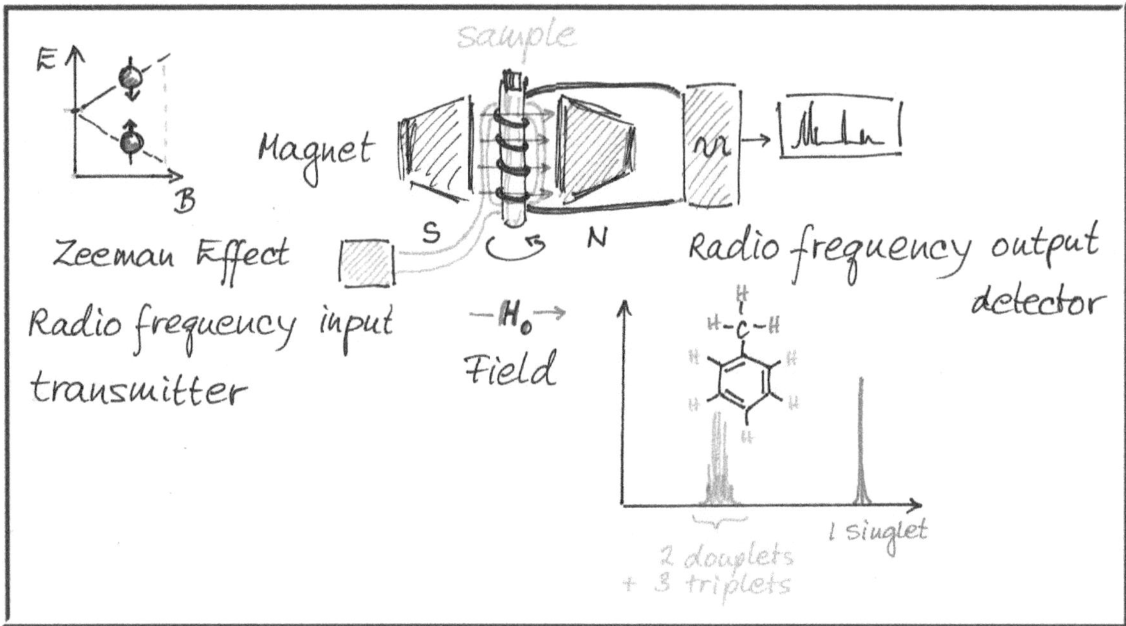


The NMR signals are based on equivalent and non-equivalent hydrogen atoms in relation to their chemical environment. All hydrogen atoms of methane for instance are equivalent having identical surrounding and therefore culminating within one signal, which when integrated accounts for 4 protons. **Integration** allows the assessment of the amount of hydrogen atoms involved in a signal (singlet) or signal group (cluster) represented by multiplets (doublet, triplet, etc.). The methane spectrum would thus be represented by a single, lone peak, far upfield, meaning with a high degree of shielding or low **chemical shift** value δ ("delta" - ppm).

**Fig.13.13**

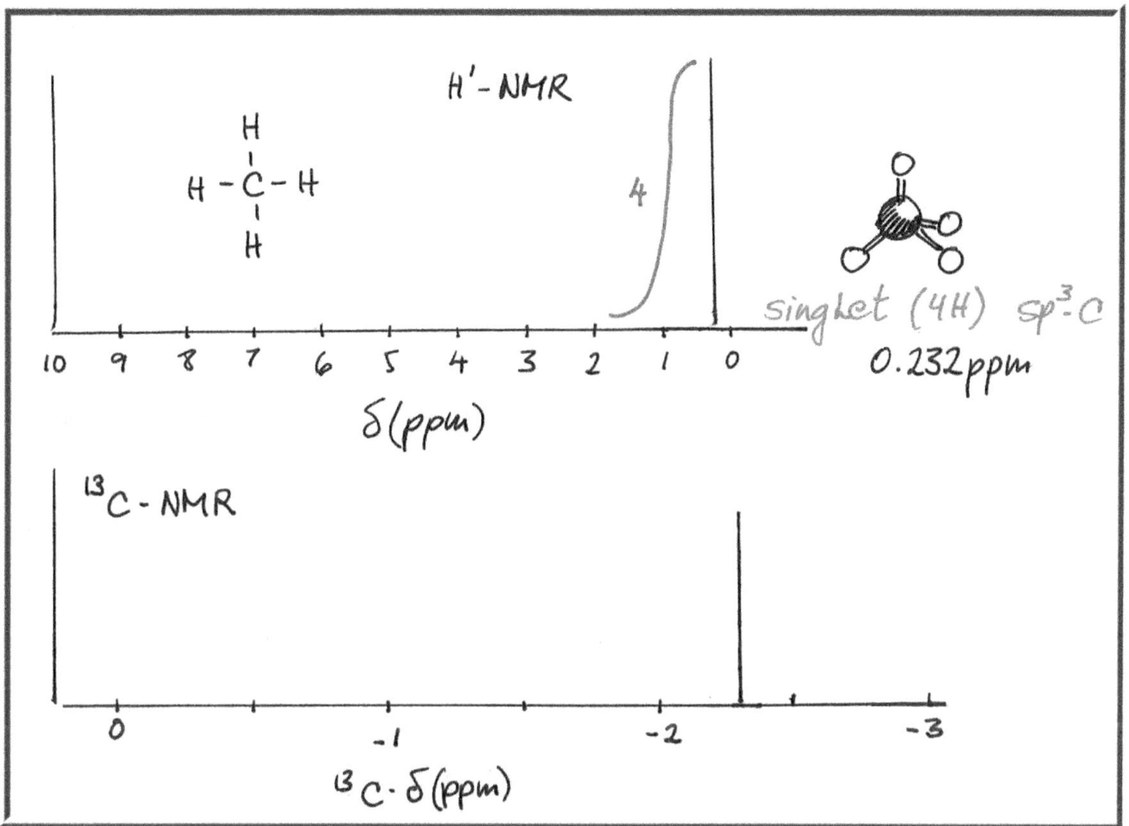


Three parameters determine the location and type of signals within a spectrum:

(1) Chemical shifts δ, measured in ppm/parts per million
(2) Spin- spin splittings and couplings
(3) Kinetic processes (reaction rates)

In order to have a zero point for the scale of chemical shift an **internal standard** is used and its signal arbitrarily set to δ=0.00 ppm. This chemical is **tetramethylsilane** (TMS), $Si(CH_3)_4$.

Groups that experience high shielding effects are positioned **upfield** of the spectrum and are characterized by low chemical shift values. Pure hydrocarbon protons are most shielded, protons connected to hetero-atoms such as those involved in hydroxyl groups (-OH) or the ones that are closer to functions containing heteroatoms such as the carbonyl (-CO-) or nitrile (-CN) groups are less shielded and their signals appear more **downfield** or at higher chemical shift values. Electronegative atoms involved in various functional groups have an electron-withdrawing quality and are therefore less shielding for neighboring hydrogen atoms. Energy transitions are higher for more shielded than for less shielded groups.

The following graph illustrates the direct relationship of **electronegativity** (EN, based on Pauling values) and **proton chemical shifts (δ).**

**Fig.13.14**

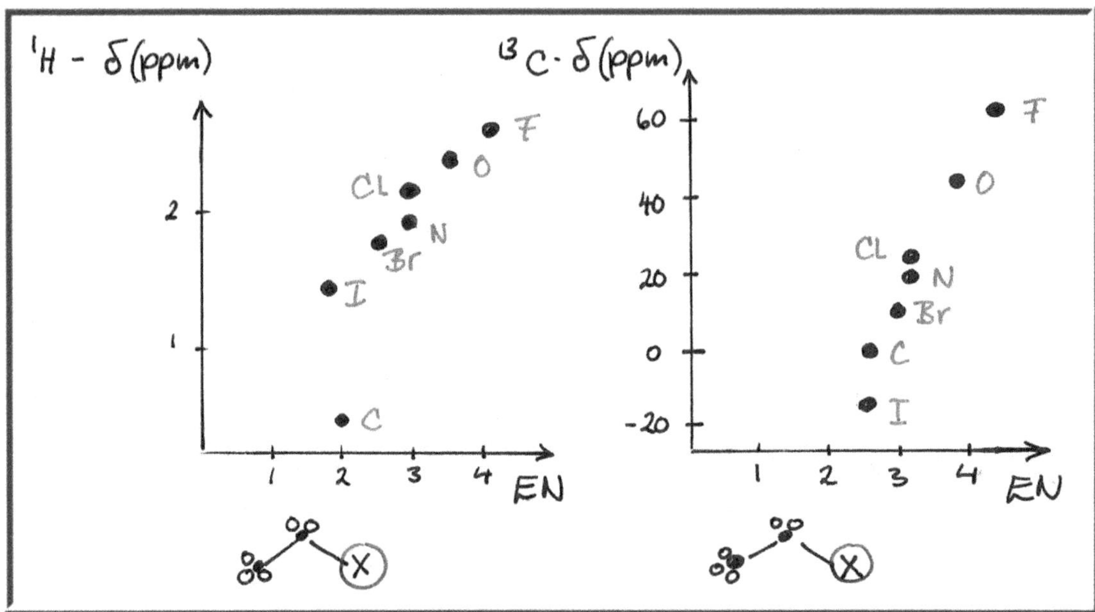


Multiple bonds as electron-withdrawing pi-systems (π-bonds) also have a de-shielding effect on attached protons. Vinyl (-CH=CH-) and aryl groups (aromatic rings) force signals downfield into higher chemical shift (δ) values.

The group most shielded is the methyl group ($-CH_3$). The group least shielded is the hydroxyl group (-OH)

One of the most significant phenomena of NMR is how non-equivalent protons interact and affect each other when they are in close proximity, meaning they are connected to adjacent atoms, especially carbon atoms. This leads to the effect of **spin-spin coupling** and the resulting **splitting of signals** into multiplets. If dealing with an ethyl group, for instance, one of the methyl protons can couple with each of the two of the neighboring methylene group protons. The possible coupling patterns are thus:

$-CH_2CH_3$ ↑↓

↑↑ (↑↓ **or** ↓↑) ↓↓ ↑↑ ↓↑ ↓↓

The left combination would be of low energy, the two middle, equivalent ones would compound to one peak of middle signal strength and the last one would be found in the high energy region of chemical shift values. The resulting peak would then be observed as a symmetrical triplet peak arrangement in the ratio 1:2:1. On the other hand, from the perspective of the methylene group, each proton can couple with three of the methyl group protons.

↑↓↑ ↓↑↓

↑↑↓ ↓↓↑

↑↑↑ ↓↓↓ ↑↓↓ ↓↓↓

So, the interaction of the two equivalent methylene protons with its three neighbors would produce a symmetric quadruplet in the ratio 1:3:3:1, found further downstream in the spectrum.

The whole spectrum of the ethyl group would then look as follows:

**Fig.13.15**

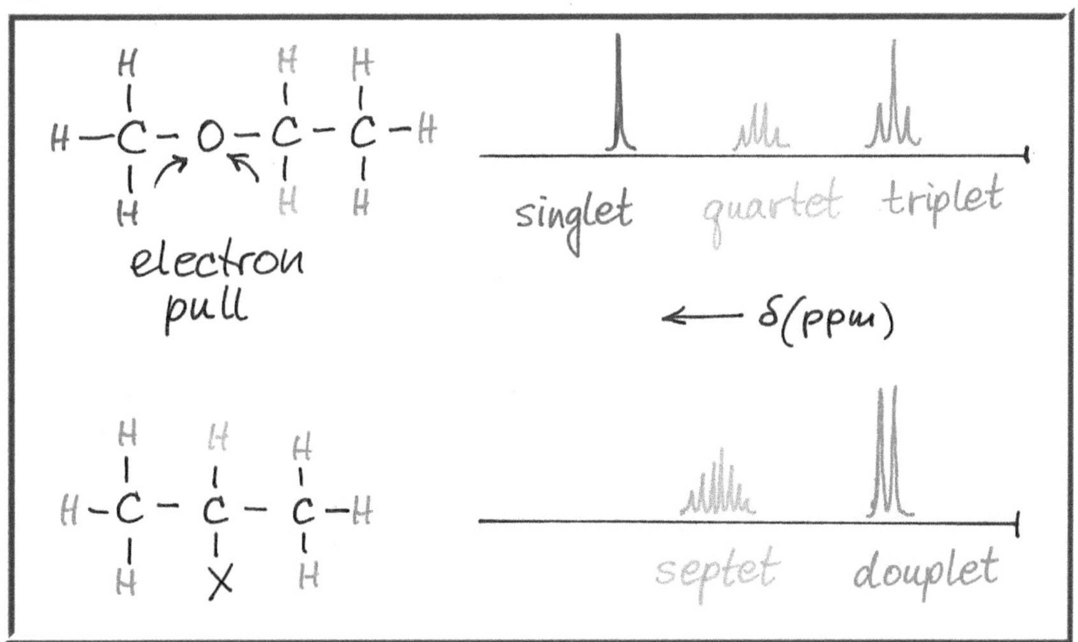


Overall a rule can result for predicting splitting patterns of proton coupling reactions, expressed as:

**M = H + n**

(M=multiplicity of signal, H = proton coupling with neighbors, and n = it's immediate, non-equivalent neighboring protons, also known as **vicinal protons**).

To apply this rule to the ethyl group, each methyl hydrogen atom coupling with its two methylene group neighbors would produce 1+2 =3 peaks or a triplet. Each methylene proton coupling with its three neighbors would add up to 1+3=4 peaks or a quadruplet.

If the molecule propane is analyzed, the two methyl groups would each produce a triplet of same energy, compounded into one in the spectrum, but the central methylene group now having six equivalent neighbors would produce a 1+6=7 multiplet or specifically a septet.

**Fig.13.16**

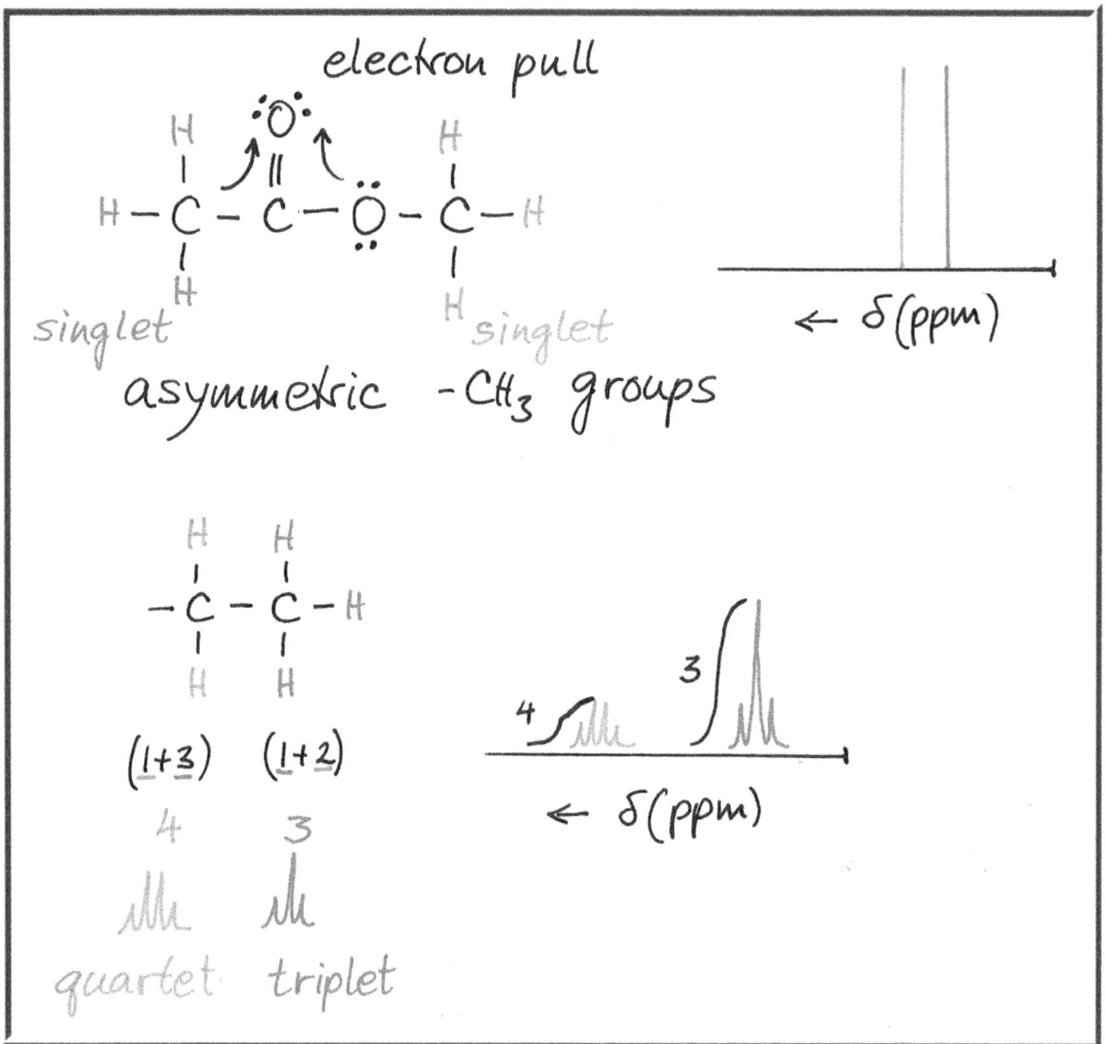


The relative heights or intensities of the multiplet signals follow the pattern of **Pascal's triangle** as is illustrated in Fig. 13.9.

So the ratios become 1:1, 1:2:1, 1:3:3:1, 1:4:6:4:1, etc.

First order splitting patterns show the smooth lines of the individual peaks. However, these can further split into further peaks just like harmonics of a main note, a result known as second order splitting. This can lead to interpretations difficulties.

**Integration** applied, appearing as a stepped line, which measures the space beneath the peaks directly correlates the relative amounts of protons involved, so the large triplet would equate to six protons and the relatively low septet only two protons.

The spacing between the major signals is attributed to the **spin-spin coupling constant J**, the ratios being unaffected by the transmitter frequency υ, whether 60 MHz (1 x $10^6$ cycles/sec), 220 MHz or other.

**$^{13}$C-NMR** looks only into the nuclear magnetic resonance of the C-13 nuclei. No coupling between non-equivalent carbon atoms is observed, probably due to the relative high mass and less flexibility within the carbon framework. The corresponding spectrum of methane $CH_4$ is also only represented by one peak (Fig. 13.11)

Propane would look as follows showing two peaks. Both methyl groups show equivalent carbon nuclei, compounding to one tall signal. The methylene carbon appears with a separate signal further downfield.

The successful interpretation of an NMR spectrum should produce the following information

(1) Number of signals pertaining to different types of protons and proton groups. Chemical environments for each involved proton of the carbon framework with and without heteroatoms.
(2) Intensity of signals relating to quantity of equivalent protons at a particular molecular site, with integration to help establish the relative ratios.
(3) Multiplicity or degree of signal splitting in relation to the number of adjacent or vicinal protons in the molecular "neighborhood".

**Example Problem 13.2**
**Evaluate the $^1$H-NMR spectra of the following molecules:**

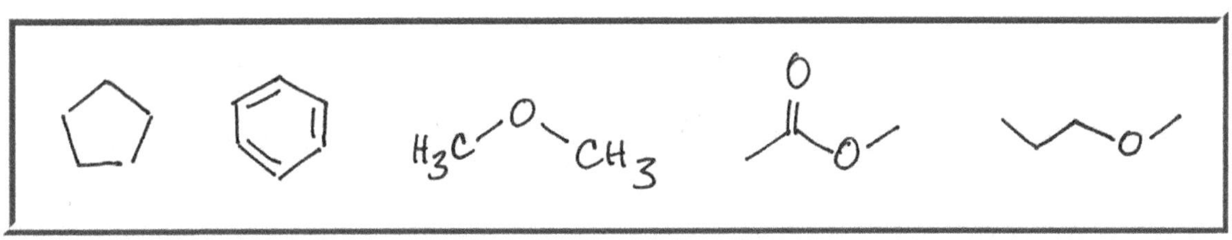


**Example Problem 13.3**
**Which of the following molecules show only one peak (singlet)?**

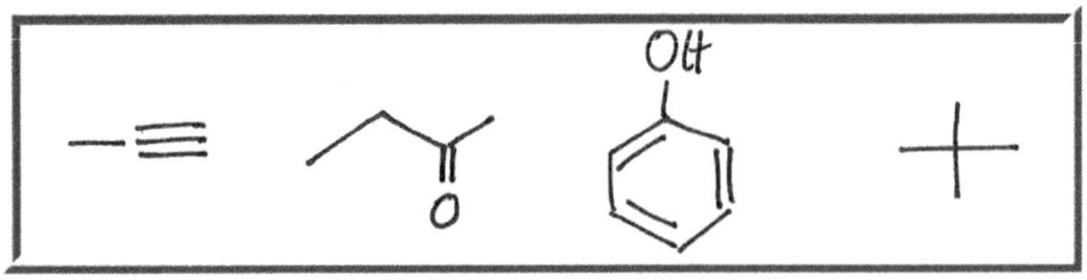


**Example Problem 13.4**
***p*-Xylene shows two $^1$H-NMR peaks at δ 2.30 and at δ 7.10 in the ratio of 3:2 or 6:4. Assign the $^1$H-nuclei to each molecular group or area.**

**Example Problem 13.5**
**Why are the hydrogen atoms on the benzene ring equivalent? When does their equivalency change?**

In order to receive a spectrum via **mass spectroscopy**, organic molecules are bombarded by a beam of medium energy electrons coming from an "electron gun", fragmenting the molecule into various smaller parts mostly in the form of ions.

**Fig. 13.17**

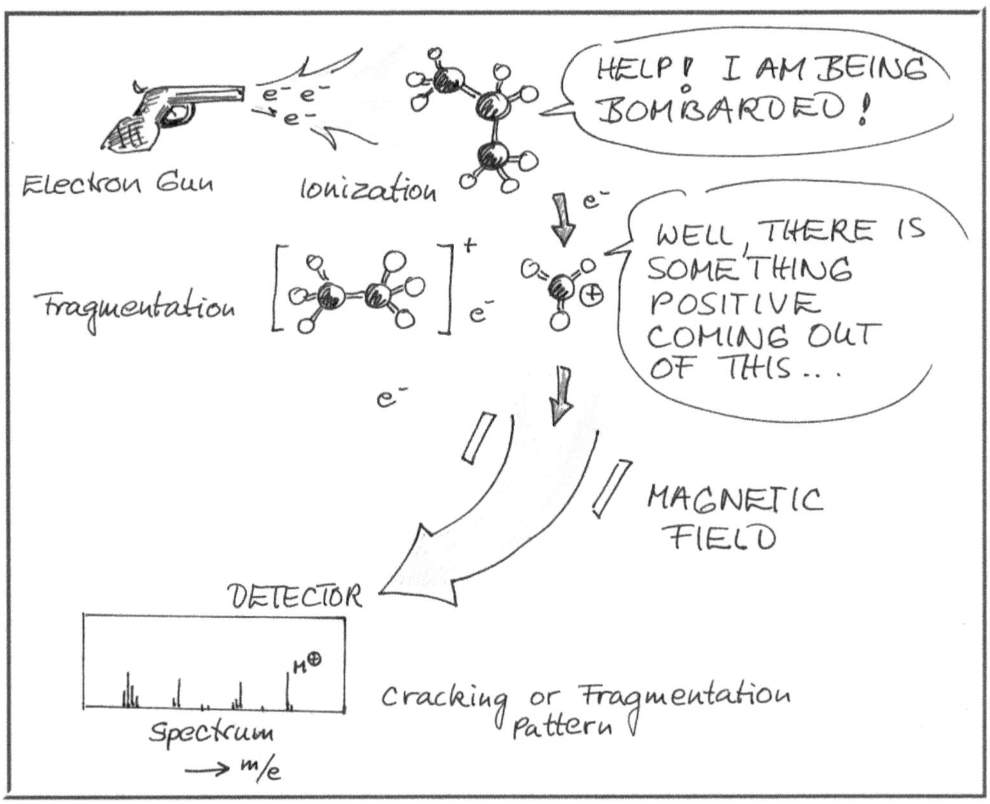

These ionized particles are then sent through a curved path surrounded by a magnetic field. Like on a racetrack the slower heavier charged species will have a harder time making the curve and will arrive at the destination point a bit later.

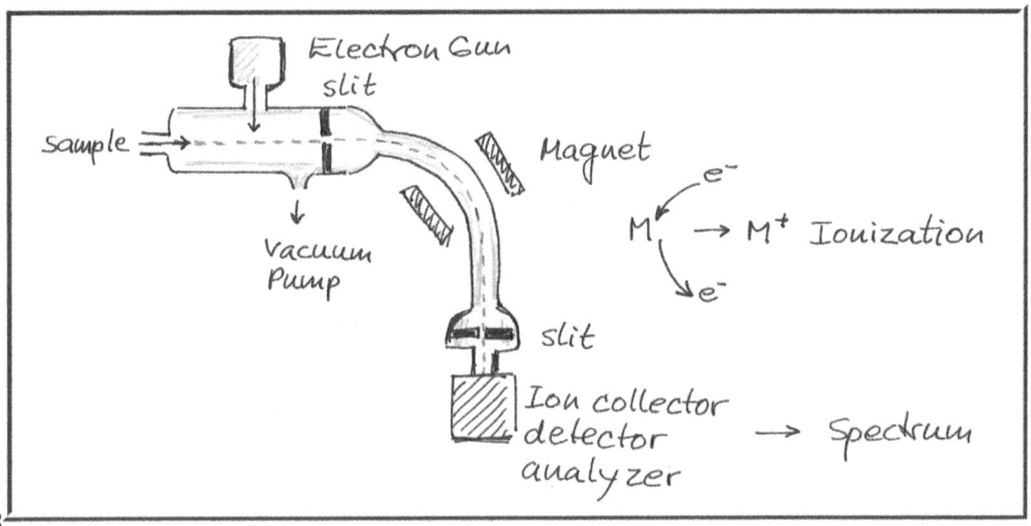


**Fig. 13.18**

An ion collector and analyzer will then a produce a type of cracking pattern based on relative e/m values (charge to mass ratios) that will look somewhat like in the following illustration:

**Fig. 13.19**

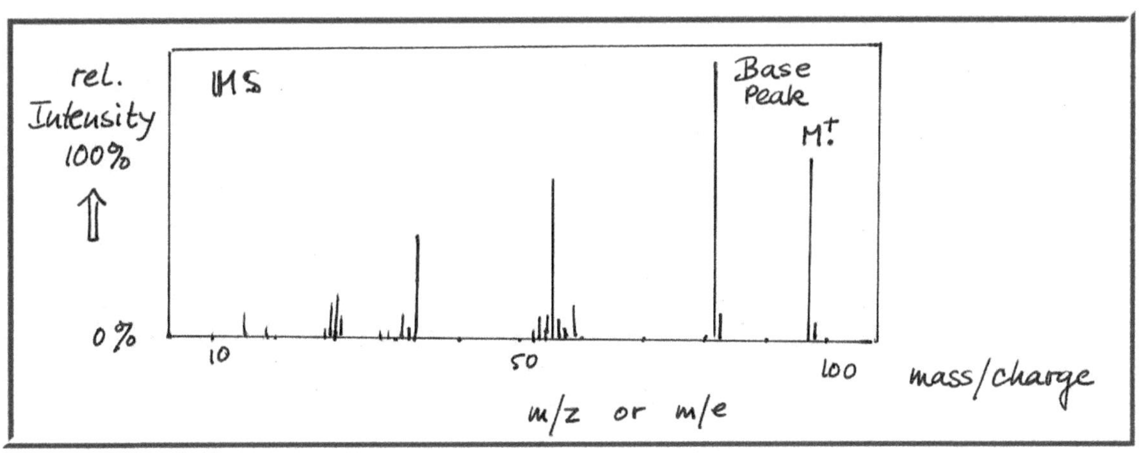


Positive ions are paramagnetic and thus are influenced and deflected (bent in their path) by magnetic fields. Neutral species are not registered.

Often the peak related to the highest mass or **highest e/m (or m/z) value** will be designated as **$M^+$** (cationic species) **or $M.^+$** (radical ionic species), representing the original parent molecule. Sometimes an $M^+$+1 peak can be present as a result of proton transfer or species containing a heavier isotope such as $^{13}C$. The highest or **tallest peak** of a spectrum represents the most abundant peak and is referred to as the **base peak**. Taller peaks are in general also the most stable, since fragmentation occurs at the weakest bonds. Any lower peaks than the molecular peak (fragmentation peaks) are called **daughter peaks**. Stable cation formation also favors certain molecular sites. Branched sites in hydrocarbons are more easily broken, not only because they appear to be weaker but due to the relative stability of the formed **carbocations**, as has been observed for many chemical relationships.

Chemical isomers produce different fragmentation patterns, so the ionization process leading to the break-down products is strictly a result of the chemical environments given by the atomic framework.

Some chemicals such as 2,2-dimethylbutane cleave so readily at the quaternary carbon (4°), that no parent ion can be observed in the spectrum.

Mass spectra can offer challenges to interpretation, since the fragmentation process does not make the system immune to follow-up chemical reactions. Sometimes unusual species occur having their own breakdown patterns and internal rearrangements of a produced molecular ion can lead to different fragmentation patterns.

**Fig.13.20**

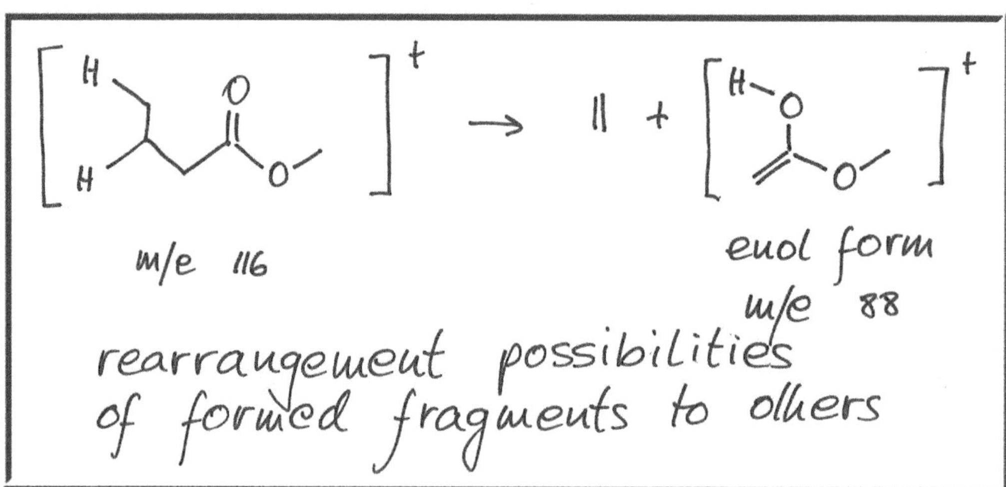


Deuterated forms of molecules, in which the hydrogen atoms are more or less substituted by deuterium atoms ($^{2}_{1}H$ or D), such as toluene, helped investigate such behavior patterns.

**Fig. 13.21**

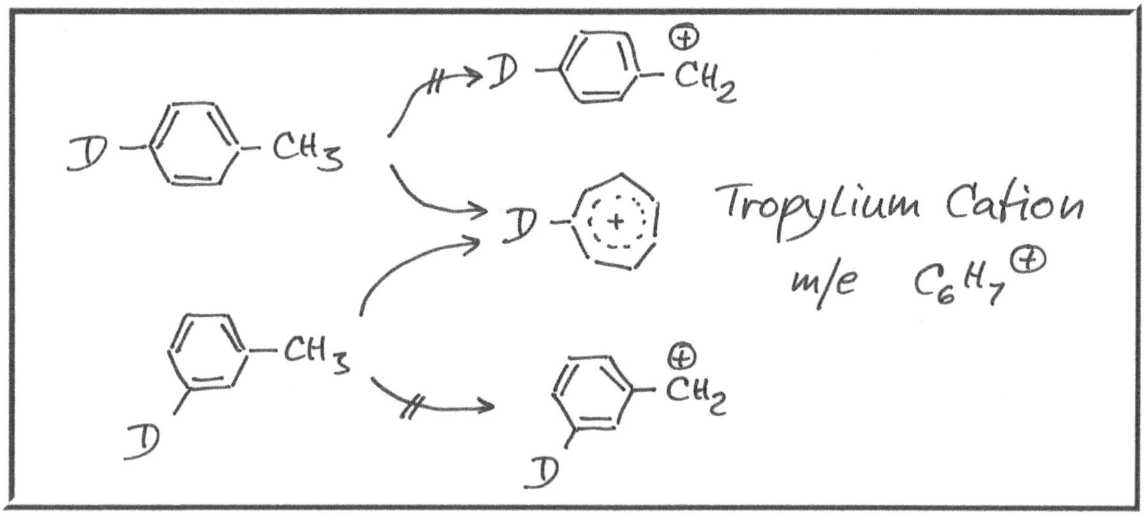


**Example Problem 13.6**
**An aromatic hydrocarbon has an MS parent peak at m/e (or m/z) at 102. It's $^{1}$H-NMR spectrum produces two peaks at δ 2.7 and δ 7.4 with relative areas in the ratio 1:5 when integrated. What is the name and structure of the analyzed molecule?**

**Example Problem 13.7**
**The mass spectrum of an alcohol ($C_3H_6O$) shows a strong daughter peak at m/e (m/z) = 31. How does it come about and is the parent molecule a primary or secondary alcohol? Explain.**

**Example Problem 13.8**
**The mass spectrum of a ketone has a parent peak at $M^{+\cdot}$ = 128 m/e (m/z). Other prominent peaks show up at m/e = 85, 71, 57 and 43. What species are involved and how are they derived from the parent molecular ion radical? What is the name and structure of the original molecule?**

**Example Problem 13.9**
**What type of $^{13}$C-NMR is expected for ethanol?**

**Raman spectroscopy** is an analytical adjunct to infrared spectroscopy. Monochromatic light from an argon laser is targeted onto the sample substance, which absorbs the light but also scatters it in various directions. The scattering pattern mirrors the interaction of the light with the molecular framework. In comparison, a blood splatter pattern can lead to detail-information of a possibly criminal or accidental event for forensic science. In Raman spectroscopy the scattering pattern is measured, analyzed by a photo-detector and delivered as spectrum similar to IR. The waves coming from the light source interact with the various bonds leading to bond vibrations and rotations, thus slightly altering or modifying the wavelengths of the out-coming waves.

**Fig. 13.22**

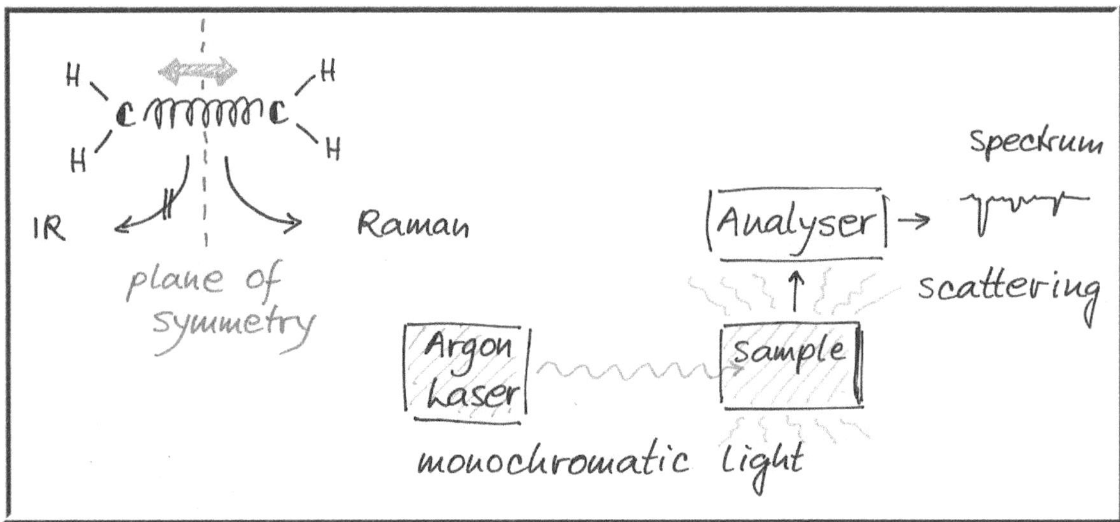


The one drawback for infrared spectroscopy is that symmetrically built molecules do not give any signal. Electrically identical groups vibrate the same and cancel out each-others signals. Raman, however produces strong signals for these molecules as is shown for the molecules 1,1,2,2-tetrachloroethene and cyclohexene, illustrated below:

**Fig.13.23**

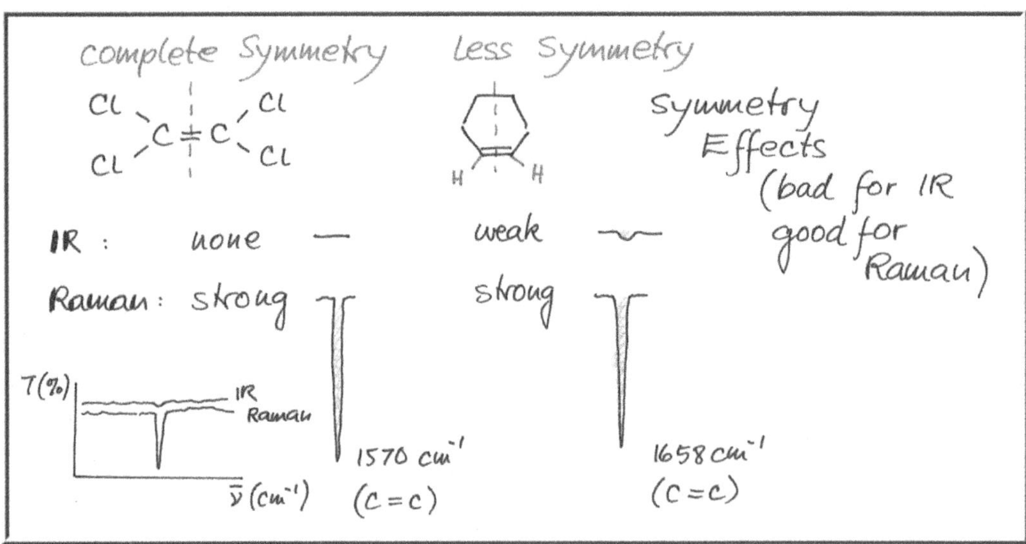

Spectra produced by actual electron transitions to higher excitation levels are referred to as **electronic spectra.** These involve **ultra-violet/visible absorption spectra (UV-vis)** and emission spectra such as **fluorescence and phosphorescence** spectra. The latter two types are complementary to absorption spectra obtained in the UV-vis region. As high frequency light with short wavelengths comes in from the high energy end of the electromagnetic spectrum and onto a target compound, the molecules will absorb photons of specific wavelengths to propel its electrons to distinct higher energy levels, depending on the type of bonds and the electro-chemical environment they are involved in. According to the **valence bond (VB)** and **molecular orbital theories (MO),** electrons from bonding atoms combine in **bonding** energy states of lower energy (**σ and π** molecular orbitals) and, if already filled, pair up in so-called **anti-bonding** energy states of higher energy levels (**$\sigma^*$ and π*,** "sigma star and pi star" molecular orbitals) than in the ground state, which actually is the non-excited state. **Non-bonding energy levels** are designated by the letter **n**.

**Fig. 13.24**

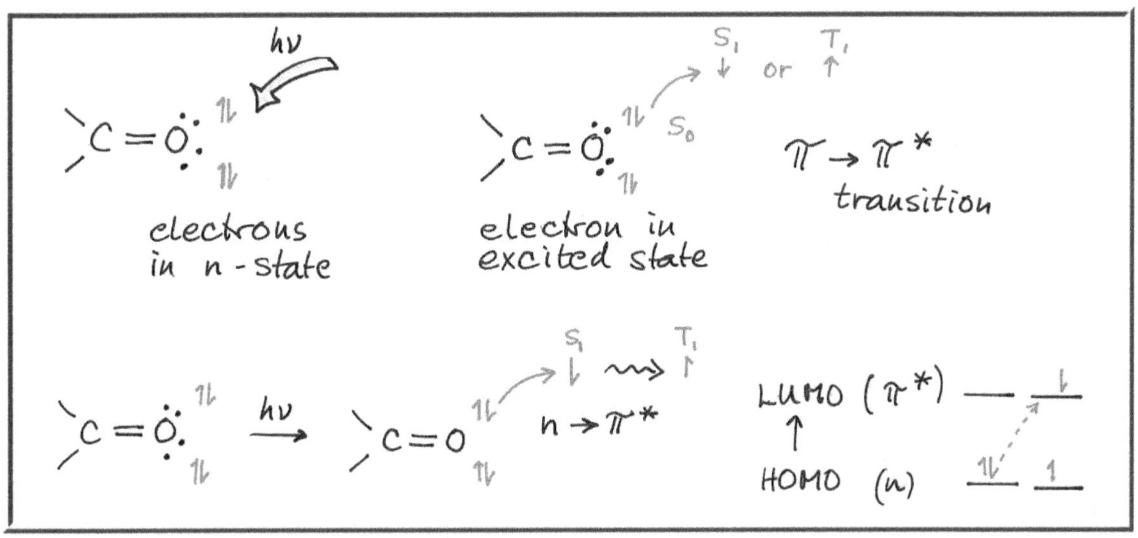


The spatial overlap of atomic orbitals (s, p and hybrids sp, $sp^2$, $sp^3$, mainly involved in carbon based molecules) leads to the extended molecular orbitals (σ for linear, coaxial overlaps and π for parallel or side-way overlaps)

**Fig.13.25**

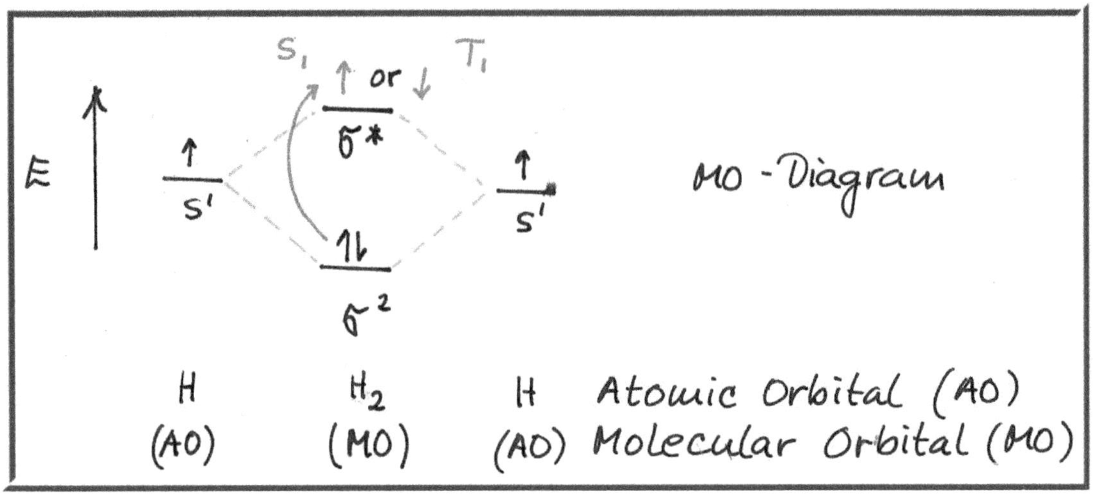


Molecular orbital theory applied to the formation of the diatomic oxygen molecule allows to establish the bond order and to make assessments about the magnetic properties of a molecule..

**Fig. 13.26**

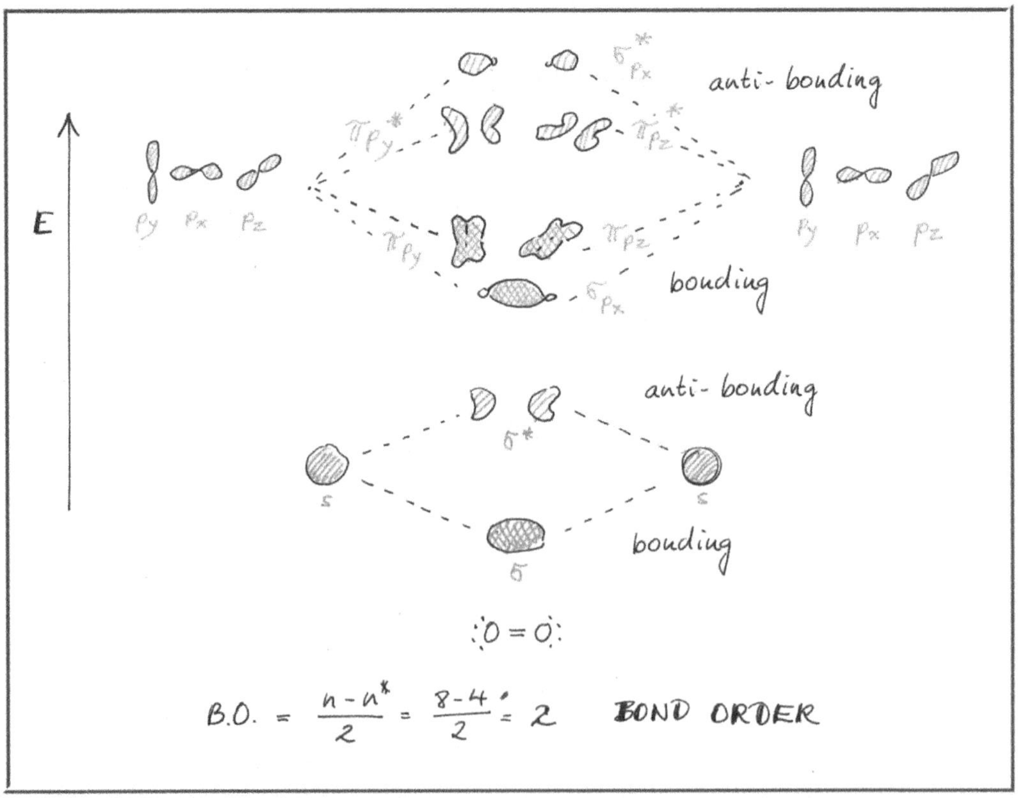

Oxygen, for instance, is a paramagnetic substance, verified by the two antiparallel spins of the two electrons occupying two separate degenerate (Hund's rule) antibonding molecular orbitals

**Fig. 13.27**

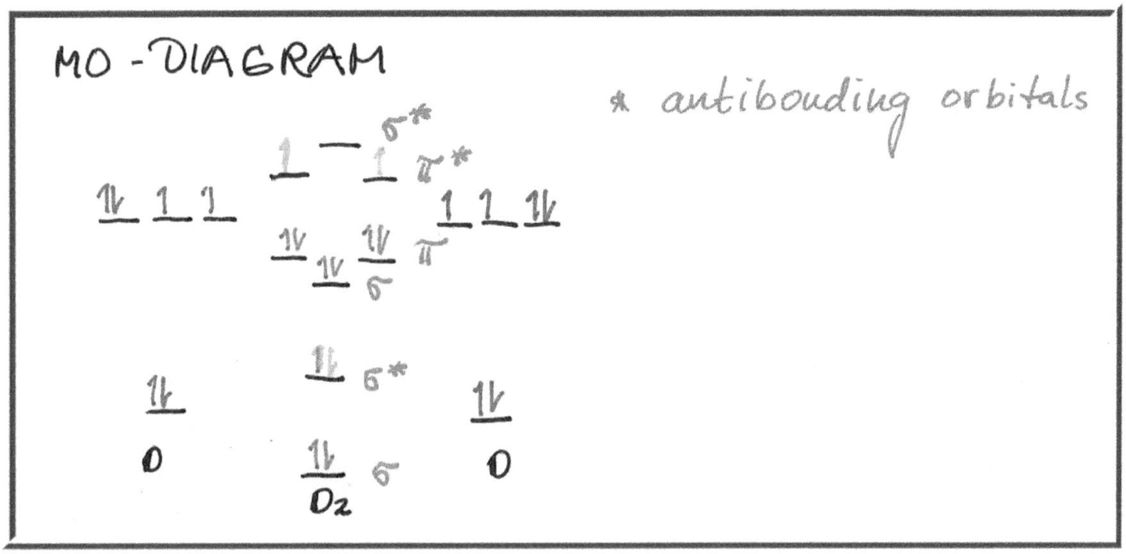


The types of observed electron transitions to higher energy levels are:

**n→σ*, n→π*; π →σ*, π→π*; σ→σ*, σ→π***

The energy ladder for these transitions is illustrated below:

**Fig.13.28**

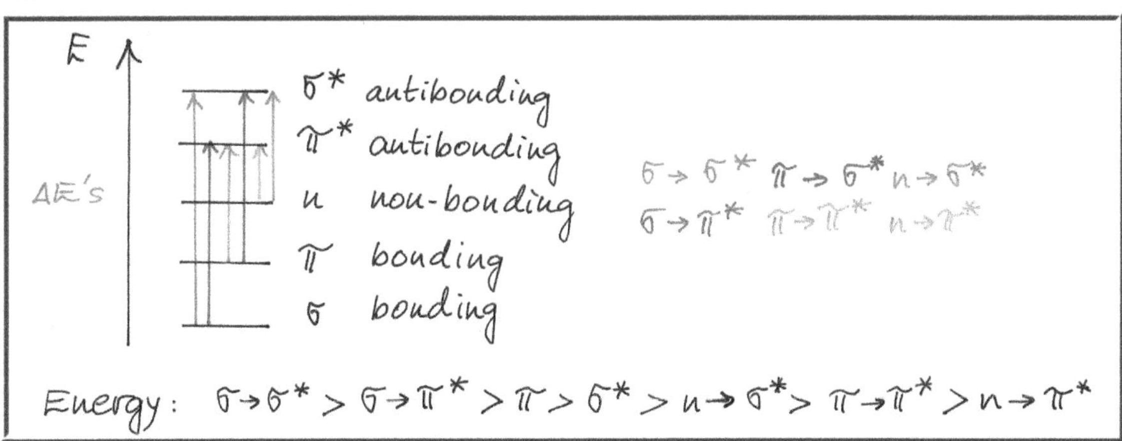

The higher the gap between the transition-states the higher the energy absorbed and the shorter the wavelength ($\lambda$, “lambda”, measured in nm) of the observed absorption peak. The intensity of the peak is qualified by the molar extinction coefficient ($\varepsilon$, “epsilon”), which is derived from the equation $\varepsilon = A/cl$ ($A = \log(I/I_o)$; according to Lambert-Beer law absorbance is the decadic logarithm of the ratio transmitted light to incident light; c = concentration of the solution and l = thickness of the sample tube or path-length of the cell).

The maximum peak in a UV-vis spectrum is usually designated by $\boldsymbol{\lambda_{max}}$ **and** $\boldsymbol{\varepsilon_{max}}$.

**Fig.13.29**

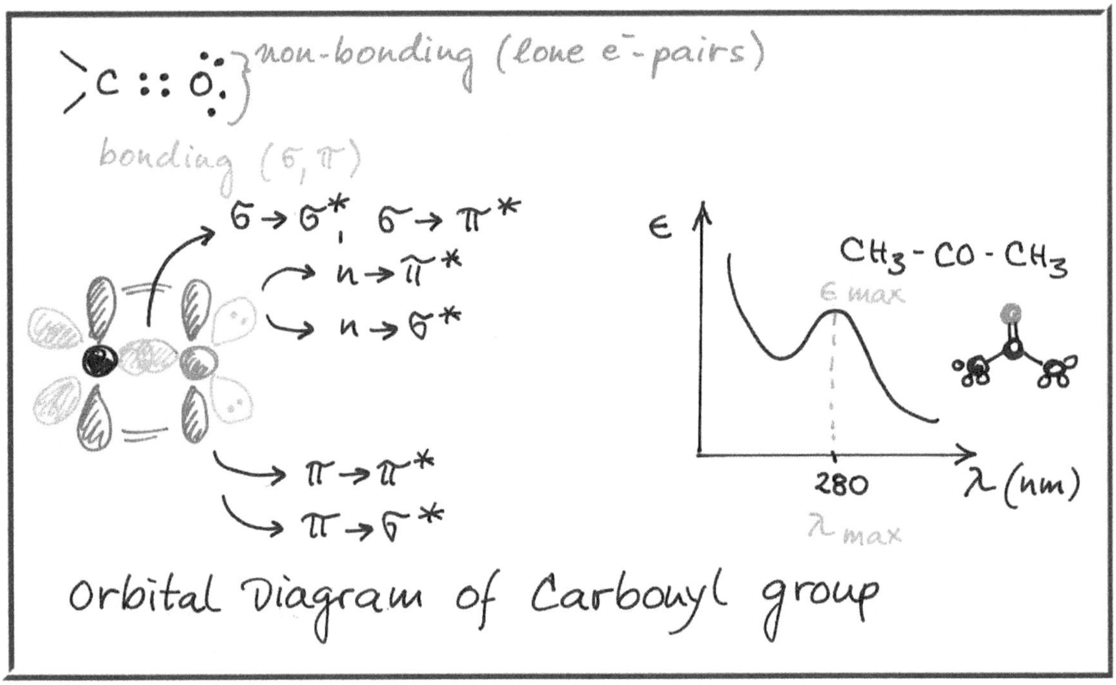


Methyleneamines have similar transitions as the carbonyl compounds

**Fig. 13.30**

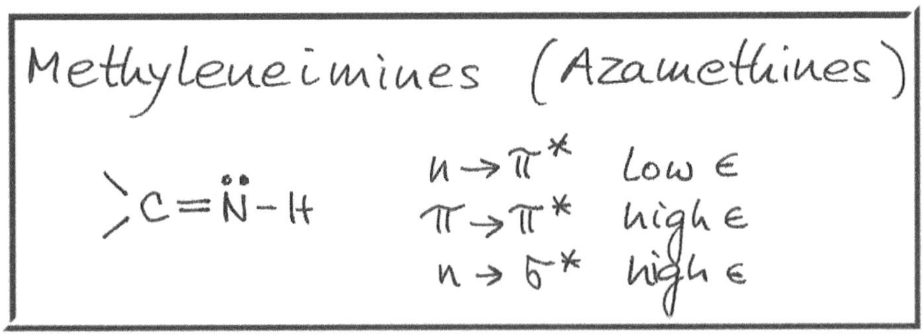

Strong $\pi \rightarrow \pi^*$ transitions are observed for any molecular groups containing multiple bonds with or without heteroatoms. Conjugation leads to bathochromic shifts, in which the absorption peak leads to higher wavelengths and higher intensities.

**Fig.13.31**

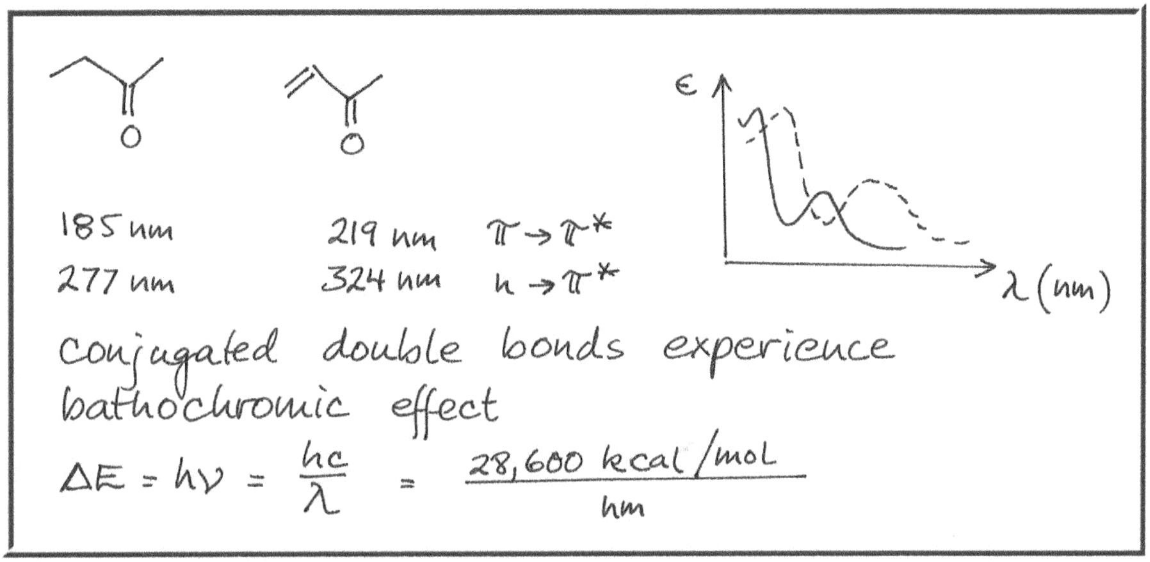


Orbitals, in which electrons are paired ( ↑↓ or ↓↓/↑↓) are called **singlet** states, which can be the case for ground levels as well as excited states (**$S_0$, $S_1$** etc.). However, paired excited orbitals can be split into more than one, singularly-filled, unpaired orbitals (↓ or ↑), which are now referred to as **triplet** states (**$T_1$** etc.) The transition from an excited $S_1$ state to an excited $T_1$ state is known as **intersystem-crossing**, which is radiation-less. Excitation to either $S_1$ or $T_1$ leads to longer bond lengths, but $T_1$ states occupy lower energy levels and are thus more stable due to Hund's rule (degenerated orbitals are first singularly filled and then doubled up).

Once an electron is elevated to a higher level by absorption ($S_0 \rightarrow S$), it has a few transition options. If it doesn't dissociate completely, it can after a short interval of non-radiative vibrational relaxation (a type of delayed reaction), return to one of the ground states via $S_1 \rightarrow S_0$ transition producing **fluorescence** as emission radiation. From $S_1$ it also has the option of crossing over radiation-less via intersystem-crossing to $T_1$, which represents another delayed reaction, and fall back into one of the ground levels of $S_0$ via **phosphorescence** as emitted radiation ($T_1 \rightarrow S_0$).

**Fig.13.32**

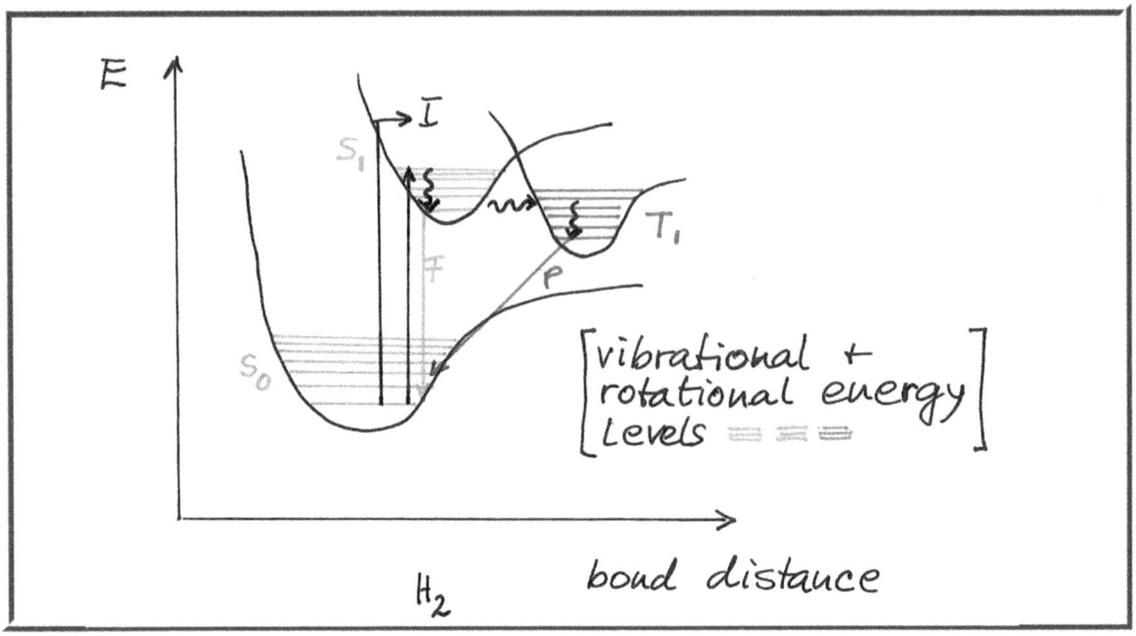


**Example Problem 13.9**
**Assign the electron transitions for the molecule 2-methyl-2-penten-4-one that produces the following UV-vis spectrum:**

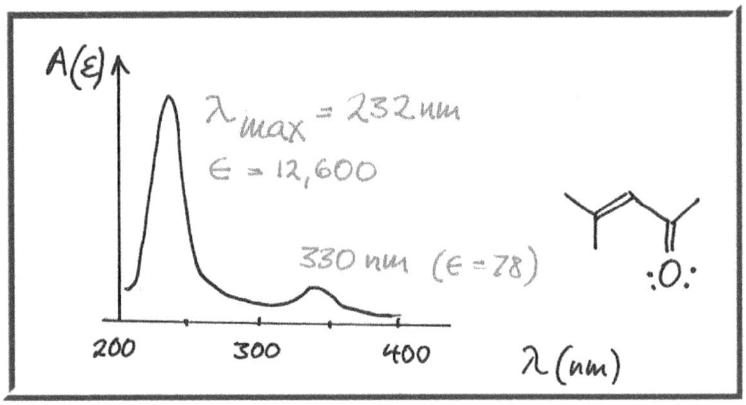


All these spectroscopic methods indirectly point to the most probable molecular structure if used in conjunction. However, the real molecule is never seen and its structure can only be assumed. Many molecules are very complex in their structure and spectroscopic data can be ambiguous and prone to interpretational errors. If successful crystallization of the compound leads to a mono-crystal, then **X-ray crystallography** can be applied, which produces diffraction patterns that via Fourier transfer analysis can actually lead to a viable image, derived from point groups and electron density maps with the help of especially designed computer programs, representing the true structure of the molecule.

**Fig. 13.33**

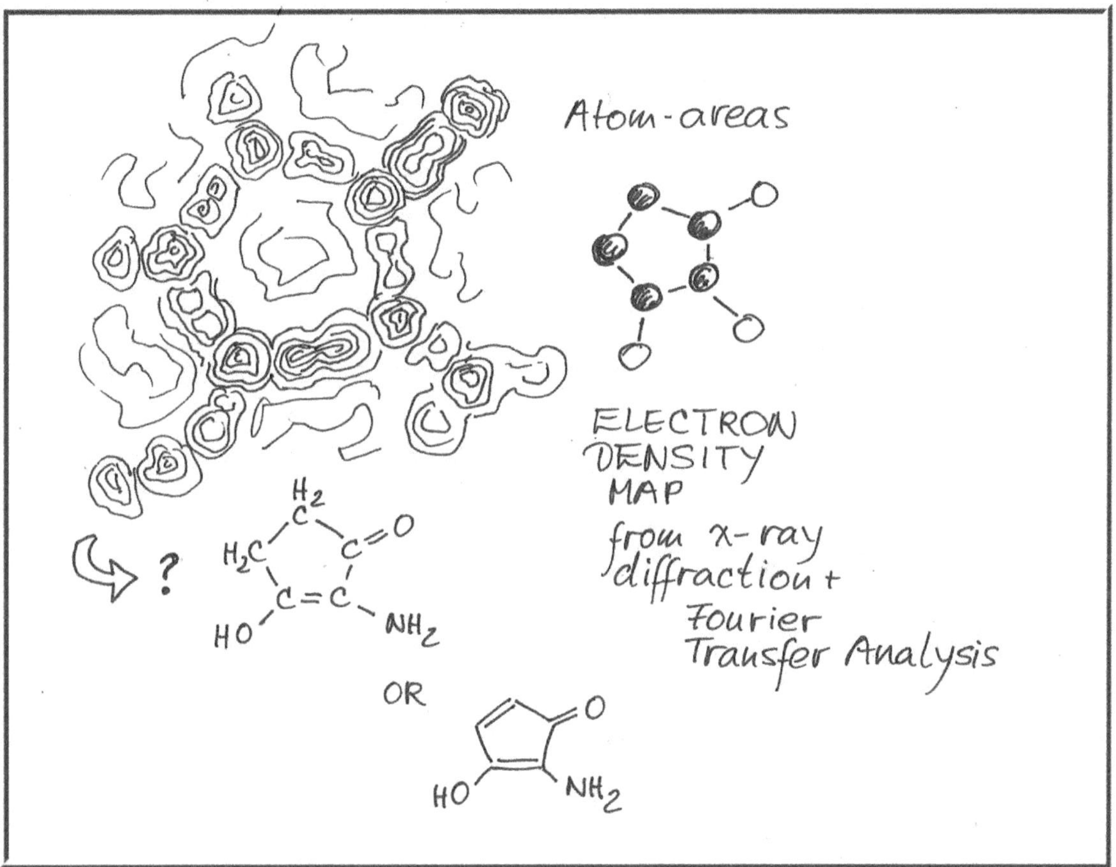


The wavelengths of x-rays are about the length of interatomic distances and thus allow refraction patterns to be produced. Assemblage of various patterns derived from different penetration levels of the mono-crystal and mathematical analysis (Fourier transfer etc.) of the "package" lead to an approximation of the actual bonding structure concerning the targeted substance in question.

**Fig. 13.34**

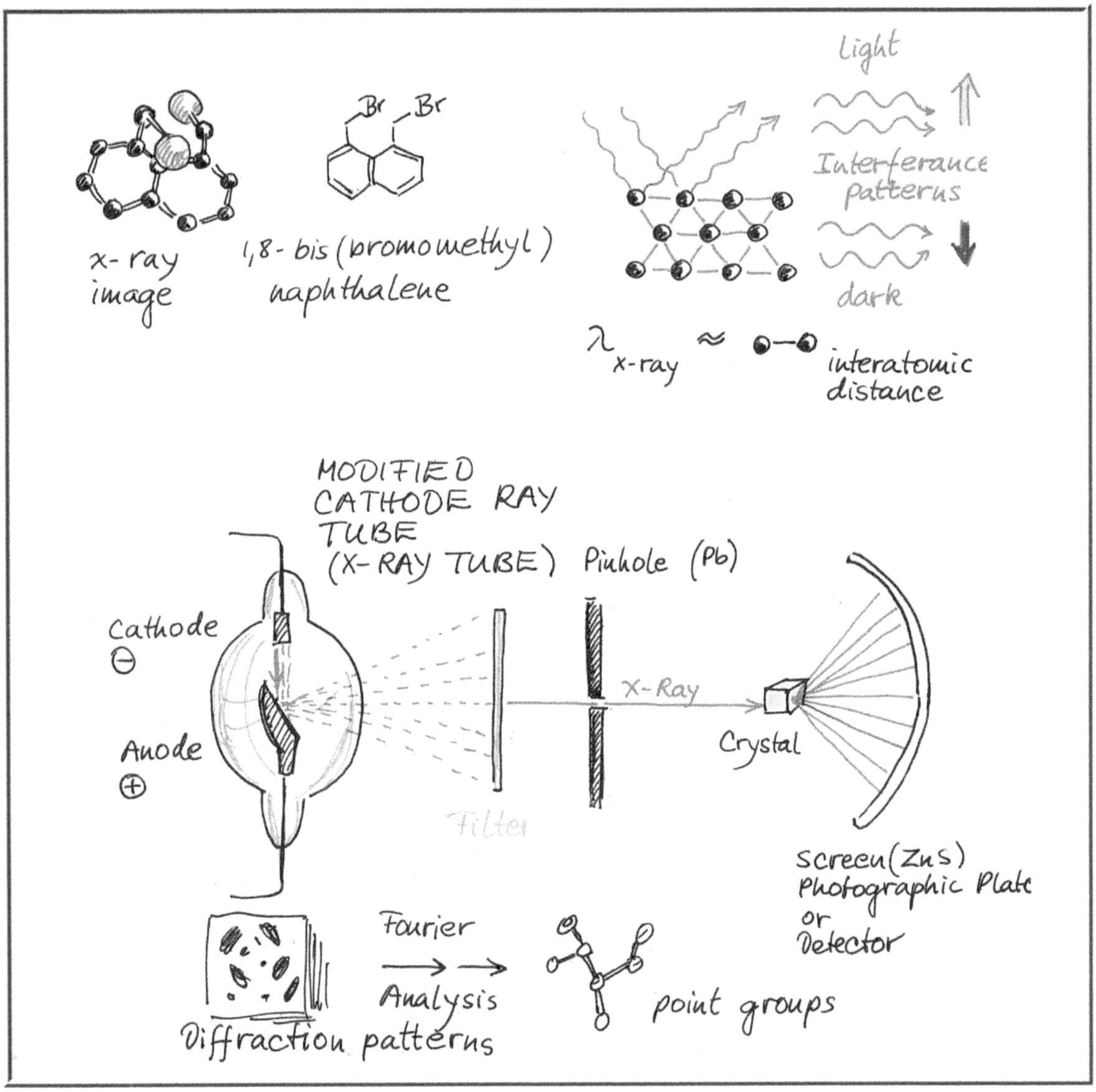


Many structural details can be elucidated. For instance an x-radiogram determined bromomethyl groups attached to a naphthalene nucleus to be oriented in opposite positions of the ring.

Highly structured and symmetrical molecules, such as those belonging to the class of phthalocyanins give obvious refraction patterns as shown below. X-ray analysis actually proved that the large rings within the crystal matrix are actually tilted against each-other.

**Fig. 13.35**

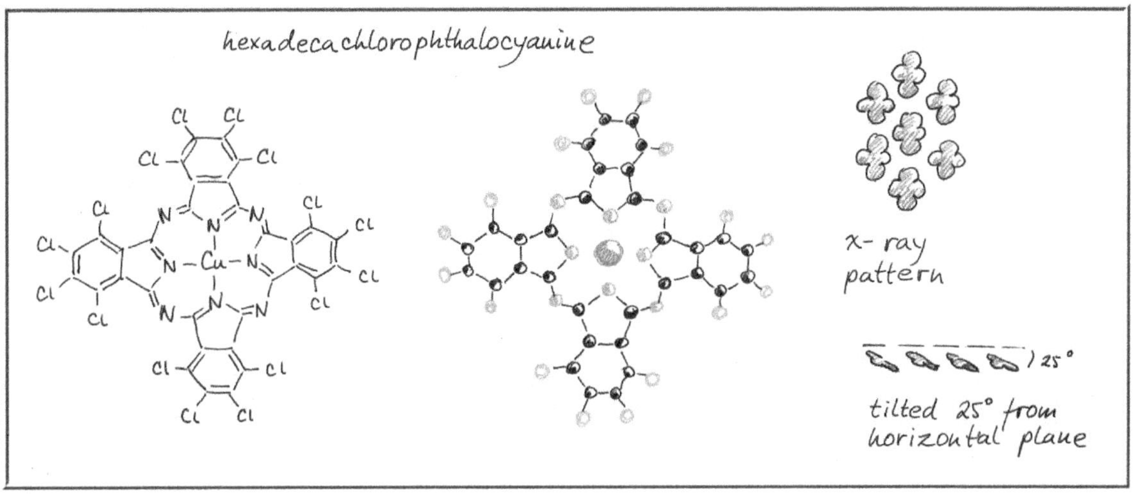


The following image shows an overview of most of the techniques applied for the structure elucidation for purified target molecule:

**Fig. 13.36**

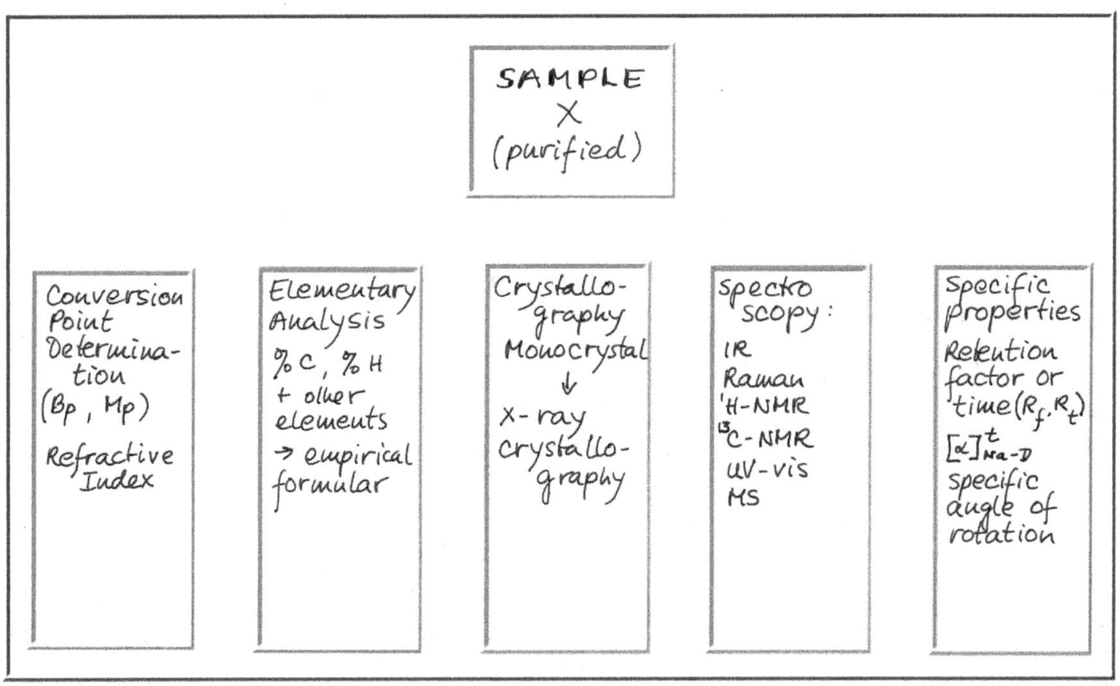


Chemists often have to deal with quite complex molecules such as the venom **bathrachotoxin**, found in natural sources like in the poison dart frogs.

**Fig. 13.37**

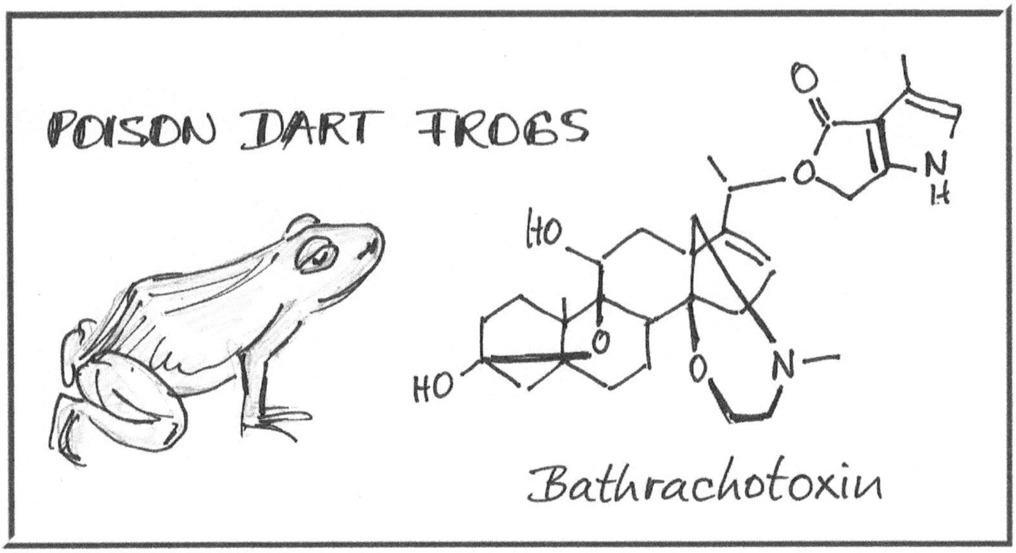


Besides the boiling point and density of liquids, a most indicative property for the identification of a liquid is its **refractive index n**. Light travels at different speeds (velocity when considered a vector) through different media. The maximum speed of c = $3.0 \times 10^8$ m/sec is attained when light travels through a vacuum. Once it hits a medium made up of materials of a certain density it is slowed down due to interaction with the particles that make up the denser medium. The deceleration effect also leads to the "bending" of light to tighter (smaller) angles at the interface than the angle of the incoming light. This phenomenon is called **refraction.** The refractive index n is the ratio of the angle of incoming light to that of the outgoing light when passing through the interface of two media or simply the ratio of the angle of an incident beam of light to the angle of the refracted beam of light, given by the formula:

$$n = \frac{\sin\theta}{\sin\phi} \quad \text{(sin theta/sin phi)}$$

The refraction of light is illustrated by the following image:

**Fig. 13.38**

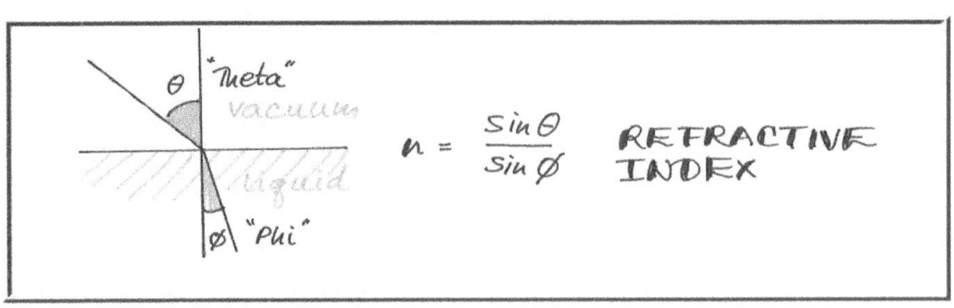

The refractive index is dependent on temperature and inversely proportional. The higher the temperature the lower the refractive index becomes, mainly due to increased particle motion having an effect on the speed of light.

The most commonly used refractometer that measures the refractive index of liquid media is the Abbé refractometer (sometimes referred to Abbé-Laue), shown in a simplified scheme below:

**Fig13.39**

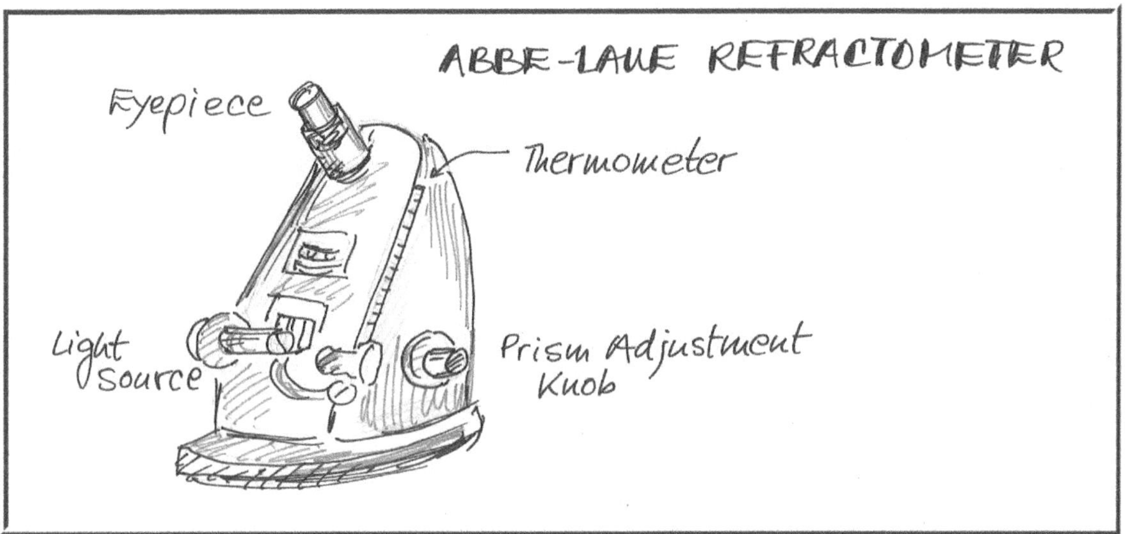


It contains a thermostat that keeps the measuring temperature usually at $20^{O}$**C**. Since the light also comes from a sodium vapor lamp utilizing the D-line of its spectrum as monochromatic light source such as is the case for polarimetry (specific angles of rotation [$\alpha^{t}{}_{D}$]), the refractive index is similarly expressed as:

$$\mathbf{n^{t}{}_{D}}$$

When measured at $20^{O}$C, the values are specifically listed as $\mathbf{n^{20}{}_{D}}$. For most liquids the values fall in the range between 1.3 and 1.7, the values given to the fourth decimal space. Water has a refractive index of 1.3300.

However, when the measuring temperature is different, a correctional factor must be considered, in order to compare with literature values usually provided as measurements. For higher temperatures a factor 0.0004 times every degree above the standard of $20^{O}$C must be subtracted from the measurement or added to the reference value. If a value of $n^{25}{}_{D}$ = 1.5263 was obtained and the literature value for the same chemical is listed as $n^{20}{}_{D}$ = 1.5283, then 5(0.0004) = 0.0020 must be added to the measured value or subtracted from the literature value, provided the experimental temperature was higher than the standard.

$$n^{20}_D = n^t_D + (t-20)(0.0004) \text{ or } n^t_D = n^{20}_D - (t-20)(0.0004)$$

The liquid to be measured is placed evenly as a homogeneous film between two prisms. With certain manual adjustments the viewer while looking through an eyepiece (ocular) attempts to bring light and dark fields evenly together, simultaneously focusing two crosshairs in the center between the two fields. With the chromatic adjustment accomplished, the refractive index can be correctly documented.

**Fig. 13.40**

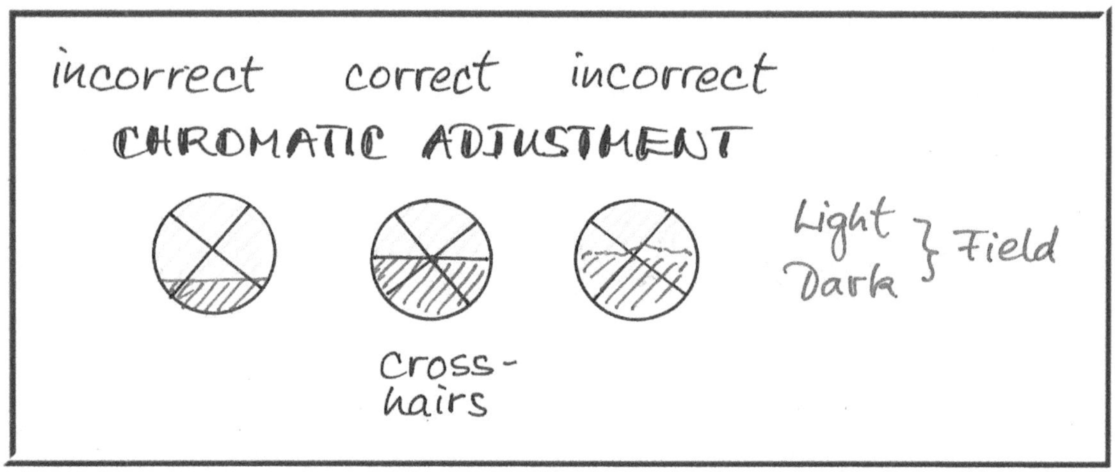

# EXPERIMENTS – Ch. 13

## *Analytical Methods*

**Key Terms and Concepts:** Wave characteristics, electromagnetic spectrum, wavelength ($\lambda$ "lambda" in units of length: nm, μm etc.) and frequency ($\nu$, "nuy" in units of cycles per time: Hz, $cm^{-1}$ etc.), $c=\lambda\nu$, wavenumber $\overline{\nu}$ ($cm^{-1}$), microwave, infrared, Raman spectroscopy, molecular bond movements (vibrations, rotations), electron transitions and ultraviolet/visual, fluorescence and phosphorescence spectroscopy, magnetic alignments of atoms and nuclear magnetic resonance spectroscopy ($^{1}H$, $^{13}C$ NMR), ionization and fragmentation of molecules and mass spectroscopy (MS), x-ray diffraction analysis of mono-crystals, polarimetry, refractometry

### I Infrared Spectroscopy of Synthesized Compounds (Acetyl Salicylic Acid, Diels-Alder Adduct, *n*-Butylbromide and *tert*-Butylchloride)

### Chemicals:

*Aspirin*
*Diels Alder adduct*
*Methanol or ethanol*
*Tert-butylchloride*
*n-butylbromide*

### Equipment and Materials:

*Infrared card*
*Infrared spectrometer (IR or FTIR)*
*Spot plate, Pasteur pipette or dropper with bulb*

### Procedure:

1. Dissolve a bit of sample compound in methanol or ethanol and spread the solution onto the film of an IR card.
2. Evaporate the solvent and insert the card into the measuring slot of the instrument.
3. Record the spectrum.
4. Repeat the procedure for the two liquids by placing the sample between two NaCl prisms, in a liqid holding device and inserting the sample carrier in the proper slot of the spectrometer.
5. Record the spectra.

### Observations and Results:

| Sample: | Main peaks ($cm^{-1}$) | Corresponding molecular group |
|---|---|---|
| | | |
| | | |

| | | |
|---|---|---|
| | | |
| | | |

## Questions and Problems:

1. Interpret the IR-spectra of each sample tested.
2. Can any impurities be detected in the samples?

## II 60 MHz $^{1}$H-NMR of synthesized Compounds (Acetyl Salicylic Acid, Diels-Alder Adduct, *n*-Butylbromide and *tert*-Butylchloride

### Chemicals:

*Aspirin*
*Diels Alder adduct*
*Various solvents (non-deuterated)*
*Deuterated solvents ($CDCl_3$ etc.)*

### Equipment and Materials:

*NMR-measuring tube*
*Pasteur pipettes*
*Glass wool or cotton*
*Spot plate*

### Procedure:

1. Pre-test the solubility of the solid sample compounds using a spot plate to found the best deuterated solvent for NMR measurement.
2. Dissolve 10-50 mg of solute in 0.9 ml deuterated solvent such as $CDCl_3$ and a trace of the internal standard TMS if required and transfer via a Pasteur pipette into the NMR tube.
3. Place the tube into the proper compartment of the instrument and record the spectrum.
4. Repeat the measurement for the two liquids, which can be placed neat into the NMR tube.
5. Measure and record the spectra

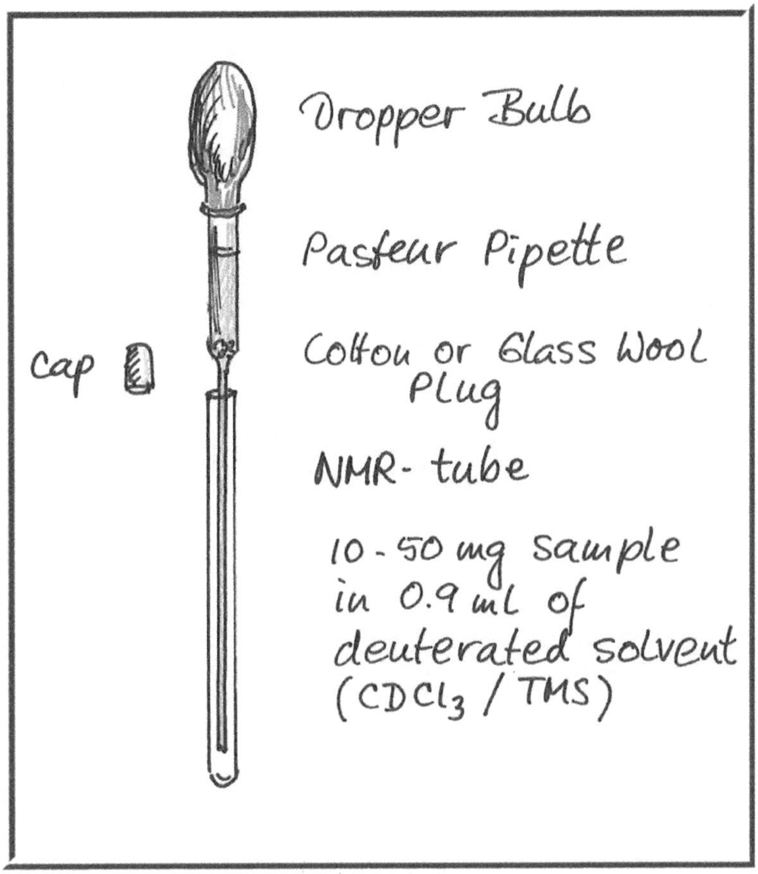


**Observations and Results:**

| Name of Chemical: | Chemical shifts of signals (δ): | Multiplicity: | Integration result (rel. number of protons): | Possible groups: |
|---|---|---|---|---|
| | | | | |
| | | | | |
| | | | | |
| | | | | |

**Questions and Problems:**

1. Interpret the NMR-spectra of each sample tested.

2. Can any impurities be detected in the samples?
3. What are the main spectral differences between the two butyl halides tested?

## III UV-vis Spectroscopy of Fluorescein and Eosin and an isolated Plant Pigment Lycopene

### Chemicals:
*Fluorescein*
*Eosin*
*Lycopene (tomato paste)*

### Equipment:
*UV-vis spectrometer*
*Glass cuvettes*

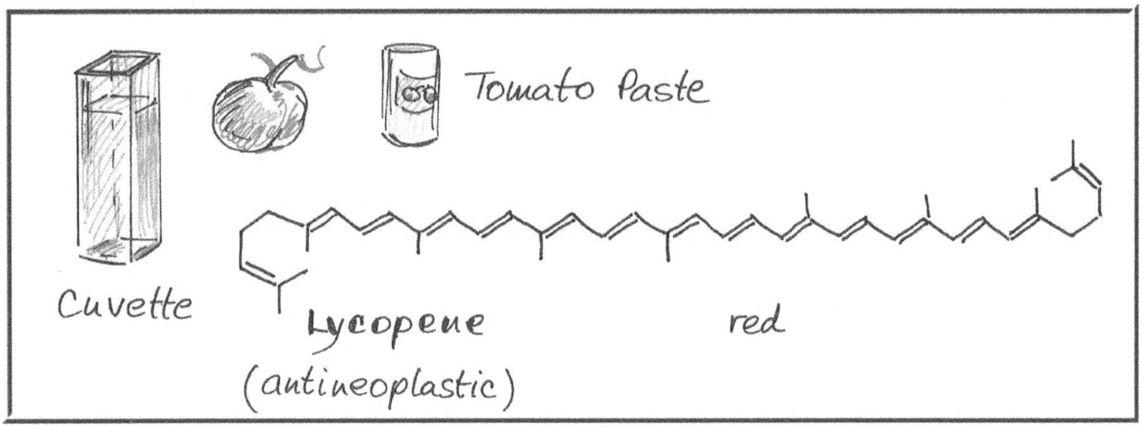


### Procedure:
1. Extract a sample of tomato paste with 2-3 mL of hexane, filter through a fluted filter paper and dilute the red solution containing the lycopene with more solvent to produce a barely visible color of the solution.
2. Fill the glass cuvette ¾ full with the solution. Fill the reference cuvette with pure solvent and measure the spectrum in the UV-vis spectrometer.
3. Clean out the cuvettes and repeat the measurements for fluorescein and eosin using aqueous alkaline solutions barely or very faintly colored. Measure the UV-vis spectrum.
4. Prepare new samples containing acidified aqueous solutions of the two dyes. Measure again and compare results.
5. If the spectra are either too thin (peaks too flat)or exceeding the dimensions of the graph paper (peaks too large), the concentration of the samples must be adjusted accordingly.

## Observations and Results:

| Chemical: | Main Absorption bands $\lambda_{max}$ (nm) | Corresponding Extinction Coefficient ($\varepsilon$) | Shift with changing pH (alkaline) | Shift with changing pH (acidic) |
|---|---|---|---|---|
| | | | | |
| | | | | |
| | | | | |

## Questions and Problems:

1. What are the main characteristics of the measured spectra?
2. What electron transitions can be correlated to the wavelengths of the absorption bands?
3. Lycopene is a conjugated polyene. How does it structurally differ from β-carotene? Which is the more conjugated system and how does it play a role in a bathochromic (going to longer wavelengths) or hypsochromic (going to shorter wavelengths) shift of the spectrum.
4. How does the pH affect the absorption spectra of fluorescein and eosin? What happens to the molecular structure if it becomes protonated?
5. What is a chromophore or chromophoric group within a molecule?
6. What is the difference between a benzoid and a quinoid system within a molecule? How do they affect absorption characteristics?

# IV Determination of the Refraction Index for the Synthesized Compounds *n*-Butylbromide and *tert*–Butylchloride

## Chemicals:

*tert-Butylchloride*
*n-Butylbromide*

## Equipment and Materials:

*Refractometer (Abbé – Laue)*
*Dropper with bulb or Pasteur pipet*

## Procedure:

1. Clean the analyzing prisms of the refractometer with ethanol.
2. Place an even homogeneous layer of the sample via dropper or pipet over the surface of the lower prism. Move the top prism onto the lower one with the sample in between.

3. Perform the chromatic adjustment by aligning light and dark fields around the cross and record the measuring temperature and given refractive index.

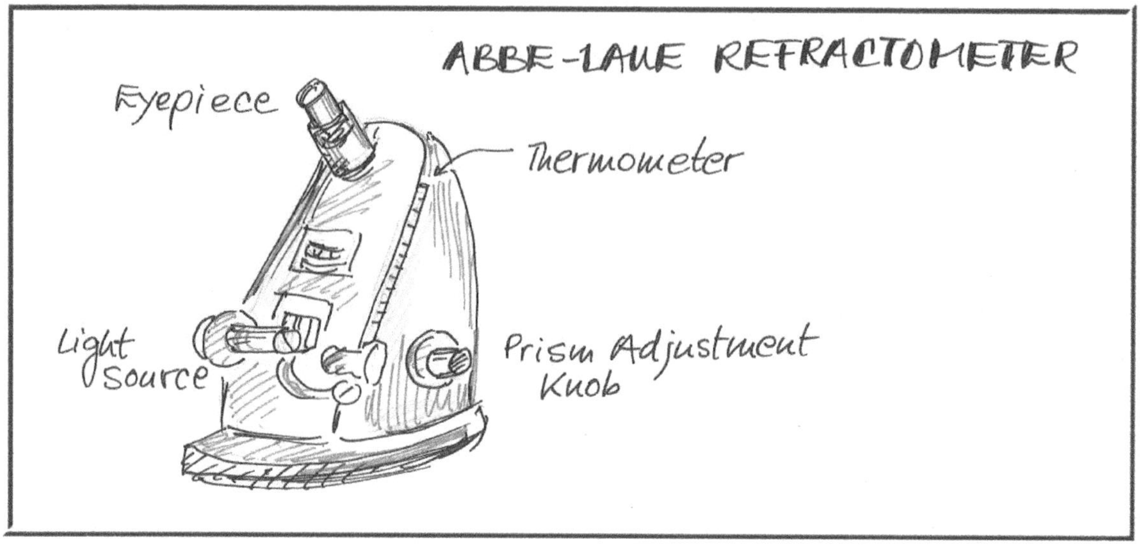


**Observations and Results:**

| Chemical: | Refractive index $n^t_D$: | Temperature (°C) | Literature Value ($n^{20}_D$) |
|---|---|---|---|
| | | | |
| | | | |

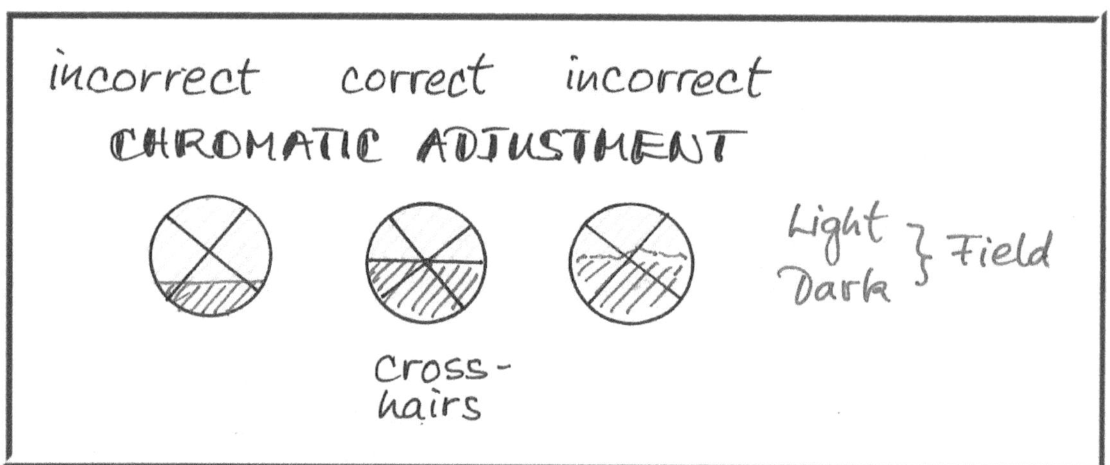

## Questions and Problems:

1. Did the experimental data correlate with the literature values?
2. What type of impurities could affect the readings?

## V Mass Spectroscopy of Natural Compounds (Research) and Molecular Categorization of the most Abundant Fragment Peaks (m/e values)

### Chemicals:

*None*

### Equipment and Materials:

*Hand-outs of various MS spectra (known and unknown)*

### Procedure:

1. Analyze the hand-out spectrum or spectra.
2. Find the molecular (parent) peak and the base peak (tallest peak).
3. Find and identify the most pronounced peaks for various fragments.
4. Propose a fragmentation scheme for the parent molecule reflected by the spectrum.

### Observations and Results:

| Sample MS spectrum | Molecular Formula and molar mass (g/mol) | Parent Peak $M^{+\cdot}$ (m/e): | Base Peak (m/e) | Fragment Peaks (m/e): | Relative Peak intensity |
|---|---|---|---|---|---|
| | | | | | |
| | | | | | |
| | | | | | |
| *Unknown* | | | | | |

### Questions and Problems:

1. What were the main characteristics of each analyzed spectrum.
2. What are the proposed paths for each fragmentation pattern?
3. What was the identity of the unknown, for which the molecular formula and molar mass were given?

www.ingramcontent.com/pod-product-compliance
Ingram Content Group UK Ltd.
Pitfield, Milton Keynes, MK11 3LW, UK
UKHW051453290926
15471UKWH00008B/1191